# FOUNDRY PROCESSES
Their Chemistry and Physics

# General Motors Research Laboratories Symposia Series

1986 Seymour Katz, Craig F. Landefeld, eds., *Foundry processes: Their chemistry and physics*, Plenum Press, New York, 1988.
1985 J. A. Bennett, M. E. Botkin, eds., *The optimum shape: Automated structural design*, Plenum Press, New York, 1986.
1984 L. Evans, R. C. Schwing, eds., *Human behavior and traffic safety*, Plenum Press, New York, 1985.
1983 M. S. Pickett, J. W. Boyse, eds., *Solid modeling by computers: From theory to applications*, Plenum Press, New York, 1984.
1981 R. Hickling, M. M. Kamal, eds., *Engine noise: Excitation, vibration and radiation*, Plenum Press, New York, 1982.
1980 G. T. Wolff, R. L. Klimisch, eds., *Particulate carbon: Atmospheric life cycle*, Plenum Press, New York, 1982.
1980 D. C. Siegla, G. W. Smith, eds., *Particulate carbon: Formation during combustion*, Plenum Press, New York, 1981.
1979 R. C. Schwing, W. A. Albers, Jr., eds., *Societal risk assessment: How safe is safe enough?* Plenum Press, New York, 1980.
1978 J. N. Mattavi, C. A. Amann, eds., *Combustion modeling in reciprocating engines*, Plenum Press, New York, 1980.
1978 G. G. Dodd, L. Rossol, eds., *Computer vision and sensor-based robots*, Plenum Press, New York, 1979.
1977 D. P. Koistinen, N.-M. Wang, eds., *Mechanics of sheet metal forming: Material behavior and deformation analysis*, Plenum Press, New York, 1978.
1976 G. Sovran, T. A. Morel, W. T. Mason, eds., *Aerodynamic drag mechanisms of bluff bodies and road vehicles*, Plenum Press, New York, 1978.
1975 J. M. Colucci, N. E. Gallopoulos, eds., *Future automotive fuels: Prospects, performance, perspective*, Plenum Press, New York, 1977.
1974 R. L. Klimisch, J. G. Larson, eds., *The catalytic chemistry of nitrogen oxides*, Plenum Press, New York, 1975.
1973 D. F. Hays, A. L. Browne, eds., *The physics of tire traction*, Plenum Press, New York, 1974.
1972 W. F. King, H. J. Mertz, eds., *Human impact response*, Plenum Press, New York, 1973.
1971 W. Cornelius, W. G. Agnew, eds., *Emissions from continuous combustion systems*, Plenum Press, New York, 1972.
1970 W. A. Albers, ed., *The physics of opto-electronic materials*, Plenum Press, New York, 1971.
1969 C. S. Tuesday, ed., *Chemical reactions in urban atmospheres*, American Elsevier, New York, 1971.
1968 E. L. Jacks, ed., *Associative information techniques*, American Elsevier, New York, 1971.
1967 P. Weiss, G. D. Cheever, eds., *Interface conversion for polymer coatings*, American Elsevier, New York, 1968.
1966 E. F. Weller, ed., *Ferroelectricity*, Elsevier, New York, 1967.
1965 G. Sovran, ed., *Fluid mechanics of internal flow*, Elsevier, New York, 1967.
1964 H. L. Garabedian, ed., *Approximation of functions*, Elsevier, New York, 1965.
1963 T. J. Hughel, ed., *Liquids: Structure, properties, solid interactions*, Elsevier, New York, 1965.
1962 R. Davies, ed., *Cavitation in real liquids*, Elsevier, New York, 1964.
1961 P. Weiss, ed., *Adhesion and cohesion*, Elsevier, New York, 1962.
1960 J. B. Bidwell, ed., *Rolling contact phenomena*, Elsevier, New York, 1962.
1959 R. C. Herman, ed., *Theory of traffic flow*, Elsevier, New York, 1961.
1958 G. M. Rassweiler, W. L. Grube, eds., *Internal stresses and fatigue in metal*, Elsevier, New York, 1959.

# FOUNDRY PROCESSES
# Their Chemistry and Physics

Edited by

## Seymour Katz and Craig F. Landefeld

*General Motors Research Laboratories*
*Warren, Michigan*

Library of Congress Cataloging in Publication Data

International Symposium on Foundry Processes, their Chemistry and Physics (1986: Warren, Mich.)
  Foundry processes, their chemistry and physics.

  (General Motors Research Laboratories symposia series)
  "Proceedings of an International Symposium on Foundry Processes, their Chemistry and Physics, sponsored by General Motors Reserach Laboratories, held September 21–23, 1986, in Warren, Michigan."–Verso t.p.
  Includes bibliographical references and index.
  1. Founding–Congresses. I. Katz, Seymour. II. Landefeld, Craig F. III. Title. IV. Series.
  TS228.99.I6  1986                    671.2                    88-5806
  ISBN-13:978-1-4612-8292-1    e-ISBN-13:978-1-4613-1013-6
  DOI: 10.1007/978-1-4613-1013-6

Proceedings of an International Symposium on Foundry Processes:
Their Chemistry and Physics, sponsored by General Motors Research Laboratories,
held September 21–23, 1986, in Warren, Michigan

© 1988 Plenum Press, New York
Softcover reprint of the hardcover 1st edition 1988
A Division of Plenum Publishing Corporation
233 Spring Street, New York, N.Y. 10013

All rights reserved

No part of this book may be reproduced, stored in a retrieval system, or transmitted
in any form or by any means, electronic, mechanical, photocopying, microfilming,
recording, or otherwise, without written permission from the Publisher

# FOREWORD

For a number of years it has been a General Motors Research Laboratories custom to hold a symposium on a subject which is new and emerging, and to invite the best people in the world in that subject to come together to talk to each other.

Initially, I had some difficulty in regarding foundry processes as a new and emerging subject. Copper alloys have been in foundry practice for about six thousand years. Foundrymen working with those alloys have been recognized, as such, for nearly all that time. Iron has a much shorter history, probably only three or four thousand years.

So what's new?

What is new is that a subject which has always been so complex and so difficult that it could only be a craft skill, with bits and pieces of knowledge and bits and pieces of insight, has begun to yield to new abilities to solve very complex problems.

We do this now because we can handle great amounts of data by computational means, using new and more complicated theoretical treatments than we could deal with before. In fact, we have a new technology with which we can attack these terribly difficult problems. Thus, foundry processing is becoming a new subject because new things can be done with it.

This symposium provided an opportunity to exchange views on the new aspects and on old aspects of the subject. We are delighted to have been able to provide this opportunity, and are proud to present these Proceedings as a record of the interesting and enlightening exchanges that occurred here.

*Dr. Robert A. Frosch*
*Vice President*
*General Motors Research Laboratories*

# PREFACE

These Proceedings are a compilation of papers and discussions presented at the International Symposium, Foundry Processes: Their Chemistry and Physics, held September 21—23, 1986 in Warren, Michigan. This Symposium was the thirty-first in a series sponsored by the General Motors Research Laboratories (GMR) for the purpose of exploring subjects important to the automotive industry and to the technical community at large.

These symposia have generally dealt with new and emerging areas of technology. This symposium, however, examines a subject with ancient beginnings from the perspective of modern science and engineering.

This change in focus has become necessary as the demand has escalated for castings with more accurate dimensions, lighter weight, higher strength, fewer defects, and lower cost. To meet these demands, improvement of foundry processing is increasing and relevant technology is being developed which can quicken this pace.

To achieve these goals, this symposium brought together an international group of experts on foundry science and allied areas to discuss new ideas and insights for foundry processing. To allow full discussion in the available time, the subjects considered were confined to liquid metal processing from melting to casting. This area was chosen because considerable relevant science and technology is emerging that promises to have major impact on product quality and cost. The symposium did not consider melt solidification, a subject that falls within the subject scope but is so broad that it could not be adequately treated as a portion of this two-day program.

The symposium covered three major subject areas: (1) production of liquid metals, (2) composition control, which includes alloy addition, tramp element removal and continuous chemical analysis, and (3) sources and control of casting defects, focusing primarily on gas porosity and non-metallic inclusions. To provide cohesiveness to the range of subjects discussed, the first paper is a critical overview of foundry liquid metal processing.

In preparing for the symposium, the organizers were confronted with the enviable situation of having a larger number of potential speakers than could normally be accommodated in the two-day symposium. In order to include as many of these valuable contributions as possible, four speakers presented shorter papers. These were Professors Fray (Paper 7a), Fruehan (Paper 8a), Loper (Paper 11a) and Wieser (Paper 16a). Professors D. R. Gaskell (Purdue University) and K. S. Goto provided prepared discussions of Papers 3, 7, and 7a.

Many people contributed to the planning and implementation of this symposium. Professors R. J. Fruehan (Carnegie Mellon University) and J. Szekely (Massachusetts Institute of Technology), served as consultants for the symposium, and were instrumental in developing the program.

A debt of gratitude is due our able Session Chairmen, Professors A. McLean (University of Toronto), K. W. Lange (University of Aachen), D. G. C. Robertson (University of Missouri-Rolla) and K. S. Goto (Tokyo Institute of Technology).

S. A. Worth (GMR Technical Information Department), the third member of the GMR triumvirate, took meticulous care of all the physical arrangements for the symposium as well as publication of these Proceedings. The creative artwork for the symposium was provided by R. Berube and S. E. McWilliams (GMR Technical Information Department), and secretarial support by F. A. Kukula and M. D. Aschmetat.

Thanks are due to those who worked behind the scenes to provide the participants with the setting that facilitated active and enlightening discussion. These GMR staff members include: J. J. Bommarito, K. Ernst, K. D. Gardels, C. Herhager, G. A. Kruger, J. Lane, D. T. Maher, W. L. Serveney, S. R. Tiderington, M. R. Tomlin and G. H. Tucker.

Finally, special thanks are due to members of GMR management, J. D. Caplan, C. S. Tuesday, G. H. Robinson, and L. R. Buzan, for providing guidance and support.

*Seymour Katz*
*Craig F. Landefeld*
*Metallurgy Department*
*General Motors Research Laboratories*

# CONTENTS

**SESSION I: FUNDAMENTALS**
    Chairman: A. McLean, University of Toronto

Paper 1 . . . . . . . . . . . . . . . . . . . . . . . . . . . . . . . . . . . . . . . 1
A Critical Overview of Liquids Metal Processing in the Foundry
    S. Katz, B. Tiwari, General Motors Research Laboratories

Paper 2 . . . . . . . . . . . . . . . . . . . . . . . . . . . . . . . . . . . . . . 53
Physicochemical Phenomena of Mechanisms and Rates of Reactions in Melting, Refining, and Casting of Foundry Irons
    E. T. Turkdogan, United States Steel Research Laboratories

Paper 3 . . . . . . . . . . . . . . . . . . . . . . . . . . . . . . . . . . . . . 101
The Capacities and Refining Capabilities of Metallurgical Slags
    I. D. Sommerville, University of Toronto

Paper 4 . . . . . . . . . . . . . . . . . . . . . . . . . . . . . . . . . . . . . 135
Partition of Alloying Elements in Freezing Cast Irons and its Effect on Graphitization and Nitrogen Blow Hole Formation
    A. Kagawa, T. Okamoto, Osaka University

**SESSION 2: PRODUCTION OF LIQUID METALS**
    Chairman: K. W. Lange, Technische Hochschule-Aachen

Paper 5 . . . . . . . . . . . . . . . . . . . . . . . . . . . . . . . . . . . . . 163
Chemical Processes and Heat Loss in Cupolas
    C. F. Landefeld, General Motors Research Laboratories

Paper 6 .................................................. 193
The Modelling of Fluid Flow Phenomena in Foundry Operations
  J. Szekely, Massachusetts Institute of Technology

Paper 7 .................................................. 219
Electrochemical Sensing of Carbon, Oxygen, and Silicon in Iron Melts
  A. R. Romero, K. Ichihara, H.-J. Engell, D. Janke, Max-Planck-Institut
  für Eisenforschung

Paper 7a ................................................. 243
Possible Uses of Sensors in the Aluminum Foundry Industry
  D. J. Fray, University of Cambridge

Paper 8 .................................................. 261
Fluid Flow and Mass Transfer in Gas Stirred Ladles
  S. Asai, I. Muchi, Nagoya University; M. Kawachi, Aichi Steel

Paper 8a ................................................. 293
Two Phase Mass Transfer in Gas Stirred Ladles
  R. J. Fruehan, S-H Kim, Carnegie-Mellon University

## SESSION 3: PURIFICATION OF LIQUID METALS
            Chairman: D. G. C. Robertson, University of Missouri-Rolla

Paper 9 .................................................. 303
The Principles of Gas and Powder Injection for Iron Refining
  G. A. Irons, McMaster University

Paper 10 ................................................. 333
Physico-chemical Aspects of Ladle Desulfurization of Iron and Steel
  H. Gaye, C. Gatellier, P. V. Riboud, IRSID

Paper 11 ................................................. 357
Thermodynamic Aspects of Removing Impurity Elements from
Carbon-Saturated Iron
  N. Sano, University of Tokyo

Paper 11a ................................................ 375
The Effect of Bismuth in Gray Cast Iron and the Chemistry of its
Neutralization with Rare Earth Metals
  C. R. Loper, Jr., University of Wisconsin-Madison

Paper 12 ................................................. 393
Chemical Inpurities in Aluminum
  J. H. L. vanLinden, R. E. Miller, R. Bachowski, Alcoa Laboratories

## SESSION 4: CASTING DEFECTS
>Chairman: K. S. Goto, Tokyo Institute of Technology

Paper 13 . . . . . . . . . . . . . . . . . . . . . . . . . . . . . . . . . . . . . . 411
The Thermodynamics and Kinetics of Gas Dissolution and Evolution from Iron Alloys
> R. J. Fruehan, Carnegie-Mellon University

Paper 14 . . . . . . . . . . . . . . . . . . . . . . . . . . . . . . . . . . . . . . 427
Formation of Porosity During Solidification of Cast Metals
> R. D. Pehlke, University of Michigan

Paper 15 . . . . . . . . . . . . . . . . . . . . . . . . . . . . . . . . . . . . . . 447
On the Detection, Behavior and Control of Inclusions in Liquid Metals
> R. I. L. Guthrie, McGill University

Paper 16 . . . . . . . . . . . . . . . . . . . . . . . . . . . . . . . . . . . . . . 467
Metal Refining by Filtration
> D. Apelian, K. K. Choi, Drexel University

Paper 16a . . . . . . . . . . . . . . . . . . . . . . . . . . . . . . . . . . . . . 495
Filtration of Irons and Steels
> P. F. Wieser, Case Western Reserve University

Symposium Participants . . . . . . . . . . . . . . . . . . . . . . . . . . . . 513

Subject Index . . . . . . . . . . . . . . . . . . . . . . . . . . . . . . . . . . 521

# A CRITICAL OVERVIEW OF LIQUID METAL PROCESSING IN THE FOUNDRY

### S. KATZ and B. L. TIWARI
*General Motors Research Laboratories*
*Warren, Michigan 48090*

## ABSTRACT

A need for understanding surrounds many aspects of foundry processing of liquid metals. Without this information it has been difficult to optimize foundry processes. This overview of foundry processing, from melting to casting, attempts to highlight areas that would particularly benefit from improved understanding. In some cases general approaches to the acquisition of needed information are suggested in other areas only the need is stated.

Major emphasis was placed on examining the inefficiencies in cupola melting; tramp element removal (ferrous and nonferrous) and the particular need for generic methods; the origins of gas defects and possible approaches for prevention; the generation and removal of solid inclusions; and finally, the development of sensors for process control and quality assurance.

## INTRODUCTION

The work collected in this volume attempts to inject new insight and ideas on foundry processes that deal with metal in the liquid state. Each of the subjects was carefully selected to address important issues, from melting to casting. The purpose of this paper is to provide cohesiveness to the volume by examining foundry processes, pointing out areas where understanding is needed to move foundry technology forward.

Many of the subjects discussed here are further addressed in the papers that follow. The subjects fall into three major categories, (1) production of liquid metals, (2) composition control, which includes alloy addition, tramp element removal and continuous chemical analysis, and (3) sources and control of casting defects which focuses primarily on gas porosity and non-metallic inclusions.

## TABLE 1
### Production Costs—Gray Iron.[4]

| Cupola Melting $/Tonne | | Induction Melting $/Tonne |
|---|---|---|
| 14.50 - 20.00 | Coke | — — |
| 0.00 - 3.50 | Oxygen | — — |
| 1.00 - 2.00 | Electricity | 46.00 - 92.50 |
| 7.00 - 13.00 | Metal Loss | 1.00 - 4.50 |
| — — | Carbon | 1.50 - 6.50 |
| 5.00 - 11.00 | Lining, Flux | 2.50 - 5.00 |
| 6.00 - 12.00 | Holding Furnace | — — |
| 33.50 - 61.50 | Total | 51.90 - 108.50 |

## PRODUCTION OF LIQUID METALS

### General

A wide range of melting furnaces are employed by the foundry industry to prepare liquid metal for castings. To identify promising areas for research and development it is only necessary to examine the performance of these furnaces according to the types of energy used for melting, i.e., combustion or electric.

Combustion-based furnaces are much less energy efficient than electric furnaces.[1,2] They also have larger associated losses of base metal and alloy by oxidation, due to exposure of the liquid metal to the gaseous combustion products.[2,3] Despite the technological disadvantages of melting with energy obtained by combustion, lower energy costs make it more economical to melt by these processes.[1,4] This is illustrated by recent data, given in Table 1, comparing the cost of cast iron produced in an electric induction furnace and in a coke fired cupola.[4] A similar situation exists in the production of liquid aluminum where costs for melting in gas-fired reverberatory furnaces are lower than in electric induction furnaces.[3]

Since combustion-based melting furnaces, despite their inefficiencies, produce metal at lower cost, the inefficiencies can be viewed as opportunities to further reduce liquid metal cost and improve melt quality. In this regard, a major opportunity exists in cupola melting.

### Cupola Melting

**General.** The cupola is the most important melting furnace in the ferrous foundry industry, producing over 70% of the cast iron tonnage. Transfer of combustion heat to iron in a cupola is potentially efficient because of the countercurrent gas-metal flow, but actual energy efficiency is generally poor. The principal

## TABLE 2
### Distribution of the Combustion Energy Among Cupola Outputs.

| Output | Percent of Input Energy |
| --- | --- |
| Iron | 20 - 45 |
| Slag | 2 - 7 |
| Off-gas sensible heat | 5 - 30 |
| Off-gas latent heat | 30 - 55 |
| Conduction through wall | 5 - 25 |

shortcomings of cupolas are: (1) incomplete combustion, (2) incomplete transfer of sensible heat, (3) expensive fuel (coke), (4) metal oxidation, and (5) inability to use the least expensive scrap (borings and turnings). Evidence suggests that significant progress can be made toward overcoming these shortcomings.

**Energy Efficiency.** The heat required to produce liquid iron at 1500°C from iron and steel scrap can be supplied by combustion of 4 wt % coke (based on the weight of metal charged) to $CO_2$. Considering the current industry-wide coke consumption is about 13 wt %, there clearly is a large margin for improvement.

The distribution of the heat of combustion among the various outputs from the cupola is given in Table 2. The percentages are based on the heat of combustion to $CO_2$ as 100%. It is clear from the data that achieving impressive fuel savings is not an utopian goal. The average amount of heat from combustion that is transferred to iron is about 33%. By simply limiting the heat losses in the last three categories to the practical minimum values given, energy efficiency almost doubles and fuel consumption is cut proportionately. Of perhaps equal importance, lower fuel rates mean equivalently higher melt rates. Thus, fewer cupolas could produce current tonnages.

Conceptually the heat loss most amenable to reduction is the sensible heat in the cupola off-gas. The results of a recent energy audit of cupola operations, shown in Figure 1,[5] suggest that an average of 20% of the heat of combustion is lost as off-gas sensible heat. Considering that the temperature of the gas entering the cupola preheat zone is about 1200°C, the average off-gas temperature of 600°C (see Figure 1) represents only a 50% reduction in the sensible heat content of the off-gas in its passage through the cupola preheat zone. It is clear that the heat loss is related to burden height. However, significant differences exist at the same burden height (see data at 3 meters), suggesting large differences exist in cupola heat transfer conditions.

A repercussion of poor heat transfer in the preheat zone is that benefits from

*References pp. 43–51*

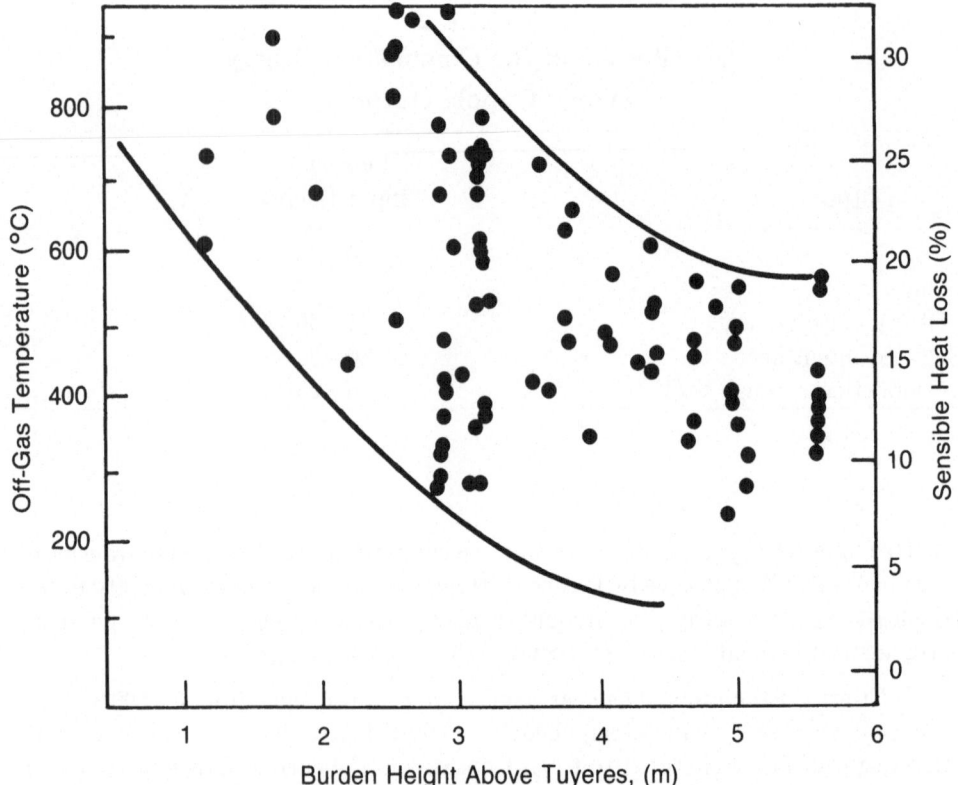

Figure 1. Measured off-gas temperatures for thirty cupolas as a function of burden height above the tuyeres.[5] The approximate sensible heat loss as a percent of the heat of combustion to $CO_2$ is given on the scale to the right.

other energy saving strategies are reduced. This is illustrated in Figure 2.[6] Here, much of the energy made available by decreasing the latent heat losses resulting from increasing the coke diameter was dissipated as sensible heat due to poor heat transfer conditions in the preheat zone. Thus, only a relatively small fraction of the newly available heat was usefully transferred to iron.

The preheat zone heat transfer problem in cupolas has not been adequately recognized. For example, the failure of coke sizing to increase iron temperature, as in Figure 2, has at times been attributed to coke.[6,7] Other studies have shown, however, that this is not the case.[8-10] The evidence thus indicates that foundries need to examine the heat transfer conditions in the cupola preheat zone more carefully and must develop strategies to recover greater amounts of heat. Helpful measures are listed in Table 3. The acceptable options may differ from foundry to foundry. What is needed to facilitate improvements is an effective mathematical model of cupola combustion and heat transfer. For this application, much could be learned even from a one-dimensional steady-state model.

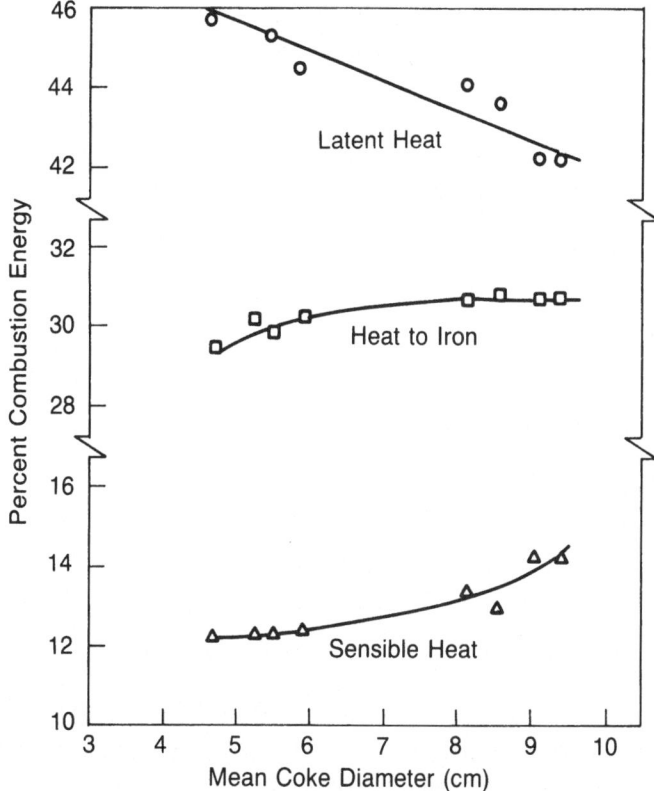

Figure 2. The distribution of combustion energy to $CO_2$ between iron and the sensible and latent heat of the off-gas for a cold blast, 74 cm internal diameter, lined cupola.[6]

Measurements of shell heat losses have shown the benefits of insulated vs. uninsulated (water cooled) shells.[11,12] The curves given in Figure 3 were generated by an extensive correlation.[11] On the basis of the potential heat savings suggested here, there was a trend to retrofit uninsulated cupolas with refractory linings. Where applied, this has produced substantial energy savings and attendently higher melt rates.[13] However, refractory costs have a large offsetting affect on cost savings, and as a result, a general change to lined cupolas has not occurred.

This situation suggests the value of examining whether conduction through the wall can be significantly reduced without the use of refractory. One idea is to use the periphery of the coke bed as insulation. This would require techniques to impede the normal preferential gas flow along the cupola wall.

The largest energy loss in cupola operation is the latent heat produced by $CO_2$ gasification of the coke bed (Boudouard reaction). The loss can be minimized by limiting the area of contact between coke and the cupola atmosphere as by increasing the size of coke (Figure 2) or by reducing the height of the coke bed. The

*References pp. 43–51*

## TABLE 3
### Conditions to Improve Preheat Zone Heat Transfer.

A. Increase available metal surface
    Higher iron to coke ratio
    Smaller metallic charge materials
    Denser coke

B. Reduce gas velocity
    Conical stack
    Oxygen enriched blast

C. Increase length of preheat zone
    Taller stack
    Lower melt zone
    Hot Blast
    Oxygen enriched blast

latter is illustrated in Figure 4 where the reduction in bed height was obtained by reducing the weight of coke per charge.[14] The effective amount of energy saved at the lower coke rates is less than that suggested by the figure because a fraction of the energy saved must be reintroduced (e.g., higher blast temperature) to counteract the lower iron temperatures that result. Since latent heat losses are lower at lower coke rates, all fuel savings obtained by reducing wall conduction and sensible heat losses also contribute to lowering the latent heat loss.

Applying the ideas outlined above will not reduce the latent heat loss below about 30%, as this requires even more effective shielding of coke from the cupola atmosphere. Conceptually high levels of separation might be achieved by coating the coke with a layer of high melting material such as lime.

The problem of latent heat loss can be totally circumvented by eliminating the coke bed. This has been achieved in the "cokeless cupola" which melts with hot gasses produced by efficient combustion of methane in external burners.[15] The bed in this cupola consists of ceramic spheres rather than coke. Of even greater effectiveness would be melting with a plasma heated inert gas (no alloy oxidation loss). Studies have shown that plasma cupola melting is feasible and effective,[16] however, the cost of electricity is an important factor in its applicability.

Another means for recovering latent heat is oxidation of CO in the cupola gas with air. This can be performed on the cupola off-gas (recuperation[17]) or inside the cupola (secondary air[18] and divided blast[19]). All three of these methods are in current use but none has achieved its full potential. Current recuperative installations recover relatively small fractions of the available heat, usually only

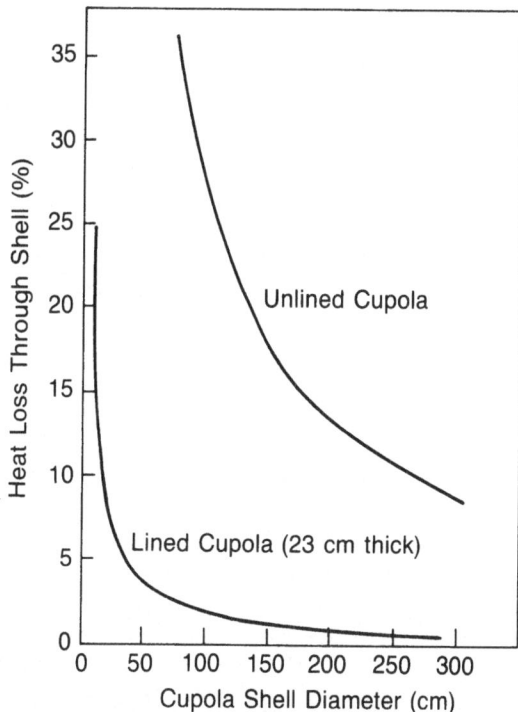

Figure 3. Heat loss through the shell of an unlined and lined (23 cm of refractory) cupola as a function of shell diameter.[11] Heat loss is expressed as a percent of the heat of combustion to $CO_2$.

enough to preheat the blast air. Full energy recovery appears to be limited by capital costs for installations with higher temperature capability and for devices to utilize the heat (electricity or steam generation equipment, etc.). Conceptually, oxidation of CO in the cupola shaft is appealing because heat can be transferred to the descending charge. Unfortunately, the secondary air process introduces air near the top of the burden and the divided blast introduces it so close to the primary tuyeres that the amount of CO that can be converted is limited due to regasification.

The major energy savings that are possible by reducing the latent heat stored in the cupola off-gas encourages further efforts to find acceptable solutions. A suggested approach is the use of mathematical modeling of cupola combustion and heat transfer to optimize the benefits of the secondary air and divided blast processes.

A promising concept for saving fuel that is not currently commercially exploited is heat generation in the cupola well (hearth). It has been reported that 2% oxygen directly injected into the cupola well raised iron temperature 70°C while the same amount of oxygen supplied in the normal manner to the blast produced only a 10°C rise in the iron temperature.[14] Clearly, generation of heat in the absence of the nitrogen diluent provided better transfer of heat to the iron. Trading the 70°C

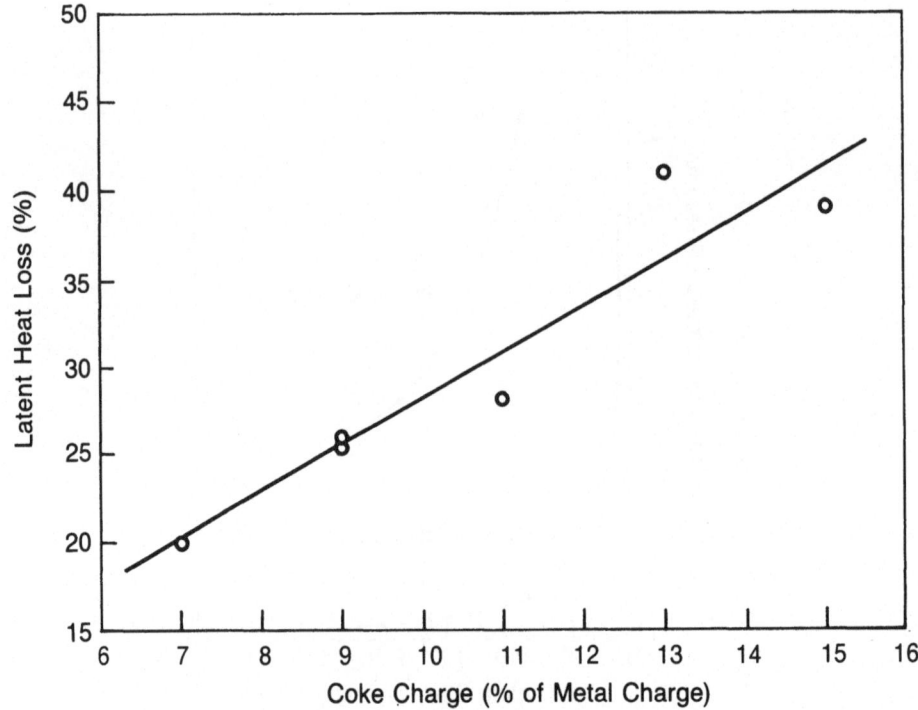

Figure 4. The relationship between latent heat loss of the off-gas and the amount of coke in the cupola charge for a cold blast, 74 cm internal diameter, lined cupola.[14]

temperature rise for a reduced coke rate resulted in the use of 22% less fuel.

Recent work with 1.7% Al additions to the cupola charge[20] has shown an analogous, but larger, temperature effect (90°C rise) from heat liberated by the reaction

$$4/3\underline{Al} + SiO_2 = \underline{Si} + 2/3Al_2O_3$$

in the slag layer in the cupola well (Figure 5). Based on the oxygen injection work,[14] trading the 90°C temperature rise for a lower fuel rate could reduce coke rates by 35-40%. This, however, is yet to be experimentally established. Unlike the results with oxygen injection, aluminum additions provide other important benefits: elimination of silicon and manganese oxidation loss and better desulfurization.[20] Thus, aluminum additions could prove to be an extremely important innovation in cupola operation.

**Cupola Fuel.** Foundry coke costs 50% more than blast furnace coke. It is produced from more expensive coals and requires longer furnace heating times to produce the large-size coke needed for efficient combustion. Desirable improvements are listed in Table 4. Achieving the indicated changes appears more likely via the formcoke route than by altering the current slot oven process. For example, the

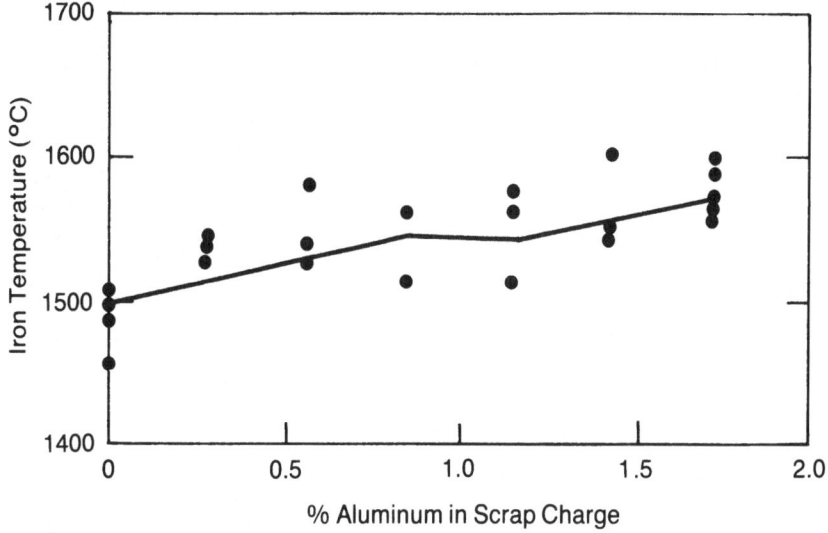

**Figure 5.** Average iron tap temperature as a function of the amount of aluminum added to the charge of a lined, 315°C hot blast, 48 cm internal diameter cupola.[20]

**TABLE 4**
**Improvements Desired in Foundry Coke.**

- Lower cost
- Larger size
- Lower porosity
- Serve as alloy source
- Serve to hold fine metallic materials

most effective way to reduce off-gas latent heat is by increasing the size of coke (see Figure 2). For the widest cupolas, 30 cm diameter coke might effectively be employed. Slot oven coke is limited by oven size to 20 cm and the mean size is considerably smaller (<15 cm). Formcoke is of uniform size and does not have the same furnace limitations.

Coke with lower porosity than that contained in current slot oven coke (~50%) could make a significant contribution towards reducing off-gas sensible heat. For example, reducing coke porosity from 50% to 25% would increase the iron/coke volume ratio in the cupola preheat zone by 20% (Figure 6).

To achieve large coke size, premium high caking coals are required in the manufacture of slot oven coke. This is not a requirement for formcoke because of the use of binders. As a result, use of low cost coal chars is an option available

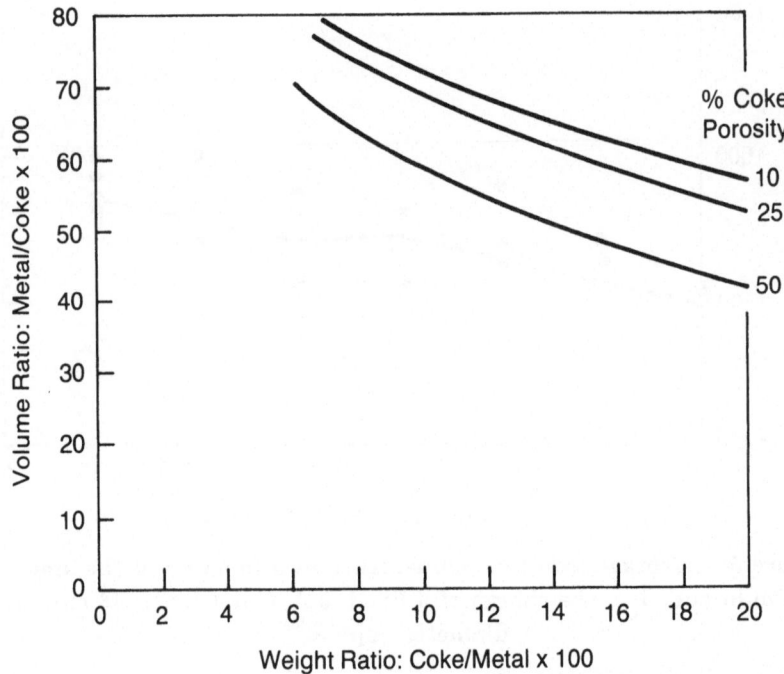

Figure 6. The effect of internal coke porosity and weight ratio, coke: metal, on the volume ratio, metal: coke, in the cupola burden.

for reducing the cost of formcoke. The viability of the idea is in part contingent on whether the high intrinsic reactivity of low cost chars[21] would lead to excessive coke gasification via the Boudouard reaction. Current foundry data is unclear on this point,[10,22] making further work in this area necessary.

Because the amount of coke used in a cupola is small relative to the iron melted, carbothermically reduced coke ash is not a significant source of silicon as it is for blast furnace iron.[23] Thermodynamically, however, carbothermic reduction of ores added to formcoke is a viable concept, not only for producing silicon, but other important cast iron alloy additions such as manganese, chromium, copper and nickel. The practicality of *in situ* alloy production rests on kinetic considerations since the estimated time coke temperatures exceed 1500°C is only about 0.5 h. Further work in this area is needed to establish feasibility.

Formcoke could also serve as a vehicle for utilizing small size scrap, such as borings and turnings. These very low cost materials are not ordinarily charged to cupolas because the melting efficiency is low due to their entrainment in the off-gas.[16,24] The $60/t cost differential between borings or turnings and normal size scrap makes it attractive to pursue means for incorporating significant amounts of these materials into the formcoke blend. This product would have the added advantage that the coke would serve to protect the fine metallic materials from oxidation.

**Alloy Elements.** Considering the need to produce iron within narrow composition limits, it is difficult to comprehend how poorly understood are the details of the chemical processes that control iron composition. In general, the carbon concentration of liquid iron increases during melting while the concentrations of other alloying elements decrease. We have gained a superficial understanding of the chemical processes involved with carbon pickup,[25] silicon loss,[20,26] and desulfurization.[27,28] However, these offer little help in predicting the important changes in liquid metal composition that occur as a result of differences in scrap composition, charge material size or changing combustion and heat transfer conditions in the cupola. It is the unpredictable changes in these conditions that produce the fluctuations in iron composition that characterize cupola operation.

Gaining a better understanding of the chemical processes in the cupola is the key to better control of iron composition, reduced oxidation losses, utilization of low cost charge materials, and improved desulfurization. Developing this information is not a simple task as we do not even have sound evidence of where within the cupola the important reactions occur. Evidence of this type can best be obtained by direct measurements of the conditions inside the cupola, either through the use of probes or by analysis of the contents of a cupola whose chemical processes were frozen by quenching. Unfortunately, earlier cupola quenching experiments[29,30] did not extract much information about chemical processes.

A procedure that could yield profound improvements in cupola operation is the addition of metallic aluminum to the charge. The benefits in higher iron temperatures and the anticipated fuel savings have already been described. In addition, aluminum additions prevent alloy oxidation loss (Figures 7 and 8) and reduce sulfur levels in the iron.[20] Further work is needed to optimize the process and to develop strategies for dealing with the residual aluminum content of the iron.

A less radical procedure that merits attention is the reduction, by carbon (coke), of $SiO_2$ and $MnO$ from slag in the cupola well. Almost no efforts have been made to control alloy losses by this method. As indicated in Figure 9, the thermodynamics of $SiO_2$ reduction at cupola operating temperatures are particularly favorable for acid slags. Calculations in Figure 9 are based on unit carbon and CO activity and silicon activity, $a_{Si} = 7 \times 10^{-4}$, relative to pure liquid silicon. The latter corresponds to the silicon activity in liquid gray and ductile base iron (3.5-4.0% C; 2.0-2.5% Si). To maximize the time available for the carbon-slag reaction, the slag layer should be deep. Objections concerning heat losses attendant with increasing the depth of the well[31] to accommodate deeper slag layers can be overcome by better insulation.

**Sensors.** The availability of sensors for continuously monitoring iron composition could greatly alter the philosophy and costs in iron melting. Currently, the type and quantity of metallic charge materials and coke used are primarily constrained not by cost but by the need to produce iron with acceptable composition. Composition sensors could uncouple the melting and the composition adjustment

*References pp. 43–51*

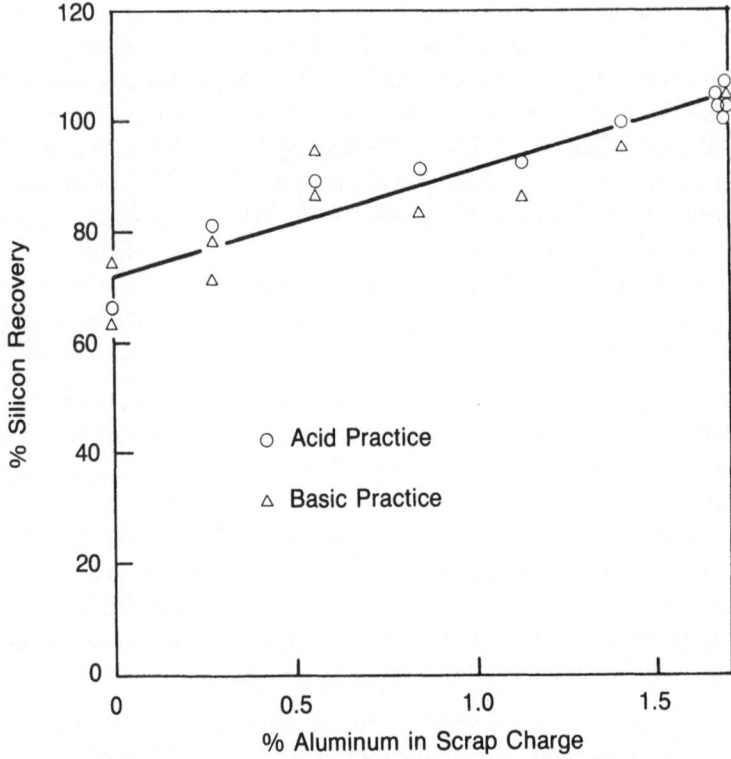

Figure 7. Percent of charged silicon that is recovered in the cupola iron as a function of the amount of aluminum added to the charge of a lined, 315°C hot blast, 48 cm internal diameter cupola.[20]

processes. Thus, melting would be performed on the basis of minimum cost and composition would be adjusted by continuous ladle additions. The status of composition sensors for iron melts is discussed in a later section.

A sensor that could greatly enhance cupola operation is one that measures the surface to volume ratio of the metal charge. Metal charge size is probably the most uncontrolled variable in cupola melting operations. Yet, charge size has a profound effect on composition and temperature of the iron produced (Figure 10), and fuel efficiency (Figure 11).

With a suitable surface to volume ratio sensor for controlling the blend of metallic charge materials used, the necessary quality of the iron could be maintained with the highest possible fuel efficiency. Two possible ways for measuring the metal charge surface to volume ratio can be suggested. The first is monitoring the pressure drop through a section of the cupola preheat zone. The data would have to be suitably corrected for gas temperatures, blast rate and the weight ratio of iron to coke. The second method is to measure the sulfur content of the cupola off-gas. As shown in Figure 12, the fraction of the coke sulfur that is discharged in the off-gas is

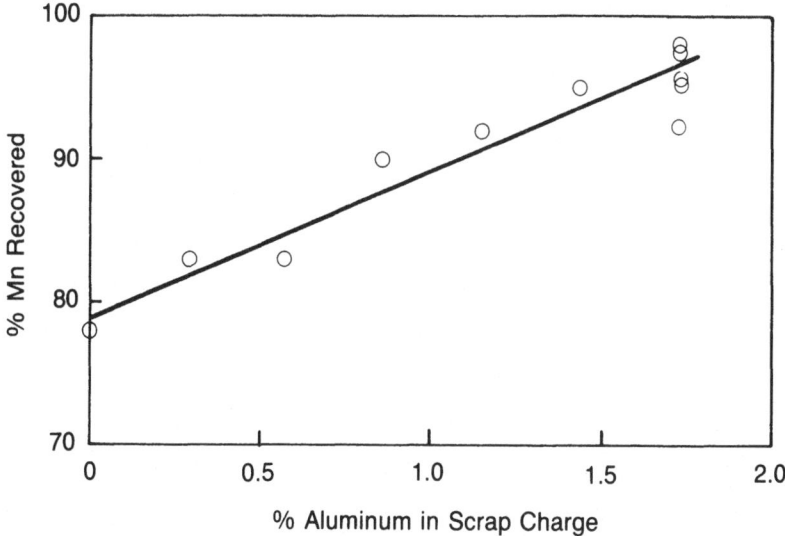

**Figure 8.** Percent of charged manganese that is recovered in the cupola iron as a function of the amount of aluminum added to the charge of a lined, 315°C hot blast, 48 cm internal diameter cupola.[20]

a function of the surface to volume ratio of the metallic charge.[11] Here, factors such as the surface area of limestone and coke bed height would have to be considered.

## COMPOSITION CONTROL

### General

With some exceptions, the basic feed material for castings is scrap. By its nature, there is uncertainty attached to the composition of scrap, both with respect to concentrations of desirable alloy elements and the presence of deleterious materials. This makes melt composition control a critical aspect of foundry operations.

Foundries, in general, are not well equipped to deal with large deviations from expected composition, nor with the presence of unexpected, undesirable impurities. Thus, the front line in composition control has been the use of quality scrap, preferably in-house scrap, and/or the use of materials of certified quality such as pig iron or secondary aluminum casting alloy. Conversely, lower scrap grades are avoided because of more poorly defined composition and greater probability of containing deleterious contaminants.

Despite all precautions and care in melting, adjustments to melt composition are usually required. Thus, an indispensable adjunct to composition control is means for rapid and accurate chemical analysis and facilities for making alloy additions. Although current analytical methods supply accurate metal composition in minutes, this is not rapid enough to uncouple the melting and composition adjustment processes involved in continuous metal production as in the cupola.

*References pp. 43–51*

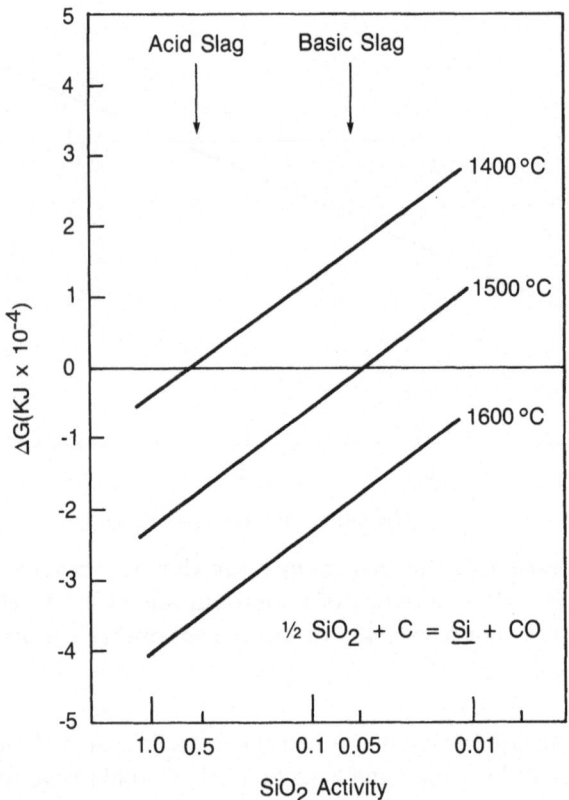

Figure 9. Free energy change for the carbothermic reduction of $SiO_2$ from slag as functions of $SiO_2$ activity and temperature.

Pressure for further work and understanding on rapid methods to control melt composition arises because the scrap system is being contaminated today at a higher rate than in previous times.[34] Steel is now commonly coated with aluminum and zinc. Aluminum-lithium alloys will soon enter the secondary aluminum system. There is a trend in both the primary metals and cast metal fields to use tailored rather than generic alloys, which brings greater variability to scrap composition. With the increase in the rate of contamination of the scrap system, the careful selection of charge materials may soon become very difficult if not too expensive. Hence, we need to learn how to deal with contaminants either by removing them or by counteracting their effects. This could be done in the foundry, or by secondary alloy producers.

Because most contamination occurs sporadically (at least today), the availability of continuous monitors for tramp elements would be beneficial to alert foundries when critical levels were exceeded. In another application, continuous melt composition analyzers used as drivers for automatic ladle additions would greatly facilitate production of tailored alloys.

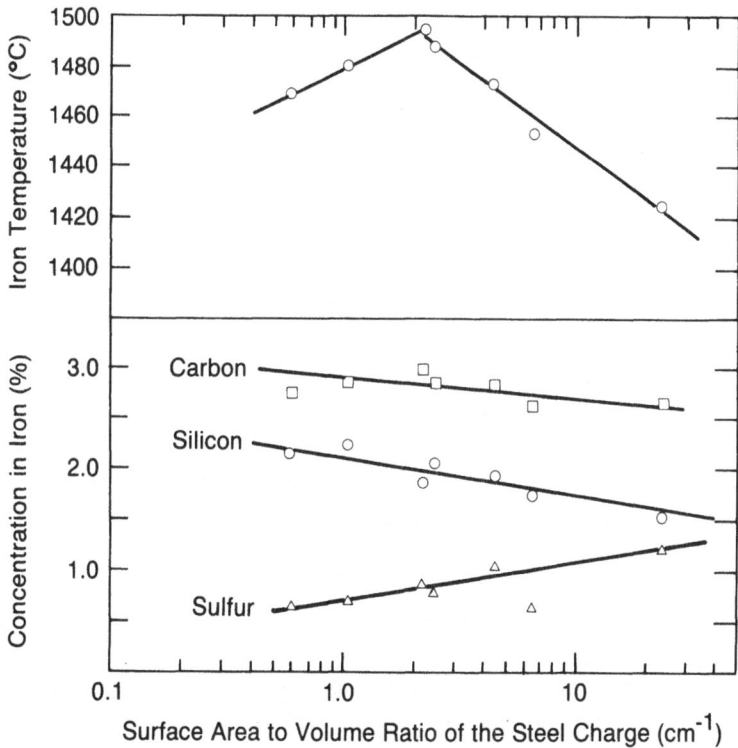

Figure 10. Effect of scrap surface area to volume ratio on iron composition and temperature. The charge consisted of equal weights steel scrap and pig iron. Only the size of the steel scrap was varied. Data was obtained from a lined, cold blast, 58 cm diameter cupola.[32]

With the increased and more varied uses being developed for ladle chemical processes, there is a need to determine which uses might appropriately be applied to foundry needs. This includes processes such as powder injection, wire feeding, melt degassing and gas shrouding. Further efforts are also needed to better understand the fluid dynamics and mass transfer in the porous plug ladles (multiple plugs) commonly used in foundry work.

### Composition Adjustments in Ferrous Melts

**Alloy Addition to Ferrous Melts.** Addition of alloy elements to cast iron poses few problems. The large degree of superheat of near-eutectic cast iron compositions obviates serious difficulties with iron freezing on the cold alloy surface. There are some problems with carbon and magnesium additions.

The difficulty with carbon additions to cast iron melts arises from ash

*References pp. 43–51*

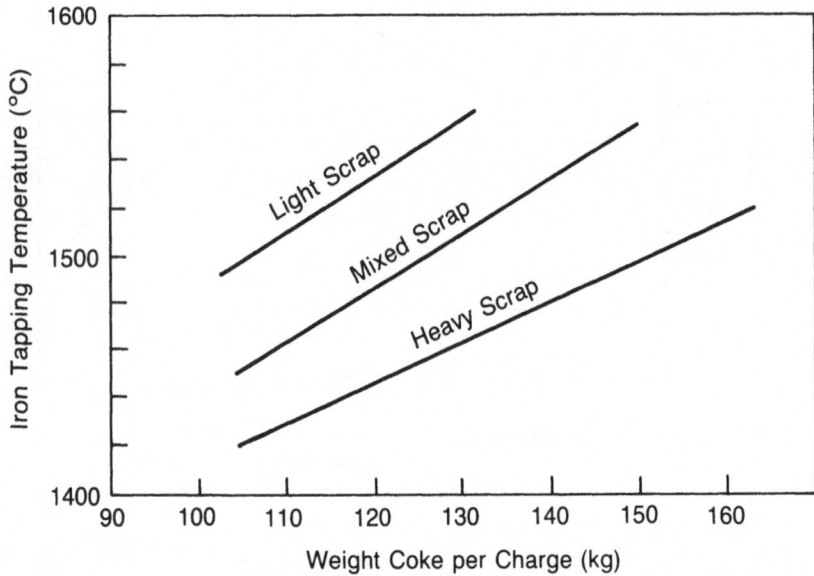

Figure 11. The effect of scrap size on cupola fuel consumption to obtain a given iron temperature. The heavy and light scrap had respective thicknesses of <2.5 cm and 2.5-10 cm. The mixed scrap contained equal weights of light and heavy scrap. The weight of the metallic charge was 907 kg.[33]

accumulation on the surface of the carbon source due to leaching of carbon. The rate of dissolution of carbon decreases with increasing ash content of the carbon source.[35] The rate is also affected by the adherence and surface coverage of the ash.[36] Similar relationships exist with respect to carbon pickup in cupola melting.[31] Means for promoting the rate of carbon dissolution from low cost, higher ash carbon sources would be worthwhile.

Difficulties with adding magnesium arise from low yields due to its volatility and the crude manner in which it is often added to the melt. Use of specially designed covered ladles obviates the problem;[37] however, they are not commonly used. Very little has been done to develop wire feeding[38] and powder injection methods[39] for foundry alloy additions. Further work in this area would be beneficial.

A common method of adding magnesium involves placing Mg-Fe-Si alloy powders in a prechamber of the casting mold.[40,41] Magnesium yields by this method are high and the late addition of magnesium prevents "fade," i.e., gradual loss of magnesium by evaporation and oxidation. The process generates considerable dross, however, which can cause serious casting defects.[42,43] The issue of the best means for adding magnesium to cast iron is still open, and better means are still being sought.

Alloy additions to steel melts require attention because the low degree of superheat causes a layer of iron to freeze on the alloy surface, thus delaying

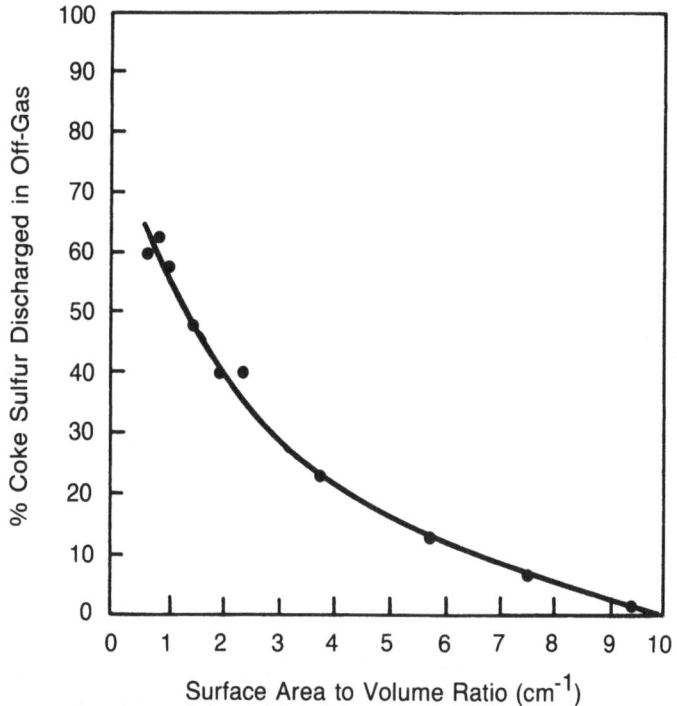

Figure 12. The percent of coke sulfur discharged in the cupola off-gas as a function of the surface area to volume ratio of the metallic charge. Data was taken from several sources.[11]

dissolution.[44,45] Additions of dense alloys, such as Fe-Mo and Fe-Nb, to ladles presents other problems because they tend to settle to the bottom of the ladle where dissolution and mixing proceeds slowly. Stirring, careful alloy sizing, and addition of elements that produce an exothermic reaction with iron[46] are used to minimize the problem. The need for further work in this area is not great.

**Purification of Cast Iron Melts.** Removal of elements from liquid iron or steel is difficult, which accounts for the importance attached to the prevention of contamination. Nevertheless, the removal of sulfur for ductile iron manufacture is imperative and similar efforts may soon be necessary for aluminum, and dissolved gases. Each of these topics is discussed briefly. Finally, techniques for dealing with large classes of impurities will be discussed.

*Desulfurizing Cast Iron.* Almost all base iron for ductile iron manufacture is desulfurized by calcium carbide ($CaC_2$) in a ladle reactor.[47] The reagent has proved so effective that the subject of iron desulfurization has drawn little recent research interest. The situation is changing because of rising environmental concerns about $CaC_2$. A number of approaches are currently under study to solve the problem. Calcium carbide manufacturers are working on proprietary methods to increase the degree of reaction of $CaC_2$, thereby minimizing the residual $CaC_2$ in spent

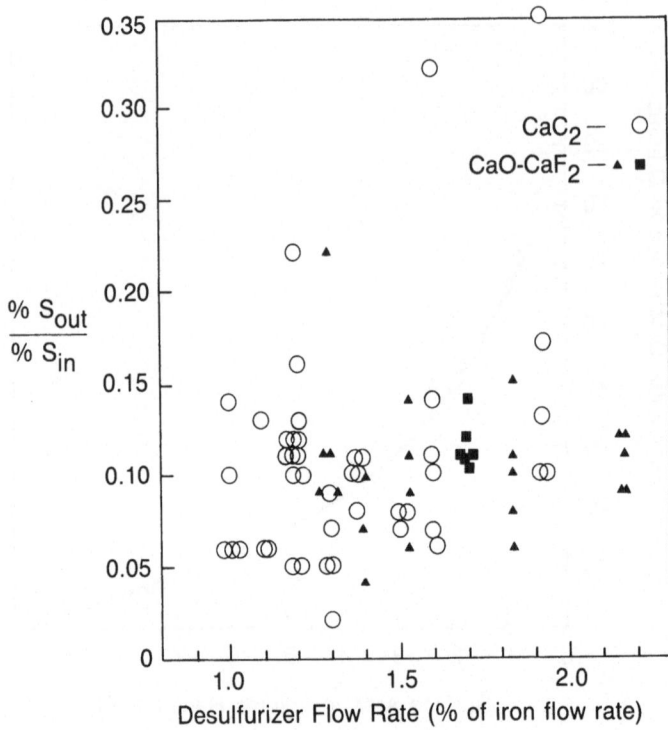

Figure 13. A comparison of plant data showing the equivalent effectiveness of CaO-CaF$_2$ and CaC$_2$ for continuous desulfurization.[52]

slag. They are also trying to provide means that promote more rapid oxidation of CaC$_2$ in the spent slag. An attractive method for achieving more complete reaction and oxidation of the spent CaC$_2$ is powder injection.[48,49]

Alternate desulfurizers are also being considered. Lime (CaO) reacts too slowly[50]; however, CaO-CaF$_2$ desulfurizers have been able to match the desulfurization rates of CaC$_2$ in laboratory[51] and foundry tests.[52] Foundry results are shown in Figure 13. The ability to achieve high desulfurization rates with CaO-CaF$_2$ is due to the unique two phase (liquid-solid) structure of the desulfurizer particles (Figure 14) which prevents the development of mass transfer limiting films, as occurs with CaO.

Further work on cast iron desulfurization is certain as the industry has a strong incentive to resolve this issue. Other promising desulfurization agents include slags in the systems, CaO-Al$_2$O$_3$-CaF$_2$, CaO-SiO$_2$-CaF$_2$ and CaO-MgO-SiO$_2$-Al$_2$O$_3$.[53-55] The search for slags with suitable sulfide capacities has been considerably simplified by recently developed correlations with optical basicity[56] and the development of computer programs for calculating oxide activity in complex slags.[55]

*Dealuminizing Cast Iron.* In many respects industry acceptance of improved

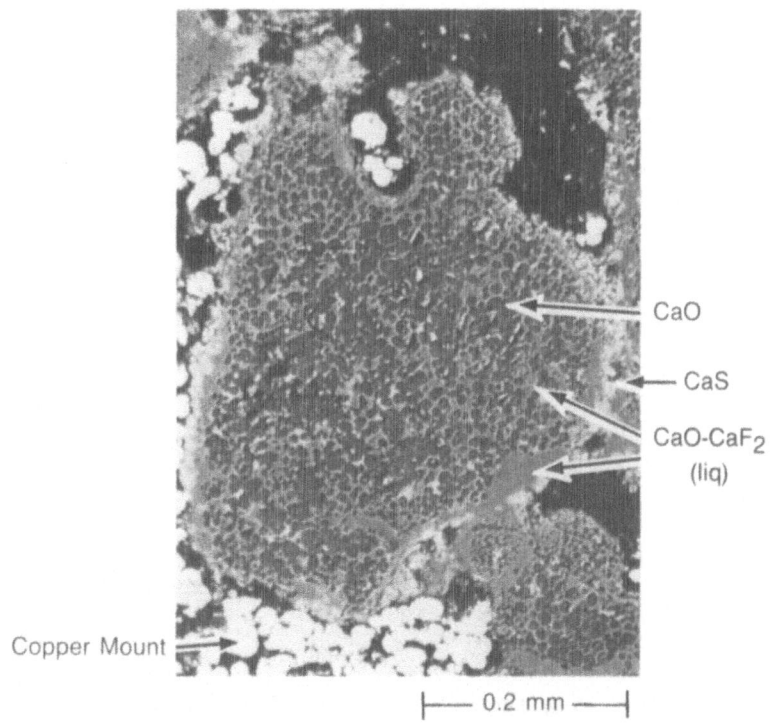

Figure 14. Electron probe microanalyzer, backscatter electron image of a cross-section of a spent desulfurizer particle (~0.5 mm diameter). The continuous, light gray phase is $CaO-CaF_2$, with $SiO_2$ and S. It is liquid at desulfurization temperatures. The darker gray phase is CaO. The white phase seen within the $CaO-CaF_2$ at the surface of the particles is CaS that presumably precipitated on cooling.

cupola performance through additions of aluminum to the charge depends on the availability of an effective, low cost method for dealuminizing the iron. The foundry industry has very serious concerns about aluminum's role in promoting gas defects[57] and its inherent drossing problems. It is thus incumbent to either remove aluminum or demonstrate how to successfully process aluminum-bearing cast iron. An effective process to dealuminize iron would also be an asset in dealing with problems from aluminum contaminated scrap, particularly as it affects induction melted iron where the opportunities to oxidize an aluminum contaminant are limited.

Work in the past[58] has shown that air or FeO will oxidize aluminum from cast iron melts. It is expected that the concurrent loss of silicon and carbon could be severe in these cases. A more appealing concept is the selection of an oxidant that does not react with carbon or silicon, e.g., CO or $SiO_2$. For dilute aluminum-bearing melts, where aluminum oxidation is likely governed by mass transfer in the iron, FeO and $SiO_2$ should dealuminize at comparable rates. This is illustrated in

*References pp. 43-51*

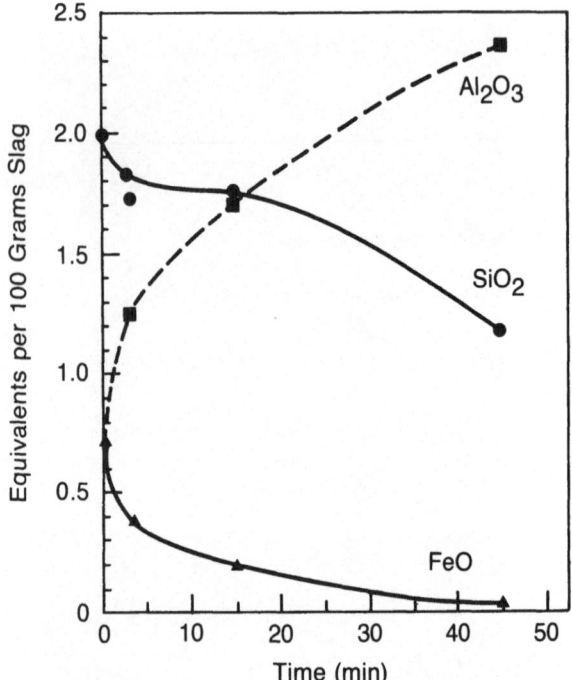

**Figure 15.** Changes at 1450°C in the concentrations of FeO, $SiO_2$ and $Al_2O_3$ in a slag in contact with an aluminum-bearing cast iron (1.5% Al). Concentration is given in units of equivalents per 100g slag.

Figure 15 which gives the concentration changes with time for FeO, $SiO_2$ and $Al_2O_3$ in a slag (initial composition, 25% FeO, 30% $SiO_2$, 35% CaO, 10% $Al_2O_3$) in contact with an aluminum-bearing cast iron (1.5% Al). The rates of $Al_2O_3$ formation are seen to be nearly the same for the periods 3-15 min and >15 min where, respectively, FeO and $SiO_2$ were separately the predominant oxidant. Other systems based on selective $Al_2O_3$ or $AlCl_3$ formation need to be investigated to establish the best and most practical means for dealuminizing iron.

*Degassing Cast Iron.* Another serious problem in iron castings is gas defects. Although these defects have probably always existed in cast iron, the recent emphasis on thin section thickness of castings, fatigue resistance and casting appearance have raised gas defects to a priority problem. As a result, there is considerable discussion about the need for degassing cast iron.

As will be shown in the last section, the cause of gas defects is not at all understood and there is a question even as to whether the gas defects originate from gases dissolved in the iron. Nevertheless, degassing of aluminum and steel effectively reduces gas defects.[59,60] Thus, an examination of methods and the potential benefits of degassing cast iron is worthwhile.

Under all conditions measured, cast iron appears saturated with respect to

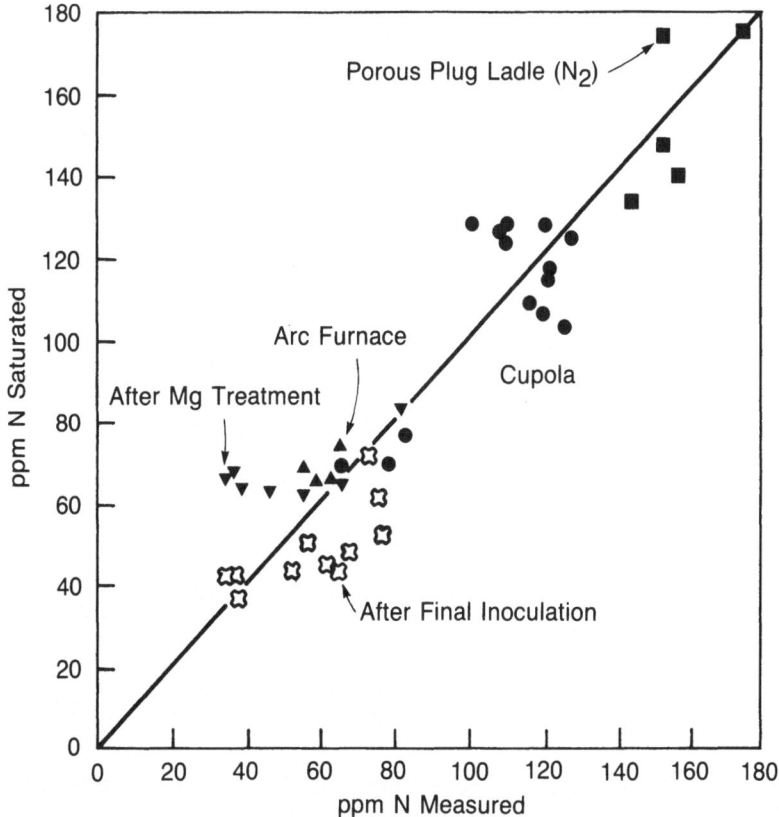

Figure 16. Apparent saturation of cast iron with nitrogen at different stages between melting and casting. The different nitrogen levels observed mainly reflect differences in iron temperatures.[61,62]

$N_2$ (see Figure 16) and in equilibrium with 0.1-0.2 atm $H_2$ (data from references 61,62). To what extent the gas content would have to be reduced to prevent defects has never been accurately determined. This might depend on mold conditions as these could affect gas redissolution rates. Definitive work on these subjects is needed. Studies are also needed to establish effective methods for degassing. Potential methods for degassing are: devices that produce small argon bubbles in aluminum melts,[63] injection of limestone, slag extraction of nitrogen[64-66] and vacuum degassing.[67] Selection of slags for assessment of nitrogen extraction has been simplified by recent correlations of nitrogen capacity and optical basicity.[68]

**Generic Solutions to Tramp Element Problems in Cast Iron.** Many elements have an effect on cast iron microstructure. To deal effectively with potential contamination by each element, on an individual basis, is impractical. Thus, it is useful to consider whether there are generic treatments that can be employed.

*References pp. 43-51*

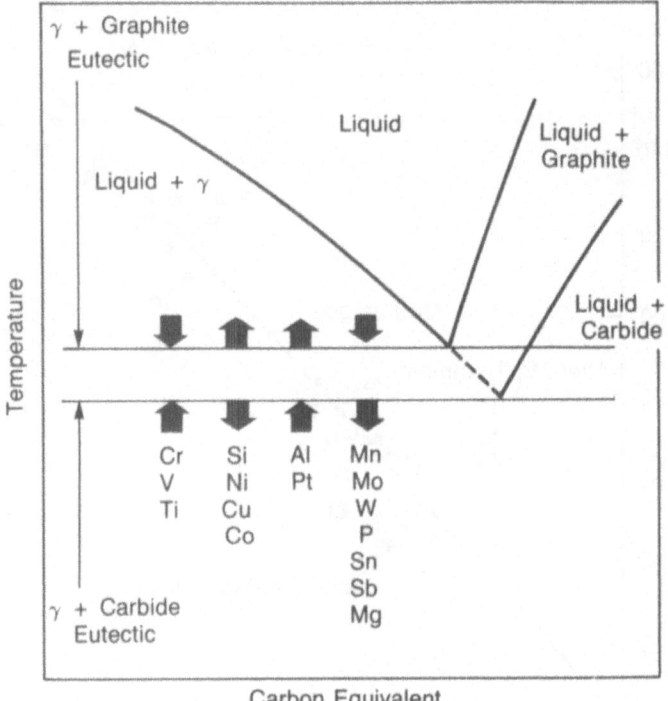

Figure 17. The effects of alloy elements on the stable and metastable Fe-C eutectic temperatures.[69]

The microstructure and properties of cast metals are mainly dictated by alloy composition and cooling rate. Two general classes of composition effects can be distinguished. The first is a bulk concentration effect in which the elements affect the thermodynamic stability of phases. This is illustrated in Figure 17 which shows the respective effects of elements on the stable and metastable Fe-C eutectic temperature.[69] Given that some undercooling always takes place, elements that increase the separation of the two eutectic temperatures will favor graphite-austenite formation. Those elements that bring the respective temperatures closer together will favor $Fe_3C$-austenite. To exert a noticeable influence, the elements must be present in relatively high concentration, usually >0.5%.

The second category is trace element effects, which largely affect graphite growth kinetics in ductile iron and eutectic cell growth in gray iron.[70] Such elements are surface active; they include S, As, Se, Sb, Te, Pb and Bi. The elements can be present in extremely low concentrations and still exert a major influence on the cast iron morphology.

Concerning the bulk concentration effect, recent work[71] has shown that the behavior illustrated in Figure 17 is predictable from thermodynamics and that

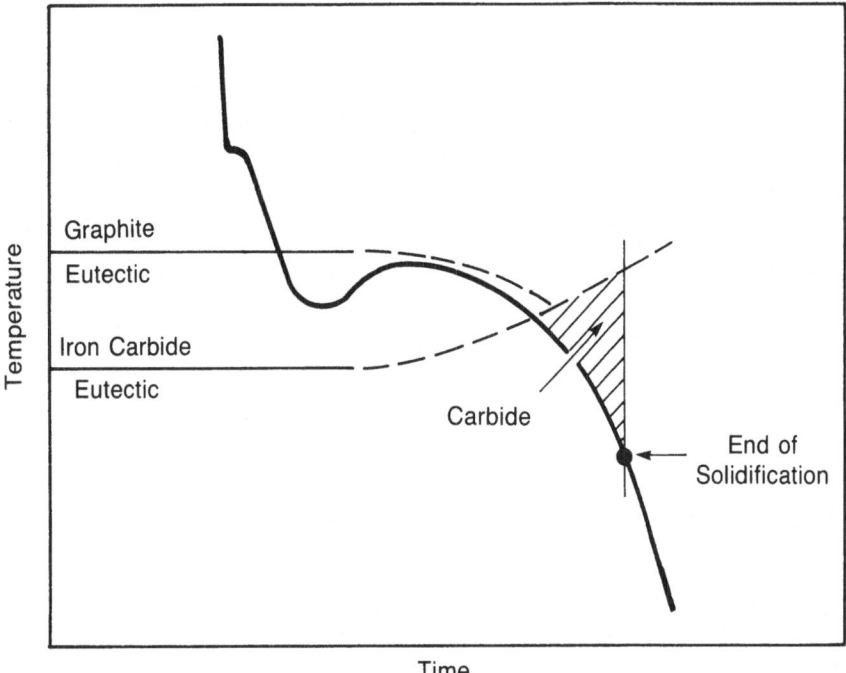

Figure 18. Thermal analysis plot of solidifying cast iron containing a carbide promoting impurity with an austenite-liquid partition ratio <1. Shown schematically are the changes in the stable and metastable eutectic temperatures during solidification.[72]

the contributions of individual elements are additive. Thus, one element can offset the influence of another. This picture is complicated by the fact that the solid-liquid partition ratios of alloy elements differ widely, with graphite-forming elements concentrating in austenite while carbide-forming elements concentrate in the liquid. Thus, as shown in Figure 18,[72] the effect of a carbide forming element is often observed late in the solidification process as a result of its continued concentration in the liquid phase.

Despite the increased complexity brought about by differences in the solid-liquid partition ratios, calculations can now be made to determine the extent that one element's effect can neutralize another. Extension of this work could lead to strategies for dealing with this class of contaminants.

Concerning the trace element effect, it has been shown that the deleterious effects of the surface-active elements in modifying the nodular graphite structure can be neutralized by very small additions of rare earth elements to the melt.[73] This is vividly illustrated by the series of photomicrographs in Figure 19 showing transformation of the spherical graphite structure of a ductile iron (Figure 19a) to flake graphite by a 0.05% Bi addition (Figure 19b) and the subsequent restoration of the spherical graphite structure by the further addition of 0.001% rare earth (Figure 19c). The mechanism of this remarkable effect is unknown at present. It is known that the technique can be effectively applied to contamination by a number of elements, Ti, Pb, Sb, Bi and Te, and the effect extends to austenite decomposition.[70] Thus, it is a generic process. Further work on understanding and extending this process is important.

Counteracting the effects of unwanted bulk elements or trace, surface-active elements by suitable elemental additions, as just discussed, is appealing because it does not require physical separation of the undesirable material from the melt. However, chemical separations beyond sulfur, aluminum and dissolved gas might ultimately become necessary, or perhaps even desirable if it makes possible the utilization of low cost, contaminated scrap.

For economic reasons the separations currently considered involve evaporation or extraction into slag. Tramp elements potentially removed by evaporation are hydrogen, nitrogen, sulfur, zinc, cadmium and bismuth.[74] By adding chlorine to the melt to form volatile chlorides, tin, zinc, titanium, zirconium, manganese and lead can be separated from cast iron without adversely affecting silicon and carbon.[75-77]

Slag separations include: vanadium and niobium into $Na_2O$-$SiO_2$ slags;[78] copper and tin into $Na_2SO_4$[77-79] or $Na_2S$-FeS slags;[80-82] and tin, lead, arsenic, antimony and bismuth into Ca-$CaCl_2$.[54,83] The above slags are all corrosive and all contain volatile and/or oxidizeable components, making near term application unlikely, unless carried out in a very large, specialized facility.

**Fluid Dynamics of Porous Plug Ladles.** The porous plug ladle (Figure 20) has proved valuable for desulfurizing continuously-flowing cast iron melts. It could also serve effectively to dealuminize or degas iron, to add carbon or silicon, etc. However, a better understanding of fluid flow and mass transfer in such reactors is needed to optimize their effectiveness. For example, it was found in developing a desulfurization process using CaO-$CaF_2$ that important changes had to be made to the system that was used with $CaC_2$. The changes made included the number and arrangement of porous plugs, the gas flow rates, and the slag depth.

Although pertinent information concerning fluid flow and mass transfer has already been developed for gas-stirred systems,[84-90] and generalized kinetic models of degassing[91,92] and desulfurization[49,89,93] are developed, further information is needed for effective application of the reactor to specific foundry tasks. The three-phase flow (iron, gas, and desulfurizer or other material) and use of multiple plugs clearly constitute a system of greater complexity than has been studied thus far. Work to characterize fluid flow and mass transfer with multiple gas plumes is progressing,[94] but much more work is needed. The need for research is especially great for the continuous reactor, because it is inherently less effective than the batch reactor.

**Purification of Steel Melts.** Steel foundries benefit from the extensive work sponsored, for many years, by the major steel companies. Techniques for desulfurization, dephosphorization and deoxidation of steel are well developed. Gas defects

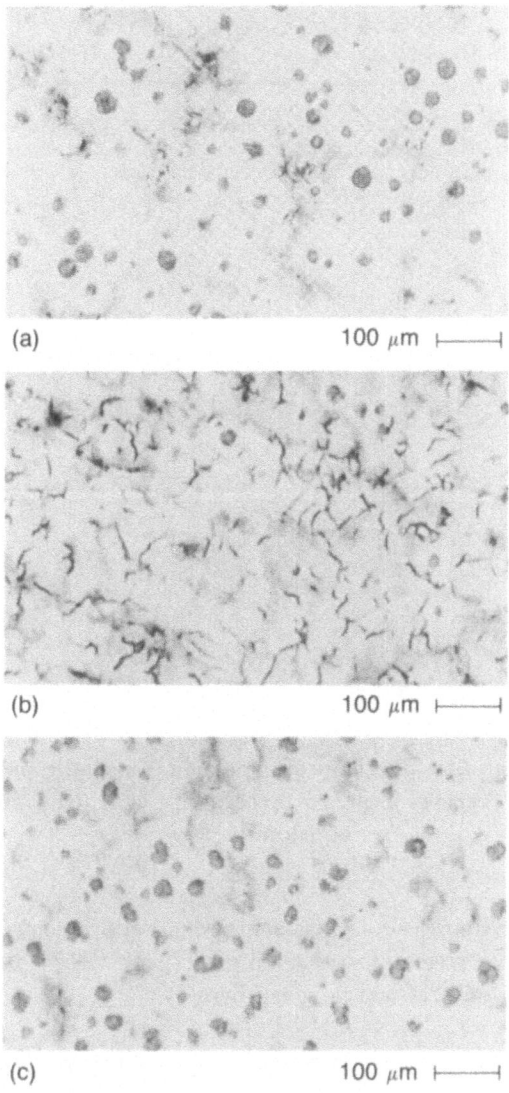

Figure 19. Photomicrographs illustrating tramp element neutralization. (a) nodular iron with no contaminants added, (b) same iron, transformed to flake iron by addition of 0.05% Bi, (c) restoration of the nodular structure by the addition of 0.001% rare earth to the bismuth contaminated iron.[73]

are less a problem with steel because gas evolution during decarburization of the melt effectively purges both hydrogen and nitrogen. Reabsorption of gaseous impurities from the atmosphere must be prevented, particularly because the solubility of $H_2$, $N_2$ and $O_2$ are much greater in steel than in cast iron.[95] Oxidizeable impurities are rather easily removed from steel; however, the non-oxidizeable impurities such

*References pp. 43–51*

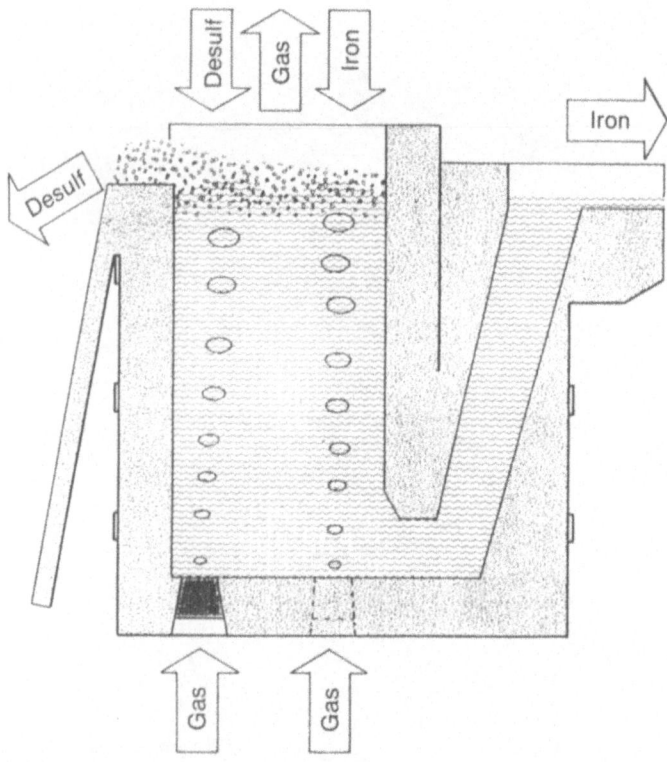

Figure 20. Schematic diagram of a porous plug ladle used for continuous desulfurization. Arrows indicate three phase (gas, iron and desulfurizer), continuous flow.

as copper and nickel represent serious contamination problems. Tramp element removal is an important steel industry research activity[74,80,81,96] and thus no further consideration of the subject will be given here.

## COMPOSITION ADJUSTMENTS—ALUMINUM MELTS

### General

Aluminum foundries serving the aerospace industry have made great progress in producing high quality, complex castings with superior mechanical properties.[97] Reasons for this success lies in the use of refined aluminum and pure alloy metals as charge materials and the use of induction furnace melting, degassing and filtration to prevent defects. Thus, technology to produce quality castings is available, albeit at high cost.

In higher volume aluminum casting production, where casting property requirements are less exacting than in aerospace applications, competitiveness dictates the use of large amounts of less costly recycled aluminum alloy (augmented by home scrap), prepared by secondary smelters from aluminum scrap.[3,98] The purchase of recycled aluminum alloy, as opposed to producing it in-house from purchased scrap, is a consequence of the high potential for serious alloy contamination which exists for aluminum alloys as a result of the wide range of elements (Cu,Si,Mg,Zn,Sn) that serve as the principal alloy addition. Because of this diversity, extensive facilities are required for preparing secondary aluminum alloys. Such facilities are beyond the capacities of many foundries. As a result, this function is served by a separate industry.

Because of the intermediary role served by the secondary smelters, foundries must generally contend only with adjustments in the composition of important trace elements such as: sodium and strontium whose concentrations vary because they are volatile and readily oxidized; or phosphorous, titanium and boron which physically separate from the melt because they are present as insoluble intermetallic compounds; or hydrogen which is present as a result of the reaction of water vapor with aluminum. With the exception of hydrogen, trace element compositions tend to be lower than desired due to the various subversive losses. Thus, it is desirable to make alloy additions late in the melt processing cycle. The practice of making late alloy additions would be greatly facilitated by the availability of composition sensors for these elements. A sensor to detect undesirably high concentrations of hydrogen would also be an important asset to foundry operations. In this case, facilities and techniques would also be needed to reduce the hydrogen concentration to the desired levels.

The situation outlined is somewhat different for the larger aluminum foundries, mostly automotive, where the favorable economics of using larger proportions of scrap are obtained by having the foundry assume the role of the secondary refiner. In this case, in addition to problems outlined above, the foundry must contend with the variety of types of aluminum scrap and the incompatibilities of certain combinations of trace elements, e.g., phosphorous or antimony with sodium and strontium.[3] The tactics developed by the large foundries to accomplish these goals are: (1) use of well characterized scrap, often obtained from subsidiaries that process wrought aluminum; (2) use of casting alloys which have rather wide permissible ranges for many of the principle aluminum alloying elements. (e.g., alloy 390.2, used for die casting, has liberal ranges for silicon, iron, copper and magnesium); (3) employment of purification processes used by secondary refiners. These include processes for removing very electropositive and volatile elements (Na,Ca,Sr,Li,Mg) and melt filtration and degassing. Finally, (4) dilution of off-specification melts with primary aluminum.

The technological needs of foundries that melt high proportions of scrap are (1) methods for better separation of the various aluminum alloy types and (2) methods for removing electropositive elements that avoid the use of chlorine, for which there are growing environmental objections. The technological needs, common to all aluminum foundries are (1) strategies to deal with hydrogen contamination and (2) composition sensors for important alloy elements. These four subjects will be examined below.

*References pp. 43–51*

**Scrap Separation.** The cost of scrap decreases as its composition becomes more uncertain. Utilization of mixed scrap, to any extent, would be difficult without separation according to alloy type since only magnesium among the major alloying elements can easily be removed from aluminum. The only effective method currently available for dealing with excessive levels of major alloying elements (except magnesium) is dilution with costly high purity alloy or primary aluminum.

The development of effective scrap separation techniques would be useful, even on a gross levels, such as dividing scrap according to cast *vs.* wrought or die cast *vs.* sand cast. A method based on melting point differences has been developed for separating cast from wrought aluminum.[99] Of much greater utility would be separations based on chemical type. The best available example of this is separation of the low magnesium, high manganese alloy used for bodies of beverage cans from the high magnesium, low manganese alloy used for the lids.[100] In this case only two alloys are involved and over a billion pounds of used beverage cans are recycled annually. Clearly, complex separation schemes are economical when large quantities of scrap can be processed. Whether any foundry has sufficient demand to afford such processes is doubtful.

**Purification of Aluminum Melts.** Potential methods for removing unwanted elements from aluminum include selective oxidation, chlorination and fluorination, electrochemical separation, evaporation and intermetallic compound formation. Primary and secondary smelters currently employ chlorination to remove lithium, sodium, magnesium, calcium and strontium.[101] The approximate theoretical minimum concentration to which elements can be reduced by chlorination is given in Table 5.[102] Surveying the data in the table it is seen that only magnesium among the major alloy elements, can be affected by these treatments. Important elements absent from these tables are: copper, silicon, tin, manganese, nickel, zinc and iron. Intermetallic compound formation has potential for removing silicon and iron[103] and a number of important trace elements. For example, the possible removal of sodium and strontium as phosphides is suggested by the poisoning of sodium and strontium modification in hypoeutectic Al-Si alloys when phosphorous is present.[3] Since sodium and strontium are relatively easily removed by other means, perhaps a more useful employment of this reaction would be for the removal of phosphorous. Similarly the poisoning of sodium and strontium modification by antimony, which is becoming a considerable concern to foundries that are remelting purchased scrap, can be taken as an indication that the former can be used to remove antimony. Antimony is added to aluminum at levels of 0.1% for the same purpose as sodium and strontium. Its concentration might also be significantly reduced by evaporation.[102] The control of titanium by precipitation as the boride is another potentially useful intermetallic reaction if appreciable amounts of $TiO_2$ (pigments) are present in the scrap. Care must be exercised in the removal of elements as intermetallic compounds since even a small excess of the added elements can have strong effects on alloy microstructure. It is clear that concentration sensors for the critical elements would greatly facilitate melt purification.

*Magnesium Removal.* As indicated, only magnesium, among the major alloy

## TABLE 5

Calculated Minimum Concentration of Alloy Elements in Aluminum Achievable by Chlorination at 727°C.[102] Based on $Cl_2$ (1 atm) + $x\underline{M}$ = $M_xCl_2(g)$ where, $p_{M_xCl_2} = p_{AlCl_3}$

| Element | Concentration |
|---------|---------------|
|         | ppm           |
| Ca      | $5 \times 10^{-7}$ |
| Na      | $7 \times 10^{-4}$ |
| Li      | $1 \times 10^{-2}$ |
| Mg      | 2             |
| Be      | 70            |

elements, is easily removed from aluminum (usually referred to as demagging). This, however, can be commercially important to foundries that melt scrap because it enables the production of low concentration magnesium alloys, such as 380 and 319, from high magnesium content scrap.

Since magnesium oxide and chloride are more stable than the respective aluminum compounds, magnesium can be selectively removed from molten aluminum by controlled oxidation. At present a chlorination process is most widely used by secondary smelters and foundries for demagging casting alloys.[104,105] In this process, magnesium is selectively reacted with $Cl_2$ to form $MgCl_2$ which is removed as a dross. While the process is reasonably efficient, it creates serious environmental conditions in the plant.[105] In addition, the $MgCl_2$ dross is hygroscopic and poses a disposal problem.

A more promising route for demagging aluminum is electrolytic refining.[106,107] This process, in comparison to chlorination, recovers magnesium in a more valuable form (metallic) with no apparent environmental problems. In the process, shown in Figure 21, magnesium is electrolytically oxidized from the molten aluminum alloy anode and concurrently deposited, by reduction, on the inert cathode at the top of the cell. Because magnesium has the lowest density in the system (see Figure 21), it floats on the electrolyte enabling separation from the aluminum. The electrolytic process has been demonstrated successfully in a 250 kg semi-plant scale furnace and appears to have potential for commercial application.

*Hydrogen Removal.* The deleterious effect of hydrogen porosity on the quality of aluminum castings is well known.[59] Porosity arises because a large difference exists between the solubilities of hydrogen in molten and solid aluminum. As shown in Figure 22, the hydrogen solubility in aluminum decreases from 0.7 to 0.02 ml /100 g Al in going from the liquid to the solid phase at the melting temperature. Coarse porosity has been observed when hydrogen levels in the melt exceed 0.1

*References pp. 43–51*

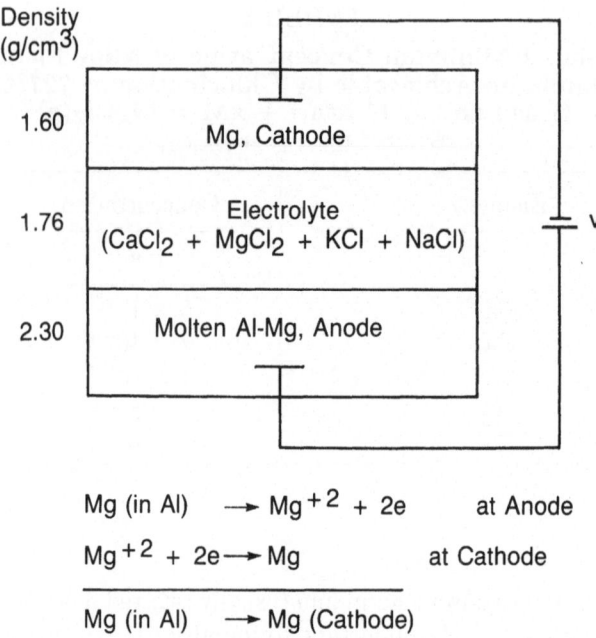

Figure 21. Schematic representation of an electrolytic demagging process.[107]

ml/100 g Al.[103] Interestingly, in castings where strength is not a criterion, hydrogen pinhole defects are often purposely produced to counter solidification shrinkage.[3] Factors that affect the amount and distribution of porosity are alloy composition, inclusion content and cooling rate.[109]

The most common source of hydrogen in aluminum melts is the reaction between molten aluminum and water vapor, forming $Al_2O_3$ and dissolved hydrogen according to the reaction:

$$2Al + 3/2H_2O = Al_2O_3 + 3\underline{H}$$

Factors that influence this reaction are (1) the imperviousness of the $Al_2O_3$ film on the melt surface,[110,111] (2) the amount of $H_2O$ present, (3) mass transfer rates between air and metal and (4) the solubility of hydrogen in the melt.

There is little data reported on the solubility of hydrogen in aluminum alloys.[102] Data for hydrogen in liquid Al-Cu and Al-Si, obtained with a Sievert's apparatus, has been reported.[112] Very recently an attempt was also made to predict hydrogen solubility in aluminum alloys using a thermodynamic liquid solution model.[113] It is clear that there is a need for more solubility measurements.

Minimizing the exposure of liquid aluminum to moisture is a first step towards preventing serious hydrogen gas defects. Charge preheating removes water and more recent furnace designs such as wet bath and electric radiation reverberatory furnaces

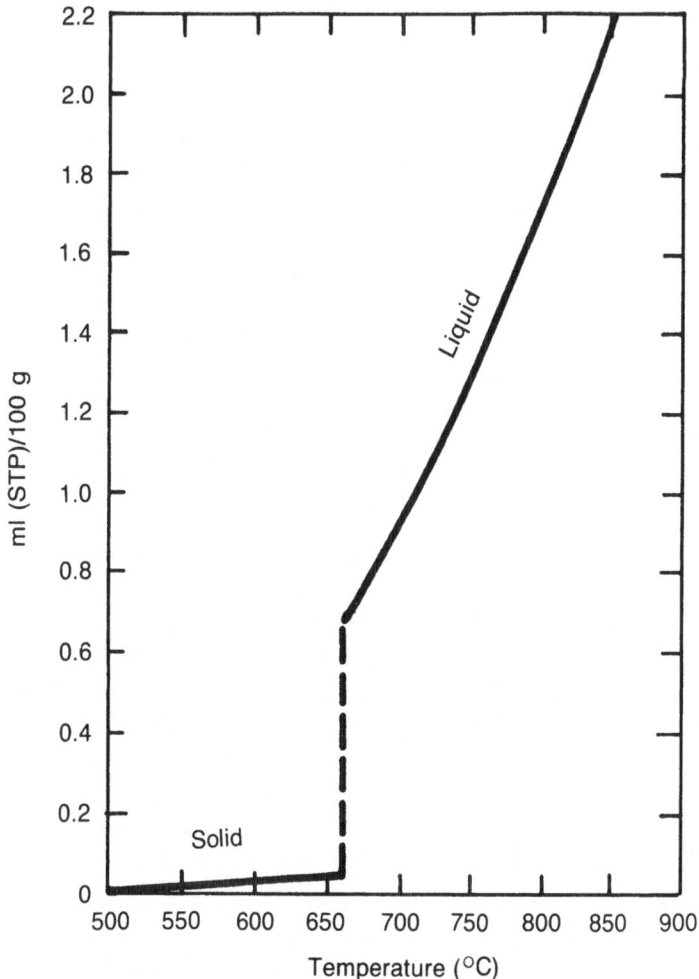

Figure 22. Solubility of hydrogen in liquid and solid aluminum under 1 atm $H_2$ pressure.[3]

minimize exposure to the atmosphere.[3] Despite best efforts to minimize hydrogen entry to the melt, hydrogen scavenging of the melt is often required to produce high quality castings.

Commercial in-line melt treatment systems effectively remove hydrogen by purging with inert gas and/or chlorine.[101] Foundry problems with dissolved hydrogen are mainly due to failure to utilize this equipment. Lower cost degassing processes would be useful for foundries. Also needed is a method to continuously monitor the hydrogen content of the melt and warn of excessive hydrogen levels. This analytical technique would also be useful in maintaining proper hydrogen levels when using precipitated gas as a countermeasure for solidification shrinkage.

*References pp. 43–51*

## Continuous Chemical Analysis of Melts

**Ferrous Melts.** As indicated earlier, beneficial changes in cupola operations are potentially feasible if the concentrations of important elements, i.e., carbon, silicon and sulfur can be continuously determined. Carbon[114] and silicon[115] concentrations can be obtained from oxygen activity measurements obtained under thermodynamically well-defined conditions (temperature and oxide activity). Silicon has also been determined by direct measurement of the activity of the metallic constituent.[116] Measurements of this nature in cast iron melts indicate the concept is feasible; however, the durability of commercial cells is not adequate. Concepts for better designed cells are now available.[117,118] Perhaps a more serious problem is the accuracy of the determinations. To be practical, the analysis must be accurate to ±0.1% C or Si. For cast iron base compositions (about 3.5% C, 2.0% Si) at 1400-1600°C, the respective changes in cell voltage for a 0.1% change in carbon and silicon are 6 and 3 mV. Achieving this level of accuracy represents a major challenge.

Electrochemical analysis appears best suited for elements with lesser requirements for accuracy. A good candidate is the determination of aluminum in cast iron, in conjunction with aluminum additions to the cupola. Iron produced with aluminum additions to the cupola contains 0.1-0.5% Al. It is likely that dealuminization to levels of 0.01-0.02% Al will be necessary to make the iron compatible with current processing. The ability to determine aluminum in cast iron in the needed concentration range has been demonstrated using oxygen sensors.[119] Electrochemical analysis of sulfur, hydrogen and nitrogen in cast iron and steel would also be very useful. However, no suitable electrolytes are known. Further work on electrolytes and durable sensor design is desirable.

**Aluminum Melts.** Due to lower temperature of operation and thus greater availability of electrolytes, the prospects for developing electrochemical methods of analysis for aluminum alloy melts are better than for cast iron. Prospective elements for analysis include hydrogen, trace elements used for grain refinement and eutectic modification (Ti,B,Na,Sr,P,Sb), magnesium to control demagging processes, and lithium which may present serious contamination problems when Al-Li aerospace components enter the scrap system.[103]

At present, with the exception of hydrogen, no commercial methods for rapid or continuous measurement of the above mentioned elements are available. Two relatively rapid techniques for hydrogen have been developed.[120] The first is the Straube-Pfeiffer test used by many foundries, which measures the highest pressure at which a hydrogen bubble forms in an aluminum melt. The second is the Telgas method which involves equilibration of a small quantity of nitrogen with hydrogen in a liquid aluminum sample and subsequently measuring the $H_2$-$N_2$ ratio. Neither of these methods of hydrogen analysis is as rapid and convenient as electrochemical sensing.

The development of electrochemical concentration cells for specific elements in aluminum is currently an active area of research.[121-126] However, no cell has

## TABLE 6
### Elements for Which Electrochemical Concentration Cells are Being Developed.

| Element | Electrolyte | Reference Material | Reference |
|---|---|---|---|
| H | $CaH_2$ | $CaH_2/Ca$ | 121,122 |
| Li | $Li_3PO_4/Li_4SO_4$ | $Li_4Ti_5O_{12}/TiO_2$ | 123 |
|  | $Na\beta\text{-}Al_2O_3$ | $Li_3AlF_6/LiF$ | 124 |
| Na | $Na\beta\text{-}Al_2O_3$ | $Na_3AlF_6/NaF$ | 124 |
|  | $Na\beta\text{-}Al_2O_3$ | $\alpha + \beta - Al_2O_3$ | 125 |
| Mg | $MgCl_2/CaCl_2$ | Mg | 126 |

reached commercialization. The elements for which cells are being developed are listed in Table 6 along with the electrolytes and the reference materials used.

## SOURCES AND CONTROL OF CASTING DEFECTS

### General

The complexity of the total casting process provides numerous opportunities for a wide range of defects. Of concern here are defects whose origins are traced to the liquid metal. These are gas porosity, non-metallic inclusions and microstructural imperfections arising from improper melt nucleation or composition control (including tramp elements). The subject of composition control was treated earlier.

### Gas Porosity Defects

**General.** Solution of this problem is an important foundry industry goal in efforts to achieve greater casting quality. Gas porosity defects have been classified in three categories (1) reaction generated defects, (2) mechanically trapped gas and, (3) precipitated dissolved-gas.[95] Reaction generated defects are large gas bubbles, up to several centimeters in diameter, produced by reaction of the melt with exogenous solid or liquid material. The most important example of this defect is CO "blowholes" produced by reaction of dissolved carbon in cast iron or steel with FeO-rich particles that are inadvertently present in the mold cavity.[95,127] Mechanically trapped gas defects, also manifested as large bubbles, are produced by physically trapping mold gases. In both cases the problems are well understood; the only difficulty is in implementing preventive measures.

The large size of the reaction-generated and mechanically-trapped defects usually renders the cast product useless. Thus, their occurrence necessitates immediate remedial action in the plant. The defect produced by precipitation of dissolved gas, called pinholes, is insidious as the gas bubbles are much smaller (1-4 mm diameter) and, as seen in Figure 23, the defect occurs just below the casting surface. In this

*References pp. 43–51*

**Figure 23.** A cross sectional view of a pinhole defect in cast iron showing its proximity to the casting surface. The width of the defect is 1.6 mm, the wall is lined with a graphite film, which is seen peeling away on the right hand side of the defect. There is a small hole connecting the defect to the casting surface. The ends of the dendrite arms are seen protruding into the pinhole cavity.

case, the defect is undetected until it is machined. In many cases neither the source nor the solution to the problem is apparent, so the defects can persist for long periods.

**Precipitation of Dissolved Gas.** The pinhole defect in aluminum castings, discussed earlier, represents one of the few cases where the source of the problem, dissolved hydrogen, is known. The CO pinhole defect in steel castings is another example where the basis for the imperfection is rather well understood.[60,95,109] In this case the source of the defect is dissolved oxygen which concentrates in the liquid phase in the course of solidification. Depending on the carbon concentration of the melt, the solubility limit for CO can ultimately be exceeded. Processes to deoxidize steel and to prevent reoxidation are sufficiently developed[128] to obviate serious CO defects in a carefully run foundry operation. On-line oxygen analysis is also available to assist the foundryman in preventing this defect.[129] However, the cost of analysis is relatively high because foundry heats are generally small and the commercially available electrochemical oxygen sensors cannot be used for more than one measurement. An oxygen sensor capable of repeated use would be a benefit. Another need is on-line carbon analysis to complement the oxygen measurement. A promising route is electrochemical oxygen activity measurements under conditions of carbon-oxygen equilibrium.[114]

An area that is in need of much understanding is "hydrogen-nitrogen" pinhole defect in cast iron and steel. Although there is a consensus that precipitation of hydrogen and nitrogen are the causes of the defect, the mechanism by which this occurs is in much doubt. Further, there is considerable conflict in existing data on the conditions for both producing and preventing the defect.[57,130,131] It is possible that the defect is produced by more than one mechanism.

The occurrence of pinhole imperfections in castings is perceived to be associated with conditions with high hydrogen and/or nitrogen levels initially in the iron, although this has not been well documented. In fact, attempted correlations between hydrogen concentration and pinholing have been disappointing.[57,130] In the case of steel[132] there are examples where large amounts of dissolved gas did not produce pinholes and, conversely, where pinholes were observed when gas concentrations were initially not very high.

As indicated in Figure 16, cast iron melts are usually saturated with respect to atmospheric nitrogen. Hydrogen levels are generally equivalent to equilibrium with ~0.1 atm $H_2$. Thus the combined partial pressures of the two gases is about 1 atm. To account for the formation of pinholes, the decreasing solubility of both gases with decreasing melt temperature is invoked, as well as the general insolubility of gases in solid iron. Recent work has shown the latter assumption is invalid for nitrogen in cast iron.[133,134] The strong influence of carbon on the activity coefficient of nitrogen in the melt results in greater nitrogen solubility in austenite than in the melt. Hydrogen solubility in the melt is only slightly greater than in austenite.[135] Thus, the possibility for producing supersaturated conditions during solidification seems limited.

Further doubt on the extent of supersaturation attained in a cooling cast iron melt is provided by the data in Figure 16 which shows that the nitrogen concentration in iron, prior to casting, e.g., after inoculation, is lower than the level produced on melting, due to the generally lower melt temperatures. Thus, foundry conditions appear favorable for maintaining nitrogen at equilibrium levels. Considering the preceding information there are reasonable grounds to suspect suggestions that precipitation of subsurface gas bubbles are due to gas absorbed prior to casting.

A more plausible argument for precipitated gas as the primary source of pinhole imperfections can be made if consideration is given to the possibility that local, near-surface supersaturation can occur due to interactions between the gas saturated melt and the mold atmosphere. This would acocunt for the subsurface location of the imperfections and the reported influence of mold conditions on pinhole formation.[130] For example, serious pinhole problems are associated with resin bonded sands containing large concentrations of organic nitrogen compounds. It can be argued in this case that with the iron melt already saturated with nitrogen, the only way to increase nitrogen further is to introduce it at higher than unit activity, a condition met by the nitrogen bearing resins.[136]

Another potential source of local supersaturation is hydrogen present in the

*References pp. 43–51*

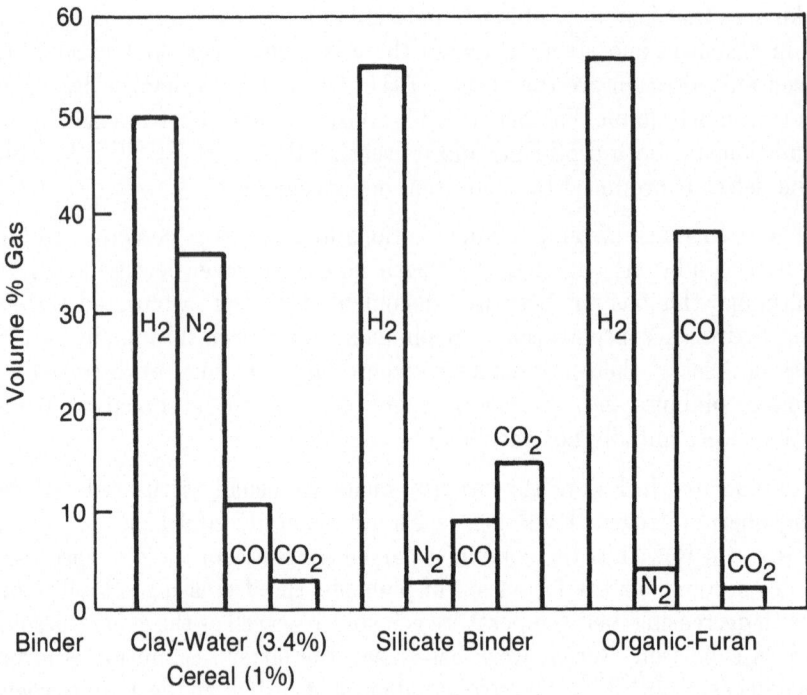

Figure 24. Mold gas compositions with different sand binders.[137,138]

mold atmosphere. The data in Figure 24 show the hydrogen concentration in the mold atmosphere is considerably higher than the equilibrium amount in the incoming iron. Thus, as long as bubbles are not nucleated, further dissolution of hydrogen can occur during mold filling. The well documented pinhole promoting effect of moisture in the mold[130] could be accounted for on this basis. Further, the inhibition of pinholes by additions of finely divided, high volatile bituminous coal (sea coal) to the molding sand[130] might be ascribed to carburization of the melt surface by the volatile organic materials that are generated in the mold. This would increase the activities of hydrogen and nitrogen near the surface and thereby reduce the tendency for further gas dissolution. Consistent with this, relatively non-volatile organic sand additives do not display this beneficial effect.[130]

The notion of local supersaturation as the source of pinhole imperfections can account for a significant amount of the data available on the subject. However, reasonable arguments can also be made for gas entrainment as the primary pinhole mechanism.[132,139] The location of the imperfections, the promotion of the effect with increasing iron turbulence[130] and the role of surface active elements[130] can support this mechanism. Even the characteristic structure of hydrogen-nitrogen pinholes, i.e., a graphite lined bubble with a graphite depleted zone in the adjacent iron, can be produced with entrained gas bubbles.[140]

It is clear there is a need for definitive experimentation to establish the pinholing mechanisms. Existing data provides great confusion. There is general agreement that certain variables affect pinhole formation. However, in many cases beneficial and deleterious effects are attributed to the same variable, e.g., carbon concentration, iron temperature, and solute elements that are either surface active, nitride forming (aluminum in particular) or oxide forming. References for all these effects are given in reference 130.

## Non-Metallic Inclusions

**General.** Since the bulk of cast metal produced is for relatively low stress applications and castings are extensively machined thus freeing the surface of imperfections, relatively simple steps have been sufficient to prevent objectionable levels of inclusions in castings. Today, with the biggest growth sector of the casting business in the area of high strength applications, e.g. ductile iron, and with competitiveness leading to greater efforts to produce thinner section and near net shape castings, inclusions are less tolerable.

**Ferrous Melts.** The main sources of inclusions in castings are mold and refractory materials and reactive metal (Al,Mg,Ca,Si) oxides and sulfides. To prevent these materials from contaminating castings, foundries generally depend on skimming and decanting to separate dross from liquid metal during pouring. Gate and runner systems in the mold are designed to minimize metal turbulence so as to facilitate floatation and trapping of larger particles. Efforts are also made to minimize contact between air and liquid metal by using tapered sprues to prevent air aspiration. Finally, to prevent sand particle contamination, uniformly distributed and effective sand binders are used.

Compared to other cast metals, cast iron has an additional advantage with respect to inclusions. Its dross, which essentially is $SiO_2$, can be minimized by control of melt temperature. Above a critical temperature, which depends on the carbon and silicon activity in the melt, carbon is oxidized in preference to silicon. This is illustrated in Figure 25.[141] This effect is due to the change in the relative stability of CO and $SiO_2$ in the temperature range encountered in cast iron production. The free energy change for the reaction

$$2\underline{C} + SiO_2 = \underline{Si} + 2CO$$

is plotted as a function of temperature in Figure 26 for the three major cast iron types. The carbon and silicon activities in these melts increase in the order malleable < gray < ductile iron. The data in Figure 26 were obtained assuming the activities of $SiO_2$ and CO were unity. From this data, the temperature required to prevent drossing ($SiO_2$ formation) falls between 1375 and 1425°C, depending on the type of iron. Observed temperatures are in good agreement with the equilibrium values.[43,142]

For low ductility casting alloys, e.g., gray iron or aluminum-silicon, further steps to reduce the level of inorganic contaminants in castings would have to be justified on the basis of improved surface finish and/or improved machinability. Melt

*References pp. 43–51*

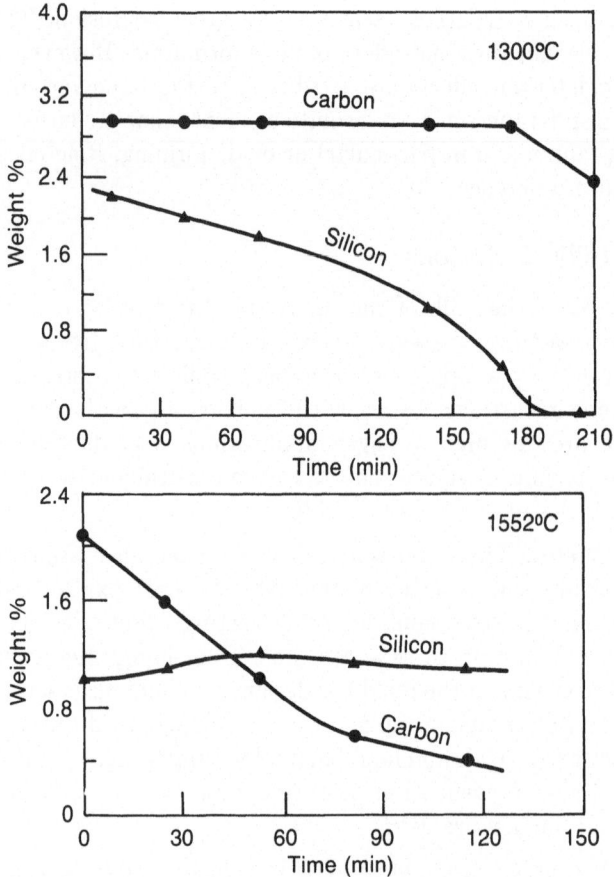

**Figure 25.** Changes in the carbon and silicon concentrations of a cast iron held at 1300 and 1552°C in air.[141]

filtration appears attractive for this application as it is effective for the removal of large particles (see Figure 27) and it occurs immediately before mold filling. Gas sparging of melts, as developed for steel,[144] might also be effective and, perhaps, less costly. However, it might be necessary to confine its use to times prior to inoculation, as it might cause unacceptable agglomeration of the nuclei purposely generated in the inoculation process.

Stokes law arguments suggest that the rise velocity of particles that could be injurious to surface appearance (>100 μm) are in an attractive range for removal by floatation (see Table 7). The sophisticated mathematical models of fluid flow now available might be applied to design of runners and gates to optimize floatation, thus making sparging and filtration unnecessary. Similarly, mold filling models might provide needed insight into minimization of mold erosion, which is another important source of inclusions.[146]

For high strength applications, e.g., ductile iron and steel, the admissible

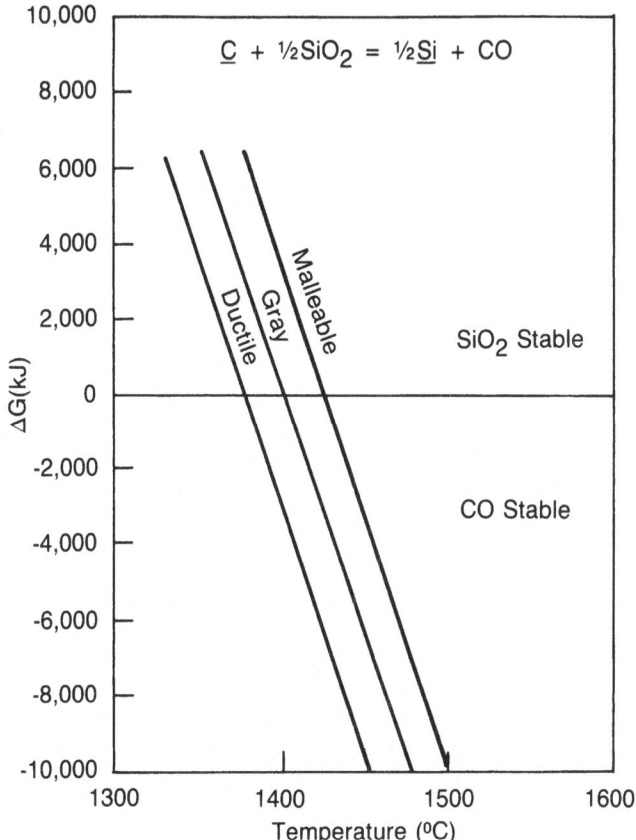

Figure 26. Relative stability of $SiO_2$ and CO in contact with different types of cast iron as a function of iron temperature.

particle size range for inclusions is at least an order of magnitude smaller than the acceptable size for the low ductility alloys. For ductile iron, the graphite nodules are in the range 10-40 μm. Thus, it is desirable to keep inclusion size <10 μm. Studies of dross formation in ductile iron[43] show that most of the dross material falls into three size categories, (1) agglomerate particles with diameters ~100 μm, (2) thread-like stringers up to 500 μm in length and (3) fine particles <10 μm in diameter, which are the building blocks for the stringers and agglomerates. In this case, removal of the larger particles might be sufficient to maintain inclusion size in nodular irons at <10 μm.

For cast steel melts, conditions similar to those for ductile iron seem to exist; the major source of dross is deoxidation which generally produces small particles (<10 μm) that subsequently agglomerate.[147] Here, in addition to filtration, the sparging techniques developed for inclusion removal in primary steel manufacture should be effective.

**Aluminum Melts.** The problem of non-metallic inclusions is more serious

*References pp. 43–51*

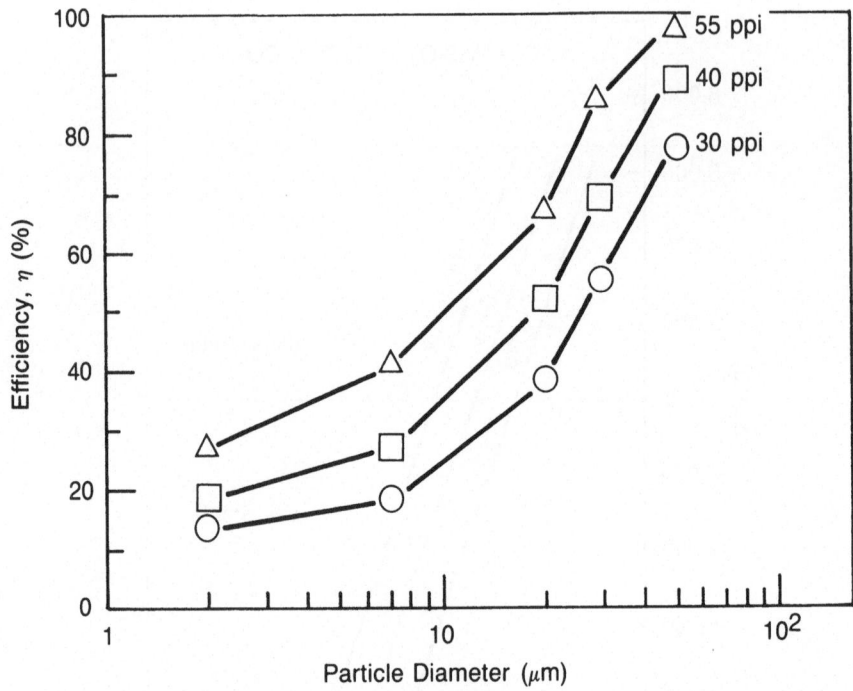

**Figure 27.** Increasing filtration efficiency as a function of particle size for three filter size openings. Filter opening size decreases in the order 30 ppi > 40 ppi > 55 ppi.[143] (ppi = pores per inch)

TABLE 7
Effect of Particle Size on Rise Velocity.[145]

| Particle Size ($\mu$m), | Rise Velocity (cm/sec), |
|---|---|
| 1 | $10^{-4}$ |
| 10 | $10^{-2}$ |
| $10^2$ | 1 |

for aluminum than ferrous melts because aluminum is more easily oxidized, and entrained oxides are more difficult to separate due to the smaller differences in densities between oxide and melt. Inclusions present in aluminum melts are mostly oxides of aluminum and magnesium and spinel. Also present are grain refiner (TiB$_2$), fluxing agents (NaCl,MgCl$_2$) and furnace lining refractory particles. The oxide film that develops on the surface of aluminum melts gives rise to entrained non-metallic particles if the film is seriously disrupted. If undisturbed, the Al$_2$O$_3$ film offers protection from further oxidation. However, the presence of alkali and alkaline earth elements, even in small amounts, increases the permeability of the film, which in turn increases both melt oxidation[110] and inclusion level.

The steps taken to prevent hydrogen pickup and dross formation will also minimize inclusions, e.g., avoiding exposure of the melt to air, minimizing turbulence during melting and handling of liquid alloy, use of furnaces designed to prevent ingress of air and introducing alkali and alkaline earth metals (eutectic modifiers) as late in the process as possible. In addition, covering fluxes will reduce gas absorption and dross formation by serving as a barrier between the melt and the atmosphere.[3]

For quality castings, removal of inclusions is essential even though precautions were taken during melting. In aluminum foundries, filtration is often employed in melting furnaces, pouring ladles[3] and in the inline melt treatment systems for removal of oxidizeable elements and hydrogen.[101] The most recent technological step taken by aluminum casters is filtration of individual castings.[143,148] This attractive idea is also being adopted by iron and steel casters.[149-152]

**Filtration.** The emergence of melt filtration as an important foundry tool necessitates further understanding of the filtration process. Among the areas requiring further attention are: the effectiveness of different filter geometries, filter efficiencies as a function of particle size at realistic metal velocities, the effect of filter composition on particle adherence, the relative efficiency of filters to remove liquid dross, and the effect of changes in metal velocity, due to filter blockage, on cast metal properties.

**Determining Metal Cleanliness.** The inclusion population in liquid cast metals could be more easily controlled if reliable on-line methods for measuring the number and size of inclusions were available. Standard methods, such as chemical analysis, quantitative metallography and volumetric analysis, are too time consuming for on-line application. Other techniques have been developed that are more rapid but their application has largely been confined to production of aluminum sheet and foil. In two of the techniques, UPILM (Ultrasonic Probe in Liquid Metals)[153] and 4M (Mansfield Molten Metal Monitor),[154] an ultrasonic signal is passed through the melt with the reflected signal having characteristics related to the number, size and shape of the inclusions. Most recently, a direct measurement technique, LiMCA (Liquid Metal Cleanliness Analysis),[155] was developed which relates the fluctuations in the apparent resistivity of the melt, moving through a very narrow orifice (150-300 $\mu$m), to the number and size of inclusions present. Material problems have made it difficult to readily adapt these techniques to iron and steel, although such attempts are currently being made.[156]

A number of uses for particle measuring devices can be envisioned, in addition to cleanliness inspection prior to casting. Examples are: determining optimum filtration conditions (filter size, pore structure, etc.) for individual castings as well as fundamental studies of liquid metal filtration; developing optimal runner systems to separate inclusions; determining that melts are properly inoculated (nucleated); and fundamental studies of nucleation phenomena during solidification. The latter

## TABLE 8
### Compositions of Some Graphite Inoculants.[70]

| % Si | % Ca | % Al | % Mn | % R.E. | % Zr | % Ti | % Ba or Sr |
|---|---|---|---|---|---|---|---|
| 80 | 1.6 | 0.9 | — | — | — | — | — |
| 75 | 1.0 | 1.3 | — | — | — | — | — |
| 63 | 1.7 | 1.0 | 5.8 | — | 6.6 | — | 2.8 |
| 64 | 2.7 | 0.9 | 6.2 | — | 6.4 | — | — |
| 47 | — | — | — | — | 39.0 | — | — |
| 55 | 6.4 | 0.9 | — | — | — | 9.1 | — |
| — | — | 0.5 | — | — | — | 68.0 | — |
| 38 | — | — | — | 24.0 | — | — | — |
| 29 | 0.3 | 0.3 | — | 50.0 | — | — | — |
| 35 | — | — | — | 26.0 | — | — | — |
| 61 | 2.3 | 1.2 | 11.0 | — | — | — | 5.2 |
| 77 | — | 0.4 | — | — | — | — | 0.6 |

is a particular need for cast iron where, despite numerous studies,[70] there remains a critical need for further understanding.

**Graphite Nucleation in Cast Iron.** Unlike aluminum alloys which are nucleated by rather well defined intermetallic compounds, graphite in cast iron is nucleated by a large number of materials (Ca,Sr,Ba,C,Ti,Mn,Zr and rare earths). Numerous nucleating alloys (inoculants), with various combinations of these elements, are sold commercially (see Table 8). Foundries generally test the individual materials to determine which works best for their castings. Since the underlying conditions that give preference to one inoculant or another are not understood, it is difficult to guarantee favorable nucleating conditions and it is unlikely that optimum conditions are often achieved. Despite a great deal of previous work in this area,[70] a device that measures the number and size of inclusions could provide very useful, new information on the roles of the various inoculant elements and their relation to conditions in the melt. Further, effects of temperature and changes in number and size of nuclei with time (inoculant fade) could be studied as well as "memory" effects[157–158] where the form of carbon in the scrap has an apparent influence on the type of carbon in the solidified casting.

## SUMMARY

A need for understanding surrounds many aspects of foundry processing of liquid metals. Without this understanding it has been difficult to optimize foundry operations. As a result, this area stands as a fertile ground for major process improvements. This paper has attempted to outline some of the areas that need serious attention.

## ACKNOWLEDGMENTS

The authors acknowledge C. F. Landefeld and J. N. Johnson of the General Motors Research Laboratories for their valuable discussions that helped to orient the ideas presented.

## REFERENCES

1. Midland-Ross Corporation, *Foundry Industry Scoping Study*, Center for Metals Production, CMP Report 86-5, November 1985.
2. F. Neumann, "Comparative Study of Cast Iron Melting in Cupolas and Induction Furnaces," *Electrowärmer International, 29*, (October 1971), 552–564.
3. *Aluminum Casting Technology*, American Foundrymen's Society, Des Plaines, IL, 1986.
4. L. Plutshack, "The Value of Yield Improvement," Paper #2, *90% Yield Seminar: Gating and Risering of Iron Castings*, Foseco Inc., Cleveland, OH, 1983.
5. R. D. Warda, E. F. Darke and H. P. Guindon, "The True Thermal Efficiencies of Thirty Cupolas," *AFS Trans, 89*, (1981), 719–730.
6. J. P. Graham, W. J. Pater, F. Dunn and H. J. Leyshon, "The Effects of Coke Size and Coke Type on Cupola Performance," *The British Foundryman*, (May 1962), 203–218.
7. H. J. Leyshon, M. J. Selby and J. Briggs, "Is Large Coke Really Necessary?" *Foundry*, (February 1974), 52–54.
8. R. Cairns, "The Effect of Using Cokes of Different Size and Varying Ash Contents on the Operation of a Single and Divided Blast Cupola," *The British Foundryman*, (November 1985), 449–452.
9. F. Danis and M. Decrop, "Influence of Blast Input, Coke Size and Melting Coke Ratios on Cupola Performance," *Foundry Trade Journal*, (March 20, 1958), 319–325, (March 27, 1958), 351–358.
10. S. Katz, "Fuels," *Cupola Handbook*, Chapter 8, American Foundrymen's Society, Des Plaines, IL, (1984), 101–128.
11. N. P. Lillybeck, "Influence of Fuel Costs on Cupola Design and Operation," *Modern Casting Tech. Report*, No. 7515, American Foundrymen's Society, Des Plaines, IL, 1975; reprinted from *Efficient Use of Fuels in Metallurgical Industries*, Inst. of Gas Technology.
12. C. F. Landefeld, "Thermal Power Losses to the Cupola Shell and Tuyeres," *AFS Trans, 93*, (1985), 383–88.
13. F. M. Degner and F. T. Kaiser, "Increasing Cupola Energy Efficiency," *AFS Trans, 88*, (1980), 609–14.
14. H. J. Leyshon and M. J. Selby, "Oxygen in Conventional and Divided-Blast Cupola Operations," *BCIRA Report*, No. 1257, March 1977.
15. R. T. Taft and H. R. Perkins, "Cokeless Before the 1980's," *Foundry Trade Journal*, (Aug. 17, 1978), 471–485.

16. S. V. Dighe, W. H. Provis, B. Buczkowski, W. J. Peck and A. D. Karp, "Plasma-Fired Cupola: An Innovation in Iron Foundry Melting," *AFS Trans, 94*, (1986), 323–334.

17. J. P. Vanderhoeck, "Blast Conditioning—II Hot Blast," *Cupola Handbook*, American Foundrymen's Society, Des Plains, IL, (1984), 199–206.

18. H. Choi, "Additions of Supplementary Air to Form an Oxygenated Zone in a Cupola," Thesis, Pennsylvania State University, March 1978.

19. H. J. Leyshon, "The Divided-Blast Cupola and its Development," *Conference on Cupola Operation*, American Foundrymen's Society, Des Plaines, IL, (1980), 247–261.

20. S. Katz and V. R. Spiranello, "Effect of Charged Aluminum on Iron Temperature, Silicon Recovery and Desulfurization in an Iron-Producing Cupola," *AFS Trans, 92*, (1984), 161–172.

21. L. R. Radovich, P. L. Walker, Jr. and R. G. Jenkins, "Importance of Carbon Active Sites in the Gasification of Coal Chars," *Fuel, 62*, (1983), 849–856.

22. S. Katz, "Defining the Coke Properties that Directly Affect the Energy Efficiency of Cupolas," Paper 22, *49th Int. Foundry Congress*, Chicago, IL, April 1982.

23. B. Ozturk and R. J. Fruehan, "Formation of SiO(g), and SiS(g), from Coke," *Iron and Steel Maker* (July 1987), 43–48.

24. R. D. Burlingame, "Metallics for Cupola Melting," *Cupola Handbook*, American Foundrymen's Society, Des Plaines, IL, (1984), 79–82.

25. S. Katz and C. F. Landefeld, "A Kinetic Model for Carbon Pickup in the Cupola: A Step Beyond the Levi Equation," *AFS Trans, 93*, (1985), 209–214.

26. C. F. Landefeld and W. J. Peck, "The Relation Between Silicon Loss and Metallic Silicon in the Cupola Charge," *AFS Trans, 91*, (1983), 1–6.

27. S. Katz and H. C. Rezeau, "The Cupola Desulfurization Process," *AFS Trans, 87*, (1979), 367–376.

28. S. Katz and C. F. Landefeld, "Desulfurization," *Cupola Handbook*, American Foundrymen's Society, Des Plaines, IL, (1984), 351–364.

29. N. E. Rambush and G. B. Taylor, "A New Method of Investigating the Behavior of Charge Material in an Iron-Foundry Cupola and Some Results Obtained," *Foundry Trade Journal*, (November 8, 1945), 197–212, 229–235, 253–260.

30. H. W. Lownie, D. E. Krause and C. T. Greenidge, "How Iron and Steel Melt in the Cupola," *AFS Trans, 60*, (1952), 766–774.

31. M. J. Selby, "Developments in Cupola Design and Operation," Paper No. 9, *45th Int. Foundry Congress*, Budapest, Hungary, October 1978.

32. S. Carter and R. Carlson, "Some Variables in Acid Cupola Melting," *AFS Trans, 62*, (1954), 267–281.

33. N. H. Keyser and W. L. Kann, Jr., "The Effect of Size of Scrap on the Tapping Temperature of a Cupola," *AFS Trans, 66*, (1958), 397–398.

34. R. H. Nafziger, A. D. Hartman, R. F. Farrell and R. D. Burlingame, "Trends in the Quality of Ferrous Scrap as Reflected in Iron Castings: 1981-1984," *AFS Trans, 94*, (1986), 417–426.

35. O. Angeles, G. H. Geiger and C. R. Loper, Jr., "Factors Influencing Carbon Pickup in Cast Iron," *AFS Trans, 76*, (1968), 3–11.

36. C. R. Loper, Jr., S. L. Liu, S. Shirvani and T. H. Whitter, "The Dissolution of Carbon in Cast Iron Melts As Studied Using Commercial Carbon Raisers and Experimental Materials," *AFS Trans, 92*, (1984), 323–337.

37. S. C. Clow, "The Effect and Control of Sulfur in Iron," *AFS Trans*, *86*, (1978), 401–410.

38. J. W. Robison, Jr., "Ladle Treatment with Steel-Clad Metallic Calcium Wire," Paper 35, *Scaninject III*, MEFOS, Lulea, Sweden, 1983.

39. P. Binder, W. Pulvermacher, G. Stolte and J. Rushe, "Stream Degassing and Ladle Degassing," *Ironmaking and Steelmaking*, *13*, (1986), 267–275.

40. M. Remondino, F. Pilastro, E. Natale, P. Costa and G. Peretti, "Inoculation and Spheroidizing Treatments Directly Inside the Mold," *AFS Trans*, *82*, (1974), 239–252.

41. O. Smalley, "Treatment of Nodular-Graphite Iron by the Inmold Process," *Foundry Trade Journal*, (September 25, 1975), 423–430.

42. R. W. Heine and C. R. Loper, Jr., "Dross Formation in the Processing of Ductile Cast Iron," *AFS Trans*, *74*, (1966), 274–280.

43. P. K. Trojan, P. J. Guichelaar, W. N. Bargeron and R. A. Flinn, "An Intensive Investigation of Dross in Nodular Cast Iron," *AFS Trans*, *76*, (1968), 323–333.

44. R. I. L. Guthrie, "Addition Kinetics in Steelmaking," *Proceedings 35th Electric Furnace Conf.*, Iron and Steel Soc. of AIME, Warrendale, PA, (1977), 30–41.

45. S. A. Argyropoulos and R. I. L. Guthrie, "The Dissolution of Titanium in Liquid Steel," *Met Trans*, *15B*, (1984), 47–58.

46. S. A. Argyropoulos, "Dissolution Characteristics of Ferroalloys in Liquid Steel," *Iron and Steel Maker*, (November 1984), 48–57.

47. W. A. Henning, "Efficiency in Desulfurization Practices: Committee 5L Report," *AFS Trans*, *94*, (1986), 815–821.

48. L. R. Farias and G. A. Irons, "A Unified Approach to Bubble-Jetting Phenomena in Powder Injection into Iron and Steel," *Met Trans*, *16B*, (1985), 211–225.

49. L-K. Chiang, G. A. Irons, W-K Lu and I. A. Cameron, "The Kinetics of Desulfurization of Hot Metal by Calcium Carbide Injection," *Process Technology Proceedings 5th Int. Iron and Steel Congress*, Iron and Steel Soc. of AIME, Warrendale, PA, (1986), 441–454.

50. F. Leclerq, J. P. Reboul, C. Gatellier, A. Chevaillier, P. Gugliermina and A. Dufour, "Hot Metal Desulfurization by Injection of Lime," *Scaninject III*, MEFOS, Lulea, Sweden, 1983.

51. C. F. Landefeld and S. Katz, "Kinetics of Iron Desulfurization by $CaO$-$CaF_2$," *Process Technology Proceedings 5th Int. Iron and Steel Congress*, Iron and Steel Soc. of AIME, Warrendale, PA, (1986), 429–440.

52. S. Katz and C. F. Landefeld, "Plant Studies of Continuous Desulfurization with $CaO$-$CaF_2$-C," *AFS Trans*, *93*, (1985), 215–228.

53. A. Moriya, T. Nagahata, K. Ieda, K. Ichihara and M. Ishikawa, "Steel Quality Improvement by Flux Injection," Paper 32, *Scaninject III*, MEFOS, Lulea, Sweden, 1983.

54. N. Sano, "Thermodynamic Aspects of Removing Impurity Elements from Carbon-Saturated Iron," *Symposium on Foundry Processes: Their Chemistry and Physics*, General Motors Research Laboratories, Warren, MI, September 21–23, 1986.

55. M. Faral and H. Gaye, "Metal Slag Equilibria," *Second Int. Symp. on Metallurgical Slags and Fluxes*, H. A. Fine and D. R. Gaskell, eds., The Metallurgical Society of AIME, Warrendale, PA, (1984), 159–179.

56. I. D. Sommerville, "The Measurement, Prediction and Use of Capacities of Metallurgical Slags," *Scaninject IV*, MEFOS, Lulea, Sweden, 1986.

57. B. Hernandez and J. F. Wallace, "The Mechanisms of Pinhole Formation in Gray Iron," *AFS Trans*, *87*, (1979), 335–348.

58. M. Svilar and J. F. Wallace, "Removal of Aluminum from Gray Cast Iron to Reduce Pinholes," *AFS Trans*, *86*, (1978), 421–430.

59. G. K. Sigworth, "A Scientific Basis for Degassing Aluminum," *AFS Trans*, *95*, (1987), 73–78.

60. J. M. Svoboda, "Behavior of Gases in Steel Castings," *Steel Founders' Research J.*, No. 9, (First Quarter 1985), 10-26.

61. M. Robinson, "Nitrogen Levels in Ductile Iron: AFS Committee 12H Report," *AFS Trans*, *87*, (1979), 503–508.

62. S. Katz, Unpublished data.

63. A. J. Clegg, "Aluminum Degassing Practice," *Int. Molten Aluminum Processing*, American Foundrymen's Society, Des Plaines, IL, (1986), 369–380.

64. K. Schwerdtfeger and H. G. Shubert, "Solubility of Nitrogen in $CaO$-$Al_2O_3$ Melts in Graphite Crucibles at 1600°C," *Arch. Eisenhüttenwes*, *45*, (1974), 649–655.

65. K. Schwerdtfeger and H. G. Schubert, "Solubility of Nitrogen and Carbon in $CaO$-$Al_2O_3$ Melts in the Presence of Graphite," *Met Trans*, *8B*, (1977), 535–540.

66. K. Schwerdtfeger, W. Fix and G. H. Schubert, "Solubility of Nitrogen in $CaO$-$SiO_2$-$Al_2O_3$ Slag in the Presence of Graphite at 1450°C," *Ironmaking and Steelmaking*, (1978), No. 2, 67–71.

67. "Secondary Steelmaking," Metallurgical Society, London, 1978.

68. I. D. Sommerville, "The Capacities and Refining Capabilities of Metallurgical Slags," *Symposium on Foundry Processes: Their Chemistry and Physics*," General Motors Research Laboratories, Warren, MI, September 21–23, 1986.

69. J. F. Janowak, R. B. Gundlach, G. T. Eldes and K. Röhrig, "Technical Advances in Cast Iron Metallurgy," *AFS International Cast Metals J.*, (December 1981), 28–42.

70. J. F. Wallace, "Effects of Minor Elements on the Structure of Cast Irons," *AFS Trans*, *83*, (1975), 363–377.

71. A. Kagawa and T. Okamoto, "Theoretical Calculations of Eutectic Temperature and Composition in Iron-Carbon Ternary and Multicomponent Alloys," *J. Mat. Sci.*, *22*, (1987), 643–650.

72. R. B. Gundlach, J. F. Janowak, S. Bechet and K. Röhrig, "On the Problems with Carbide Formation in Gray Cast Iron," *Third Int. Symp. on the Physical Metallurgy of Cast Iron*, Sodertalje, Sweden, August 29-31, 1984.

73. J. C. Sawyer and J. F. Wallace, "Effects of Neutralization of Trace Elements in Gray and Ductile Iron, Part 2," *AFS Trans*, *79*, (1971), 386–404.

74. D. N. Pocklington, "Removal of Residuals from Liquid Steel-Practical Limitations and Potential Techniques," *Ironmaking and Steelmaking*, *12*, (1985), 289.

75. R. B. Coates and H. J. Leyshon, "Refining Molten Iron by Chlorine Treatment," *BCIRA Journal*, *11*, (July 1963), 451–457.

76. R. N. Andrews, J. B. Andrews and C. A. Andrews, "Removal of Excess Manganese from Gray Iron Using Polyvinylchloride as a Chlorine Source," *AFS Trans*, *92*, (1984), 505–514.

77. R. E. Brown, H. V. Makar and R. J. Divilo, "Refining Molten Iron by Sulfide-Forming Slags and Chlorination," *Reports of Investigations*, 8065, U.S. Bureau of Mines, Washington, DC, (1975).

78. F. Tsukihashi, A. Werme, F. Matsumoto, A. Kasahara, M. Yukinobu, T. Hyodo, S. Shiomi and N. Sano, "Thermodynamics of the Soda Slag System for Hot Metal Treatment," *Second Int. Symp. on Metallurgical Slags and Fluxes*, H. A. Fine and D. R. Gaskell, eds., The Metallurgical Society of AIME, Warrendale, PA, (1984), 89–106.

79. H. V. Makar and R. E. Brown, "Upgrading Copper-Contaminated Ferrous Scrap by Treatment with Sodium Sulfate," *AFS Trans*, 82, (1974), 45–54.

80. X. Liu and J. H. E. Jeffes, "Effect of Sodium Sulfide on Removal of Copper and Tin from Molten Iron," *Ironmaking and Steelmaking*, 12, (1985), 293–294.

81. T. Okazaki and D. G. C. Robertson, "Removal of Tramp Elements: Mathematical Modelling," *Ironmaking and Steelmaking*, 12, (1985), 295–298.

82. T. Imari and N. Sano, *Tetsu-to-Hagane*, 72, (1986), S962.

83. M. Köhler and H-J. Engell, "Partition Equilibria of Tramp Elements Between Iron Melts and Calcium-Halide Slags," *Second Int. Symp. on Metallurgical Slags and Fluxes*, H. A. Fine and D. R. Gaskell, eds., The Metallurgical Society of AIME, Warrendale, PA, (1984), 483–496.

84. J. Szekely, T. Lehner and C. W. Chang, "Flow Phenomena, Mixing and Mass Transfer in Argon Stirred Ladles," *Ironmaking and Steelmaking*, (1979), No. 6, 285–293.

85. T. DebRoy and A. K. Majumdar, "Predicting Fluid Flow in Gas-Stirred Systems," *J. Metals*, (November 1981), 42–48.

86. M. Sano and K. Mori, "Fluid Flow and Mixing Characteristics in Gas Stirred Molten Metal Bath," *Trans ISIJ*, 23, (1983), 169–175.

87. D. Mazumdar and R. I. L. Guthrie, "Mixing Models for Gas Stirred Metallurgical Reactors," *Met Trans*, 17B, (1986), 725–733.

88. S. Asai, T. Okamoto, J. C. He and I. Muchi, "Mixing Time of Refining Vessels Stirred by Gas Injection," *Trans ISIJ*, 23, (1983), 43–50.

89. N. El-Kaddah and J. Szekely, "Mathematical Model for Desulfurization Kinetics in Argon-Stirred Ladles," *Ironmaking and Steelmaking*, (1981), No. 6, 269–278.

90. G. M. Marrone and D. J. Kirwan, "Mass Transfer to Suspended Particles in Gas-Liquid Agitated Systems," *AIChE Journal*, 32, (1986), 523–525.

91. G. K. Sigworth and T. A. Engh, "Chemical and Kinetic Factors Related to Hydrogen Removal from Aluminum," *Met Trans*, 13B, (1982), 447–460.

92. T. DebRoy, N. H. El-Kaddah and D. G. C. Robertson, "Mixed Transport Control of Gas-Liquid Metal Reactions," *Met Trans*, 8B, (1977), 271–277.

93. S. Ohguchi and D. G. C. Robertson, "Kinetic Model for Refining by Submerged Powder Injection: Part I Transistory and Permanent Contact Reactions," *Ironmaking and Steelmaking*, 11, (1984), 262–273.

94. S. H. Kim and R. J. Fruehan, "Physical Modeling of Liquid/Liquid Mass Transfer in Gas Stirred Ladles," *Met Trans*, 18B, (1987), 381–389.

95. R. W. Monroe, "Gas Holes in Iron and Steel Castings," *Steel Founders' Research J.*, No. 3, (Third Quarter 1983), 5–12.

96. C. Bodsworth, "Technological Means for Removal of Tramp Elements," *Ironmaking and Steelmaking*, 12, (1985), 290–292.

97. E. Bossing, "Aluminum Aerospace Castings — 25 Years in Review," *International Molten Aluminum Processing*, American Foundrymen's Soc., Des Plaines, IL, (1986), 1–30.

98. *Recycled Metals in the 1980's*, National Association Recycling Industries, New York, NY, 1982.

99. D. Montagna and H. V. Makar, "Method for Wrought and Cast Aluminum Separation," *U.S. Patent*, 4,330,090, May 18, 1982.

100. K. A. Bowman, "Alcoa's Used Beverage Can (UBC), Alloy Separation Process," *Recycle and Secondary Recovery of Metals*, P. R. Taylor, H. Y. Sohn and N. Jarrett, eds., The Metallurgical Soc. of AIME, Warrendale, PA, (1985), 429–444.

101. D. V. Neff, "Impurity Control in Aluminum Alloy Melting Processes Using the Gas Injection Pump," *International Molten Aluminum Processing*, American Foundrymen's Soc., Des Plaines, IL, (1986), 341–368.

102. G. K. Sigworth and T. A. Engh, "Refining of Liquid Aluminum — A Review of Important Chemical Factors," *Scandinavian J. Metallurgy*, *11*, (1982), 143–149.

103. J. H. L. VanLinden, R. E. Miller and R. Bachowski, "Chemical Impurities in Aluminum," *Symposium on Foundry Processes: Their Chemistry and Physics*, General Motors Research Laboratories, Warren, MI, September 21–23, 1986.

104. B. L. Tiwari, "Demagging Processes for Aluminum Alloy Scrap," *J. Metals*, *34*, No. 9, (1982), 54–58.

105. M. R. Smith, "The Present Status of the Derham Process for Magnesium Removal in Secondary Aluminum Smelting," *Conservation and Recycling*, *6*, No. 1–2, (1983), 33–40.

106. B. L. Tiwari and R. A. Sharma, "Electrolytic Removal of Aluminum from Scrap Aluminum," *J. Metals*, *36*, No. 7, (1984), 41–43.

107. B. L. Tiwari, B. J. Howie and R. M. Johnson," Electrolytic Demagging of Secondary Aluminum in a Prototype Furnace," *AFS Trans*, *94*, (1986), 385–390.

108. P. D. Hess and G. K. Turnbull, "Effects of H on Properties of Al Alloys," *Hydrogen in Metals*, I. M. Berstein and A. W. Thompson, eds., ASM, Metals Park, OH, (1974), 277–87.

109. K. Kubo and R. D. Pehlke, "Mathematical Modeling of Porosity Formation in Solidification," *Met Trans*, *16B*, (1985), 359–365.

110. C. N. Cochran, D. L. Belitskus and D. L. Kinosz, "Oxidation of Aluminum-Magnesium Melts in Air, Oxygen, Flue Gas and Carbon Dioxide," *Met Trans*, *8B*, (1977), 323–332.

111. W. Theile, "The Oxidation of Aluminum and Aluminum Alloy Melts," *Aluminum*, *38*, (1962), 707–715.

112. W. R. Opie and W. J. Grant, "Hydrogen Solubility in Aluminum and Some Aluminum Alloy," *Trans AIME*, *188*, (1950), 1237–1241.

113. R. Y. Lin and M. Hoch, "Solubility of Hydrogen in Molten Aluminum Alloys," TMS-AIME Annual Meeting, New Orleans, LA, 1986.

114. A. R. Romero, J. Harkki and D. Janke, "Oxygen and Carbon Sensing in Fe-O-C and Fe-O-C-X Melts at Elevated Carbon Contents," *Steel Research*, *57*, (1986), 636–686.

115. S. C. Ghorpade, R. W. Heine and C. R. Loper, Jr., "Oxygen Probe Measurements in Cast Irons," *AFS Trans*, *83*, (1975), 193–198.

116. K. Ichihara, D. Janke and H-J. Engell, "A New Silicon Sensor for Hot Metal Measurements," *Steel Research, 57,* (1986), 166–187.

117. D. Janke, "Basic Considerations on the Design of Oxygen Probes for Continuous Measurements in Steel Melts," *Arch. Eisenhüttenwes, 54,* (1983), 259–266.

118. K. Nagata, N. Tsuchiya, M. Sumito and K. S. Goto, "Oxygen Potentials in Liquid Pig Iron and Slag and Analysis of Reactions in the Blast Furnace by Means of Affinities of the Reactions," *Tetsu-to-Hagane, 68,* (1982), 2271–2278.

119. S. Katz, D. E. McInnes, D. L. Brink and G. A. Wilkinson, "Determination of Aluminum in Malleable Iron from Measured Oxygen," *AFS Trans, 88,* (1980), 835–844.

120. P. D. Hess, "Methods for Determining Hydrogen in Aluminum Alloys," *Light Metals 1972,* The Metallurgical Society, AIME, Warrendale, PA, (1972), 367–85.

121. R. Gee and D. J. Fray, "Instantaneous Determination of Hydrogen Content in Molten Aluminum and Its Alloys," *Met Trans, 9B,* (1978), 427–430.

122. B. L. Tiwari, unpublished data.

123. P. C. Yao and D. J. Fray, "Determination of the Lithium Content of Molten Aluminum Using a Solid Electrolyte," *Met Trans, 16B,* (1985), 41–46.

124. A. A. Debreuil and A. D. Pelton, "Probes for the Continuous Monitoring of Sodium and Lithium in Molten Aluminum," *Light Metals 1985,* The Metallurgical Society of AIME, Warrendale, PA, (1985), 1197–1205.

125. R. J. Brisley and D. J. Fray, "Determination of the Sodium Activity in Aluminum and Aluminum-Silicon Alloys Using Sodium Beta Alumina," *Met Trans, 14B,* (1983), 435–440.

126. B. L. Tiwari and B. J. Howie, "Electrochemical Probe for Measuring Magnesium Concentration in Molten Aluminum," U.S. Patent 4,601,810, July 22, 1986.

127. E. F. Ryntz, Jr., R. E. Schroeder, W. W. Chaput and W. O. Rassenfoss, "The Formation of Blowholes in Nodular Iron Castings," *AFS Trans, 91,* (1983), 139–144.

128. R. J. Fruehan, *Ladle Metallurgy Principles and Practice,* Iron and Steel Society of AIME, Warrendale, PA, 1985.

129. E. T. Turkdogan and R. J. Fruehan, "Review of Oxygen Sensors for Use in Steelmaking and of Deoxidation Equilibrium," *CIM Quarterly,* Vol. II, (1972), 371–379.

130. S. F. Carter, W. J. Evans, J. C. Harkness and J. F. Wallace, "Factors Influencing the Formation of Pinholes in Gray and Ductile Iron," *AFS Trans, 87,* (1979), 245–268.

131. R. V. Naik and J. F. Wallace, "Surface Tension-Nucleation Relations in Cast Iron Pinhole Formation," *AFS Trans, 88,* (1980), 367–388.

132. F. Chen and J. Keverian, "Effect of Nitrogen on Subsurface Pinholes in Steel Castings," *AFS Trans, 74,* (1966), 281–289.

133. A. Kagawa and T. Okamoto, "Partition of Nitrogen in Hypo-Eutectic and Nearly Eutectic Iron-Carbon Alloys," *Trans Japan Inst. of Metals, 22,* No. 2, (1981), 137–143.

134. A. Kagawa and T. Okamoto, "Behavior of Nitrogen in Solidifying Cast Iron," *Trans Japan Foundrymen's Society, 2,* (1983), 12–15.

135. A. Kagawa and T. Okamoto, "Partition of Alloying Elements in Freezing Cast Irons and Its Effect on Graphitization and Nitrogen Blowhole Formation," *Symposium on Foundry Processes: Their Chemistry and Physics,* General Motors Research Laboratories, Warren, MI, September 21–23, 1986.

136. S. Yamada, H. Kubota and E. Kato, "Gas Defects in Steel Castings Caused by Phenol Urethane Cold Box Cores," *Imono*, 57, No. 3, (1985), 23–27.

137. C. Locke and R. L. Ashbrook, "Nature of Mold Gases, A Review," *AFS Trans*, 80, (1972), 91–104.

138. W. D. Scott and C. E. Bates, "Decomposition of Resin Binders and the Relationship Between Gases Formed and the Casting Surface Quality," *AFS Trans*, 83, (1975), 519–524.

139. E. T. Turkdogan, "Physicochemical Phenomena of Mechanisms and Rates of Reactions in Melting, Refining and Casting of Foundry Irons," *Symposium on Foundry Processes: Their Chemistry and Physics*, General Motors Research Laboratories, Warren, MI, September 21–23, 1986.

140. S. Yamamoto, Y. Kawano, Y. Murakami, B. Cheng and R. Ozaki, "Producing Spheroidal Graphite Cast Iron by Suspension of Gas Bubbles in Melts," *AFS Trans*, 83, (1975), 217–226.

141. R. W. Heine, "Oxidation-Reduction Principles Controlling the Composition of Cast Irons," *AFS Trans*, 59, (1951), 121–138.

142. R. W. Heine and C. R. Loper, Jr., "Dross Formation in the Processing of Ductile Cast Iron," *AFS Trans*, 74, (1966), 274–280.

143. L. J. Gauckler, M. M. Waeber, C. Conti and M. Jacob-Duliere, "Ceramic Foam for Molten Metal Filtration," *J. Metals*, (September 1985), 47–51.

144. E. T. Turkdogan, "Ladle Deoxidation, Desulfurization and Inclusions in Steel — Part I: Fundamentals," *Arch Eisenhüttenwes*, 54, No. 1, (1983), 1–10.

145. A. Nicholson and T. Gladman, "Non-Metallic Inclusions and Developments in Secondary Steelmaking," *Ironmaking and Steelmaking*, 13, No. 2, (1986), 53–69.

146. K. Yonekura, Y. Yamammoto, M. Nakamura, M. Nakamichi, H. Yoshioka and M. Ohashi, "Finding the Source of Sand Inclusions Using Tracing Sands," *AFS Trans*, 94, (1986), 277–284.

147. W. O. Philbrook, "Oxide Inclusions in Steel (A), Oxygen Reactions with Liquid Steel," *Int. Metals Rev.*, L. H. VanVlack, ed., ASM, Metals Park, OH, (1977), 187–201.

148. D. Apelian, C. E. Ekert, R. Mutharasen and R. E. Miller, "Refining Molten Aluminum by Filtration Technology," *Refining and Alloying of Liquid Aluminum and Ferro-Alloys*, T. A. Engh, S. Lyng and H. A. Oye, eds., Norwegian Inst. Technology, Trondheim, 1985.

149. A. M. Arzt, "Filtration of Ferrous Metals," *Modern Casting* (March 1986), 24–26.

150. S. Ali, R. Mutharasen and D. Apelian, "Physical Refining of Steel Melts by Filtration," *Met Trans, 16B*, (1985), 725–742.

151. A. Aubrey, J. Brockmeyer and P. F. Wieser, "Dross Removal from Ductile Iron with Ceramic Foam Filters," *AFS Trans*, 93, (1985), 171.

152. A. Ilhan, I. Dutta, J. Brockmeyer and P. F. Wieser, "Cast Steel Quality Improvement by Filtration with Ceramic Foam Filters," *AFS Trans*, 93, (1985), 177.

153. D. A. Doutre, "The Development and Application of a Rapid Method of Evaluating Molten Metal Cleanliness," PhD Thesis, McGill University, 1984.

154. T. L. Mansfield, "Molten Aluminum Quality Measured with Reynolds 4M$^{TM}$ System," *Light Metals 1984*, The Metallurgical Society of AIME, (1984), 1305–1328.

155. R. I. L. Guthrie and D. A. Doutre, "On-Line Measurement of Inclusions in Liquid Metals," *Int. Seminar on Refining and Alloying of Liquid Aluminum and Ferro-Alloys*, T. A. Engh, S. Lyng and H. A. Oye, eds., Norwegian Inst. Technology, Trondheim, 1985.

156. R. I. L. Guthrie, "On the Detection, Behavior and Control of Second Phase Particles in Liquid Metals," *Symposium on Foundry Processes: Their Chemistry and Physics*, General Motors Research Laboratories, Warren, MI, September 21–23, 1986.

157. W. J. Williams, "A Study of Some Metallurgical Factors Influencing Chill and Mottle Formation in Gray Iron," *BCIRA Journal of Research and Development*, *4*, (1952), 403.

158. T. Wolverson, "Ultra-Light Iron Castings for the Motor Industry," *British Foundryman*, *1*, (August 1957), 395.

## DISCUSSION

**R. C. Creese** *(West Virginia University)*

Please elaborate on the nature of problems foundries are experiencing in desulfurizing with calcium carbide?

**A.** The problem is calcium carbide reacts with water to form acetylene which is explosive. Foundries, particularly when desulfurizing continuously, cannot take the chance of adding a quantitative amount of calcium carbide. Enough extra calcium carbide must be added to insure that under almost any condition, the desired low sulfur concentrations will be achieved. As a result, the spent slag contains considerable amounts of calcium carbide. When this is disposed of, it creates problems like fires and explosions and people don't like to work with it. In addition, various government agencies, State and Federal, have required, that the spent calcium carbide slag be treated as a hazardous material. Disposal of the spent desulfurizer in a prescribed manner is quite expensive.

**A. McLean** *(University of Toronto)*

One of the things that GM has been involved in over the past two or three years has been the application of plasma as an energy source in the cupola. You referred to coke and energy transfer, would you comment on where plasma sits in that scenario?

**A.** Ordinarily, the cupola is a coke-fired device. In the case of the plasma cupola, the melting energy is obtained from a combination of electricity and coke. The fact that electricity is employed means an expensive energy component is introduced and something that is very special must be achieved to make this worthwhile. One thing that such a cupola can do is melt very fine metallic charge material, like borings. In a coke-fired cupola operation, loose, small pieces of metallic material will be blown out of the cupola by the high velocity blast. But in a plasma cupola, using a considerable amount of energy from electricity, the total amount of gas that is generated is relatively small. As a result, this very inexpensive charge material can be melted without major losses due to entrainment in the gas stream. The plasma cupola concept is economical when melting such material. There is a lot of work going on now that is sponsored by Westinghouse and Electric Power Research Institute. They are looking into other benefits of this concept, such as *in-situ* production of silicon and better desulfurization. They have formed a consortium with a group of foundries. As a result, the information that they generate is not generally available. I am not really familiar with the status of these programs.

# PHYSIOCHEMICAL PHENOMENA OF MECHANISMS AND RATES OF REACTIONS IN MELTING, REFINING, AND CASTING OF FOUNDRY IRONS

### E. T. TURKDOGAN

*USS, a Division of USX Corporation*
*Technical Center*
*Monroeville, Pennsylvania 15146*

## ABSTRACT

Mechanisms and rates of metallurgical reactions are discussed in terms of physicochemical phenomena occurring in melting, refining, and casting of foundry irons. Solidification characteristics of foundry irons and some of the casting problems are shown to be strongly influenced by the interfacial energy-related phenomena. Rate-controlling reaction mechanisms are discussed in relation to the desulfurization of cupola iron in a gas-stirred ladle (batch process) or a reactor (continuous process). Examples are given to show the variations in the rate-controlling reaction mechanisms with the type and scale of the experiments on gas-slag-metal reactions. A brief commentary is made on the rate of gasification of coke and its effect on the cupola performance. The state of gas-slag-metal reactions in the cupola is shown to be similar to that in the iron-blast furnace.

In relation to the subjects discussed in this paper, the following technical data are compiled in the Appendix: (1) solute interaction coefficients in liquid iron alloys; (2) solubilities of Mg, MgS, and TiC in liquid iron alloys; and (3) surface tension of liquid cast iron.

## INTRODUCTION

We have come a long way during the past few decades in developing a better understanding of the complexity of rates of reactions in high-temperature metallurgical processes. The principles of the rates of heat and mass transfer, interfacial reaction kinetics, nucleation, and growth of reaction products are well established through theoretical considerations of individual rate processes for specified conditions. However, the experimentally observed rates of interfacial reactions in gas-slag and metal-slag systems are often difficult to interpret in terms of the limiting rate laws or combinations thereof for mixed rate control.

Unless the experimental data reveal a physicochemical phenomenon that has a decisive influence on the rate of a reaction, there is little to be gained from an annotated list of references to the so-called rate measurements. In a recently published book by the author,[1] selected examples of rate measurements are given to demonstrate various phenomenological aspects of the rate-controlling processes, which are not artifacts of some experimental circumstance.

With the objective of assisting the technology implementation in the foundry industry, the rate phenomena are discussed in this paper with specific examples of practical importance:

1. Interfacial energy-related phenomena in foundry iron castings, with particular reference to the mechanism of (i) formation of pinholes in castings and (ii) nodulation of graphite in ductile iron.

2. Rate-controlling reaction mechanisms in the desulfurization of cupola iron in gas-stirred melts.

3. Rate of gasification of coke and its effect on the cupola performance.

4. The state of gas-slag-metal reactions in the cupola.

In relation to the subjects discussed in this paper, the following technical data are compiled in the Appendix:

1. Solute interaction coefficients in liquid iron alloys.
2. Solubilities of Mg, MgS, and TiC in liquid iron alloys.
3. Surface tension of liquid cast iron.

## INTERFACIAL ENERGY-RELATED PHENOMENA IN FOUNDRY IRON CASTINGS

In the refining of liquid metals and their solidification, the rates of interfacial reactions between a liquid metal and a second phase, the latter being a gas, slag, refractory oxide, or graphite, are often influenced by the interfacial energy-related phenomena. For example, a spontaneous interfacial turbulence may occur when a surface active solute is transferred across the liquid metal/slag interface, and thus, markedly affect the rate of the reaction. Another consequence of interfacial turbulence is the preferential erosion of a solid at the three-phase junction, the so-called "flux-line attack," as for instance, at the refractory oxide/slag/metal junction. Then, there is the familiar retarding effect of surface active oxygen and sulfur on the rates of gas-metal reactions.

In this section of the paper, I will discuss two special cases of interfacial energy-related phenomena: (i) the formation of subsurface pinholes in foundry iron castings, and (ii) the role of interfacial energy in the nodulation of graphite in ductile iron.

## MECHANISM OF FORMATION OF SUBSURFACE PINHOLES IN FOUNDRY IRON CASTINGS

The formation of subsurface pinholes in foundry irons, cast in green sand molds, is known to be a consequence of mold-liquid metal interaction, resulting in the entrapment of gas bubbles ($H_2$, $N_2$, or $CO$) at the initial stages of solidification. The cause of pinhole formation in castings, that is of major concern to foundrymen, has been investigated extensively since the mid-1950s; the references cited[2-7] are selected examples of the earlier investigations.

The pinholes, usually < 3 mm diameter, are located just beneath the surface of the casting. The inner surface of the pinhole in gray and ductile irons is coated with a thin layer of graphite, and the surrounding matrix is usually graphite free. Aluminum contents of over 0.01% promote pinhole formation; however, at concentrations above 0.2% Al, there is no pinhole formation. As would be expected, the elimination of moisture or resin constituents such as urea, $(NH_2)_2CO$, and hexamine, $(CH_2)_6N_4$, sharply reduces the extent of pinhole formation. Also, carbonaceous mold additions, such as 5 to 6% seacoal or 2% pitch, are known to be effective in reducing pinholing in gray and ductile irons. Other additions, such as iron oxide and sodium silicate mold washes, also reduce pinholing.

An attempt is made in the following discussion to identify criteria for the formation of pinholes in foundry iron castings.

**Release of Gas Bubbles**—The volatile matter in the sand mold is the source of gas which forms bubbles in the liquid metal. These form most easily at the crevices of the mold in contact with the liquid metal. The criterion for bubble growth at crevices is given by the following thermodynamic relationships for contact angles $\theta < 90°$ and $\theta > 90°$.

For $\theta < 90°$:

$$\frac{0.02\sigma}{P_0 + P_f} \sin\theta > r > \frac{0.02\sigma}{P_g - (P_0 + P_f)} \tag{1}$$

For $\theta > 90°$:

$$-\frac{0.02\sigma}{P_0 + P_f} \cos\theta > r > \frac{0.02\sigma}{P_g - (P_0 + P_f)} \sin\theta \tag{2}$$

where $\theta$ is the contact angle between the liquid metal and the mold, $r$ is the pore radius at the mold-metal interface, $\sigma$ the surface tension of the liquid metal, $P_g$ the gas pressure in the pores of the mold, $P_0$ the atmospheric pressure, and $P_f$ the ferrostatic pressure.

For the values of $\sigma = 900$ mNm$^{-1}$, $P_g = 3.2$ bar, $P_0 + P_f = 1.2$ bar, the bubble release will be sustained by pore sizes in the range 9 to 15 $\mu$m for $\theta = 90°$ and 5.8 to 11.5 $\mu$m for $\theta = 140°$. For a given excess gas pressure in the pores, the lower the surface tension, the smaller would be the pore sizes for the release of bubbles at the mold-liquid metal interface. The mold coating that reduces the pore size and/or the contact angle $\theta$, will shift the critical value of $\sigma$ to a lower level for the onset of pinholing.

*References pp. 83–84*

When the gas bubble reaches a critical size before the onset of solidification, it is released into the melt. At low gas-flow rates in inviscid liquids, the bubble volume, $V_d$, at detachment is given by[8]

$$V_d = \frac{2\pi r \sigma}{\rho g} \qquad (3)$$

where $\rho$ is the density of the liquid metal, and $g$ the gravitational acceleration. Davidson and Schuler[8] derived the following theoretical equation for bubble volume at detachment for isothermal and constant flow conditions:

$$V_d (\text{cm}^3) = 0.022 G^{1.2} \qquad (4)$$

where the gas-flow rate $G$ is in units of $\text{cm}^3\text{s}^{-1}$. The ratio $G/V_d$ is the bubble frequency, $f$, per second

$$f = 45.45 G^{-0.2} \qquad (5)$$

In the present context, the gas-flow rate $G$ is that generated from the volatile matter in the mold.

The gas bubbles released from the mold into the liquid metal, prior to the onset of solidification, decrease in size with decreasing $\sigma^2$ as noted from the relations in Equations 2 and 3. Also, noting that the velocity of rise of small bubbles in a liquid medium is proportional to the square of the bubble diameter, liquid cast iron of lower surface tension would be expected to be more susceptible to pinholing during solidification in a sand mold. This theoretical deduction has, in fact, been substantiated by the experimental observations of Wallace and co-workers,[9,10] who found that the propensity to pinholing in green sand mold castings increases with decreasing surface tension of the liquid cast iron.

Wallace and co-workers also found that the threshold value of surface tension for pinhole formation increases from a low value of about 400 $\text{mNm}^{-1}$ for white iron at 1400°C to a high value of about 1100 $\text{mNm}^{-1}$ for ductile iron, the gray iron having an intermediate critical surface tension for the onset of pinholing. The threshold value of surface tension is also a function of solidification rate. For example, Naik and Wallace[10] showed that ductile iron cast in a thin section, 6.4 mm thick pad, (representing fast solidification) had no pinholes, while that cast in a thicker section, 19 mm, (representing slow solidification) developed pinholes.

The mold additions which affect the extent of pinholing, may be explained, in part, by the theoretical relation in Equation 2. The mold wash or dressing when exposed to the temperature of the liquid cast iron may reduce the average pore size at the mold surface, thus lower the threshold $\sigma$ for the onset of pinholing. Also, if the mold coating lowers the contact angle $\theta$, the threshold $\sigma$ will be lowered as indicated by Equation 2. In this connection, reference may be made to the papers by Selcuk and Kirkwood[11] and by McSwain, et al,[12] who measured the contact angle between liquid cast iron and polycrystaline graphite at 1200°C. For Mg-treated cast iron (very low sulfur activity), the contact angle $\theta = 140°$ decreases to $\theta = 85°$

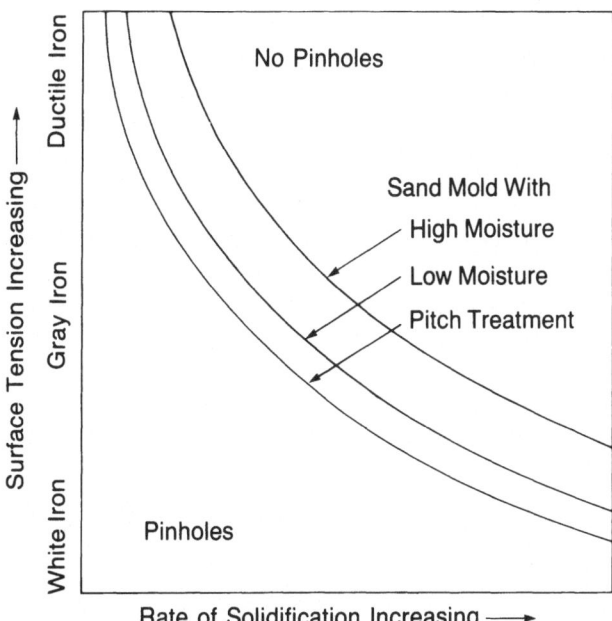

Figure 1. Schematic Representation of Conditions for Subsurface Pinhole Formation in Foundry Iron Castings in Sand Molds.

with the addition of 0.05% S. If the contact angle of liquid cast iron with a carbon substrate is less than that with the sand mold, the beneficial effect of carbon coating of the mold on reducing pinholing may be explained as described above on the basis of Equation 2.

Based on the foregoing theoretical considerations that are substantiated by experimental observations, the criteria for the onset of pinhole formation during the early stages of solidification of cast iron may be depicted by the relation between surface tension and rate of solidification as shown schematically in Figure 1 for the sand mold of high- and low-gas content and for that treated with a carbonaceous material.

Small additions of aluminum or titanium to cast iron with modifiers and inoculants have long been known to intensify the pinhole formation.[13,14] However, no satisfactory physicochemical explanation has so far been given for this intriguing phenomenon. As discussed below, this phenomenon may be a consequence of the associated chemisorption of a nitride or a carbide, resulting in lowering of the surface tension, hence, rendering the cast iron more susceptible to pinholing.

**Effect of Aluminum**—Hernandez and Wallace[9] have shown that small additions of aluminum markedly affect the surface tension of liquid cast iron. Their surface tension data are given in Figure 2a for gray iron with the average base composition: 3.3% C, 2.2% Si, 0.7% Mn, 0.1% S, and 0.1% P. The surface tension reaches a minimum at about 0.1% Al, then increases with increasing concentration

*References pp. 83-84*

of Al. At 0.4% Al, the surface tension appears to level off at about 770 mNm$^{-1}$, which is about 120 mNm$^{-1}$ higher than that for the Al-free gray iron. Within the range 0.01 to 0.25% Al, where $\sigma < 600$ mNm$^{-1}$, the gray iron is known to be susceptible to pinholing.

Without interaction with other solutes, the aluminum at low concentrations should have no measurable effect on the surface tension of completely deoxidized iron alloys as in cast irons. Noting that air-melted cast irons invariably contain nitrogen in the range 40 to 80 ppm, the observed effect of Al on $\sigma$ may well be attributed to the interaction of dissolved Al and N in liquid cast iron. As shown in the Appendix, in cast iron containing, for example, 60 ppm N, the nitrogen activity would be 0.02, which lowers the surface tension by 112 mNm$^{-1}$. Because of the relatively strong interaction between Al and N, there can be N-induced Al chemisorption at the surface of the cast iron, resulting in further lowering of the surface tension.

By extrapolating the experimental data of Evans and Pehlke[15] to lower temperatures, the solubility product of AlN in liquid iron alloys at 1400°C is calculated to be

$$\frac{[a_{Al}][a_N]}{a_{AlN}} = 5.75 \times 10^{-3} \tag{6}$$

where the solute activities $a_{Al}$ and $a_N$ are derived from the alloy composition as described in the Appendix. The variation of $\sigma$ with $a_{AlN}$ is shown in Figure 2b; the values of $a_{AlN}$ are for 60 ppm N assumed to be present in the cast iron used in the experiments of Hernandez and Wallace. Had the AlN remained in solution in a supersaturated-metastable state, the surface tension would have continued to decrease in the range $a_{AlN} > 1$. It is interesting to note that the surface tension reaches a minimum at about $a_{AlN} = 1$, corresponding to $\sim 0.1\%$ Al. An increase in $\sigma$ with increasing concentration of Al> 0.1 percent suggests a decrease in the concentration of dissolved N because of the precipitation of AlN when $a_{AlN} > 1$. With decreasing nitrogen activity, the N-induced chemisorption of Al will also decrease, resulting in a further increase in the surface tension of the metal. At sufficiently high concentration of Al, hence, low activity of N, the surface tension will be for the N-free gray iron and not affected further by Al in solution, as indicated by the experimental data in Figure 2.

In relation to the chemisorption of complex species, reference may be made to the surface tension of Fe–3% C–Cr melts measured by Whalen, et al.[16] The surface tension of pure liquid iron is not affected by carbon and lowered slightly by chromium, thus

$$\sigma_{Fe-Cr} = \sigma_{Fe} - 8[\%Cr] \tag{7}$$

Yet, in the ternary alloys Fe–3% C–Cr, the surface tension at 1350°C decreased by 350 mNm$^{-1}$ at 19 percent Cr; with further addition of Cr, the surface tension increased. From a thermodynamic analysis of these data, Belton[17] showed that the surface tension minimum is indicative of associated adsorption of CrC.

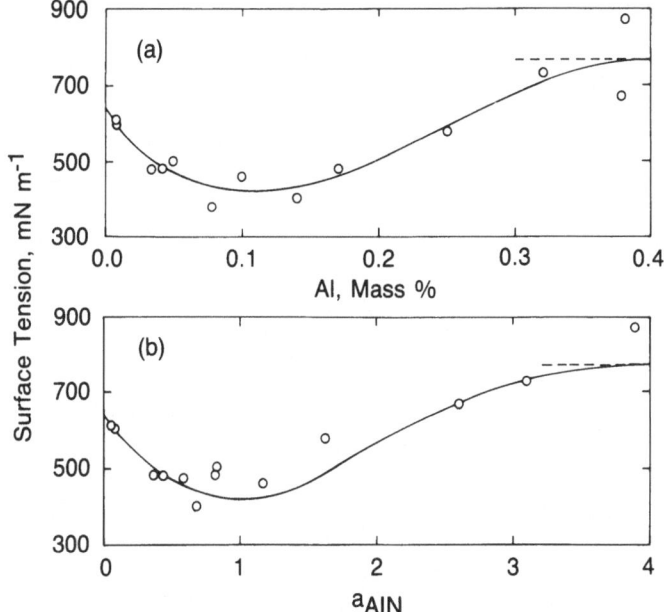

Figure 2. Surface Tension of Liquid Cast Iron at 1340–1430°C Using Data of Hernandez and Wallace:[9] (a) Effect of Al Content; (b) Effect of AlN Activity Calculated for 60 ppm N Assumed to be Present in the Liquid Cast Iron (3.3% C, 2.2% Si, 0.7% Mn, 0.1% S and 0.1% P).

To test the validity of the proposed associated chemisorption of AlN species on the surface of liquid iron alloys, the surface tension measurements should be made at known activities of nitrogen in liquid cast iron containing aluminum up to about 1%.

**Effect of Titanium**—The fixation of dissolved nitrogen by titanium as TiN would be expected to alleviate the N-induced chemisorption of aluminum, and thus reduce the tendency to pinholing. Yet, as shown by Dawson,[14] the extent of pinhole formation in cast iron increases with increasing concentration of titanium (up to 0.17% Ti investigated). In their surface tension measurements, Hernandez and Wallace found that the addition of titanium lowered the surface tension of liquid cast iron. Their data plotted in Figure 3a are for the base composition: 3.3% C, 1.7% Si, 0.6% Mn, 0.1% S, and 0.1% P.

At the concentrations of Ti up to about 2%, the titanium has no effect on the surface tension of pure liquid iron. Therefore, the observed effect of Ti on the surface tension of liquid cast iron may be attributed to the associated chemisorption of complex species Ti(N,C). The activities of TiC in liquid cast iron are calculated using the the solubility data given in the Appendix. There is much scatter in the surface tension data plotted in Figure 3b as a function of TiC activity. The dotted

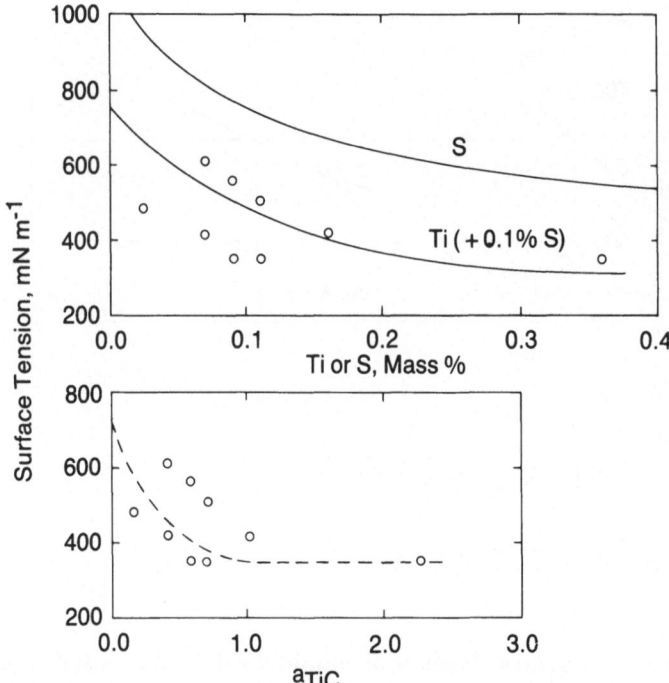

Figure 3. Surface Tension of Ti-Bearing Liquid Cast Iron at 1380–1430°C From Data of Hernandez and Wallace[9] for the Base Composition: 3.3% C, 1.7% Si, 0.6% Mn, 0.1% S and 0.1% P.

curve is drawn to reach a plateau at $\sigma \sim 300$ mNm$^{-1}$ when $a_{TiC} \to 1$. Because of high concentration of carbon in the cast iron, a small amount precipitated as TiC at $a_{TiC} > 1.0$ will hardly alter the extent of associated chemisorption of TiC. Consequently, the surface tension will remain unchanged at a low level; hence, the pinholing will persist even at relatively high concentrations of titanium. More accurate surface tension data are needed to confirm the validity of the proposed associated chemisorption of TiC species on the surface of liquid cast iron.

## MECHANISM OF GRAPHITE NODULATIONS WITH MODIFIERS AND INOCULANTS

Many years of research efforts were devoted to the understanding of the mechanism of graphite nodulation with the addition of modifiers and inoculants to cast iron. The early researches resulting from many controversial points of view have been disclosed in later publications of which a selected few are cited in the present discussion.[11,12,18–20]

**Criteria for Spheroidal Graphite Growth**—In both hypo- and hyper-eutectic iron, i.e., for carbon equivalents, CE, below or above the eutectic at

4.26% CE, the nodular graphite forms in direct contact with the liquid iron. Subsequently, the austenite precipitates in the carbon-depleted zones surrounding the graphite spherulites, thus resulting in a nodular structure with a spherical shell of austenite surrounding the graphite nodule.

From their study of nucleation and growth of graphite spherulites, and other well substantiated experimental evidence, the following conclusions have been drawn by Hunter and Chadwick[18] concerning the formation of spheroidal graphite:

1. A graphite spherulite develops only in an undercooled melt.

2. Most of the eutectic graphite is precipitated on spherulites which are growing non-cooperatively in the melt, i.e., austenite and graphite are growing non-contiguously and independently.

3. A graphite spherulite originates in the melt as a minute flake of graphite; during its growth this flake undergoes interfacial breakdown with crystal multiplication to give an approximately radial array of crystallites as shown schematically in Figure 4.

4. Within each crystallite, the basal planes are in general initially oriented towards the radial direction, but as growth proceeds they become aligned approximately parallel to the periphery.

**Modifiers for Ductile Iron**—A high interfacial energy between graphite and liquid iron is an *a priori* condition for the nodular growth of graphite. This condition is fulfilled by the addition of about 1% of a magnesium-ferrosilicon alloy (known as the modifier containing about 45% Si, 8% Mg, 1.5% Ca, 1% Al, and 0.06% Ti) to cast iron previously desulfurized to less than about 0.01% $S$. In some practices, rare-earth containing modifiers are used. The magnesium or rare-earth in the modifier forms a stable sulfide, hence reduces the activity of dissolved sulfur, resulting in a marked increase in the interfacial energy between graphite and liquid iron, as demonstrated by the measurements of contact angle and surface tension.[11,12] As shown by McSwain *et al*,[12] at low sulfur activities the interfacial energy for the basal plane of graphite is lower than that for the prism plane. It is for this reason that in the spheroidal (low surface area) growth, enforced by the high interfacial energy, the basal plane ultimately becomes essentially parallel to the periphery of the graphite spherulite.

Prolonged holding of modified liquid cast iron prior to casting results in a low grade nodular-cast structure because of the familiar Mg or Ce fade caused by vaporization, air oxidation, and reaction with the ladle refractory lining. The consequences of Mg and Ce fade are well demonstrated in the experimental work of Selcuk and Kirkwood[11] who measured the contact angle and surface tension of modified melts held *in vacuo* to enhance Mg and Ce fading. This is demonstrated in Figure 5 by the plot of $\sigma$ against $a_S$. As the total Mg content of the metal decreased from 380 ppm to 80 ppm, the surface tension decreased from about 1500 to 1100 mNm$^{-1}$, and the contact angle decreased from about 140° to 128°, indicating a decrease in the graphite-melt interfacial energy. The graphite in the casting at

*References pp. 83-84*

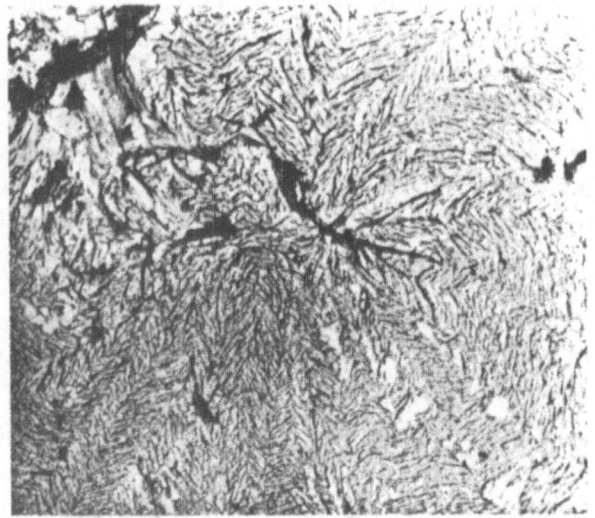

a. Replica of the diametral section of a graphite spherulite     x6500

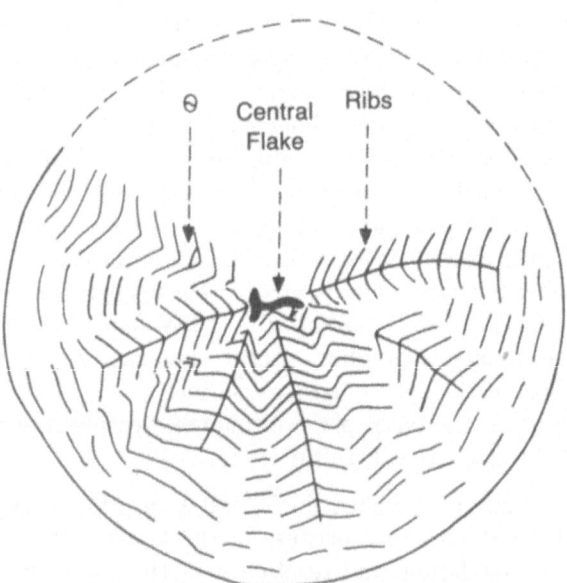

b. Schematic representation of the main features observed on the diametral sections of graphite spherulites

Figure 4.   Reproduced From the Paper by Hunter and Chadwick.[18]

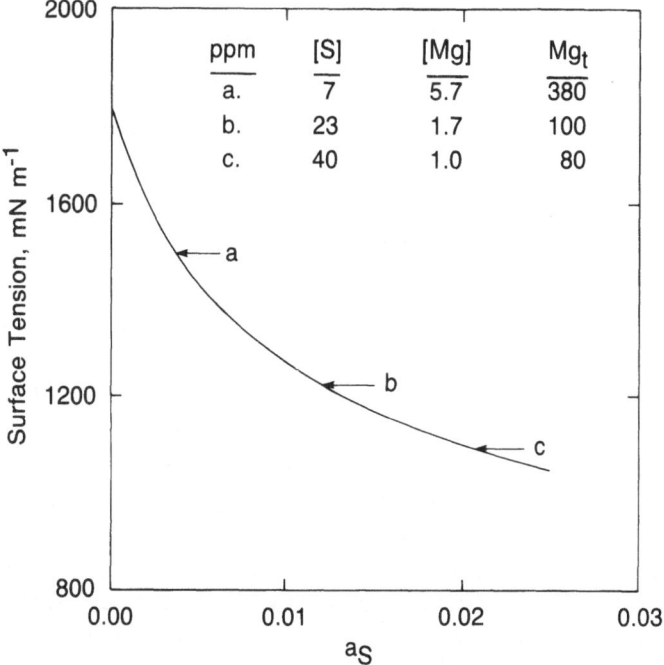

Figure 5. Concentrations of Dissolved [S] and [Mg], for MgS Saturation, Estimated from the Surface Tension Data of Selcuk and Kirkwood[11] for Liquid Cast Iron at 1200°C, Subsequent to Mg Treatment.

(a) was completely nodular and at (c) flaky; at (b) a mixture of nodular and flaky graphite. Similar observations were made in an earlier investigation by Khrapov.[21]

The numbers in parenthesis in Figure 5 are calculated concentrations of dissolved [S] and [Mg] in equilibrium with MgS dispersed in the melt computed from the solubility and activity data given in the Appendix. It appears from the measured values of surface tension that most of the Mg in the metal, determined by chemical analysis for total magnesium $Mg_t$, must have been in the form of finely dispersed particles of of MgS and MgO some of which float out of the melt with increasing holding time.

**Nucleation of Graphite in Gray and Ductile Iron**—The post inoculation of gray and ductile iron, immediately before or during pouring into a mold, is essential for the nucleation of graphite in the undercooled liquid at a temperature that is above the Fe-Fe$_3$C eutectic temperature, so that the formation of cementite would be suppressed completely.

The type of additives and the manner of inoculation remained a closely guarded secret in the foundry industry for a long period of time, well into the early parts of the 20th Century. Reference may be made to a paper by Patterson and Lalich[22] for a colorful dicsourse on the progress in cast iron inoculation from

*References pp. 83-84*

## Typical Compositions of Inoculants in Mass Percent, the Balance Being Iron

| Si    | Mn    | Al      | Ca      | Ba      | Sr      | Zr  | RE   |
|-------|-------|---------|---------|---------|---------|-----|------|
| 60/65 | 9/11  | 1.0/1.5 | 1.5/3.0 | 4/6     |         |     |      |
| 60/65 | 5/7   | 0.8/1.3 | 0.6/0.9 | 0.6/0.9 |         | 5/7 |      |
| 36/40 |       | 0.5 max | 0.5 max |         |         |     | 9/11 |
| 73/78 |       | 0.5 max | 0.1 max |         | 0.6/1.0 |     |      |

the time of the Iron Age to the late 1970s. Typical compositions of inoculants (ferrosilicon-based alloys) for gray and ductile iron are in the accompanying table.

It has long been thought that the graphite crystallites nucleate on oxide and sulfide particles that are formed by the addition of inoculants. However, it was only during the past decade that a better understanding of graphite nucleation materialized by detailed microstructural studies using scanning electron microscopy, energy dispersive x-ray analysis, and combined electron microscope-microanalyser, EMMA-3. In particular, reference may be made to the papers of Jacobs *et al*,[20] on the identification of heterogeneous nuclei for graphite spheroids in chill-cast iron. Figure 6, reproduced from their paper, is a typical example of a graphite nodule with a central particle on which the graphite nucleated. As is shown in Figure 7, the matrix particles extracted from the metal sample reveal in more detail the makeup of the nucleus on which the graphite flake crystallizes. The hexagonal-shaped platelet, ~ 1 $\mu$m diameter and ~ 0.3 $\mu$m thick, has a duplex character. The central particle, ~ 0.05 $\mu$m diameter, was identified as a mixed sulfide, (Ca, Mg)S, the outer region a spinel-type oxide containing Mg, Al, Si, and O. From these detailed observations, Jacobs *et al*, concluded that the sulfide particles formed (with the addition of inoculants) present crystalline faces upon which the platelet of spinel can grow epitaxially. The oxide platelet presents two crystalline faces to the carbon-supersaturated melt upon which the basal plane of graphite can grow epitaxially. The graphite flakes thus nucleated will grow as (i) flakes in gray iron (low interfacial energy) or (ii) spheroids in ductile iron (high interfacial energy).

## RATE CONTROLLING REACTION MECHANISMS IN LADLE DESULFURIZATION OF BLAST FURNACE AND CUPOLA IRONS

**Process Variables Affecting Rate of Desulfurization of Iron**—The physicochemical aspects of the sulfur reaction in gas-slag-metal systems are known well enough that the subject need not be reviewed here. Instead, the attention is drawn to certain unique features of process variables which affect the rate of desulfurization of liquid iron alloys by lime-based slags, or by injection of lime-based desulfurizing agents.

The extent of desulfurization of liquid iron alloys, controlled by the reaction,

$$(CaO) + [S] \rightarrow (CaS) + [O] \tag{8}$$

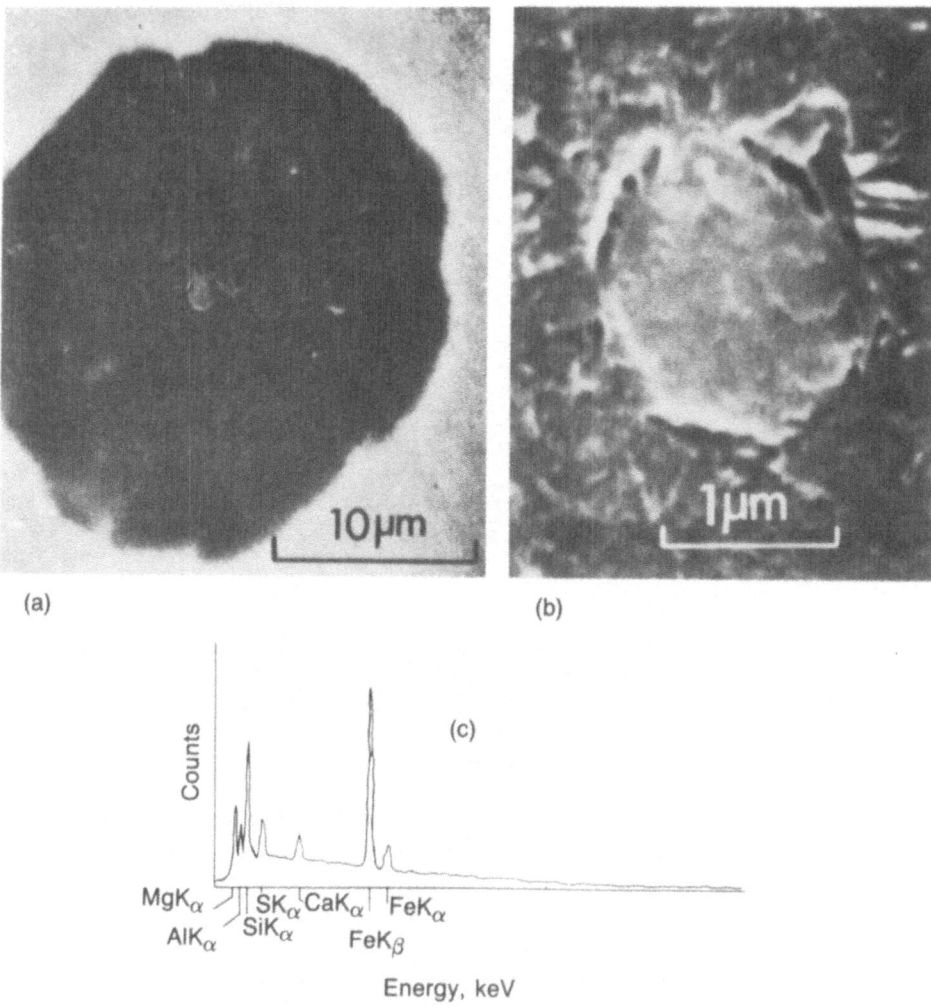

Figure 6. Diametrical Section of Graphite Nodule in Cast Iron Examined Under SEM. After Jacobs et al.[20]
 a) Secondary Electron Image
 b) Central Particle Shown in (a)
 c) X-ray Spectrum From the Central Particle.

becomes greater with slags of high sulfur capacity, e.g., $CaO$-$CaF_2$ and $CaO$-$Al_2O_3$, and with liquid metals of high sulfur activity and low oxygen activity, as in hot metal and foundry iron. As is seen from an example of an early study[23] in Figure 8, the addition of Si and particularly Al increased the rate of desulfurization of carbon-

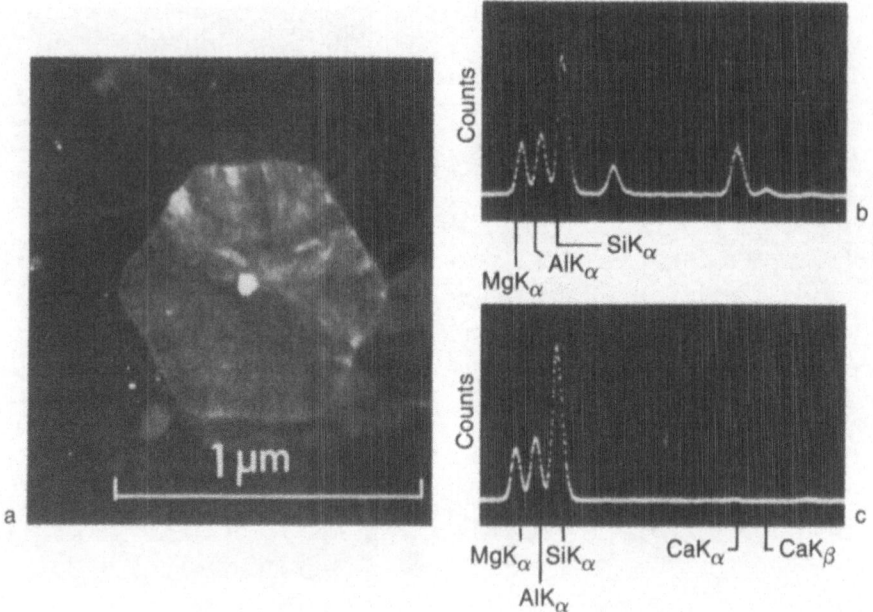

a. Dark field micrograph
b. X-ray spectrum from the central particle
c. X-ray spectrum from the outer region of the hexagonal platelet

Figure 7. EMMA-3 Analysis of a Matrix Particle Extracted From Ductile Iron. After Jacobs et al.[20]
a) Dark Field Micrograph
b) X-ray Spectrum From the Outer Region of the Hexagonal Platelet
c) X-ray Spectrum From the Central Particle.

saturated liquid iron with molten calcium aluminate ($CaO/Al_2O_3 = 50/50$).*

Although the lime is good desulfurizer chemically, the solid reaction products hinder the rate of desulfurization by the reaction

$$(2x + 2)CaO(s) + [Si] + 2[S] \rightarrow 2xCaO \cdot SiO_2(s) + 2CaS(s) \tag{9}$$

because of slow diffusion of ions through the solid coatings of CaS and $2CaO \cdot SiO_2$ around the CaO particles. As is seen from the experimental data of Landefeld and Katz[24] in Figure 9,** an addition of 10% $CaF_2$ greatly increases the rate of

---

\* Nominal conditions in these experiments were: 1505°C; 8 g of C-saturated iron containing 1.08% S, and 4 g of calcium aluminate ($CaO/Al_2O_3 = 50/50$).
\*\* Nominal conditions in these experiments were: 1450°C; 2% of CaO + flux mixture; 13.4 kg of iron alloy initially containing 0.1% S and 0.4 to 0.5% Si; and stirring of the melt with $N_2$ bubbling at $1 \times 10^{-3} - 6 \times 10^{-3}$ m$^3$/min.

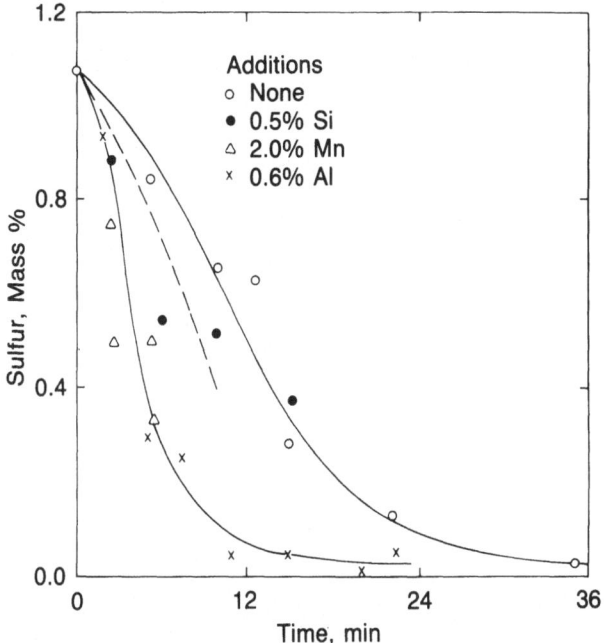

**Figure 8.** Desulfurization of Graphite-Saturated Iron (8g) at 1505°C With a Calcium Aluminate Slag (4g) With and Without the Additions of Si, Mn or Al. After Turkdogan et al.[23]

desulfurization with CaO because of the fluxing of the solid reaction products with $CaF_2$. The CaO-saturated aluminosilicate melt reacts faster than CaO, but slower than molten $CaO$-$CaF_2$ (saturated with CaO).

Small additions of Al to the metal will also increase the rate of desulfurization with CaO. In the reaction

$$CaO(s) + 2/3[Al] + [S] \rightarrow 1/3(Al_2O_3) + (CaS) \quad (10)$$

the alumina generated will react with CaO and CaS forming a liquid reaction product, hence make more CaO available for further reaction with sulfur. This expectation has been confirmed by the experimental work of Kor,[25] as demonstrated by a few examples of his experimental data in Figure 10.* It is seem that an addition of 0.1% Al to the metal increases the rate of desulfurization of iron by lime more than that achieved by using a mixture of 90 percent CaO and 10 percent $CaO \cdot Al_2O_3$. We see from the comparison of the experimental data in Figure 10 that the rate of desulfurization achieved with the mixture of 90% CaO and 10%

---

* Nominal conditions in these experiments were: 1510°C; 15 kg iron either saturated with carbon or containing 2 percent C, both containing about 0.07% S; 225 g of the desulfurizing mixture; and mechanical stirring with a graphite blade rotating at 100 rpm.

*References pp. 83–84*

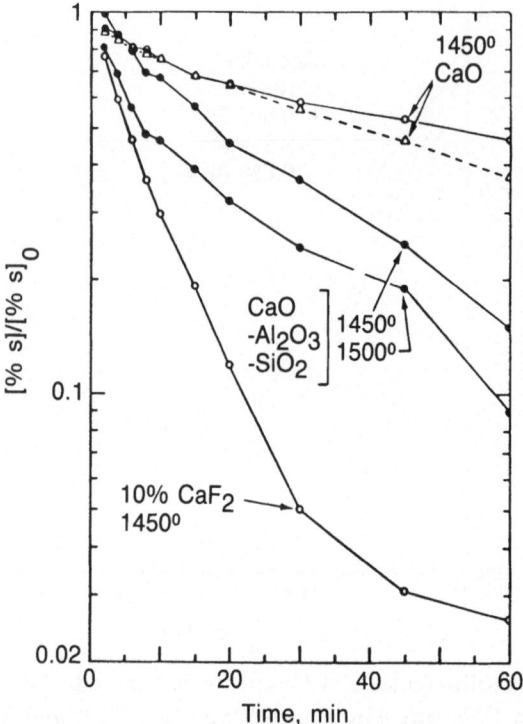

Figure 9. Desulfurization of Graphite-Saturated Iron (13.4 kg) With CaO and CaO-Saturated Fluoride or Aluminosilicate Slag (270 g) in $N_2$-Stirred Melts ($2.75 \times 10^{-3}$ m$^3$/min). After Landefeld and Katz.[24]

$CaF_2$ is similar to that with CaO when the metal initially contains 0.1% Al. The data in Figure 11 are for graphite-saturated iron and iron containing 2% C with or without the addition of aluminum. The effect of carbon content of iron on its rate of desulfurization with CaO is as would be expected.

For the same type of slag in these laboratory experiments, the rate of desulfurization in the mechanically-stirred melt (Figure 10) is about two-thirds of that in the gas-stirred melt at $2.75 \times 10^{-3}$ m$^3$/min (Figure 9).

It is all too clear from these experimental observations that the presence of a fluxing agent is a necessary condition for effective desulfurization of liquid iron alloys with lime. Because of the adverse effect of aluminum on the castability of the foundry irons causing pinholes, no aluminum-bearing additives can be made to enhance the rate of desulfurization with CaO; $CaF_2$ is an obvious choice as a fluxing additive.

The apparent rate constants obtained from laboratory experiments with small (a few grams) or large (several kilograms) mass of melts should be interpreted with due caution, because of the dependence of the rate controlling reaction mechanism

# PHYSICOCHEMICAL PHENOMENA

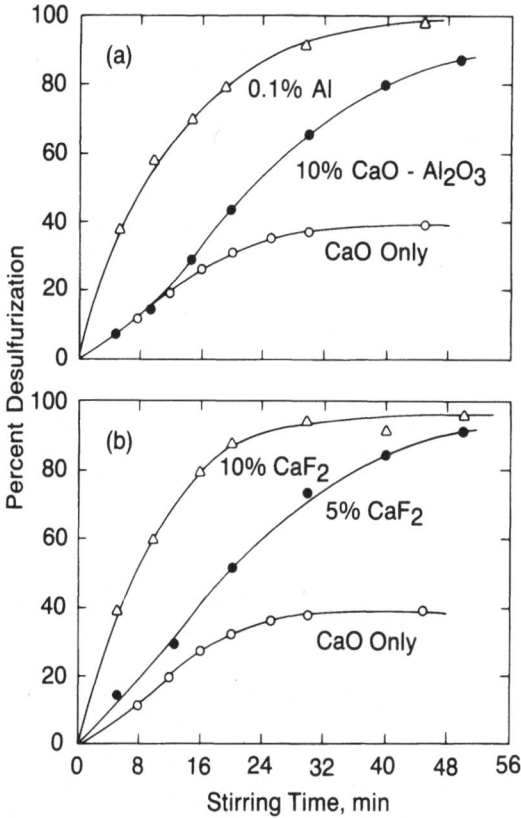

Figure 10. Desulfurization of Carbon-Saturated Liquid Iron at 1510°C With CaO, CaO-Al$_2$O$_3$ or CaO-CaF$_2$ Mixtures, While Melt Being Stirred Mechanically. After Kor.[25]

on the size of the melt and the experimental technique employed, as demonstrated by the following examples.

From the experimental results of the earlier work of Turkdogan *et al*,[23] reproduced in Figure 8, the average apparent rate constants are: $k' = 0.05$ min$^{-1}$ for no Al addition and $k' = 0.14$ min$^{-1}$ for 0.6% Al addition to the metal. In these small scale experiments with 8 g metal and 4 g slag, each layer was about 6 mm thick. If the rate of desulfurization were controlled by diffusion of sulfur in the metal, the measured desulfurization rate would correspond to an apparent diffusivity of about $3 \times 10^{-4}$ cm$^2$/s, which is an order of magnitude higher than the diffusivity of sulfur in graphite-saturated liquid iron. The observed high rate of desulfurization of small iron samples may well be due to the interfacial turbulence caused by the transfer of surface active sulfur from metal to slag. This phenomenon was well demonstrated experimentally by Kozakevitch *et al*,[26] and subsequently by others.

*References pp. 83–84*

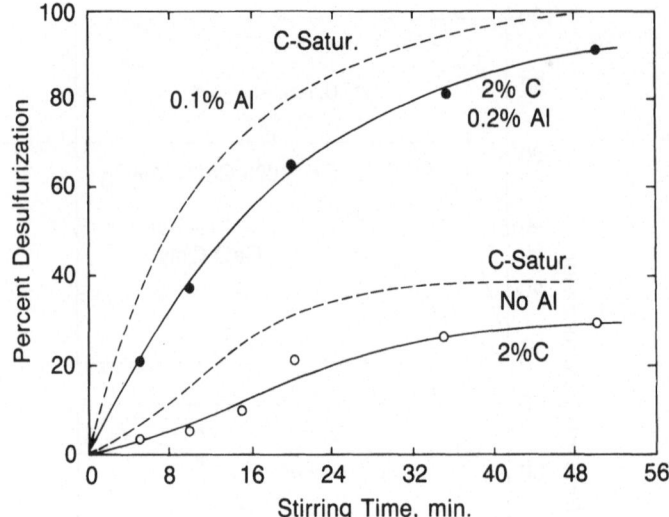

**Figure 11.** Effect of Addition of Al on the Rate of Desulfurization of Liquid Iron (2% C and C-Saturated) at 1510°C With CaO, While Melt Being Stirred. After Kor.[25]

For mass-transfer controlled slag-metal reactions, under the conditions of (i) reaction equilibrium at the slag-metal interface, (ii) no dispersion of slag and metal, and (iii) no solute-induced interfacial turbulence, the apparent rate constant $k$ can be estimated from the following theoretical relation for a given temperature and pressure.

$$k(\text{cm/s}) = (bD\dot{Q})^{1/2} \tag{11}$$

where $\dot{Q}$(cm/s) is the volume flux of gas bubbles across the interface, $D$ the solute diffusivity in the metal, and $b$ a constant (for liquid metals $b = 120$ cm$^{-1}$ as determined experimentally by Subramanian and Richardson[27]). Equation 11 is based on the surface renewal or penetration theory of Higbie[28] for gas-stirred liquids with uniform distribution of gas bubbles crossing the interface.

For the volumetric gas-flow rate, $V_o$, measured at atmospheric pressure and room temperature (20°C), and the nominal area of the slag-metal interface, $A$, Equation 11 for the reaction temperature of $T(K)$ at atmospheric pressure is transformed to

$$k(\text{cm/s}) = \left(bD\frac{TV_0}{293A}\right)^{1/2} \tag{12}$$

The diffusivity of S in carbon-saturated liquid iron is $D = 3 \times 10^{-5}$ cm$^2$/s at 1400–1500°C; inserting this value of $D$ and $b = 120$ cm$^{-1}$ in Equation 12, the following relation is obtained for $k$ at 1450°C.

$$k(\text{cm/s}) = 0.146(V_0/A)^{1/2} \tag{13}$$

where $V_0$ is in cm$^3$/s and A in cm$^2$.

In their desulfurization experiments with gas-stirred melts at 1450°C, Landefeld and Katz[24] found that the apparent rate constant was approximately proportional to the square root of the volume gas-flow rate, in general accord with the theoretical relation in Equation 11. In their crucible assembly holding 13.4 kg metal, the nominal area of the slag-metal interface was $A = 175$ cm$^2$. From their rate measurements with the gas-flow rates of up to $6 \times 10^{-3}$ m$^3$/min, the following relation is obtained for desulfurization with CaO - 10% CaF$_2$ slag

$$k(\text{cm/s}) = 0.044(V_0/A)^{1/2} \tag{14}$$

which is about one-third of that in Equation 13 derived from the theoretical considerations. The experimental value of $k$ being somewhat smaller than that calculated from the theoretical Equation 11 may be due, in part, to the presence of a large volume fraction of solid CaO in the slag layer that would hinder the rate of transfer of sulfur from metal to slag.

Because of the high sulfur capacities of CaO-saturated molten calcium aluminate and CaO-CaF$_2$ containing some SiO$_2$, the reverse sulfur reaction would be negligibly small, and consequently, a simplified form of the rate equation can be used

$$\ln \frac{[\%S]}{[\%S]_0} = -\frac{k}{\ell}t \tag{15}$$

where $\ell$ is the depth of liquid metal and the subscript 0 indicates the initial sulfur content of the metal. For a truly mass transfer controlled slag-metal reaction, the mass-transfer coefficient as defined by Equations 11 and 15 should be independent of the type and composition of the molten slag, provided it has a high sulfur capacity. Yet, we see from the experimental results of Landefeld and Katz in Figure 9 and those of Kor in Figure 10 that the rate of desulfurization with CaO-saturated liquid aluminate is about one-third of that for CaO-saturated liquid fluoride slag. One major difference, however, is in the liquid volume fraction in these CaO-saturated slags, e.g., for the mixtures used, the aluminate slag is about 27% liquid and the fluoride slag about 50% liquid.

Two experimental data points in Figure 12a for 0.27 and 0.50 liquid volume fractions of slag are compared with that calculated from Equation 13 for a completely liquid slag and the experimental gas flow rate of $V_0/A = 0.26$ cm/s. The results indicate the rate of desulfurization of gas-stirred liquid iron by a CaO-saturated slag could be increased by decreasing the amount of excess solid CaO, i.e., by increasing the liquid volume fraction of the ladle slag.

In relation to the foregoing discussion, reference should be made to the work of Coon[29] who investigated the rate of desulfurization of foundry iron (500 kg mass) with lime-fluorspar mixtures (2% of the metal charge) in a ladle fitted with a porous plug for gas (air) stirring. The highest rate of desulfurization was obtained with the addition of 5% CaF$_2$; the rate decreased at higher concentrations of fluorspar in the CaO-CaF$_2$ mixture. This finding is contrary to that obtained in smaller scale experiments (with low-Si irons) discussed above. It should be noted, however, that

*References pp. 83–84*

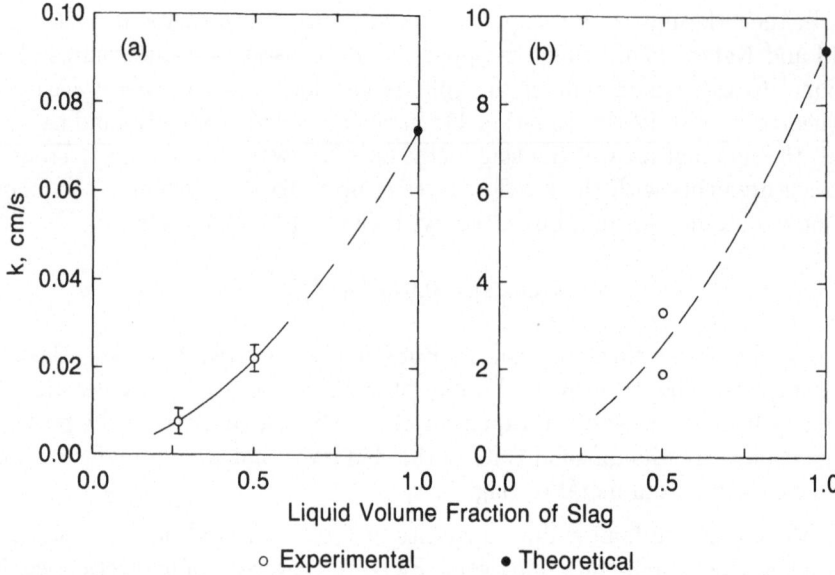

Figure 12. Rate Constant for Desulfurization of Foundry Iron With Lime-Saturated Slag in Gas-Stirred Melts is Related to the Liquid Volume Fraction of the Slag: (a) Bench-Scale Experiments[24] With 13.4 kg Melt and $V_0/A = 0.26$ cm/s; (b) Plant Continuous Desulfurization[30] in a Reactor With 3500 kg Melt and $V_0/A = 1.41$ cm/s.

in the experiments of Coon with 2% Si-foundry iron stirred with air injection, the loss of silicon from the metal during the desulfurizing treatment increased with increasing proportion of fluorospar in the $CaO\text{-}CaF_2$ mixture, accompanied by an increase in the slag volume. At present, the author can offer no explanation as to why such a sequence of events has led to the lowering of the rate of desulfurization at high contents of $CaF_2$.

**Rate of Desulfurization Under Practical Conditions**—As is seen from the plant data in Figure 13, compiled by Kor,[25] the extent of desulfurization of hot metal (initially containing 0.03 to 0.10% S) increases with an increase in the amount of injected material. In these plant trials, little or no difference could be seen between the desulfurizing capabilities of the $CaC_2$-Mg and CaO-Mg mixtures. However, an addition of about 3% Al noticeably increased the desulfurizing capability of the CaO-Mg mixture. The observed beneficial effect of Al is due to (i) the lowering of oxygen activity of the hot metal and (ii) the fluxing of Ca(Mg)O and Ca(Mg)S with the alumina generated *in situ* in the melt.

Foundry iron has to be desulfurized to low levels of sulfur (< 0.010% S) in the production of ductile and compacted-graphite iron. Although calcium carbide is an effective desulfurizer, alternative methods of desulfurization are being explored in the foundry industry, because of the fume emission during $CaC_2$ injection and

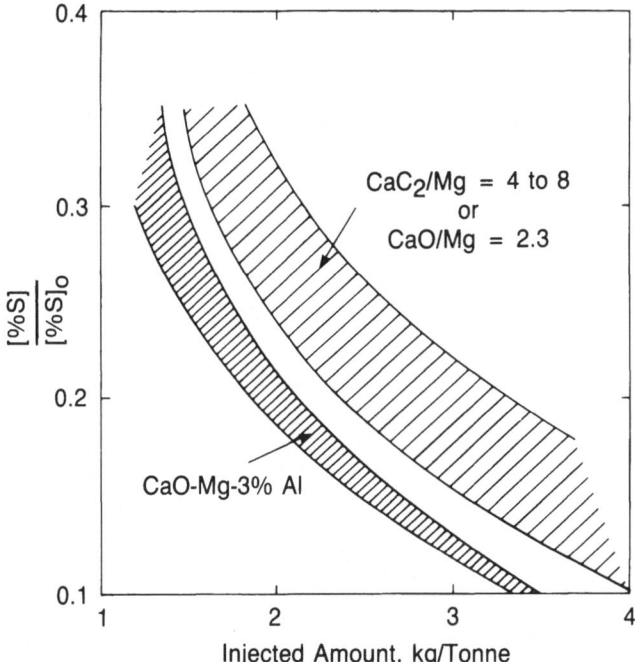

Figure 13. Desulfurization of Hot Metal in the Transfer Ladle by Injection of $CaC_2$-Mg, CaO-Mg or CaO-Mg-3% Al. After Kor.[25]

the pollution problems in the disposal of the waste product containing $CaC_2$ and CaS. The use of Mg or a CaO-Mg mixture is not favored, because it increases dross defects in castings. As mentioned earlier, aluminum or alumina containing desulfurizing mixtures cannot be used, because of contamination of the iron with aluminum, resulting in the formation of subsurface pinholes in castings.

Preference is given nowadays to desulfurization of foundry iron, in a continuous or a batch process, with a lime-fluorspar top slag in a reactor, or a ladle, fitted with porous plugs for gas stirring of the melt.

Recently, Katz and Landefeld[30] reported the results of plant trials on continuous desulfurization of foundry iron with a lime-spar top slag in a reactor (0.76 m diameter and 1.02 m deep 3.5 t melt) stirred with nitrogen flowing at the rate of about 0.38 $Nm^3$/min. The rate constant for desulfurization was found to be much greater than that obtained in laboratory experiments[24] for the same gas flow rate per unit cross-sectional area of the reactor. This vast difference in the rate constants is probably due to a more extensive slag-metal mixing when the melt is stirred with gas injection on a plant scale operation.

According to the experimental work of Ishida et al,[31] with 2.5 t liquid steel, the energy density of stirring, $\dot{\varepsilon}$, should exceed 60 $W/t^{-1}$ to insure slag-metal mixing, which corresponds to an argon flow rate of $V_o > 1$ $Nm^3$/min injected at

*References pp. 83–84*

about 3-m depth in a 200-t heat, as usually practiced in the ladle desulfurization of steel at atmospheric pressure. For gas-stirred melts corresponding to $60 < \dot{\varepsilon} < 120$ W/t, Ishida, et al, obtained the following empirical relation for the mass-transfer controlled slag-metal reaction. The exponent 2.1 for $\dot{\varepsilon}$ has no particular physical significance.

$$k'(\text{min}^{-1}) = 7.6 \times 10^{-6}(\dot{\varepsilon})^{2.1} \tag{16}$$

The rate constant $k'$ is defined in the usual manner by the relation

$$\log \frac{S - S_e}{S_0 - S_e} = -k't \tag{17}$$

where the subscripts $o$ and $e$ indicate, respectively, the initial and the equilibrium concentrations of sulfur in the metal.

For a gas-stirred liquid iron bath, the energy density of stirring, $\dot{\varepsilon}$, is given by the following thermodynamic relation based on the buoyancy energy of the injected gas

$$\dot{\varepsilon}(\text{W/t}) = 14.23 \frac{V_0 T}{M} \log \left(1 + \frac{H}{1.46 P_0}\right) \tag{18}$$

where $V_0$ the gas-flow rate (Nm$^3$/min), $M$ the mass of liquid metal (tonne), $T$ the melt temperature (K), $H$ the depth of gas injection (m), and $P_0$ the gas pressure at the surface of the melt (bar).

For the operating conditions and the dimensions of the reactor in the plant studies of Katz and Landefeld, i.e., $V_0 = 0.38$ Nm$^3$/min, $T = 1723K$, $M = 3.5$ t, $H = 1.02$ m, and $P_0 = 1$ bar, the power of stirring is $\dot{\varepsilon} = 613$ W/t for which $k' = 5.43$ min$^{-1}$, and, therefore, $k = 9.2$ cm/s. This computed rate constant for mass-transfer controlled desulfurization in the stirred melt is compared in Figure 12b with the plant data for 0.5 liquid volume fraction of slag in the reactor. Similar to the results of the laboratory experiments, the rate of desulfurization in the reactor appears to be related to the liquid volume fraction of the slag. Additional experimental data are needed with high-sulfur capacity slags of varying liquid volume fraction to confirm the relationships in Figure 12 derived from Equation 13 for laboratory scale stirred melts, and Equation 16 for plant scale operations with completely molten slags in gas-stirred melts.

## RATES OF REACTIONS IN THE CUPOLA

**Rate of Gasification of Coke and Effect on Cupola Performance—** Kinetics of gasification of various types of carbon in $O_2$, $CO_2$, and $H_2O$ have been studied extensively for over four decades. Many comprehensive review papers have been published periodically on this subject; therefore, only a brief commentary will be made here on certain aspects of coke gasification that is pertinent to the cupola practice.

Because of the porous nature of coke, and most other forms of carbon, the rate of gasification is affected in a particularly unique manner by the particle size and temperature, characterized by three limiting cases. The rate is reported usually in terms of mass fraction gasified per unit time, $d(w/w_0)/dt$, which is abbreviated here by $R_0$.

1. At low temperatures and with small size particles, there is rapid gas diffusion in and out of the pore structure, resulting in uniform internal burning for which the chemical reaction controlled $R_0$ is independent of the particle size and shape.

$$R_0 \propto \Phi'S \tag{19}$$

where $\Phi'$ is the apparent rate constant for a given temperature, pressure and gas composition, and $S$ the connected pore surface area per unit mass.

2. At higher temperatures and with larger size particles, the gas diffusion in and out of the pore structure is limited to the outer casing of the carbon particle for which the mixed controlled $R_0$ is inversely proportional to the particle (spheroidal) diameter. Therefore, for a given temperature, pressure, and gas composition,

$$R_0 \propto \frac{1}{d}\left(\frac{\Phi'SD_e}{\rho}\right)^{1/2} \tag{20}$$

where $d$ is the diameter of the spheroidal carbon particle, $\rho$ the bulk density of carbon, and $D_e$ the effective gas diffusivity in porous carbon.

3. At much higher temperatures and with much larger size particles, the rate of gasification is controlled by mass transfer in the gas-film boundary layer for which the rate is determined primarily by the particle size, the porosity of the coke bed, $\varepsilon_b$, and the gas flow rate.

$$R_0 \propto \frac{D(1-\varepsilon_b)^{1/3}}{\varepsilon_b \rho d^{4/3}} \tag{21}$$

where $D$ is the molecular $CO$-$CO_2$ interdiffusivity in $N_2$.

The apparent heat of activation for the rate of gasification controlled by the limiting case (2) is about one-half of that for complete internal burning, the limiting case (1). Because of the small temperature effect on the gas diffusivity and the mass transfer coefficient, the rate of gasification is almost independent of temperature for the limiting case (3).

Reference should be made to text books, as for example that by Szekely, et al,[32] for detailed information on the rates of reaction of gases with porous solids.

The experimental rate data in Figure 14, from the work of Turkdogan and Vinters,[33] are for the limiting case (1) for almost complete internal burning of coke granules ($\sim$ 0.5 mm diameter) in $CO_2$-$CO$ mixtures at atmospheric pressure. Similar results were obtained in subsequent work by Aderibigbe and Szekely.[34] The important feature of the rate data is the strong retarding effect of CO on the rate of gasification of coke with $CO_2$, within the regime of complete or partial internal burning. A similar kinetic effect of CO was found also with other types of carbon.

*References pp. 83–84*

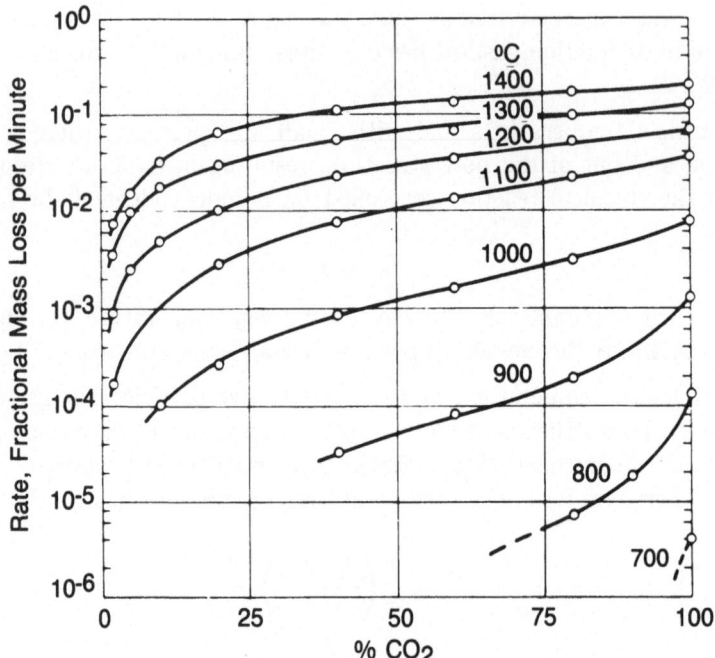

Figure 14. Effects of Temperature and Gas Composition on the Rate of Gasification of Coke Granules (~0.5mm diam) in $CO_2$-CO Mixtures at Atmospheric Pressure. After Turkdogan and Vinters.[33]

Since the operating conditions in the cupola do not favor internal burning of coke, this aspect of the rate of gasification will not be pursued further.

The experimental studies of cupola performance have revealed that the gasification of coke (5–12 cm diameter) with $CO_2$ is limited primarily to the region below the melting zone at temperatures above 1350–1500°C, depending on the cupola charge, coke size, and the rate of air blast.[35,36] From an analysis of these experimental data, Katz[37] showed that much of the gasification of coke occurring below the melting zone is controlled by interparticle mass transfer, i.e., the limiting case (3). Above the melting zone, i.e., at temperatures below 1350–1500°C, the rate of gasification drops rapidly as would be expected for the rate limiting case (2).

It follows from these observations that the coke reactivity, characterized by the parameter

$$\left(\frac{\Phi' S D_e}{\rho}\right)^{1/2} \qquad (22)$$

will not have much influence on the $CO_2$ gasification of coke, hence, on the performance of the cupola. Since the coke size has a major effect on the rate of gasification by $CO_2$, which is an endothermic reaction, the temperature profile in the cupola and the temperature of the iron will be affected by the coke size. The larger the coke size, the lower the rate of gasification, hence, the higher would be the temperature of the iron. Also, as indicated by Equation 21, the rate of gasification decreases, hence, the iron temperature increases, with an increase in the porosity, $\varepsilon_b$, of the coke bed, as achieved with uniform coke size.

An increase in the temperature of iron achieved by using uniform and large coke size in the cupola results in more carbon pickup by the iron and less loss of iron, manganese, and silicon to the slag. Reference may be made to a comprehensive review paper by Katz[37] for a detailed discussion of the properties of coke that affect the rate of gasification, the thermal efficiency, and the overall performance of the cupola.

**Rate of Carbon Dissolution in Liquid Iron**—The rate of dissolution of graphite in liquid iron has been studied by several investigators. As shown, for example, by Angeles, et al,[38] and by Mihajilovic and Marincek,[39] the rate of dissolution is controlled by mass transfer of carbon through the diffusion boundary layer at the melt-graphite interface. As noted by these and other investigators, the carbon with high ash content, as in coke, dissolves in liquid iron at a rate slower than graphite of low ash content. Of particular interest are the test results reported by Leyshon and Thibault[40] on the behavior of cokes in the cupola. They found that for a given content of fly-ash in the coke, the amount of carbon pickup by iron was lower with coke containing fly-ash of high fusion temperature. With the addition of 5% $CaF_2$ to the coke charge in the cupola, the carbon pickup by iron increased and became independent of the fusion temperature of the fly-ash in cokes of different source.

The inhibition of the carbon dissolution process by ash is due to the formation of a barrier film at the coke-liquid metal interface. However, no explanation has so far been given for the noticeable effect of the fusion temperature of fly-ash in the carbon dissolution process. As demonstrated by the following calculations, this unique behavior of fly-ash may be attributed to the formation of CO bubbles at the coke surface, under certain conditions.

Fly-ash with a high fusion temperature consists of primarily aluminum silicate of mullite composition ($3Al_2O_3 \cdot 2SiO_2$) for which the activity of silica is about 0.5 (relative to pure solid $SiO_2$). Now, let us consider the following reaction between fly-ash on the coke surface with liquid iron.

$$(SiO_2) + 2[C] = [Si] + 2CO(g) \tag{23}$$

For the foundry iron compositions and $a_{SiO_2} = .05$, the equilibrium partial pressure of CO at 1450°C is in the range 1.6 to 2.0 bar. If the total gas pressure in the cupola hearth is less than 1.6 or 2.0 bar, there is potential for the nucleation of CO bubbles at the crevices of the coke by the reaction in Equation 23. The gas bubbles retained at the coke surface will certainly hinder the carbon dissolution process. The fluxing of the fly-ash with $CaF_2$ lowers the activity of silica, hence, curtails the possibility of the formation of CO bubbles at the coke surface.

**State of Slag-Metal Reactions**—Because of the complexity of reactions occurring in the melting zone and during passage of metal droplets through the

*References pp. 83–84*

slag layer in the cupola hearth, it is difficult, if not impossible, to resolve the rate controlling mechanisms of slag-metal reactions in the cupola or the blast furnace.

According to a study of plant data by Katz and Rezeau,[41] there is close approach to slag-metal equilibrium in the cupola with respect to the coupled reactions between silicon, manganese, and sulfur. However, for reactions involving C and CO, there is no gas-slag-metal equilibrium. The state of gas-slag-metal reactions in the cupola deduced from this study of the plant data differs in some respect from that deduced by the author[42] from blast-furnace data briefly summarized below.

Plant data for slag and metal tap samples from various domestic and foreign blast furnaces (for tap temperatures of 1450–1525°C) are scattered within the shaded areas shown in Figure 15 for silicon, manganese, and sulfur distribution ratios between slag and metal. However, for a given blast furnace, the scatter in the data is usually much less and the effect of slag basicity on the distribution ratios is qualitatively in accord with the equilibrium data. That is, for any given blast furnace, the daily average plant data give the [%Si]/(%SiO$_2$) ratio that decreases with increasing slag basicity; [%Mn]/(%MnO) and (%S)/[%S] ratios increase with increasing slag basicity.

The dotted curves in Figure 15 are calculated from the known equilibrium data for the three-phase gas-slag-metal reactions for 1 bar CO (an average CO pressure in the blast-furnace hearth).[43]

$$(SiO_2) + 2[C] = [Si] + 2CO(g) \tag{23}$$

$$(MnO) + [C] = [Mn] + CO(g) \tag{24}$$

$$(CaO) + [S] + [C] = (CaS) + CO(g) \tag{25}$$

Of the three reactions considered, the distribution of silicon between metal and slag is closer to equilibrium for the three-phase gas-slag-metal reaction.

We may also consider the following two-phase coupled reactions.

$$(CaO) + [Mn] + [S] = (CaS) + (MnO) \tag{26}$$

$$(CaO) + 1/2[Si] + [S] = (CaS) + 1/2(SiO_2) \tag{27}$$

$$2(MnO) + [Si] = (SiO_2) + 2[Mn] \tag{28}$$

Blast-furnace data are compared in Figure 16 with the equilibrium relations for manganese-sulfur and manganese-silicon coupled reactions in melts saturated with graphite. For blast-furnace slags containing on the average 40% CaO, the ratio (%S)/[%S] is scattered about the equilibrium line for the reaction in Equation 26 at 1500°C.

The values of $K_{MnSi}$

$$K_{MnSi} = \left(\frac{[\%Mn]}{(\%MnO)}\right)^2 \frac{(\%SiO_2)}{[\%Si]} \tag{29}$$

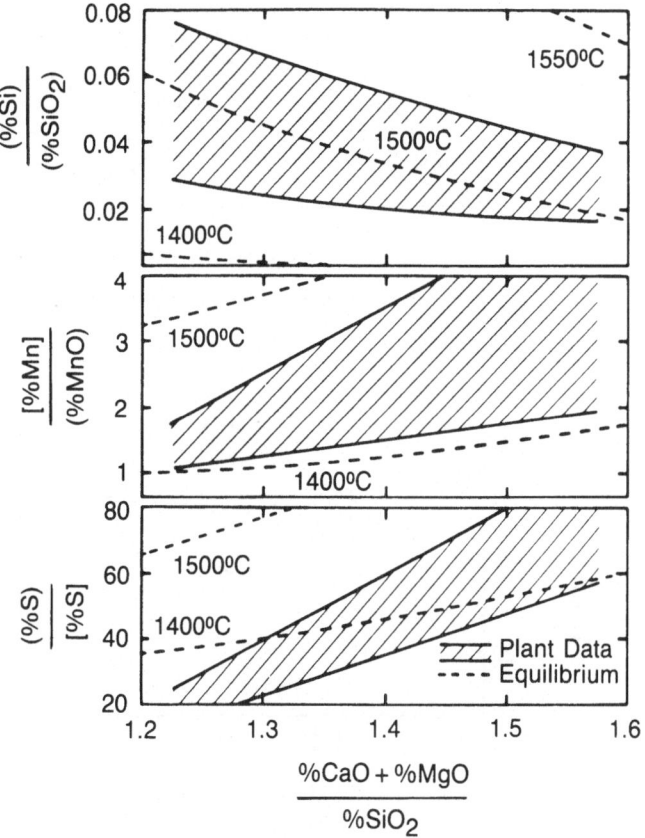

Figure 15. Blast Furnace Data are Compared With the Equilibrium Relations for Slag/Metal Distribution of Silicon, Manganese and Sulfur in Graphite-Saturated Melts at 1 bar Pressure of CO. After Turkdogan.[42]

from plant data scattered within the shaded area in Figure 16 (bottom diagram) are below the equilibrium line (dotted). This observation from the plant data may indicate that as the metal droplets pass through the slag layer, the direction of the reaction might be

$$2(MnO) + [Si] \rightarrow (SiO_2) + 2[Mn] \tag{30}$$

Such an intuitive deduction, however, may not be correct. Observed departure from equilibrium for reaction in Equation 28 may be attributed to competition between FeO and MnO in the slag for oxidation of silicon. It is a generally accepted view that with a blast-furnace burden of low basicity, the residual iron oxide in the melting zone of the bosh is higher than with a burden of high basicity. In slags of low basicity, the higher concentration of iron oxide in the upper part of the slag layer would be in competition with MnO for reaction with both silicon and carbon. Consequently, departure from equilibrium for reactions in Equations 24 and 28 is expected to be greater in slags of low basicity. The trends seen from the plant

*References pp. 83–84*

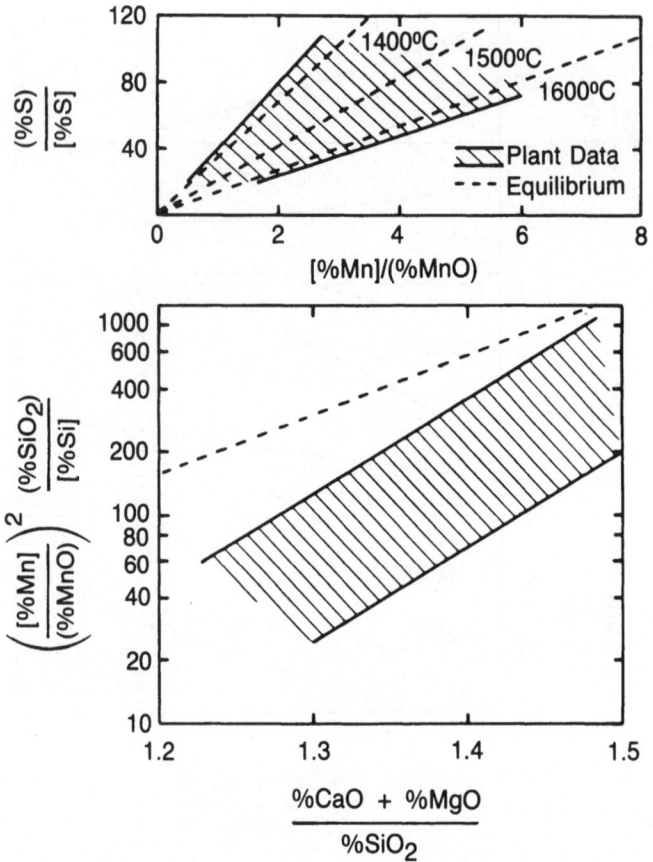

Figure 16. Blast Furnace Data are Compared With the Equilibrium Relations for Manganese-Sulfur and Manganese-Silicon Reactions in Graphite-Saturated Melts at 1 bar Pressure of CO. After Turkdogan.[42]

data in Figures 15 and 16 substantiate the deductions made from the foregoing argument.

Analysis of the cupola plant data in the manner described above for the blast furnace may shed further light on the state of gas-slag-metal reactions in the cupola hearth. From the composition ranges for basic cupola operations with tap temperatures of 1450–1530°C, given in the paper by Katz and Rezeau,[41] it is seen that the ratio (%S)/[%S] is in the range 25–50 for the slag basicity of 1.3 and increases to 40–120 at the basicity of 1.5, which are in general accord with the blast-furnace plant data in Figure 15. The ratio [%Si]/(%SiO$_2$) for basic cupolas are at the lower end of the shaded area in Figure 15a for the blast-furnace data. Indications are that the state of reactions in basic cupolas are similar to that in iron blast furnaces with respect to the departures from or closeness to the three-phase gas-slag-metal reaction equilibria.

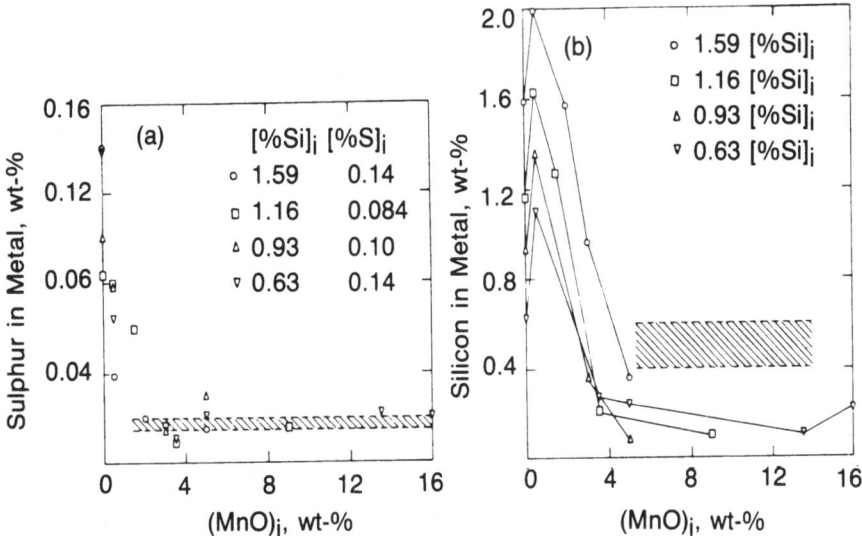

Figure 17. Change in (a) Sulfur and (b) Silicon Content of Metal Droplets Passing Through a 10 cm Deep Slag Column at 1480°C the Shaded Region is for Gas-Slag-Metal Equilibrium at 1 bar Pressure of CO. After Turkdogan et al.[43]

**Reaction of Metal Droplets in Molten Slags**—In a study by Turkdogan, et al,[43] experiments were made to simulate the events in the upper part of the blast-furnace hearth where metal droplets pass through the slag layer; a similar condition will prevail in the cupola hearth. In these experiments, solid pieces (4–5 mm diameter) of graphite-saturated iron containing silicon and sulfur were dropped on the surface of a pool of blast-furnace slag (75 g) contained in a graphite tube (20 mm diameter) at 1480°C; an argon atmosphere was maintained over the melt. Twenty pieces of metal (7 g total) were dropped one at a time at 10–20 s time intervals, and the melt was then rapidly cooled. In the experiments with each alloy, different amounts of MnO were added to the slag during melting. If there were no appreciable buoyancy effect due to moderate gas evolution, the estimated residence time of the droplets during descent in the slag column would be about 1–2 s. Despite the shortness of the reaction time, appreciable changes occurred in the composition of the metal droplets.

As is seen from the experimental data in Figure 17a, with slags containing more than 2% MnO, the metal droplets were desulfurized rapidly to the equilibrium level. Changes in the silicon content of the metal with increasing MnO content of the slag are shown in Figure 17b.

Rapid approach to the gas-slag-metal equilibrium achieved during a short time of descent of metal droplets in a 10-cm deep slag column, has never been experienced in conventional experiments with stirred or unstirred slag and metal samples contained, one on top of the other, in a crucible. With the limited experimental

*References pp. 83–84*

data available, it is not possible to resolve the kinetics of these fast gas-slag-metal reactions. In the system investigated, vapor species, such as CO, SiO, and SiS, may play an important role in reaction kinetics. Also, the diffusion of solutes in the metal droplets will be greatly enhanced by the interfacial turbulence that accompanies the transfer of a surface-active solute, such as sulfur, across the metal surface.

## CONCLUDING REMARKS

Metallurgical reactions occurring in melting, refining, and solidification of foundry irons are controlled by a variety of reaction mechanisms, depending on the type of reaction involved in the process. In the casting phase of the foundry practice, the solidification characteristics of the cast iron are greatly influenced by the surface tension of the liquid metal, in particular, the nucleation and mode of growth of graphite: flaky graphite in the production of gray iron (low-surface tension) and nodular graphite in the production of ductile iron (high-surface tension). Also, foundry iron of low-surface tension, and solidifying at a slow rate, is susceptible to pinholing in castings, caused by the interaction between the sand mold and the liquid metal. Aluminum and titanium at low concentrations lower the surface tension of liquid cast iron, which, in turn, becomes more susceptible to pinholing. This unusual solute effect on the surface tension may be attributed to the associated chemisorption of AlN and TiC species. This concept of complex chemisorption should be tested experimentally.

Large variations are noted in the results of laboratory experiments on the rates of gas-slag-metal reactions. We have seen from the selected examples given that the rate-controlling reaction mechanism, and the corresponding rate constant, depend on the scale and the type of the experiment. The results of laboratory experiments with gas-stirred melts cannot be extrapolated to plant scale operations in the ladle refining process. The rate of desulfurization of cupola iron with a lime-rich slag in a gas-stirred reactor appears to depend on the liquid volume fraction of the lime-saturated slag. Indications are that a lime-rich-slag with a high liquid fraction may sustain a higher rate of desulfurization. However, more experimental data are needed to substantiate this indication.

Rate phenomena pertaining to the reactions in the cupola are closely associated with the properties of coke that affect the rate of gasification, the thermal efficiency, and the overall performance of the cupola.

The state of gas-slag-metal reactions (with respect to the reaction equilibria) in the cupola hearth is similar to that in the blast-furnace hearth. Although there are departures from the three-phase gas-slag-metal equilibria, the manganese, silicon, and sulfur contents of the iron at tap vary with the slag composition in a manner that is predictable from equations for the reaction equilibrium constants. That is, the ratio $[\%Si]/(\%SiO_2)$ decreases with decreasing temperature and increasing slag basicity; the ratios $[\%Mn]/(\%MnO)$ and $(\%S)/[\%S]$ increase with increasing tap temperature and increasing slag basicity.

# REFERENCES

1. E. T. Turkdogan, "Physiocochemical Properties of Molten Slags and Glasses," The Metals Society, London, 1983.
2. J. V. Dawson and L. W. L. Smith, *BCIRA J. Res. & Develop.*, 1956, *6*, 226–248.
3. J. V. Dawson, *BCIRA J. Res. & Develop.*, 1959, *7*, 824–831.
4. E. Haack, *Foundry*, 1961, *89*, 80–83.
5. V. T. Burtsev, A. A. Vertman, A. M. Samarin, and G. Filipp, *Russian Castings Production*, 1965, May, 203–206.
6. F. P. H. Chen and J. Keverian, *Trans AFS.*, 1966, *74*, 281–289.
7. K. H. Tiegel, *Giesserei*, 1968, *55*, 169–174.
8. J. F. Davidson and B. O. G. Schuler, *Trans. Inst. Chem. Eng.*, 1960, *38*, 144–154; 335–342.
9. B. Hernandez and J. F. Wallace, *Trans. AFS.*, 1979, *87*, 335–368.
10. R. V. Naik and J. F. Wallace, *Trans. AFS.*, 1980, *88*, 367–388.
11. E. Selcuk and D. H. Kirkwood, *J. Iron Steel Inst.*, 1973, *211*, 134–140.
12. R. H. McSwain, C. E. Bates, and W. D. Scott, *Trans. AFS*, 1974, *82*, 85–94.
13. J. V. Dawson, *BCIRA J. Res. & Develop.*, 1962, *10*, 433–437.
14. J. V. Dawson, *Foundry*, 1973, *101*, 89.
15. D. B. Evans and R. D. Pehlke, *Trans. AIME*, 1964, *230*, 1651–1656.
16. T. J. Whalen, S. M. Kaufman, and M. Humenik, *Trans. ASM*, 1962, *55*, 778–785.
17. G. R. Belton, *Metall. Trans.*, 1972, *3B*, 1465–1469.
18. M. J. Hunter and G. A. Chadwick, *J. Iron Steel Inst.*, 1972, *210*, 117–123; 707–717.
19. G. S. Cole, *Trans. AFS*, 1972, *80*, 335–348.
20. M. H. Jacobs, T. J. Law, D. A. Melford, and M. J. Stowell, *Met. Technology*, 1974, *1*, 490–500; 1976, *3*, 98–108.
21. A. Ya. Khrapov, *Fiz. Metall. & Metallovedenie*, 1958, *6*, 281–288.
22. V. H. Patterson and M. J. Lalich, *Trans. AFS*, 1978, *86*, 33–42.
23. E. T. Turkdogan, R. A. Hancock, and J. Pearson, *J. Iron Steel Inst.*, 1955, *179*, 338–341.
24. C. F. Landefeld and S. Katz, in "5th Inter. Iron Steel Congress, Process Technology Proc.," The Iron and Steel Society of AIME, Warrendale PA, (1986), 429–440.
25. G. J. W. Kor, Private communication, U. S. Steel Corporation, Technical Center, Monroeville, PA, 1986.
26. P. Kozakevitch, G. Urbain, and M. Sage, *Mem. Sci. Rev. Métál.*, 1955, *52*, 161–171.
27. K. N. Subramanian and F. D. Richardson, *J. Iron Steel Inst.*, 1968, *206*, 576–583.
28. R. Higbie, *Trans. Inst. Chem. Eng.*, 1935, *31*, 365–389.
29. P. M. Coon, *Trans. AFS*, 1980, *88*, 471–480.
30. S. Katz and C. F. Landefeld, *Trans. AFS*, 1985, *93*, 215–228.
31. J. Ishida, K. Yamaguchi, S. Sugiura, K. Yamano, S. Hayakawa, and N. Demukai, *Denki-Seiko*, 1981, *52*, 2–8.

32. J. Szekely, J. W. Evans, and H. Y. Sohn, "Gas-Solid Reactions," Academic Press, New York, 1976.
33. E. T. Turkdogan and J. V. Vintners, *Carbon*, 1970, *8*, 39–53.
34. D. A. Aderibigbe and J. Szekely, *Ironmaking & Steelmaking*, 1981, *8*, 11–19.
35. F. Danis and M. Decrop, *Foundry Trade Journal*, 1958, March 20, 319–325; March 27, 351–358.
36. I. J. Petrovsky and A. B. Draper, "Combustion of Coke in a Cold Blast Cupola," CAES Pub. No. 562–80, Center for Air Environment Studies, University Park, PA, 1980.
37. S. Katz, in "Cupola Handbook," 5th Edition, 1984. American Foundrymen's Society, Des Plaines, Illinois.
38. O. Angeles, G. H. Geiger, and C. R. Loper, *Trans. AFS*, 1968, *76*, 629–637.
39. A. Mihajilovic and B. Marincek, *Arch. Eisenhüttenwes.*, 1973, *44*, 507–512.
40. H. J. Leyshon and M. R. Thibault, *AFS Inter. Cast Metals J.*, 1978, *3*(3), 25–38.
41. S. Katz and H. C. Rezeau, *Trans. AFS*, 1979, *87*, 367–376.
42. E. T. Turkdogan, *Trans. Iron Steel Inst. Japan*, 1984, *24*, 591–611.
43. E. T. Turkdogan, G. J. W. Kor, and R. J. Fruehan, *Ironmaking & Steelmaking*, 1980, *7*, 268–280.

# APPENDIX

**Solute Activities in Liquid Iron Alloys**—The solute concentration in mass percent is converted to the thermodynamic activity, $a_i$, defined as

$$a_i = f_i[\%i] \tag{A1}$$

where $f_i$ is the activity coefficient of the solute $i$, defined such that $f_i \to 1.0$ as the concentrations of the solutes approach zero. For dilute metallic solutions, as in low-alloy steels and foundry iron, the activity coefficient of solute $i$, affected by the alloying elements $k, l, m \ldots$, is usually approximated by the following relation.

$$f_i = f_i^i \times f_i^k \times f_i^\ell \times f_i^m \times \ldots \tag{A2}$$

The interaction coefficients, $e_i^j = \partial \log f_i^j / \partial [\%j]$, used to calculate solute activities in foundry irons are listed in the accompanying table.

### Interaction Coefficients, $e_i^j$
### Alloying Element, $j$

| Solute, $i$ | Al | C | Mn | P | S | Si |
|---|---|---|---|---|---|---|
| Al | 0.049* | 0.091 | — | — | 0.030 | 0.056 |
| C | 0.043 | 0.15* | −0.012 | 0.051 | 0.046 | 0.089* |
| Mn | — | −0.07 | 0 | 0 | −0.048 | 0 |
| N | 0.026* | 0.13 | −0.02 | 0.045 | 0.007 | 0.047 |
| S | 0.035 | 0.11 | −0.026 | 0.29 | −0.014* | 0.063 |
| Si | 0.058 | 0.20* | 0 | 0.11 | 0.056 | 0.11* |

\* Values are for 1400°C; others are for 1600°C; however, the temperature effect is small.

The values of $e_S^C$ and $e_S^{Si}$ given in the accompanying table are for C and Si concentrations below 1 percent. The following values of $f_S^j$ are to be used for higher concentrations of C and Si.

| Mass% C or Si: | 2.0 | 2.5 | 3.0 | 3.5 | 4.0 | 4.5 | 5.0 |
|---|---|---|---|---|---|---|---|
| $f_S^C$: | 1.79 | 2.14 | 2.53 | 3.05 | 3.74 | 4.56 | 5.75 |
| $f_S^{Si}$: | 1.37 | 1.50 | 1.64 | 1.78 | 1.95 | 2.10 | 2.32 |

**Solubility of Mg in Liquid Iron Alloys**—Trojan and Flinn[1] measured the solubility of Mg in liquid Fe-C alloys by equilibrating the melt with liquid magnesium under a pressure of argon. Subsequently, Guichelaar, et al,[2] made similar measurements with liquid Fe-Si-Mg alloys. Their data have been used in numerous studies to derive the equilibrium relations for the solubility of magnesium in liquid iron; however, there are some variations in the interpretation of the above mentioned experimental data. A reassessment of these experimental data is considered desirable.

*References pp. 96–97*

The equilibrium relation for the solubility of Mg (in units of mass% bar$^{-1}$) is represented by

$$Mg(g) = [Mg]$$

$$K_{Mg} = \frac{[\%Mg]f_{Mg}}{P_{Mg}} \quad (A3)$$

where $f_{Mg}$ is the activity coefficient affected by the alloying elements. In the experiments with the Fe-C-Mg melts coexistent with liquid Mg, the latter contained less than 2% Fe, therefore, the Mg vapor pressure prevailing in the reactor would be essentially the same as that for pure Mg for which the following is obtained from the data of Guichelaar *et al.*

$$\log P^\circ_{Mg}(\text{bar}) = -\frac{6778}{T} + 4.934 \quad (A4)$$

In the experiments with the Fe-Si-Mg alloys, the Mg-rich phase contained relatively large concentrations of Fe and Si. In addition to determining the miscibility gap in this system, Guichelaar, *et al*, also measured the equilibrium vapor pressure of Mg in the system by the "boiling method."

These experimental data are presented in Figure A1 as a plot of log $([\%Mg]/P_{Mg})$ versus the concentration of C or Si in the iron-rich phase. For the Fe-C-Mg system, there is a linear relation; for the Fe-Si-Mg system, a linear relation holds up to about 18% Si. From the slopes of the lines, the following interaction coefficients are obtained:

$$e^C_{Mg} = -0.15 \text{ and } e^{Si}_{Mg} = -0.046 \quad (A5)$$

The intercepts of the lines with the ordinate axis of %C and %Si $\to$ 0 give the equilibrium constant $K_{Mg}$ for pure liquid iron (undercooled). The temperature dependence of $K_{Mg}$ shown in Figure A2 is represented by

$$\log K_{Mg} = \frac{4110}{T} - 3.692 \quad (A6)$$

which gives the following free energy equation for the solution of Mg vapor in liquid iron:

$$Mg(g) = [Mg](1\%)$$

$$\Delta G_S = -78,700 + 70.67T \text{ Jmole}^{-1} \quad (A7)$$

For the quaternary melts Fe-C-Si-Mg, the activity coefficient of Mg may be approximated by the product

$$f_{Mg} = f^C_{Mg} \times f^{Si}_{Mg} \quad (A8)$$

The solubility of Mg in liquid Fe-C-Si-Mg alloys calculated from Equations A5, A6, and A8 agrees well with the measured values reported by Trojan and Flinn.[1]

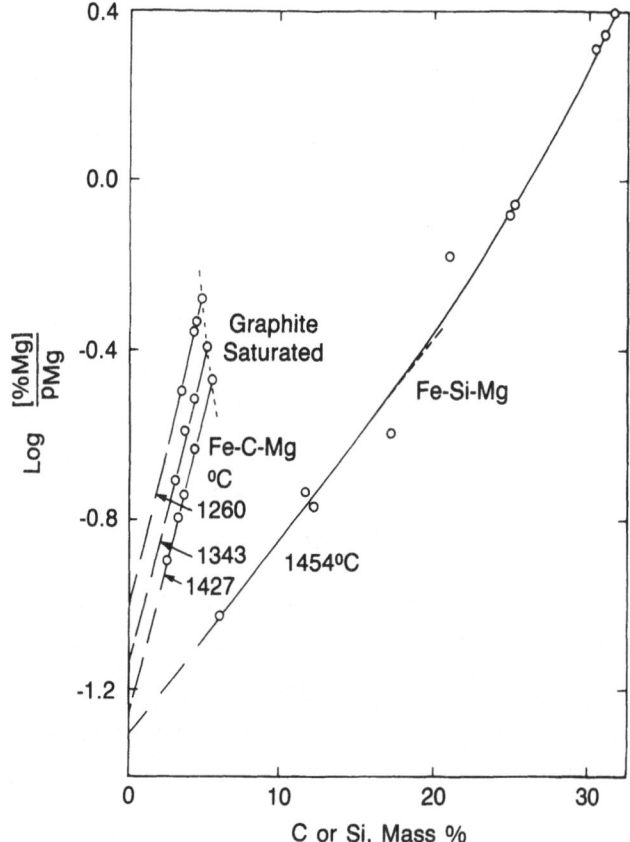

**Figure A1.** Solubility of Mg in Liquid Fe-C Alloys[1] and Liquid Fe-Si Alloys.[2]

In the experiments by Engh et al.,[3] the Mg solubility in graphite-saturated liquid iron at 1260°C was measured by equilibrating the liquid Fe-C-Mg alloys with liquid Pb-Mg alloys. The vapor pressure of Mg in the two-liquid system was derived from $P°_{Mg}$ (Equation A4) and the known activity coefficient of Mg in the Pb-Mg phase. The ratio %Mg/$P_{Mg}$(bar) — 0.553 they obtained for graphite-saturated iron compares well with 0.525 determined by Trojan and Flinn. Noticeably higher values were reported by other investigators: 0.80 by Speer and Parlee[4] and 0.70 by Irons and Guthrie.[5]

**Solubility of MgS in Liquid Iron Alloys**—From the JANAF Thermochemical Data,[6] the following equation is derived for the standard free energy of formation of MgS.

$$Mg(g) + 1/2S_2(g) = MgS(s)$$

$$\Delta G° = -539,740 + 193.05T \text{ Jmole}^{-1} \tag{A9}$$

*References pp. 96–97*

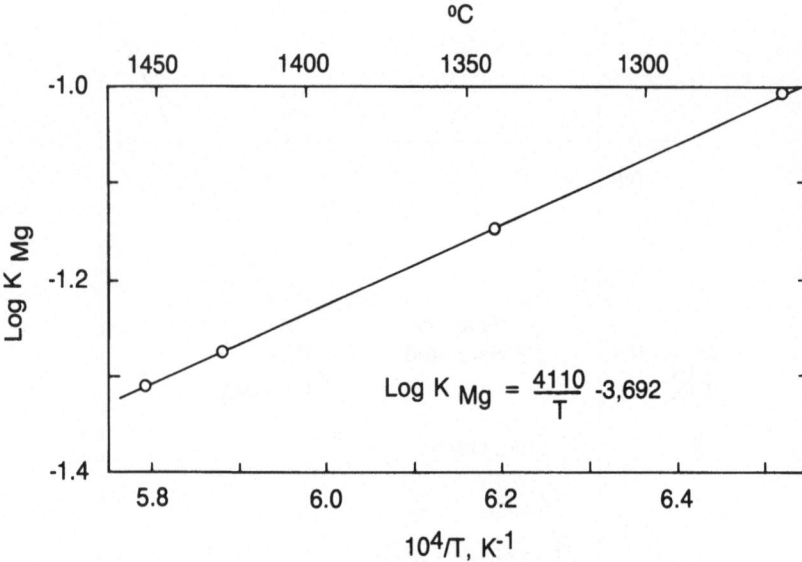

**Figure A2.** Temperature Dependence of the Equilibrium Constant for the Solubility of Mg in Liquid Iron (Undercooled), %Mg bar$^{-1}$.

The uncertainty in the value of $\Delta G°$ is estimated to be ±3 kJ. The free energy of solution of sulfur in liquid iron is known to be[7]

$$1/2 S_2(g) = [S](1\%)$$
$$\Delta G_S = -135,060 + 23.43T \text{ Jmole}^{-1} \tag{A10}$$

From Equations A7, A9, and A10, the following equilibrium relation is obtained for the solubility of MgS in liquid iron.

$$MgS(s) = [Mg] + [S]$$
$$\log K_{MgS} = \log[\%Mg][\%S]f_{Mg}f_S = -\frac{17,026}{T} + 5.168 \tag{A11}$$

where the activity coefficients $f_{Mg}$ and $f_S \to 1.0$ for dilute concentrations of Mg, C, and S dissolved in iron.

For low-alloy steels, the solubility product at 1600°C is calculated to be (% Mg] [% S] = $1.1 \times 10^{-4}$. For graphite-saturated liquid iron at 1260° with 4.4%C, hence $f_{Mg}^C = 0.22$ and $f_S^C = 4.4$, the calculated solubility product is [%Mg] [%S] = $1.2 \times 10^{-6}$, which is only slightly lower than $1.8 \times 10^{-6}$ calculated by Engh, et al.[3] However, as is seen from the plot in Figure A3, the calculated solubility product is in general accord with the experimental data of Engh et al, for graphite-saturated liquid iron at 1260°C. Some of the data points lying above the calculated line is indicative of the presence of dispersed particles of MgS in the melt; the same point of view was expressed by Engh, et al, in their paper.

PHYSICOCHEMICAL PHENOMENA

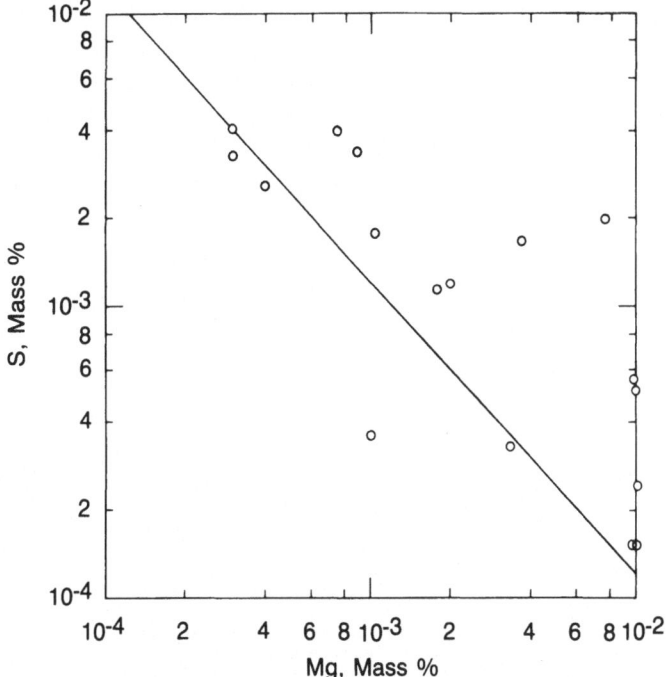

Figure A3. Sulfur and Magnesium Contents of Carbon-Saturated Liquid Iron at 1260°C from Data of Engh et al;[3] the Line is Calculated for MgS Saturation, i.e., [%Mg] [%S] = 1.2 × 10$^{-6}$.

**Solubility of TiC in Liquid Iron Alloys**—Delve[8] measured the solubility of TiC in graphite-saturated liquid iron at 1500 and 1600°C by analyzing the quenched samples for the acid soluble Ti, which is taken to represent the concentrations of dissolved Ti in the melt in equilibrium with TiC. The following results were obtained:

| °C | %C(dissolved) | %Ti(dissolved) |
|---|---|---|
| 1500 | 5.14 | 0.27 |
| 1600 | 5.42 | 0.29 |

From the experimental data on the solubility of TiC in austenite[9] and the appropriate thermochemical data, the following equation is derived for the solubility product of TiC in liquid iron alloys

$$\log \left( \frac{[\%Ti][\%C] f_{Ti} f_C}{a_{TiC}} \right) = -\frac{6160}{T} + 3.25 \qquad (A12)$$

From Equation A12, the interaction coefficients in the accompanying Table and the solubility data of Delve, the following values of $e_{Ti}^C$ are obtained.

| °C | $e_{Ti}^C$ |
|---|---|
| 1500 | −0.24 |
| 1600 | −0.20 |

*References pp. 96–97*

**Surface Tension of Liquid Iron Alloys**—The physicochemical phenomena encountered in the refining and solidification of foundry iron are often related to the surface tension of the liquid metal or the interfacial tension between liquid metal and a solid substrate, e.g., graphite or the refractory mold material. In relation to the metallurgical reaction phenomena discussed in this paper, selected surface tension data are presented here for iron-based liquid alloys pertinent to the compositions of foundry iron.

The experimental data for the surface tension of liquid iron and its binary alloys have been compiled recently by Keene.[10] For purified liquid iron, the average value of the surface tension at temperature $t(°C)$ is represented by

$$\sigma_{Fe} = (2367 \pm 500) - 0.34t, \text{mNm}^{-1} \tag{A13}$$

Keene derived the following weighted average limiting values of $\sigma$ (mNm$^{-1}$) for dilute solutions of X in Fe-X binary alloys.*

| | | |
|---|---|---|
| Fe-C: | Virtually no effect of $C$ on $\sigma_{Fe}$ | (A14) |
| Fe-Ce[11]: | $\sigma = \sigma_{Fe} - 700[\%Ce]$ | (A15) |
| Fe-Mn: | $\sigma = \sigma_{Fe} - 51[\%Mn]; \quad \partial\sigma/\partial t = -0.22$ | (A16) |
| Fe-N: | $\sigma = \sigma_{Fe} - 5585[\%N]$ | (A17) |
| Fe-P: | $\sigma = \sigma_{Fe} - 25[\%P]$ | (A18) |
| Fe-S: | (Discussed later) | (A19) |
| Fe-Si:[12] | $\sigma = \sigma_{Fe} - 30[\%Si]; \quad \partial\sigma/\partial t = -0.25$ | |
| | (Discussed later) | (A20) |

Gibbs' exact treatment of surface thermodynamics gives, for fixed unit surface area and constant temperature and pressure,

$$d\sigma = -RT \sum_{i}^{k} \Gamma_i d\ln a_i \tag{A21}$$

where $\Gamma_i$ is the surface excess concentration of the *ith* component and $a_i$ its activity. For a ternary system, Equation A21 is reduced to

$$d\sigma = -RT \left( \Gamma_2 d\ln a_2 + \Gamma_3 d\ln a_3 \right) \tag{A22}$$

Since carbon dissolved in iron has virtually no effect on the surface tension of liquid iron, for the ternary system Fe-C-S, Equation A22 is simplified to

$$d\sigma = -RT\Gamma_S d\ln a_S \tag{A23}$$

---

* When preference is given to a particular set of measurements, the reference is made to that paper.

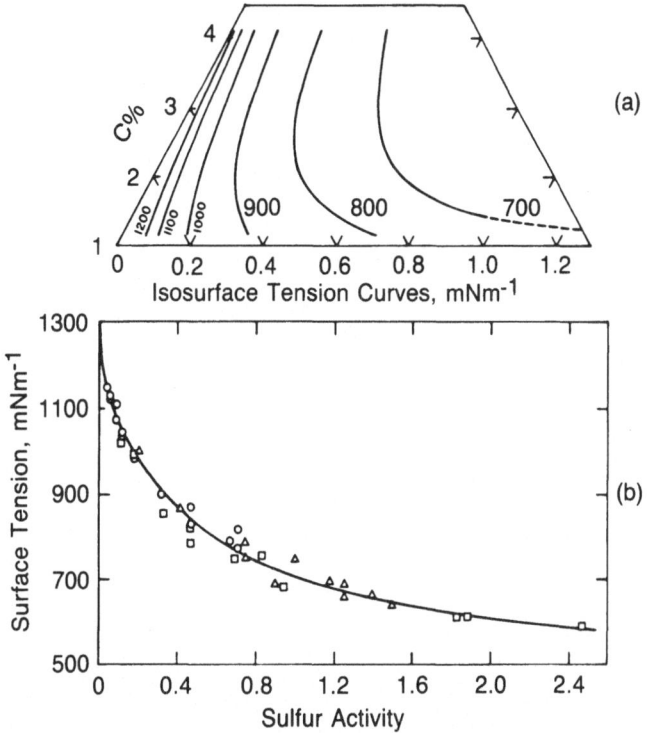

Figure A4.  Surface Tension of Liquid Fe-C-S Alloys. After Kozakevitch.[13]

As shown, for example by Kozakevitch,[13] the addition of carbon to iron-sulfur alloys lowers the surface tension (Figure A4a). In Figure A4b, the plot of surface tension against the activity of sulfur illustrates that the apparent effect of carbon on the surface tension results from the effect of carbon on the activity coefficient of sulfur dissolved in iron.

Whalen, et al,[14] measured the surface tension of liquid alloys Fe-3% C-Si at 1450°C. Their data adjusted to 1400°C, using $\partial\sigma/\partial t = -0.25$, are plotted in Figure A5a against the concentration of silicon in mass percent. The same data are plotted in Figure A5b against the activity silicon and are compared with that for the binary system Fe-Si. Considering that an uncertainty of 100 to 200 mNm$^{-1}$ in the values of $\sigma$ is not unusual in surface tension measurements by different techniques, the data in Figure A5b show that $\sigma$ varies with $a_{Si}$ in a similar manner for both the Fe-Si and Fe-C-Si melts. This similarity is as would be expected, because carbon has no effect on the surface tension of liquid iron.

The lower curves in Figure A5a are from the surface tension data of Washchenko and Rudoy[15] for liquid cast iron containing 3.5% C, 0.22% Mn, 0.04% P, 0.04% S, and 0.1–2.8% Si. The extrapolation of the curve for 1400°C to zero silicon content gives $\sigma = 1180$ mNm$^{-1}$ for the iron $-$ 3.2% C - 0.04% S alloy where the activity of sulfur is $a_S = 0.10$; this value of $\sigma$ is about 50 mNm$^{-1}$ higher than that in Figure A4 (adjusted to 1400°C).

*References pp. 96–97*

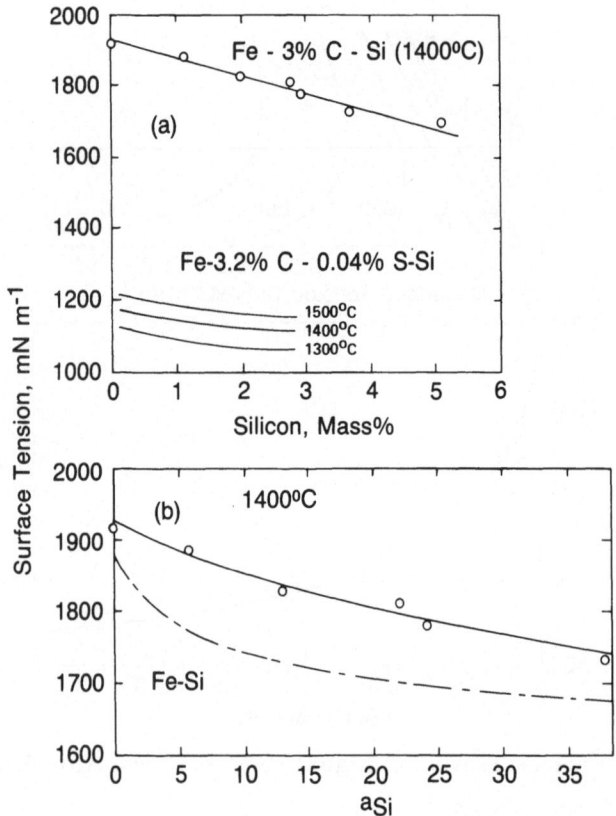

Figure A5. Surface Tension of Liquid Alloys Fe-Si[12], Fe-C-Si[14] and Fe-C-Si-S.[15]

Washchenko and Rudoy also measured the effect on sulfur on the surface tension of Fe-3.2% C-2.8% Si-S alloys. The curve $a$ in Figure A6a represents their experimental data for 1400°C plotted against the activity of sulfur. The curve $b$ is for the Fe-C-S alloys, reproduced from Figure A4 with the adjustment of $\sigma$ from 1450 to 1400°C. Since silicon also lowers the surface tension of liquid iron, the curve $a$ should have been below curve $b$, or vice versa. Assuming that the effects of $a_S$ and $a_{Si}$ on $\sigma$ are additive, the curve $c$ is calculated from curve $b$ for cast iron containing 3.2% C and 2.8% Si. The calculated values of $\sigma$ relative to the Fe-C-S alloys are 230 mNm$^{-1}$ lower than the measured values of Waschenko and Rudoy.

In the experimental work of Selcuk and Kirkwood,[16] it was found that the surface tension of liquid cast iron (not treated with Mg or Ce) increased from about 1100 to 1270 mNm$^{-1}$ with vacuum treatment at about 1200°C during which the sulfur content of the iron decreased from about 0.006 to 0.0025% S. The iron alloy used in the experiments contained 3.68% C, 2.58% Si, 0.13% Mn, 0.026% P, and 0.022% S (initial), for which the activity coefficient of sulfur is $f_S = 3.75$. Their values of $\sigma$ are scattered within the shaded area shown in Figure A6b. With the

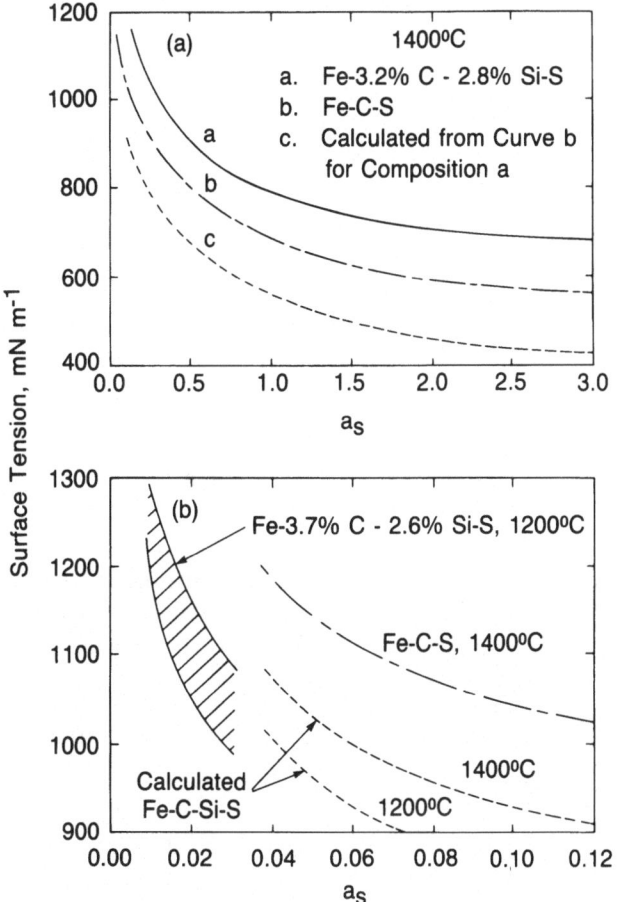

**Figure A6.** Effect of Sulfur Activity on the Surface Tension of Liquid Alloys Fe-C-S[13] and Fe-C-Si-S.[15,16]

assumption of additive effects of $a_S$ and $a_{Si}$ on $\sigma$, the dotted curves are calculated from the curve for the Fe-C-S alloys. It is interesting to note that the calculated curve for 1200°C for $a_S > 0.04$ is in line with the experimental data of Selcuk and Kirkwood for the range $0.009 < a_S < 0.022$.

In relation to a study of the formation of surface pinholes in foundry iron castings, Hernandez and Wallace[17] measured the surface tension of a series of molten cast irons with the base composition: 3.3% C, 1.0% Mn, 2.0% Si, 0.1% P, and 0.1% S. Their results for the temperatures of 1340 to 1430°C are plotted in Figure A7 as a function of $a_S$, derived from their alloy compositions. The dotted curves are calculated from the $\sigma$ versus $a_S$ curve for the system Fe-C-S with the assumption of additive effects of solutes on $\sigma$. The cast iron often contains nitrogen in the range 40 to 80 ppm. The experimental results of Hernandez and Wallace are scattered about the calculated curves (1400°C) for zero and 60 ppm N. It

*References pp. 96–97*

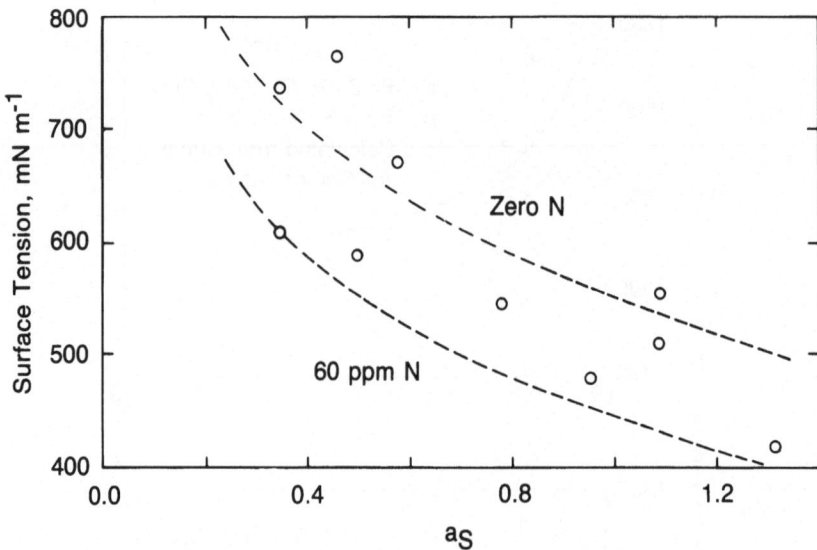

**Figure A7.** Effect of Sulfur on the Surface Tension of Molten Cast Iron at 1340-1430°C, Measured by Hernandez and Wallace;[17] Dotted Curves are Calculated as Described in the Paper.

should be noted that a negative temperature effect on the surface tension reported by these investigators is contrary to the results of other studies. That is, because of the presence of a surface active element, such as sulfur, the surface tension of foundry iron should increase with increasing temperature, as found in the work of Washchenko and Rudoy.[15]

For an approximate estimate of the surface tension of liquid foundry iron containing various alloying elements, the following linear equation may be used for a given temperature and sulfur activity.

$$\sigma_{T, a_S} = \sigma(\text{Fe} - \text{S}) - \sum \Delta\sigma_X(\text{Fe} - \text{X}) \tag{A24}$$

where $\Delta\sigma_X$ is the decrease in the surface tension of iron at a given concentration of X in the binary alloy Fe-X. In addition to S, the alloying elements which have measurable effects on the surface tension of liquid foundry iron are Si, Mn, and N; the activity of oxygen in cast iron is sufficiently low that its effect on $\sigma$ need not be considered. Using the coefficients for Mn and N in Equations A16 and A17, and the curve in Figure A5b for the alloy Fe-C-Si, Equation A24 is expanded to

$$\sigma_{T, a_S} = \sigma(\text{Fe} - \text{S}) - \Delta\sigma_{Si} - 51[a_{Mn}] - 5585[a_N] \tag{A25}$$

As is seen from the compiled data in Figure A8, the surface tension data of various investigators for the Fe-S melts are in close agreement. The points read off from the curves in Figs. A4 and A8 are plotted in Figure A9 as $\sigma$ versus log $a_S$.

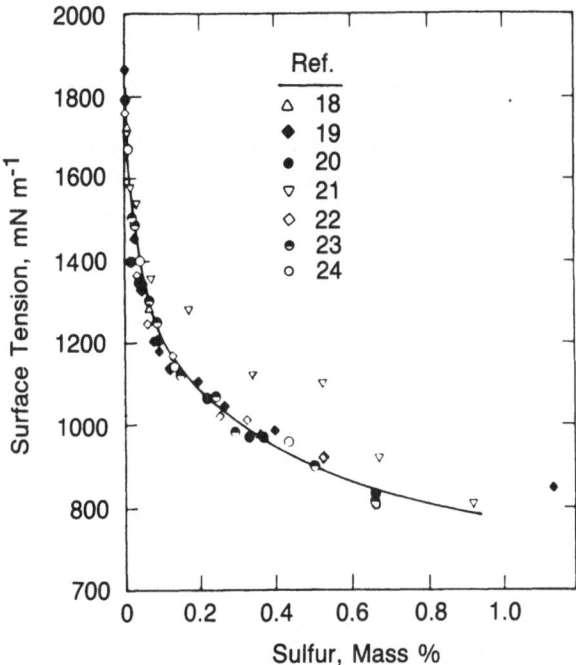

Figure A8.  Surface Tension of Fe-S Alloys at 1550-1660°C.

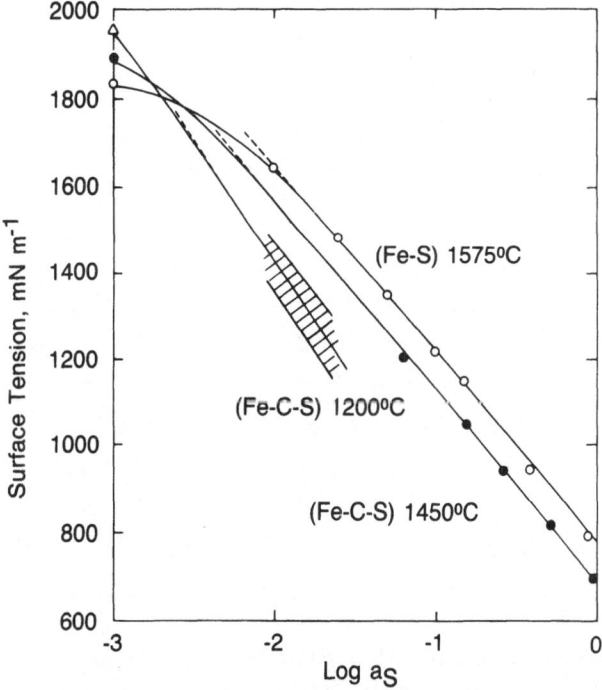

Figure A9.  Surface Tension of Fe-S and Fe-C-S Melts as Affected by Sulfur Activity.

*References pp. 96–97*

It is seen that for $a_S > 0.01$, $\sigma$ is a linear function of $-\log a_S$, a limiting case for almost complete surface coverage with chemisorbed S. The shaded area represents the data of Selcuk and Kirkwood for 1200°C, corrected for the effect of Si on $\sigma$. For their alloy composition, the activity of silicon is $a_{Si} = 35.6$ for which $\Delta\sigma_{Si} = 160$ mNm$^{-1}$ from the curve in Figure A5b. Adding this quantity to the measured values of $\sigma$ for the Fe-C-Si-S alloys gives the estimated value of $\sigma$ for the Fe-C-S or Fe-S alloys as shown by the shaded area in Figure A9. For sulfur activities $a_S > 0.01$, the effect of temperature on $\sigma$ is estimated be $\partial\sigma/\partial t(°C) = 0.65$ for the Fe-S and Fe-C-S alloys. Noting that for the Fe-Si alloys, $\partial\sigma/\partial t = -0.25$, an average value of $\partial\sigma/\partial t = 0.40$ may be used for foundry iron. It should be noted that the surface tension data of Washchenko and Rudoy give $\partial\sigma/\partial t$ in the range 0.37 to 0.50.

## REFERENCES

1. P. K. Trojan and R. A. Flinn, *Trans. ASM*, 1961, *54*, 549–566.
2. P. J. Guichelaar, P. K. Trojan, T. Cluhan, and R. A. Flinn, *Metall. Trans.*, 1971, *2B*, 3305–3313.
3. T. A. Engh, H. Midtgaard, J. C. Berke, and T. Rosengvist, *Scand. J. Metall.*, 1979, *8*, 195–198.
4. M. Speer and N. Parlee, *AFS Cast Metals Res. J.*, 1972, *8*, 122–128.
5. G. A. Irons and R. I. L. Guthrie, *Can. Metall. Quart.*, 1976, *15*, 325–352.
6. JANAF Thermochemical Tables—1974 Supplement, *J. Phys. Chem. Reference Data*, 1974, *3*, 458.
7. S. Ban-ya and J. Chipman, *Trans. AIME*, 1968, *242*, 940–946.
8. F. D. Delve, *Trans. AIME*, 1958, *212*, 183–185.
9. K. J. Irvine, F. B. Pickering, and T. Gladman, *J. Iron Steel Inst.*, 1967, *205*, 161–182.
10. B. J. Keene, "A Survey of Extant Data for the Surface Tension of Iron and Its Binary Alloys," NPL Report DMA(A)67, June 1983, UK.
11. A. F. Vishkarev, Yu. V. Kryakovskii, S. A. Bliznyukov, and V. I. Yavoiskii, *Izv. VUZ. Chern. Metall.*, 1962, *5*(3), 60–66.
12. E. S. Levin, P. V. Gel'd, and B. A. Baum, *Russ. J. Phys. Chem.*, 1966, *40*. 1455–1459.
13. P. Kozakevitch, in "Surface Phenomena of Metals," p. 223, Chem. Ind., London, 1968.
14. T. J. Whalen, S. M. Kaufmann, and M. Humenik, *Trans. ASM*, 1962, *55*, 778–785.
15. K. I. Waschenko and A. P. Rudoy, *Trans. AFS*, 1962, *70*, 855–864.
16. E. Selcuk and D. H. Kirkwood, *J. Iron Steel Inst.*, 1973, *211*, 134–140.
17. B. Hernandez and J. F. Wallace, *Trans. AFS*, 1979, *87*, 335–348.
18. F. A. Halden and W. D. Kingery, *J. Phys. Chem.*, 1955, *59*, 557–559.
19. V. Trzin-Tan, R. A. Karasev, and A. M. Samarin, *Izv. Akad. Nauk SSSR, Met i Toplivo*, 1960, (2), 49–52.
20. P. Kozakevitch and G. Urbain, *Mem. Sci. Rev. Met.*, 1961, *58*, 517–534.
21. B. V. Tsarevskii and S. I. Popel, *Fiz. Metall. i. Metallovedenie*, 1962, *13*, 125–128.
22. B. F. Dyson, *Trans. AIME*, 1963, *227*, 1089–1102.

23. H. Gaye, L. D. Lucas, M. Olette, and P. V. Riboud, in "Inter. Symp. on Interface Phenomena in Metallurgical Systems," Can. Inst. Min. Met., Hamilton, Aug. 1981.
24. K. Ogino, S. Hara, T. Miwa, and S. Kimoto, *Trans. Iron Steel Inst. Japan*, 1984, *24*, 522–531.

## DISCUSSIONS

**R. J. Fruehan** *(Carnegie Mellon University)*

I'm a little skeptical about any kind of surface tension measurement that is made in carbon saturated iron because of the overwhelming effect of sulfur. Unless the sulfur content is very low, I expect sulfur to dominate and mask the effect of other elements. For example, with carbon saturated iron containing 50 ppm sulfur, the sulfur activity would be about 0.03 because of the effect of carbon on sulfur activity. As you pointed out many years ago, with activities of 0.03, the sulfur coverage on the surface would exceed 99%. If you develop any model to describe the chemisorption of two surface active species at once, such as Ti-N, Ti-C, or Al-N, these would have to have enormous absorption coefficients to overcome the affect of sulfur. Would you like to comment on that?

**A.** Whether or not there can be another chemisorption on top of an already chemisorbed layer is quite a legitimate question. The concept of associated chemisorption in a complex alloy system proposed here is intended to make all of us think a little harder about this subject. What we see as the effect of aluminum on the surface tension is very intriguing as it is only in the presence of nitrogen that this behavior is observed. Limited data are available and as I pointed out in my presentation, additional work is needed to resolve this phenomenon.

**R. Singh** *(General Motors)*

I'm a little skeptical about the effect of aluminum and titanium on pinhole formation. I have worked with hypoeutectic and hypereutectic cast irons with section thicknesses ranging from 1.5 to 6 cm and have not observed any correlation between aluminum and titanium concentration and pinhole propensity.

**A.** Well, this morning Dr. Katz drew our attention to the conflicting evidence on the causes of pinhole formation. I took what was reported by a respected group of people at face value and tried to rationalize their observations. Whenever the surface tension of the cast iron is low, for one reason or another, the pinhole tendency becomes greater. And that does make sense in terms of bubble nucleation. So, there seems to be a tie between surface tension and pinhole tendency. The magic effect of aluminum on surface tension, based on one group of work, fits the argument I proposed. Yet you say that there are other data which contradict this effect of aluminum.

**E. Kato** *(Waseda University)*

I would like to get your thoughts on tellurium and pinholing. Tellurium has a potent effect on reducing the surface tension of iron yet it is also said to reduce pinholing. How do you account for that?

A. I don't have an answer for that. I know the paper you are referring to and again, going back to what Dr. Katz said, we do have conflicting evidence on the causes of pinhole formation. Certainly we don't have all the answers.

**R. J. Fruehan** *(Carnegie Mellon University)*

I would just like to give a possible explanation for the effect of tellurium. Whenever we add a surface active element to liquid iron, we're doing two things: we're lowering the surface tension which favors bubble nucleation, as Dr. Turkdogan has suggested. But we're also reducing the number of surface sites for gas reaction to occur on, and as I will show tomorrow, tellurium greatly reduces the rate of the chemical reaction for nitrogen on the surface. So, the formation of nitrogen molecules on the liquid iron surface is impeded by tellurium. If the rate of gas desorption was sufficiently retarded, gas bubbles could not effectively be formed.

**S. Katz** *(General Motors)*

Foundrymen generally believe that the source of the gas that forms pinholes is gas dissolved in the liquid iron. The thought is that gas precipitates during the cooling and solidification process. This belief derives from the fact that under ordinary conditions liquid cast iron, entering the mold cavity, is saturated with nitrogen and it additionally contains quantities of hydrogen. You have provided fresh input with your suggestion that the gas source for pinholes is the mold atmosphere. I think it important that we determine which of the two mechanisms is the correct one. A rather simple experiment to distinguish between the mechanisms might be to determine if pinholes can be generated in degassed iron. The potential problem here, is that the small bubbles asperated into the iron, according to your mechanism, might dissolve so rapidly that there would be no evidence that the process had occurred. Do you feel this is a realistic problem and can you suggest other approaches for establishing the source of the gas?

A. According to the studies cited in my paper, the propensity to pinholing is closely associated with gas generation in the mold as depicted in Figure 1 of my paper. As far as I am aware, the pinholing is not observed in metal mold castings. However, I agree with what you say about the casting experiments with degassed cast iron. Such experiments, using sand molds of different gas content, may help to clarify the mechanism of pinholing, provided due care is taken to prevent air bubble entrainment during casting.

**D. R. Gaskell** *(Purdue University)*

If we take at face value that, with nitrogen present in cast iron, the surface tension goes through a minimum with increasing aluminum addition, doesn't this present us with a theoretical problem? It seems in this situation the surface excess starts as a positive value but ultimately becomes negative. This theoretical difficulty leads me to question the experimental data. Would you comment on that?

**A.** In the mechanism I proposed, an increase in surface tension with increasing aluminum content ($> 0.1\%$) is attributed to a decrease in the activity of nitrogen caused by the precipitation of AlN. When most of the dissolved nitrogen is precipitated as AlN, there will be no associated chemisorption of AlN complex, hence the tension of cast iron will increase to that for the nitrogen-free melt. This is a hypothesis and I am hoping that someone will do very careful experiments at control nitrogen activities and find out if there is anything in this line of thinking.

**N.P. Lillybeck** *(Deere & Co.)*

I don't have a question but I have a comment that might divide the problem up a little bit. The statement the 0.1% Al is the point at which pinholing is a maximum applies only to the gray irons. In a ductile iron it would be some higher value.

**A.** I agree.

**R. D. Pehlke** *(University of Michigan)*

I think you've put the role of aluminum in desulfurization in very good perspective. The fact that aluminum assists in desulfurization has led some people improperly to think aluminum sulfide might be involved.

I would like to raise the following question. Have you considered that, in part, the higher desulfurization rate obtained with aluminum present, is due to its forming a slag phase which has a greater sulfide capacity and therefore a higher driving force for reaction?

**A.** The examples I cited are for desulfurization by $CaO-CaF_2$ and $CaO-Al_2O_3$. With these slags of high sulfur capacity, the aluminum assists desulfurization by lowering the oxygen activity of the metal. In the case of desulfurization by lime + aluminum injection, the alumina generated in the deoxidation reaction fluxes CaO and CaS, thus facilitates the rate of desulfurization. I am surprised to hear that there are still some people who improperly think that aluminum sulfide might be involved.

# THE CAPACITIES AND REFINING CAPABILITIES OF METALLURGICAL SLAGS

## I. D. SOMMERVILLE

*Department of Metallurgy and Materials Science*
*University of Toronto*
*Toronto, Ontario, Canada M5S 1A4*

## ABSTRACT

After reviewing the concepts of the basicity of slags and capacities of slags, the paper describes how the two can be correlated quantitatively using the optical basicity approach. The capacities discussed are those that are relevant under the highly reducing conditions typical of foundry processes, namely the sulfide, phosphate, water, nitride, cyanide and carbide capacities, and where sufficient data exists the effect of temperature on these capacities is included. The application of these correlations to the assessment of the refining capabilities of slags, and hence to the calculation of the equilibrium slag-metal distributions of sulfur, phosphorous, hydrogen and nitrogen is indicated.

## INTRODUCTION

It is convenient and practically useful to regard metallurgical slags as being composed of three types of oxygen ion:

1. Unbonded, double charged ions, $O^{2-}$ i.e., "free" oxygen ions

2. Singly bonded, singly charged ions, $O^-$ i.e., "terminal" oxygen ions

3. Doubly bonded, uncharged ions, $O^0$ i.e., "bridging" oxygen ions

Since apparently only the "free" oxygen ions participate in the slag-metal refining reactions, such as desulfurization or dephosphorization, there is considerable interest in the determination and maximization of the proportion of oxygen ions in this form. A slag in which this proportion is very low is said to be "acid," while one in which it is high is described as "basic." The more basic a slag is, the greater will be its refining capability, so slag basicity is an important parameter in refining reactions involving a slag phase.

Since the concept of basicity of slags is such a useful and important one, various different ways of expressing it have been devised over the years. The most common approach is the use of a basicity ratio or basicity index in which the basic oxides are placed on the numerator and acid oxides on the denominator. Of these, by far the most commonly used is the simple V-ratio

$$\frac{\%CaO}{\%SiO_2}$$

While this is useful as a first approximate guide, in slags containing appreciable quantities of components other than CaO and SiO$_2$ it is a very imperfect measure of basicity. Since most blast furnace slags tend to contain significant amounts of MgO and Al$_2$O$_3$, the ratio

$$\frac{\%CaO + \%MgO}{\%SiO_2 + \%Al_2O_3}$$

has found some application in industrial practice. This expression, however, implies that MgO is equivalent to CaO as a basic oxide and that Al$_2$O$_3$ is equivalent to SiO$_2$ as an acid oxide on a weight % basis, and clearly neither of these equivalences is accurate. The use of molar % or mole fractions as the concentration units in these indices, while having some appeal as being fundamentally more justifiable, can actually make these equivalences less accurate, depending on the relative atomic weights of the oxides concerned. In an attempt to correct this situation, Bell et al[1] suggested the ratio

$$\frac{N_{CaO} + 0.5 N_{MgO}}{N_{SiO_2} + 0.33 N_{Al_2O_3}}$$

where $N$ is the mole fraction of the component concerned. This expression implies that MgO is half as effective as CaO as a base on a molar basis, while Al$_2$O$_3$ is one third as effective as SiO$_2$ as an acid, again on a molar basis.

Expressed in terms of wt. % of each oxide, this ratio is

$$\frac{\%CaO + 0.7\%MgO}{0.94\%SiO_2 + 0.18\%Al_2O_3}$$

and this has indeed proved to be a very useful measure of basicity for blast furnace[2] and cupola[3] slags. A similar expression for steelmaking slags has been given[4] as

$$\frac{\%CaO + 1.4\%MgO}{\%SiO_2 + 0.84\%P_2O_5}$$

but clearly such ratios are rather arbitrary and empirical.

The other approach adopted has been the use of excess base expressions, of the form $\Sigma$ basic oxides $-\Sigma$ acid oxides, where the concentration units can be either weight %, molar % or mole fraction. This method has not been widely used in recent years.

The main problem with these expressions for both basicity index and excess base, however, is that they involve an arbitrary decision as to whether a component is a basic or acidic oxide, so that they can be assigned to the numerator or denominator in a basicity ratio, or assigned a positive or negative sign in an excess base relation, respectively. For most components the situation is clear and this poses no real problem. However, for intermediate oxides, such as $TiO_2$ for example, this can indeed pose some difficulty. A secondary problem with ratios is that the assessment of basicity becomes impossible in slags free of any recognized acid components, such as might be used in secondary processing or iron or steel. Obviously, an excess base equation could still be applied in this situation.

Over ten years ago, Wagner[5] published his thought-provoking paper on the concepts of basicity in metallurgical slags in which he suggested that a suitable parameter for quantifying the basicity of a slag is the relation:

$$B_{carb} = \frac{C_{CO_3^{2-}}}{C^*_{CO_3^{2-}}}$$

where $C_{CO_3^{2-}}$ is the carbonate capacity of an unknown slag and $C^*_{CO_3^{2-}}$ is the carbonate capacity of a reference slag. The carbonate capacity is defined as

$$C_{CO_3^{2-}} = \frac{(\text{wt}\% \, CO_3^{2-})}{p_{CO_2}}$$

A similar expression involving sulfide capacities was also envisaged, but the carbonate capacity was preferred since it was felt that the activity coefficient of the carbonate ion probably varies less than that of the sulfide ion. This is due to the fact the carbonate ion is expected to display weaker interactions with surrounding cations than the sulfide ion.

The practical difficulty in using this approach is that the number of systems for which carbonate capacity values have been determined is very limited. Subsequent work of Tokyo[6] and Toronto[7] has amply confirmed Wagner's contention that the carbonate capacity of the slag he chose as a reference slag is almost zero, so that values of $B_{carb}$ would be extremely large in highly basic slags, but obviously this difficulty can be removed simply by selecting a more basic slag as the reference slag. Other measures of basicity have also been suggested and used with varying degrees of success over limited composition ranges.

**The Concept of Optical Basicity**—An alternative approach is the use of optical basicity which has been pioneered and developed in the field of glass chemistry, by Duffy and Ingram.[8-11] In this approach, basicities are regarded in terms of the electron donor power of oxygen ions present. This can be determined experimentally by introducing trace quantities of metal ions such as $Tl^+$, $Pb^{2+}$

*References pp. 118–119*

## TABLE 1
### Optical Basicity Values Calculated by Equation (1)

| Oxide | Electronegativity of Element | Λ |
|---|---|---|
| $K_2O$ | 0.8 | 1.40 |
| $Na_2O$ | 0.9 | 1.15 |
| BaO | 0.9 | 1.15 |
| SrO | 0.95 | 1.07 |
| $Li_2O$ | 1.0 | 1.00 |
| CaO | 1.0 | 1.00 |
| MgO | 1.2 | 0.78 |
| $Al_2O_3$ | 1.5 | 0.61 |
| $SiO_2$ | 1.8 | 0.48 |
| $B_2O_3$ | 2.0 | 0.42 |
| $P_2O_5$ | 2.1 | 0.40 |
| $CO_2$ | 2.5 | 0.33 |
| $SO_3$ | 2.5 | 0.33 |

and $Bi^{3+}$, which have the electron configuration $d^{10}s^2$, and the effect is measured from spectroscopic shifts in the $^1S_o \rightarrow {}^3P_1 (s^2 \rightarrow sp)$ transition due to the electron donating power of the oxides present. This transition occurs in the u.v. region, and the greater the degree of red shift in the $^1S_o \rightarrow {}^3P_1$ frequency compared with the free (gaseous) $d^{10}s^2$ ion, the greater the basicity. The red shift can be regarded as resulting from the expansion of the outer $s$ and $p$ orbitals of the probe ion brought about by electron donation from the surrounding oxygen ions. This is called the "nephelauxetic" effect.[8,11]

The word "optical," which is rather alien to the metallurgical context, is thus derived from the fact that the basicity of the slag is measured spectroscopically in transparent slags. From a large number of such measurements on solid glasses, mainly using $Pb^{2+}$ as the probe ion, Duffy and Ingram[9] discovered that the optical basicity of an oxide, Λ, is related to the Pauling electronegativity, $x$, of the element involved by the expression

$$\Lambda = \frac{0.75}{x - 0.25} \tag{1}$$

Use of this relation allows calculation of the optical basicity for any non-transition metal oxide as illustrated in Table 1. Hence, by use of the relationship

$$\Lambda = X_{MO_x}\Lambda_{MO_x} + X_{MO_y}\Lambda_{MO_y} \cdots \tag{2}$$

it is possible to calculate the bulk or average value of Λ for a slag of any composition involving these oxides. In Equation 2, $X$ is the equivalent cation fraction, based on the fraction of negative charge "neutralized" by the charge on the cation concerned. The method of calculating $X$ and hence Λ for a slag has been given elsewhere,[12] as have effective values of Λ for the transition metal oxides, FeO, MnO, $Fe_2O_3$ and $TiO_2$, deduced from sulphide capacities of slags containing these oxides. These values cannot be measured experimentally because the slags are opaque.

Although the optical basicity scale is constructed on the basis that $\Lambda_{CaO} = 1$, and thus the $\Lambda$ values for all other components are measures of "lime character" or "lime equivalence," the values can be used in systems which do not contain lime. It should also be noted that this approach allows realistic basicity values to be ascribed to slags which contain no recognized "acid" constituents.

**The Concept of Capacities in Metallurgical Slags**—The concept of capacities of various species in slag has become popular as a way of expressing the ability of the slag to retain that species in solution at equilibrium with either a metal or a gas phase, or both. It can be used to express the solubility of gaseous species in the slag, or the relative ability of the slag to remove that species from the metal phase. Thus, capacities are particularly useful in connection with species which are usually considered as contaminants in the metal phase, such as sulphur, phosphorus, hydrogen and nitrogen. As pointed out by Elliott,[13] the concept of capacity is most appropriately applied to those species which are sparingly soluble in liquid slags under the conditions prevailing in typical metallurgical operations.

The popularity of the capacity concept is largely due to the fact that it provides a method of ranking slags in terms of their relative ability to retain any given species in solution without the necessity of assigning values to all the relevant thermodynamic quantities. While it may be argued that a certain amount of empiricism is thereby introduced there can be no doubt that this 'short-circuiting' has proved extremely useful, and allowed advances in our understanding of slag-metal and gas-slag reactions which would otherwise have had to await the determination of a great deal of fundamental thermodynamic information, particularly with regard to the slag phase.

Of the various capacities mentioned in the metallurgial literature, this review will only be concerned with six which are relevant to the reducing conditions typical of cupola operations. These are the sulphide, phosphate, water, nitride, cyanide and carbide capacities, whose definitions are given in the following sections. Where the species in question is entering the slag from the metal phase, as in the case of sulfur and phosphorous, a high capacity is desirable, but where it is entering from the atmosphere, as in the case of water vapour and nitrogen, a high capacity is undesirable, since the slag can act as a "pump," transferring deleterious elements into the metal phase.

**Sulfide Capacity**—Since the sulfide capacities of slags are usually measured using the gas-slag reaction

$$(O^{2-}) + \frac{1}{2}S_{2(g)} = (S^{2-}) + \frac{1}{2}O_{2(g)} \tag{3}$$

it is most commonly defined by the expression

$$C_S = (wt\%S)\left(\frac{p_{O_2}}{p_{S_2}}\right)^{\frac{1}{2}} \tag{4}$$

where (wt%S) is the concentration of sulfur in the slag, while $p_{O_2}$ and $p_{S_2}$ are the partial pressures of oxygen and sulfur in the gas phase. Some authors have preferred

*References pp. 118–119*

to define it with reference to the slag-metal reaction

$$(O^{2-}) + [S] = (S^{2-}) + [O] \tag{5}$$

where round brackets indicate the slag phase and square brackets the metal phase, in which case the definition becomes

$$C'_S = (wt\%S) \cdot \frac{[a_O]}{[a_S]} \tag{6}$$

The two definitions are, of course, related through the free energy of the gas-metal exchange reaction

$$\frac{1}{2}S_{2(g)} + [O] = [S] + \frac{1}{2}O_{2(g)} \tag{7}$$

**Phosphate Capacity**—The phosphate capacity is usually defined as

$$C_{PO_4^{3-}} = \frac{(wt\% PO_4^{3-})}{(p_{O_2})^{5/4}(p_{P_2})^{\frac{1}{2}}} \tag{8}$$

although some authors have preferred the form

$$C_P = \frac{(wt\% P)}{[a_P][a_O]^{5/2}} \tag{9}$$

Normally the phosphate capacity is obtained from the slag-metal distribution of phosphorus rather than a gas-slag equilibration, i.e. the reaction

$$[P] + \frac{5}{2}[O] + \frac{3}{2}(O^{2-}) = (PO_4^{3-}) \tag{10}$$

so in a sense the latter definition is the more logical choice. Obviously the two definitions are related through the free energies of solution of phosphorus and oxygen in liquid iron.

**Water Capacity**—The hydroxyl capacity has been defined as

$$C_{OH^*} = \frac{(wt\% OH^*)}{p_{H_2O}^{\frac{1}{2}}} \tag{11}$$

where the charge on the $OH^*$ group is variable depending on the basicity of the environment. Since all previous work on water solubility in slags has been virtually unanimous that the solubility is proportional to the square root of the partial pressure of water vapour, this is incorporated into the definition.

Alternatively, the term can be called the water capacity, defined as

$$C_{H_2O} = \frac{(ppmH_2O)}{p_{H_2O}^{\frac{1}{2}}} \tag{12}$$

This avoids assigning a charge to the hydroxyl group, but essentially the same effect emerges in the amphoteric behaviour of water vapour dissolved in slags. Thus, it behaves differently depending on the basicity of the slag, acting as a base (combining with $O^0$) groups in acid slags, as an acid (combining with $O^{2-}$ ions) in highly basic slags, and in an intermediate way (combining with $O^-$ ions) in slags of intermediate basicity.

**Nitride, Cyanide and Carbide Capacities**—Under oxidizing conditions, the solubility of nitrogen in slags appears to be negligible.[14] However, under reducing conditions in the presence of graphite, it can exist in solution either as nitride $(N^{3-})$ or as cyanide $(CN^-)$ ions. The equilibrium solubility is directly related to $p_{N_2}$[15–18] and inversely related to $p_{CO}$.[15,16] As the following equations indicate, this is true for the formation of both ionic species:

$$\frac{3}{2}(O^{2-}) + \frac{3}{2}C + \frac{1}{2}N_{2(g)} = (N^{3-}) + \frac{3}{2}CO_{(g)} \quad (13)$$

$$\frac{1}{2}(O^{2-}) + \frac{3}{2}C + \frac{1}{2}N_{2(g)} = (CN^-) + \frac{1}{2}CO_{(g)} \quad (14)$$

In the presence of graphite, carbon can dissolve in slags as cyanide $(CN^-)$ ions, as in Equation 14, or as carbide ions, by the reaction

$$3C + (O^{2-}) = (C_2^{2-}) + CO_{(g)} \quad (15)$$

In view of the shared ion it is convenient to discuss the solubilities of nitrogen and carbon in slags together. Because of the differing dependences on the partial pressures of nitrogen and carbon monoxide, these solubilities are most clearly compared in terms of the nitride, cyanide and carbide capacities, defined as follows:

$$C_N = (\% N^{3-}) \cdot \frac{p_{CO}^{3/2}}{p_{N_2}^{1/2}} \quad (16)$$

$$C_{CN} = (\% CN^-) \cdot \frac{p_{CO}^{1/2}}{p_{N_2}} \quad (17)$$

and

$$C_C = (\% C_2^{2-}) \cdot p_{CO} \quad (18)$$

## CORRELATION AND PREDICTION OF CAPACITIES USING OPTICAL BASICITY

**Sulphide Capacities**—As a first step, all the available sulphide capacity data at 1500°C was plotted against the optical basicities of the slags,[12] as shown in Figure 1. For the 183 points included, the correlation coefficient, $r^2 = 0.965$, so that

*References pp. 118–119*

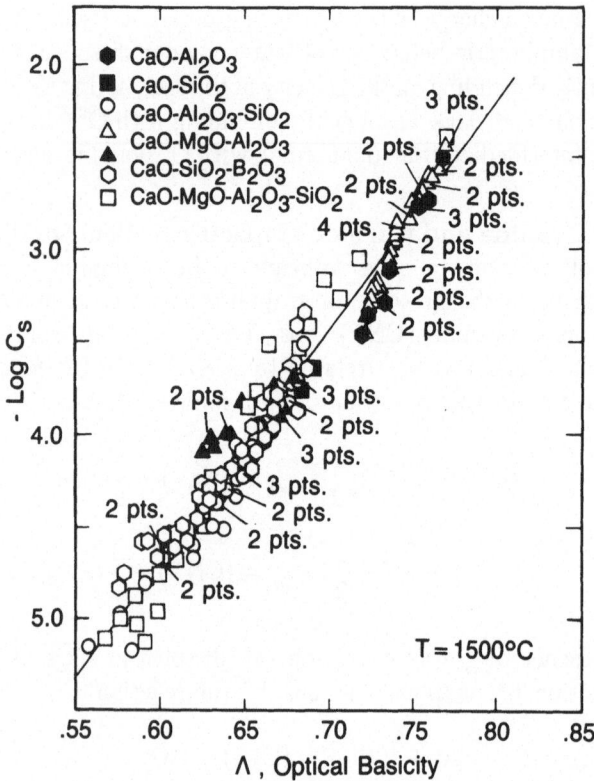

Figure 1. Relationship between log $C_S$ and optical basicity at 1500°C.

the data in the seven systems is remarkably consistent, and the relationship can be used with considerable confidence. However, since 1500°C is not a particularly useful temperature at which to have data for the field of iron and steel making, the next step was to extend this to include the effect of temperature.[19] Data at 1550°C[20] and 1650°C[21,22] permitted this, and also allowed calculation of sulphide capacity data at 1400°C, applicable to external treatment of hot metal, and at 1600°C, applicable to ladle treatment of steel. The lines for the five temperatures are compared in Figure 2, while for convenience in calculating data at any temperature in the range 1400-1700°C, a general expression has been formulated:[19]

$$\log C_S = \left(\frac{22690 - 54640\Lambda}{T}\right) + 43.6\Lambda + 25.2 \tag{19}$$

Use of equations (2) and (19) allows the calculation of $iso - C_S$ lines in any system of interest. As an example, Figure 3 shows such data for the $CaO-Al_2O_3-SiO_2$ system at least 1600°C. However, this approach allows the calculation of sulphide capacities in almost any oxide system of interest at any temperature in the range 1400-1700°C. It should be noted, however, that slags containing any appreciable amounts of soda do not seem to fit the pattern followed by all these other systems,

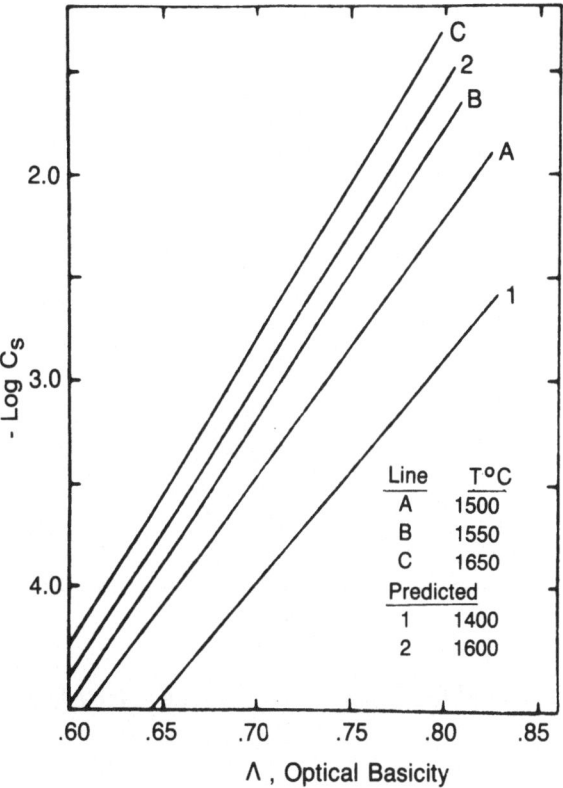

Figure 2. Relationship between log $C_S$ and optical basicity at temperatures in the range 1400 to 1650°C.

which are predominantly, though not exclusively, lime-based. The reasons for this are not understood at the present time.

**Phosphate Capacities**—As far as the author is aware, Gaskell[23] was the first to connect slag-metal phosphorus distribution data with the optical basicities of the slags concerned. This was followed by Suito and Inoue,[24,25] who plotted not only the phosphorus distribution, but also $\gamma_{P_2O_5}$ and the phosphate capacity against the slag optical basicities. The concept was also used extensively by Mori[26] in his excellent paper on phosphorus distribution, and Figure 4 is taken from his paper. At 1600°C, the equation for the excellent linear relationship shown is:

$$\log C_{PO_4^{3-}} = 17.55\Lambda + 5.72 \qquad (20)$$

so that, the phosphate capacity is clearly a very strong function of optical basicity, changing very rapidly over relatively limited ranges of basicity. To avoid any possible confusion, it should be pointed out that all these authors have used values for the optical basicities of the transition metal oxides calculated from the Pauling electronegativities of the cations, ignoring any possible complications introduced

*References pp. 118–119*

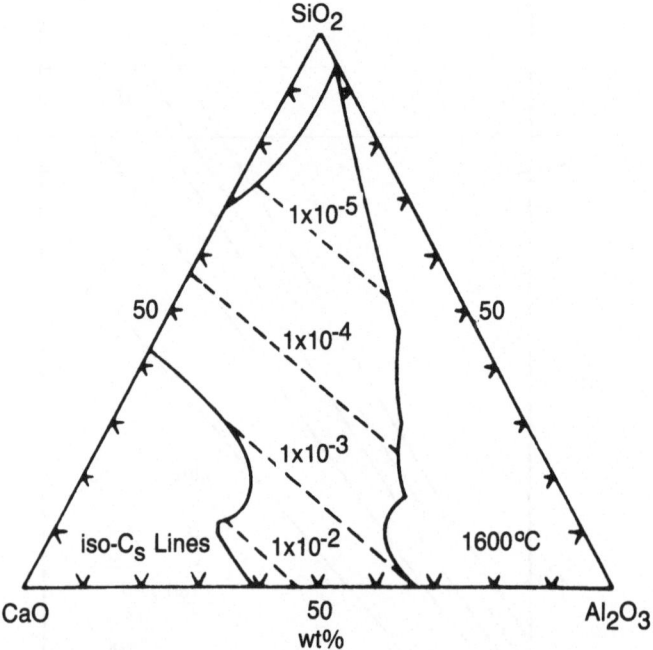

**Figure 3.** Iso-sulphide capacity lines in the CaO-Al$_2$O$_3$-SiO$_2$ system at 1600°C.

by the unfilled $d$ electronic sub-shell in these cations. Thus, the values of phosphate capacity cannot be correlated quantitatively with the values of the other capacities discussed in this review.

**Water Capacities**—A number of studies have been conducted on the water solubility or capacity in slags of the CaO-MgO-SiO$_2$ system,[27–31] with only the one study in the CaO-Al$_2$O$_3$ system.[32] In the former system the solubility data was correlated with the basicity index

$$\frac{N_{CaO} + N_{MgO}}{N_{SiO_2}}$$

where $N$ represents the mole fraction of the component concerned. Use of this index suffered from the difficulty that data for the ternary system and the two binary systems, CaO-SiO$_2$ and MgO-SiO$_2$, could not all be plotted on the same line. This problem was solved by the use of the term $(1 - a_{SiO_2})$ as a measure of basicity,[33] but while this provided a real advance, it suffers from the difficulty that the number of systems for which extensive data on the activity of silica are known is very limited. Also, obviously this parameter cannot be used in silica-free systems. Replotting this data in terms of the optical basicities of the slags provided the basis of the curve shown in Figure 5. Fresh determinations were carried out in the CaO-Al$_2$O$_3$-SiO$_2$ system, partly because of its industrial significance in connection with fluxes for ladle metallurgy and partly because higher basicities could be

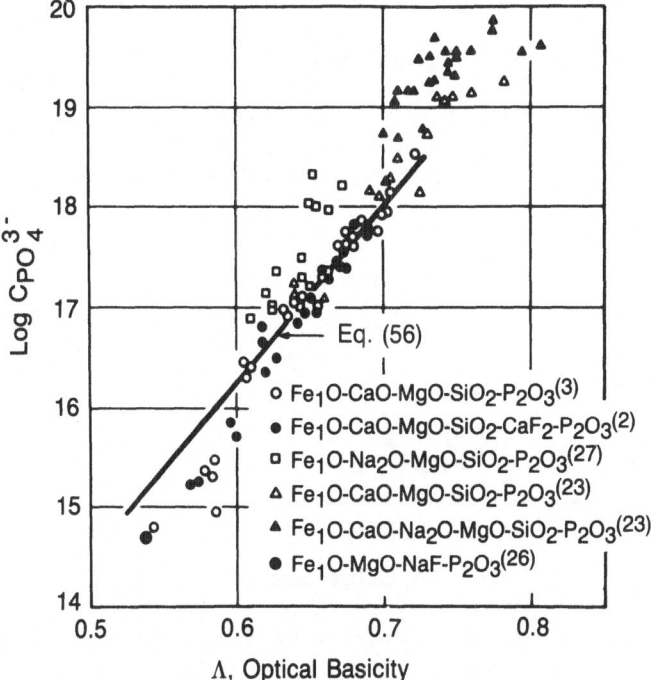

Figure 4. Relationship between log $C_{PO_4^{3-}}$ and optical basicity at 1600°C (after Mori[26]).

attained in this system than were possible in the CaO-MgO-SiO$_2$ system. This more recent data has also been included in Figure 5, which shows that all four systems can now be plotted on the one curve. The curved nature of the plot is due to the amphoteric nature of water vapour dissolved in slags, referred to earlier. The equation for this relationship is:

$$\log C_{H_2O} = 12.04 - 32.63\Lambda + 32.71\Lambda^2 - 6.62\Lambda^3 \tag{21}$$

and the present work has shown in agreement with almost all previous work in this area, that the water capacity of slags is independent of temperature in the range 1375-1600°C. The correlation coefficient, $r^2 = 0.98$, so that this curve can be used with considerable confidence to calculate water capacities in a wide range of oxide systems for which no data presently exist.

**Nitride, Cyanide and Carbide Capacities**—The most reliable and extensive studies on the dependence of the nitrogen solubility on slag composition are those of Davies and Meherali[17] on the CaO-Al$_2$O$_3$-SiO$_2$ and CaO-MgO-Al$_2$O$_3$ systems and of Schwerdtfeger and Schubert on the CaO-Al$_2$O$_3$ system. No mention was made of the cyanide ion in the former study, but in the binary system the two ions were found to display opposite dependence on slag basicity, the solubility of CN$^-$ ions increasing the solubility of N$^{3-}$ ions decreasing with increase in the

*References pp. 118-119*

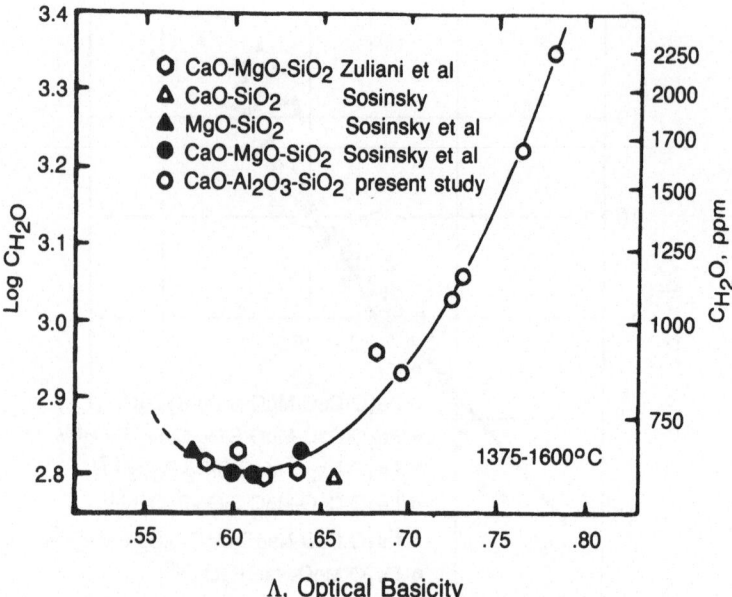

Figure 5. Relationship between $\log C_{H_2O}$ and optical basicity at temperatures in the range 1375 to 1600°C.

CaO/$Al_2O_3$ ratio. This effect was also found by Shimoo et al,[15] and appears to indicate that cyanide ions are introduced by reaction with "free" $O^{2-}$ ions, while nitride ions are introduced by reaction with oxygen ions bound in the aluminate network. Davies and Meherali[17] found similarly that the nitrogen solubilities were inversely related to slag basicity in both the systems studied, and that over the limited range of silica contents investigated, the nitrogen content was directly related to the concentration of silica, which tends to suggest that nitrogen can also substitute for oxygen ions bound in the silicate network. On this basis, Equation 13 could perhaps be more accurately written in the form:

$$3\left(O^{-}\right) + \frac{3}{2}C \times \frac{1}{2}N_{2(g)} = \left(N^{3-}\right) + \frac{3}{2}CO_{(g)} \tag{22}$$

The substitution of such a relatively highly charged ion into the network seems surprising, so that perhaps the observed effect of slag composition on nitride solubility should be explained in terms of the formation of AlN or $Si_3N_4$ in solution in the slag, or by depolymerization of the slag structure.[4]

The data of Schwerdtfeger and Schubert[34,35] at 1600°C are replotted against the optical basicities of the slag concerned in Figures 6-8, and reasonably accurate linear correlations are obtained. The data of Davies and Meherali,[17] recalculated in accordance with Equation 16, have been included in Figure 6 for comparison. Clearly there is good agreement as to the slope of the relationship between $\log C_N$ and optical basicity, but some discrepancy as to the effect of temperature, since the line through the data of Davies et al. at 1500°C almost superimposes on that of Schwertfeger et al. at 1600°C. The few points obtained by Davies et al. at 1500° and 1600°C have also been added to give some indication of the effect of temperature on the nitride capacity, although the data of Shimoo et al.[15] would suggest a much larger temperature dependence.

The fact that the two ions containing nitrogen show opposite dependence on slag basicity means that the proportion of nitrogen in the two ionic forms is a function of basicity. This is illustrated in Figure 9 for the data of Schwerdtfeger and Schubert[34] at 1600°C, and appears to indicate that the cyanide ion only occurs to any appreciable extent in slags of higher basicity, which is in qualitative agreement with the fact that Davies et al. did not report the existence of the cyanide ion in their slags.

The opposite dependences of the two ions on slag basicity also means that the total dissolved nitrogen content of slags passes through a minimum at a certain value of basicity. Up to this value the nitride ion predominates, so that the solubility decreases with increasing basicity while to the basic side of this value the cyanide ion predominates and solubility increases with further increase in basicity. This is illustrated in Figure 10 where the total nitrogen, given by the compromise expression

$$(\%N)_{tot} \cdot \frac{p_{CO}}{p_{N_2}^{\frac{1}{2}}},$$

is plotted against the optical basicity of the slags used by Davies et al.[17] and Schwertdtfeger et al.[34,35]

At any given slag composition, the fraction of nitrogen present as the nitride ion has been found to be inversely related to $p_{CO}$, and hence for carbon-saturated conditions to the oxygen potential.[15,16,34] Because the solubilities of the two carbon-containing ions both increase with slag basicity, the proportion of carbon in the two forms is virtually independent of slag composition.

In a further extension of their studies, Schwerdtfeger and Schubert[36] found that addition of 36% $CaF_2$ to a slag in which the $CaO/Al_2O_3$ ratio was one increased the cyanide and carbide capacities, while lowering that of nitride. This is consistent with the effect of $CaF_2$ on the structure of these slags, and hence on the relative proportions of network and "free" oxygen ions.

Where the nitrogen content in solution in a slag is determined by reaction with the metal phase, it is influenced by interactions occurring between nitrogen and the other solute elements in the metal. Thus, because of the positive interaction between nitrogen and silicon in solution in iron, the nitrogen content of blast furnace slag increases with the silicon content of the metal,[37] though Schwerdtfeger et al.[38] found that the slag in the blast furnace is far from equilibrium with regard to nitrogen solubility. In view of the positive interaction between carbon and nitrogen in solution in iron, the effect of increasing the carbon content of the metal would be expected to be similar.

*References pp. 118–119*

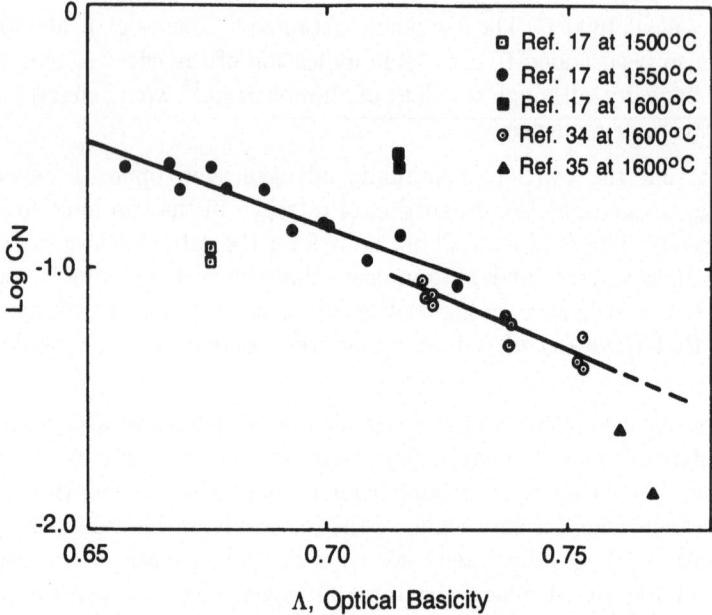

Figure 6. Relationship between log $C_N$ and optical basicity at temperatures in the range 1500 to 1600°C.

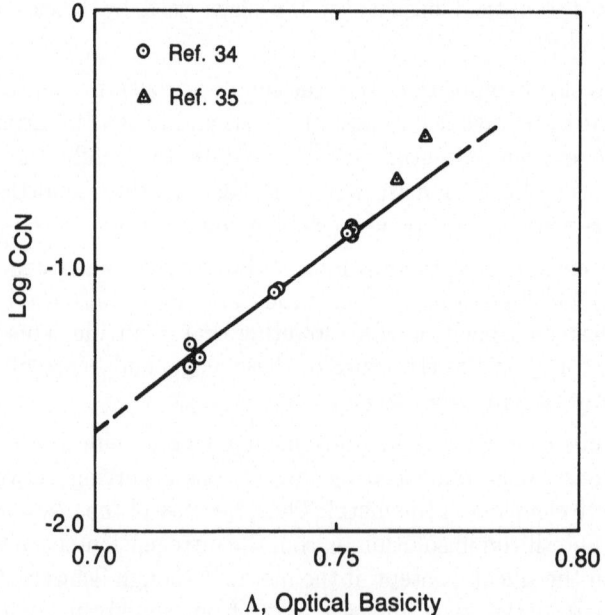

Figure 7. Relationship between log $C_{CN}$ and optical basicity at 1600°C.

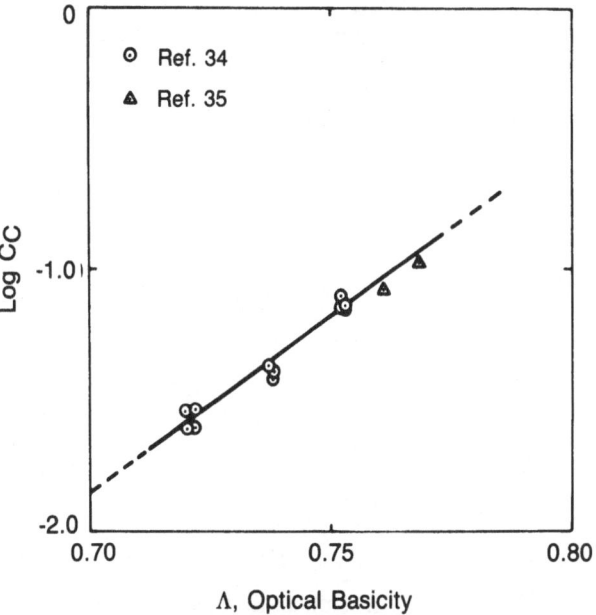

Figure 8. Relationship between log $C_C$ and optical basicity at 1600°C.

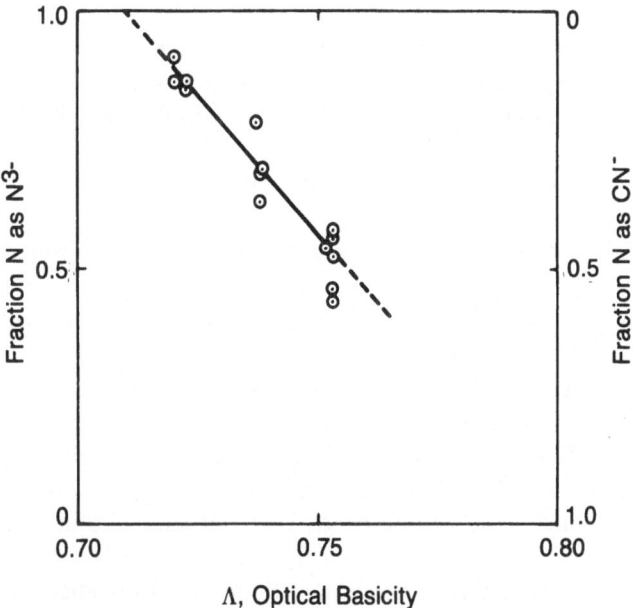

Figure 9. Fractions of nitrogen present at nitride and cyanide ions as a function of slag basicity.

*References pp. 118–119*

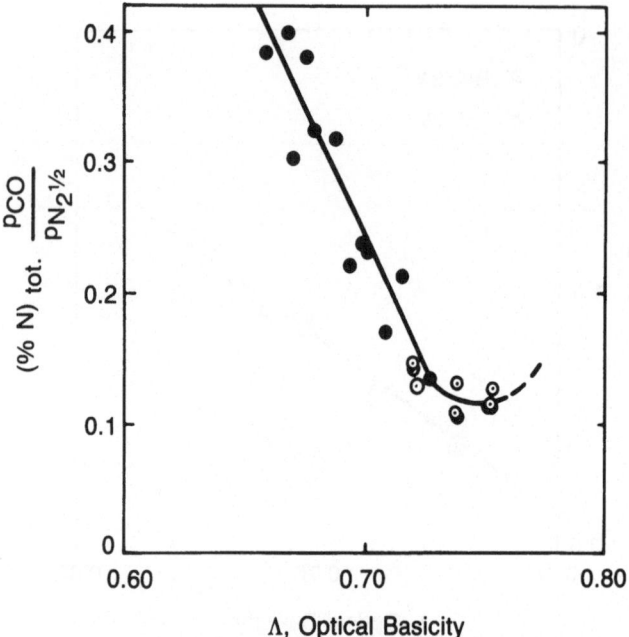

Figure 10. Relationship between the total nitrogen content of slags and their optical basicities.

**Use of Capacities of Slags**—Sulfide capacities are useful because when combined with the oxygen potential of the slag-metal system, they allow calculation of the slag-metal distribution of sulfur. The required relationship is

$$\log \frac{(\%S)}{[a_S]} = -\frac{770}{T} + 1.30 + \log C_S - \log[a_O] \qquad (23)$$

Substitution of Equation 19 into Equation 23 yields the relationship

$$\log \frac{(\%S)}{[a_S]} = \left(\frac{21920 - 54640\Lambda}{T}\right) + 43.6\Lambda - 23.9 - \log[a_O] \qquad (24)$$

For situations where $a_O$ is known Equation 24 can therefore be used to calculate the equilibrium distribution of sulphur between slag and metal for a wide range of slag compositions. Hence where the relative weights of slag and metal are known, the residual sulphur content of the metal is also obtainable. Also, the equilibrium sulphur distribution can be compared to that actually obtained in practice to give an assessment of the degree to which equilibrium is attained. For cupolas operating with acid to mildly basic slags, desulfurization appears to approach equilibrium.[3], based on assumption that $a_O$ is determined either by Mn/MnO or Si/SiO$_2$ equilibrium.

As suggested above, values of $[a_O]$ can be calculated from the known thermodynamics of the reaction controlling the oxygen potential of the slag-metal system, such as the MnO-Mn or the $SiO_2$-Si equilibrium. It should be stressed, however, that the enthalpy of the reaction controlling the oxygen potential may well alter the sign of the temperature-dependent term in Equations 23 and 24, so that while the sulphide capacity of slags increases with increase in temperature, the sulfur distribution ratio usually decreases with increase in temperature.

Phosphate capacities are similarly useful in permitting calculation of the slag-metal distribution of phosphorus for a range of slag compositions. This area has been documented by other authors cited,[23,26] and will not be elaborated further here.

Water capacities are of great importance in connection with hydrogen pick-up by iron or steel by transport through the slag layer from a humid atmosphere.[39] The extent of hydrogen pick-up increases with increase of slag basicity, as shown earlier and also with decrease of the oxygen content of the metal, since the controlling reaction is

$$(H_2O) = 2[H] + [O] \qquad (25)$$

Thus the metal is most vulnerable to hydrogen pick-up from the slag when it is in a highly reduced condition, as it is, for example, when it is tapped from a blast furnace or cupola, and the potential seriousness of this, as measured by the equilibrium hydrogen content of the metal, can be assessed from a knowledge of the slag composition and the oxygen content of the metal.

Nitrogen pick-up by the metal from the slag will apparently only be significant under reducing conditions. However, as shown earlier, the dependence on slag basicity is not so straight-forward. The risk would appear to be greatest with fairly acid or with highly basic slags, though in the case of the acid slags the rate of nitrogen transport through the slag layer would probably preclude serious contamination of the metal with nitrogen.

## CONCLUSIONS

1. The concept of capacities in slags has been shown to be useful and meaningful.
2. The sulphide, phosphate and water capacities have been shown to correlate with slag basicity. The accuracy of these correlations is sufficiently high to permit the prediction of values of these capacities.
3. The nitride, cyanide and carbide capacities have also been shown to correlate with slag basicity, but more data is required before prediction can be made about their values in other systems.
4. These capacities are of great importance in connection with slag-metal and gas-slag equilibria and hence are useful in helping to predict and control the residual contents of sulphur, phosphorus, hydrogen and nitrogen in either iron or steel.

*References pp. 118–119*

## ACKNOWLEDGEMENTS

The author wishes to thank D. J Sosinsky and A. McLean for help with some of the experimental data mentioned and for useful discussions of their significance. Appreciation is also expressed to the American Iron and Steel Institute for financial support of part of this work.

## REFERENCES

1. M. R. Kalyanram, T. B. Macfarlane and H. B. Bell, *J.I.S.I.*, London, 1960, *195*, 58–64.
2. A. S. Venkatadri and H. B. Bell, *J.I.S.I.*, London, 1969, *207*, 1110–1113.
3. S. Katz and C. F. Landefeld, *Cupola Handbook*, 5th Ed., American Foundrymen's Soc., Des Plaines, IL, 1984, 351–363.
4. E. T. Turkdogan, *Physiochemical Properties of Molten Slags and Glasses*, The Metals Society, London, 1983.
5. C. Wagner, *Met. Trans. B.*, 1975, *6B*, 405–409.
6. K. Kawahara, S. Shibata and N. Sano, *Proc. Fifth Process Technology Conference*, Washington, April 1986, 691–696.
7. D. J. Sosinsky, I. D. Sommerville and A. McLean, *Proc. Sixth Process Technology Conference*, Washington, April 1986, 697–703.
8. J. A. Duffy and M. D. Ingram, *J. Amer. Ceram. Soc.*, 1971, *93*, 6448–6454.
9. J. A. Duffy and M. D. Ingram, *J. Inor. Nuclear Chem.*, 1975, *37*, 1203–1206.
10. J. A. Duffy and M. D. Ingram, *Phys. and Chem. of Glasses*, 1975, *16*, 119–123.
11. J. A. Duffy and M. D. Ingram, *J. Non-Cryst. Solids*, 1976, *21*, 373–410.
12. D. J. Sosinsky and I. D. Sommerville, *Second Internat. Symp. on Metallurgical Slags and Fluxes*, Eds., H. A. Fine and D. R. Gaskell, Met. Soc. A.I.M.E., 1984, 1015–1026.
13. J. F. Elliott, Second *Internat. Symp. on Metallurgical Slags and Fluxes*, Eds., H. A. Fine and D. R. Gaskell, Met. Soc. A.I.M.E., 1984, 45–61.
14. H. O. Mulfinger and H. Meyer, *Glastech. Ber.*, 1963, *36*, 481–486.
15. T. Shimoo, K. Kimura and M. Kawai, *J. Japan Inst. Metals*, 1972, *36*, 723–727.
16. T. Choh, T. Hanaki, T. Kato and M. Inoue, *Trans., I.S.I. Japan*, 1973, *13*, 218–225.
17. M. W. Davies and S. G. Meherali, *Met. Trans.*, 1971, *2*, 2729–2733.
18. E. A. Dancy and D. Janssen, *Can. Met. Quart.*, 1976, *15*, 103–110.
19. D. J. Sosinsky and I. D. Sommerville, *Met. Trans. B*, 1986, *17B*, 331–337.
20. M. S. Shuja, Ph.D. Thesis, University of Strathclyde, Glasgow, 1970.
21. K. P. Abraham and F. D. Richardson, *J.I.S.I.*, 1960, *196*, 313–317
22. R. A. Sharma and F. D. Richardson, *Trans. Met. Soc. A.I.M.E.*, 1965, *233*, 1586–1592.
23. D. R. Gaskell, *Trans. I.S.I. Japan*, 1982, *22*, 997–1000.
24. H. Suito and R. Inoue, *Trans. I.S.I. Japan*, 1984, *24*, 47–53.
25. R. Inoue and H. Suito, *Trans. I.S.I. Japan*, 1985, *25*, 118–124.
26. T. Mori, Trans. *Japan Inst. Metal*, 1984, *25*, 761–771.

27. T. Fuwa, S. Ban-ya, T. Fukushima and Y. Iguchi, *Trans. I.S.I. Japan*, 1966, *6*, 225–231.
28. Y. Iguchi, S. Ban-ya and T. Fuwa, *Trans. I.S.I. Japan*, 1969, *9*, 189–194.
29. Y. Iguchi and T. Fuwa, *Trans. I.S.I. Japan*, 1970, *10*, 29–35.
30. D. J. Zuliani, M. Iwase, A. McLean and T. R. Meadowcroft, *Can. Met. Quart.*, 1981, *20*, 181–187.
31. D. J. Zuliani, M. Iwase and A. McLean, *Trans. I.S.S.-A.I.M.E.*, 1982, *1*, 61–67.
32. K. Schwerdtfeger and H. G. Schubert, *Met. Trans. B*, 1978, *9B*, 143–144.
33. D. J. Sosinsky, M. Maeda and A. McLean, *Met. Trans. B*, 1985, *16B*, 61–66.
34. K. Schwerdtfeger and H. G. Schubert, *Met. Trans. B*, 1977, *8B*, 535-540.
35. K. Schwerdtfeger and H. G. Schubert, *Arch. Eisenhüttenw.*, 1974, *45*, 649–655, 905.
36. K. Schwerdtfeger and H. G. Schubert, *Met. Trans. B.*, 1977, *8B*, 689–691.
37. E. Schurmann, F.-J. Hufnagel and A. Drevermann, *Stahl u. Eisen*, 1967, *87*, 645–654.
38. K. Schwerdtfeger, W. Fix and H. G. Schubert, *Ironmaking and Steelmaking*, 1978, *5*, 2, 67–71.
39. D. J. Zuliani, M. Hasegawa, R. A. Heard, D. J. Sosinsky and A. McLean, *Internat. Symposium on Phys. Chem. of Iron and Steelmaking*, Met. Soc. of C.I.M., 1982, I 40–46.

## DISCUSSIONS

**R. J. Fruehan** *(Carnegie-Mellon University)*

I'd like to make a couple of comments and then ask one question. I am intrigued by the nitrogen capacity data. As Iain said, the cyanide capacity increases with basicity and the nitride capacity decreases. One of the possibilities that accounts for the observed relationship between basicity and nitride capacity is the formation of silicon nitride complex in the highly acid slags.

Another point is that I'm convinced that one of the overriding things we have to know is what is controlling the oxygen potential in any metallurgical operation. The slag chemistry will tell us what the capacity of slags is, but we must also know the oxygen potential. Therefore, it's good to look at metal-slag distribution for the element and to correlate this with the capacity data to determine what is controlling the oxygen potential.

My question is: You showed very nice linear relationships between $\log C_S$ and optical basicity. If I return to the fundamental definition of sulfide capacity, it is proportional to the oxide ion activity divided by the activity coefficient of the sulfur ion. That would be the thermodynamic quantity in the slag. Why does $\log C_S$ vary linearly with basicity? I would have never expected such a simple correlation.

**A.** I think the sulfide capacity correlates with basicity because the only oxygens that are replaceable by sulfur are the ones that we call "free" oxygen. I say free because I think we really don't know how "free" is free. There are certainly oxygens that are not tied up in the network. The proportion of these oxygens is a function of basicity, or in a sense is the basicity.

As you pointed out, the term sulphide capacity is a way of grouping together several quantities, which cannot be defined or measured singly, such as the activity of the oxide ion and the activity coefficient of the sulfide ion. This grouping was done over 30 years ago by Fincham and Richardson,* and the concept of sulfide capacity has emerged as an extremely useful parameter. However, it is not clear as to exactly why there should be a linear relation between $\log C_S$ and optical basicity. It is conceivable that optical basicity is a logarithmic measure of the activity of the available or "free" oxygen ions.

**K. S. Goto** *(Tokyo Institute of Technology)*

It was a very interesting talk. I understand that the optical basicity is a very convenient parameter to express sulfide capacities and distribution of many elements between slag and metal. How is this quantity measured?

**A.** Briefly, optical basicity is evaluated by measuring the spectroscopic shift in the uv spectrum of a probe ion between the value of the free ion and the value in an oxide environment. This shift can be regarded as resulting from the expansion

---

* 1. C. J. B. Fincham and F. D. Richardson, *Proc. Royal Soc.*, 1954, A *223*, 40–62.

of the outer $s$ and $p$ orbitals of the probe ion caused by electron donation from the surrounding oxygen ions, which is related to the "nephelauxetic" effect.[8-11]

**K. S. Goto**

Can you measure chemical shifts when you have NaCl of $PbCl_2$ present?

**A.** As far as I know, these correlations have never been measured by this technique but in principle they can be. However, they have been measured by an alternative technique, the photoacoustic spectroscopic technique, which is used by Nakamura et al. at Kyushu Institute of Technology.*

**K. S. Goto**

Let's go back to the original idea of basicity. Now, basicity might be defined as the activity of the oxide ion. That means basicity can be defined only for an oxide system and cannot accommodate the presence of other ions.

**A.** Yes, but when discussing any of these nonoxide components, we always talk about their equivalent optical basicity. In other words, the fluoride or chloride will act as being equivalent to an oxide with that optical basicity. Clearly the halides have a depolymerizing effect on slag structure, and so behave in qualitatively the same way as basic oxides. Thus it is not inappropriate to say that they act as being equivalent to basic oxides.

**H. Gaye** *(IRSID)*

Have you tried to use the correlations for phosphate capacity to estimate the phosphorous distribution? I feel there is too much scatter in the data to get anything credible for practical purposes.

**A.** We have not tried to use the phosphate capacity to determine the phosphorous distribution. Certainly the phosphorous correlation is not as good as that for the sulfide data. As I stated, Gaskell[23] started the work on phosphate capacity. Then Suito and Inoue[24-25] and Mori[26] took that over. We basically have left this work to others, because we were more interested in the water capacity and so on.

There is one other point that I would like to make. When considering phosphorous distribution, the slags contain FeO and MnO. In this case it is difficult to talk in terms of a measured optical basicity, because these slags are opaque. What in fact has been done by all of the above authors is to ignore the fact that these transition metal oxides have unfilled $d$ shells, and simply calculate their optical basicities using the relationship with the Pauling electronegativity.[9] Duffy and Ingram[8-11] who originated and developed the

---

* 6. T. Nakamura, Y. Ueda and F. Noguchi, Second Internat. Symposium on Metallurgical Slags and Fluxes, Eds. H. A. Fine and D. R. Gaskell, Met. Soc. A.I.M.E., 1984, 1005–1013.

*References p. 133*

optical basicity concept, have stressed very strongly that the spectroscopic shifts, that apply are only for the $s-p$ transition. When FeO and MnO are present in the slag it may be the $d$ electrons that are responding to the electron donation, which may cause some confusion. So, they and I have never used values for FeO and MnO as calculated from Pauling's electronegativity. I mentioned that in the abstract. Thus, the data on phosphate capacities is not directly comparable with the other ones that were discussed. The photoacoustic technique yields values for the optical basicities of FeO and MnO in reasonable agreement with values deduced retroactively from sulfide capacity data.[12] And so it appears that this method can be used to measure optical basicities in slags containing transition metal oxides.

# A COMMENT OF THE EQUIVALENCE OF THE CHEMICAL AND OPTICAL BASICITIES OF SLAGS

## D. R. Gaskell
*Purdue University*

Mori's linear correlation of $\log C_{PO_4^{3-}}$ with $\Lambda$ at 1600°C,[1]

$$\log C_{PO_4^{3-}} = 17.44\Lambda + 5.72 \tag{1}$$

was derived from Suito and Inoue's[2] correlation for MgO-saturated slags containing CaO

$$\log C_P = 17.55\Lambda + \frac{29990}{T} - 23.737 \tag{2}$$

where

$$C_{PO_4^{3-}} = \frac{(\%PO_4^{3-})}{p_{O_2}^{5/4} \cdot p_{P_2}^{1/2}} \tag{3}$$

and

$$C_P = \frac{(\%P)}{[a_P][a_O]^{5/2}} \tag{4}$$

are related via[1]

$$\log C_P = \log C_{PO_4^{3-}} - \frac{21680}{T} - 1.87 \tag{5}$$

At 1600°C, Equation 3 can be expressed as[1]

$$C_{PO_4^{3-}} = 5.80 \times 10^9 \left[ \frac{(\%P_2O_5)\sum(\%i/M_i)}{\gamma_{P_2O_5}} \right]^{1/2}$$

or

$$\log C_{PO_4^{3-}} = 9.76 + \frac{1}{2}\left[\log(\%P_2O_5) + \log\sum(\%i/M_i)\right] - \frac{1}{2}\log\gamma_{P_2O_5} \tag{6}$$

where $\sum(\%i/M_i)$ is the moles of oxide in 100 g slag.

From examination of the thermodynamic data for six slag systems at 1600°C, Mori obtained the correlation[1]

$$\log\gamma_{P_2O_5} = 9.40 - 38.09\Lambda \tag{7}$$

which is shown in Figure 1. Substitution of Equation 7 into Equation 6 gives

$$\log C_{PO_4^{3-}} = 5.06 + 19.05\Lambda + \frac{1}{2}[\log(\%P_2O_5) + \log(\%i/M_i)] \tag{8}$$

which, in comparison with Equation 1, shows that the good linear correlation of $\log C_{PO_4^{3-}}$ with $\Lambda$, arises from the linear variation of $\log\gamma_{P_2O_5}$ with $\Lambda$.

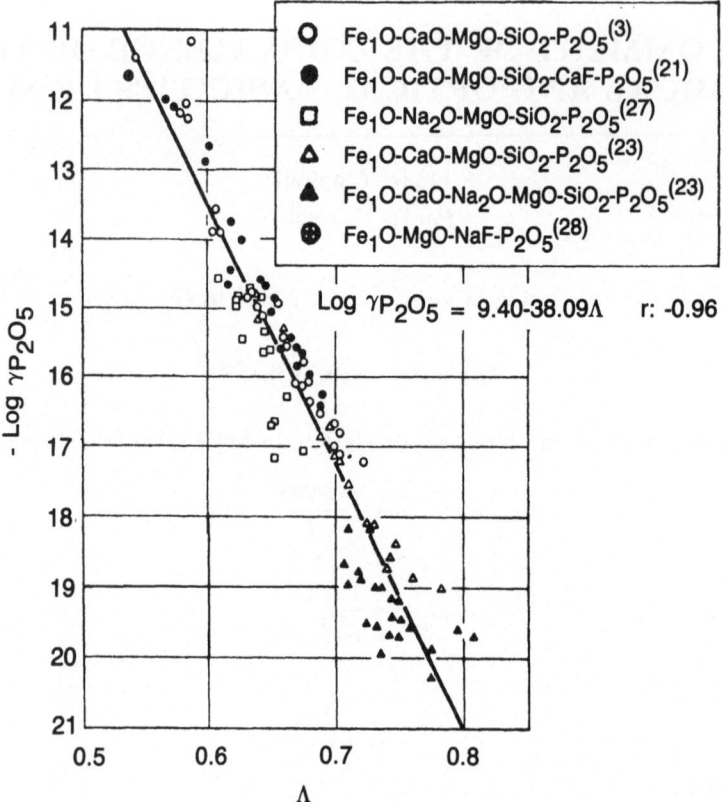

Figure 1. The variation of log $\gamma_{P_2O_5}$ with optical basicity in various slag systems at 1600°C (Mori[1]).

The observed linear variation of log $\gamma_{P_2O_5}$ with $\Lambda$ is particularly interesting in that it requires that the iso-$\gamma_{P_2O_5}$ lines in a slag system be parallel with the iso-$\Lambda$ lines. In a ternary system plotted on a Gibbs triangle, iso-$\Lambda$ lines are straight lines which radiate from a point outside the triangle. Figure 2 shows the iso-$\gamma_{P_2O_5}$ lines determined by Goto et al in the system $Na_2O$-$P_2O_5$-$SiO_2$ at 1300°C[3] in comparison with the iso-$\Lambda$ lines, and, as is seen, the two sets of lines are reasonably parallel with one another. Figure 3 shows the variation of log $\gamma_{P_2O_5}$ with $\Lambda$ in the systems $Na_2O$-$P_2O_5$ and $Na_2O \cdot P_2O_5$-$Na_2O \cdot SiO_2$ and the slight curvatures in these variations are measures of the deviations from parallelism. Figure 4 shows a comparison of the iso-$\Lambda$ lines with the iso-$\gamma_{PO_{2.5}}$ lines determined by Tsukihashi et al in the system $Na_2O$-$PO_{2.5}$-$SiO_2$ at 1200°C.[4] If iso-$\gamma_{P_2O_5}$ lines and iso-$\Lambda$ lines are parallel with one another, then, by virtue of the relationship.

$$a_{PO_{2.5}}^2 = \gamma_{PO_{2.5}}^2 X_{PO_{2.5}}^2 = a_{P_2O_5} = \gamma_{P_2O_5} \cdot X_{P_2O_5}$$

iso-$\gamma_{PO_{2.5}}$ lines are not parallel with iso-$\Lambda$ lines, as is apparent in Figure 4. However,

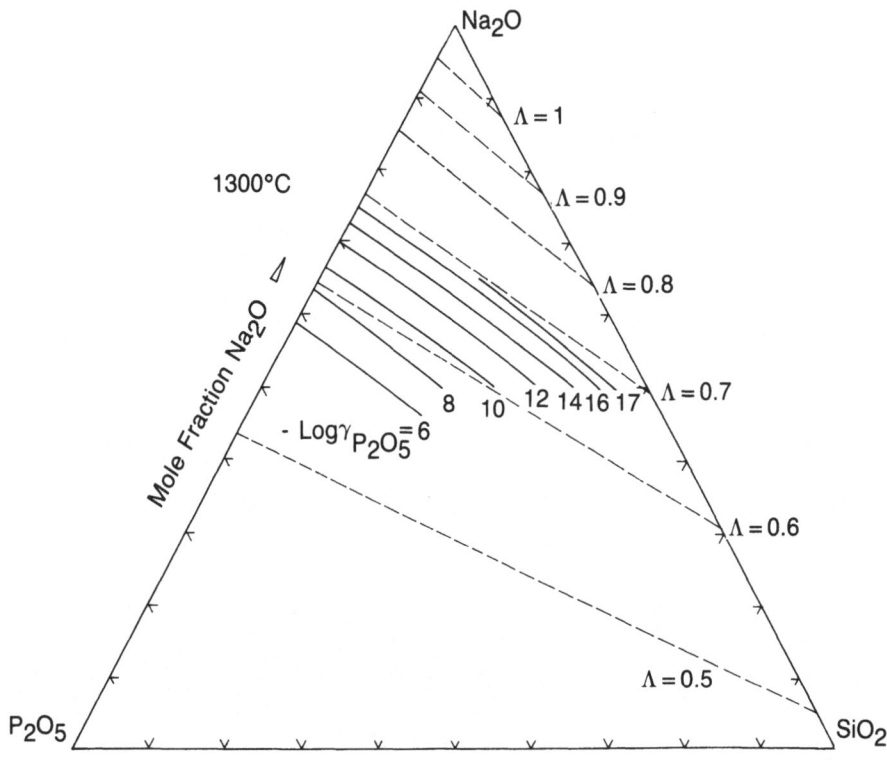

**Figure 2.** Iso-log$\gamma_{P_2O_5}$ lines and iso-optical basicity lines in the system $Na_2O$-$P_2O_5$-$SiO_2$ at 1300°C.[3]

the variation of log $\gamma_{P_2O_5}$ (calculated from log $\gamma_{PO_{2.5}}$) with $\Lambda$ along a section of the composition path between $Na_2O$ and 0.75 $SiO_2$-0.24 $PO_{2.5}$ is linear. This variation is shown in Figure 3, which also shows the discrepancy between the numerical values obtained by Goto *et al* and those obtained by Tsukihashi *et al*.

In view of the commonly-held belief that the chemical basicity of a slag is determined by the thermodynamic activity of the most basic of the oxides present in the slag, it is of interest to examine the relationship between the activity of the most basic oxide component and the optical basicity. The activities of CaO in the system CaO-$AlO_{1.5}$-$SiO_2$, determined by Rein and Chipman at 1600°C[5], and shown in Figure 5, appear to be closely parallel with the iso-$\Lambda$ lines. The variations of log $a_{CaO}$ in the sub-systems CaO-$SiO_2$, CaO-$SiO_2 \cdot AlO_{1.5}$ and CaO-$AlO_{1.5}$ with $\Lambda$, shown in Figure 6, are reasonably represented by the equation.

$$\log a_{CaO} = 12.5\Lambda - 10 \tag{9}$$

The sulfide capacity in this system, expresssed as

$$C_S = (\%S) \left[ \frac{p_{o_2}}{p_{s_2}} \right]^{1/2}$$

*References p. 133*

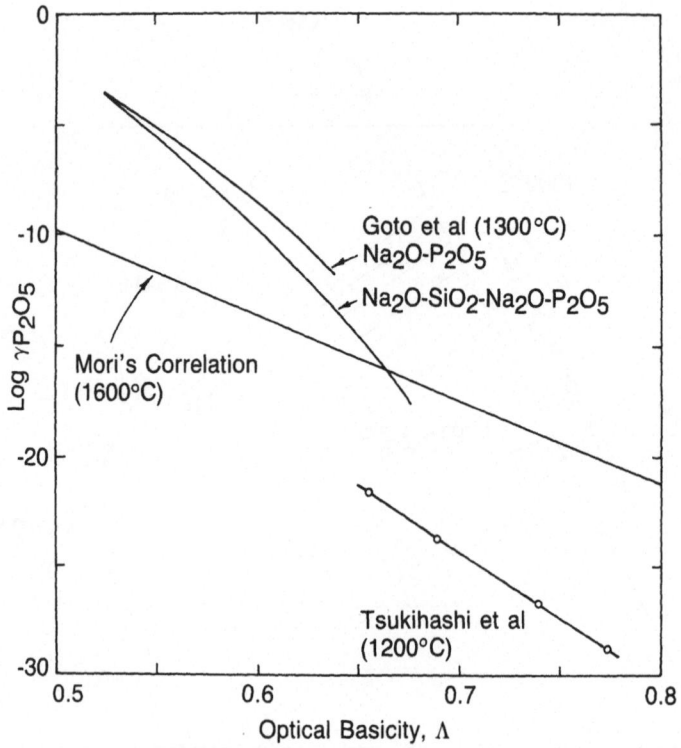

Figure 3. The variation of log $\gamma_{P_2O_5}$ with optical basicity in sodium phosphate and sodium silicophosphate melts in comparison with Mori's correlation.

can be rewritten as

$$C_s = \frac{32K \cdot a_{CaO} \cdot \sum(\%i/M_i)}{\gamma_{CaS}} \tag{10}$$

where $K$ is the equilibrium constant for the reaction

$$CaO + \frac{1}{2}S_2 = CaS + \frac{1}{2}O_2$$

The activity coefficient of CaS, $\gamma_{CaS}$, does not vary significantly with composition[6] and the sum, $\sum(\%i/M_i)$ varies in the range 1.33-1.78. Thus Equation 10 shows that the value of $C_S$ is determined by $a_{CaO}$, and the linear variation of log $a_{CaO}$ with $\Lambda$, shown in Figure 6, explains the good observed correlation of log$C_S$ with $\Lambda$. At 1600°C, $K$ has the value $4.78 \times 10^{-3}$. Substituting this, Equation 9, and the average values of $\sum(\%i/M_i) = 1.5$ and $\gamma_{CaS} = 12$ into Equation 10 gives

$$\log C_S = 12.5\Lambda - 11.72 \tag{11}$$

which is in agreement with the correlation at 1600°C of [7]

$$\log C_S = 14.42\Lambda - 13.08 \tag{12}$$

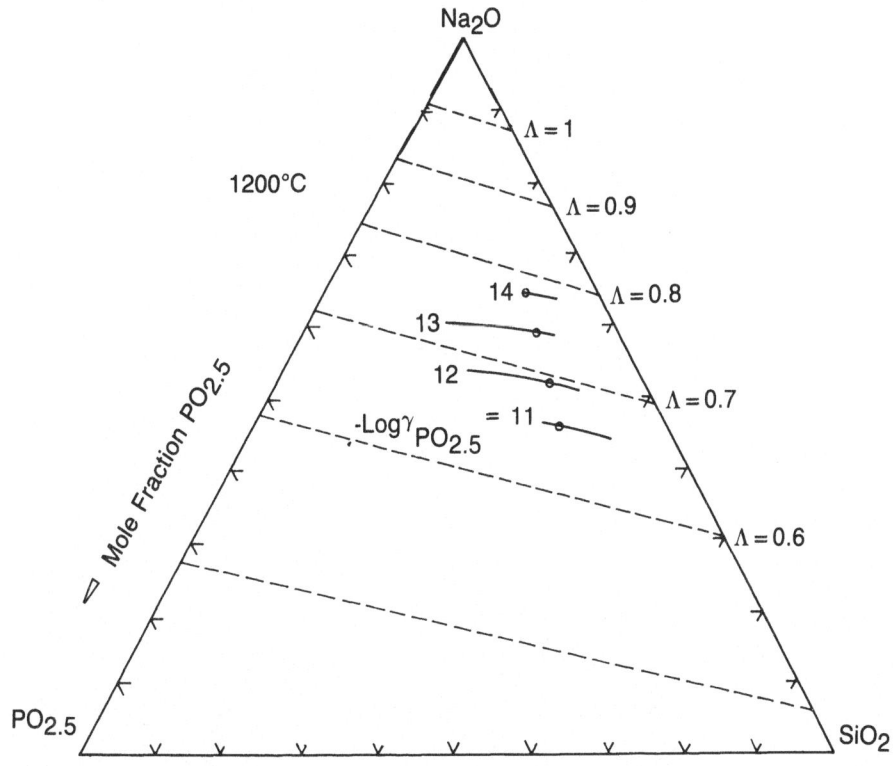

Figure 4. Iso-log$\gamma_{PO_{2.5}}$ lines and iso-optical basicity lines in the system $Na_2O$-$PO_{2.5}$-$SiO_2$ at 1200°C.[4]

With $\Lambda = 0.7$, Equation 11 gives $\log C_S = -2.97$ and Equation 12 gives $\log C_S = -2.99$.

The variation of $\log a_{Na_2O}$ with composition in the system $Na_2O$-$P_2O_5$-$SiO_2$ at 1300°C[3] is shown in comparison with the iso-$\Lambda$ lines in Figure 7, and, again, a reasonable degree of parallelism is apparent. The variations of $\log a_{Na_2O}$ in the sub-systems $Na_2O$-$P_2O_5$, $Na_2O$-$P_2O_5$-$SiO_2$ and $Na_2O$-$P_2O_5$ are shown in Figure 8. In each of the sub-systems $\log a_{Na_2O}$ is a linear function of $\Lambda$ but the slope of the line increases with increasing $X_{P_2O_5}/X_{SiO_2}$ to such an extent that the variation in the ternary cannot be represented by a single line. Thus, in this ternary system, linear variations of $\log C_S$ with $\Lambda$ can only be expected in melts of constant $X_{P_2O_5}/X_{SiO_2}$.

Similar parallelism occurs in the system $Na_2O$-$B_2O_3$-$SiO_2$, illustrated in Figure 9 as lines of constant partial molar free energy of $Na_2O$ relative to that in $Na_2O \cdot 2B_2O_3$[8] and, to a lesser extent, in the system $Na_2O$-$SiO_2$-$Al_2O_3$, shown in Figure 10.[9]

*References p. 133*

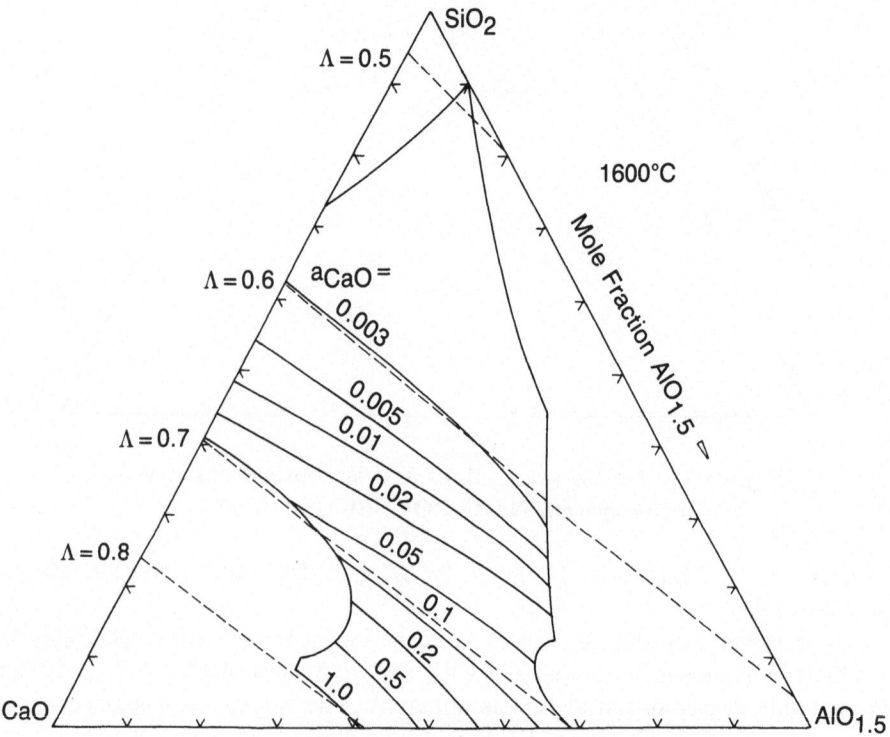

Figure 5. iso $-a_{CaO}$ lines and iso-optical basicity lines in the system $CaO$-$SiO_2$-$AlO_{1.5}$ at 1600°C.[5]

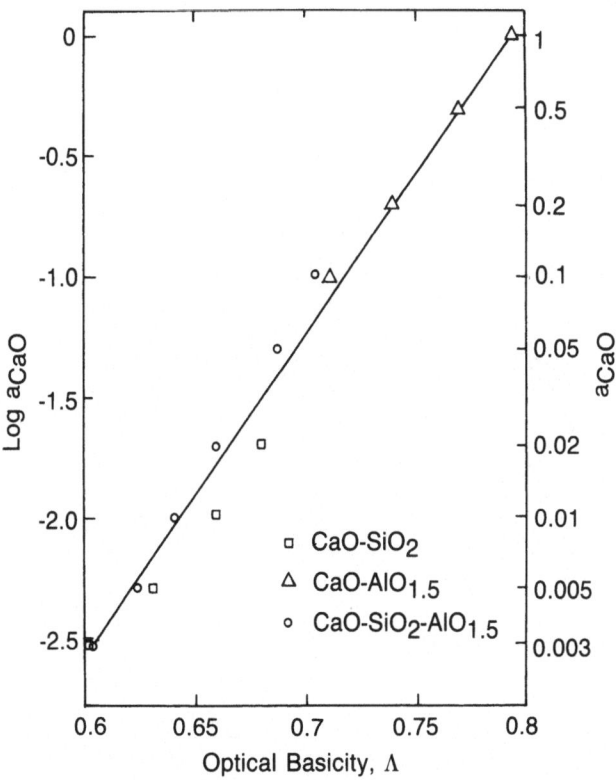

Figure 6. The variation of log $a_{CaO}$ with optical basicity in the systems $CaO\text{-}SiO_2$, $CaO\text{-}AlO_{1.5}$ and $CaO\text{-}SiO_2 \cdot AlO_{1.5}$ at 1600°C.

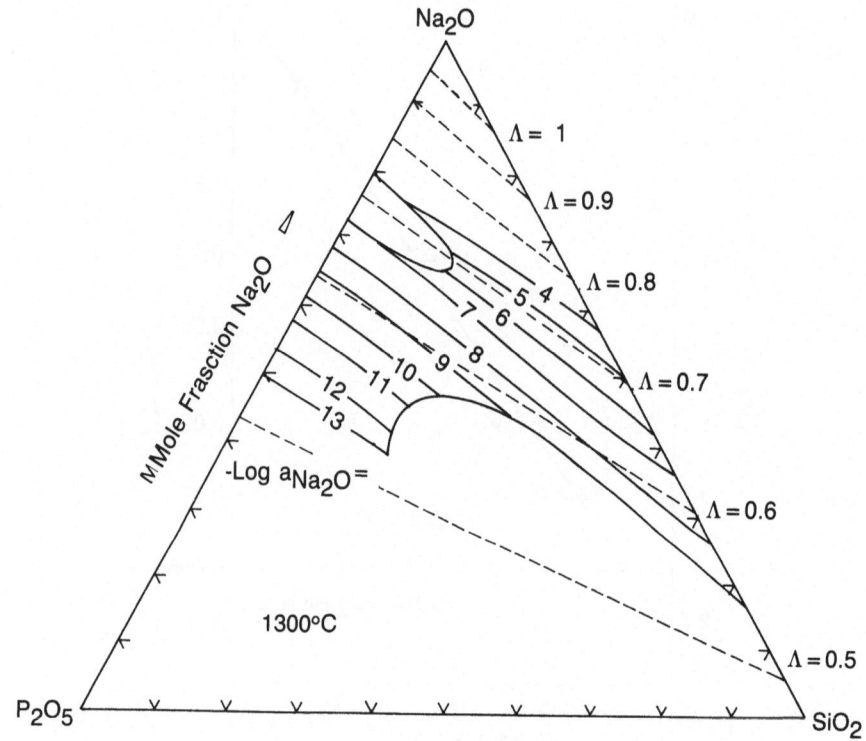

Figure 7. Iso − log $a_{Na_2O}$ lines and iso-optical basicity lines in the system $Na_2O$-$P_2O_5$-$SiO_2$ at 1300°C.[3]

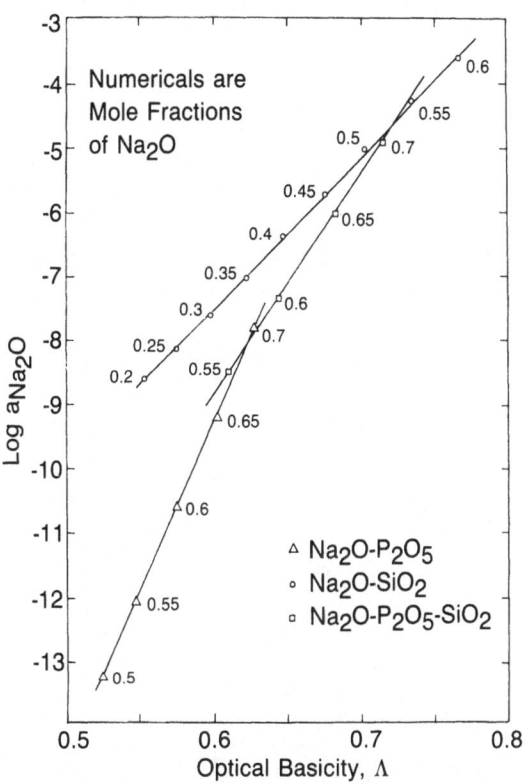

**Figure 8.** The variation of log $a_{Na_2O}$ with optical basicity in the systems $Na_2O$-$SiO_2$, $Na_2O$-$P_2O_5$ and $Na_2O$-$P_2O_5$-$SiO_2$ at 1300°C.

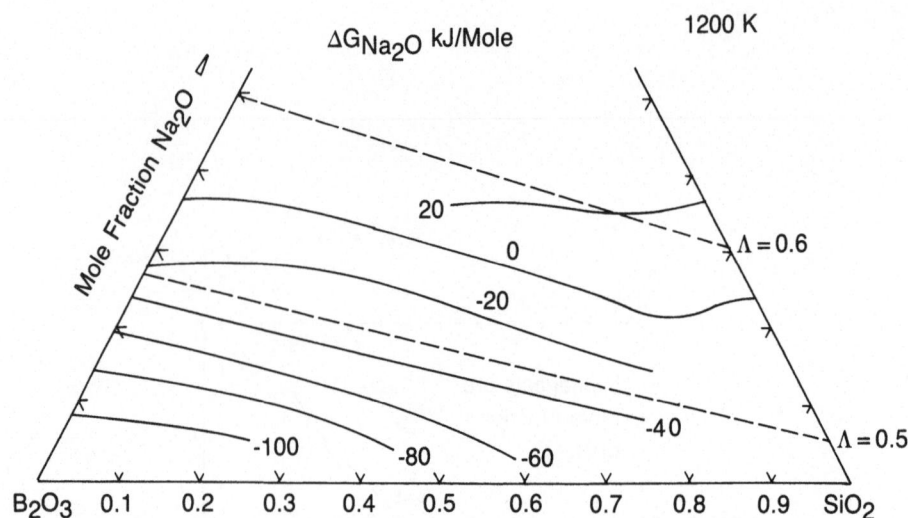

Figure 9. Constant partial molar free energy of Na$_2$O lines and iso-optical basicity lines in the system Na$_2$O-B$_2$O$_3$-SiO$_2$ at 1200K.[7]

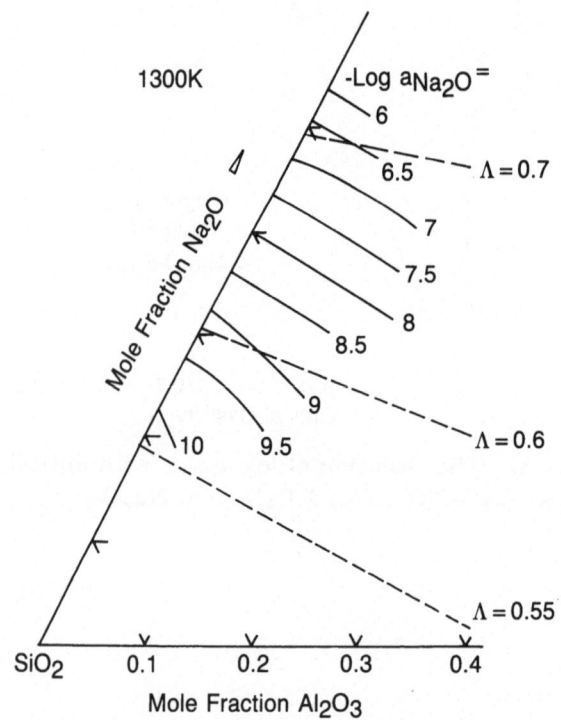

Figure 10. Iso − log $a_{Na_2O}$ lines and iso-optical basicity lines in the system Na$_2$O-SiO$_2$-Al$_2$O$_3$ at 1300K.[8]

# REFERENCES

1. T. Mori, Trans. *JIM* (1984), *25*, 761.
2. H. Suito, R. Inoue and M. Takada, *Tetsu to Hagane* (1981), *67*, 2645.
3. K. S. Goto, S. Yamaguchi and K. Nagata, *Proceedings of the Second International Symposium on Metallurgical Slags and Fluxes*, eds. H.A. Fine and D. R. Gaskell, TMS-AIME, Warrendale, Pa (1984), 467.
4. F. Tsukihashi, A. Werme, F. Matsumoto, A. Kasahara, M. Yukinobu, T. Hyodo, S. Shiomi and N. Sano, *ibid.*, 89.
5. R. H. Rein and J. Chipman, *Trans. Met. Soc. AIME* (1965), *233*, 415.
6. R. A. Sharma and F. D. Richardson, *JISI* (1962), *200*, 373.
7. D. J. Sosinsky and I. D. Sommerville, *Met. Trans. B.* (1986) *17B*, 331.
8. K. Asai and T. Yokokawa, Trans. *JIM* (1982), *23*, 571.
9. H. Itoh and T. Yokokawa, Trans. *JIM* (1984), *25*, 897.

# PARTITION OF ALLOYING ELEMENTS IN FREEZING CAST IRONS AND ITS EFFECT ON GRAPHITIZATION AND NITROGEN BLOWHOLE FORMATION

### AKIO KAGAWA and TAIRA OKAMOTO

*The Institute of Scientific and Industrial Research
Osaka University
Osaka 567, Japan*

### ABSTRACT

Partition of alloying elements, on solidification of Fe-C base melts to primary austenite and eutectic, were evaluated thermodynamically and verified experimentally. Further extension of the thermodynamic treatment yielded relationships that connect the partition coefficient of an alloy addition to its effects on the stable and metastable eutectic temperature and the propensity for solidification in the stable or metastable mode. In iron, alloy partition coefficients and the derivative quantities were shown to exhibit a periodicity with respect to atomic number. The values are also strong functions of carbon concentration, but are relatively unaffected by moderate additions of other elements (up to a few percent).

The case of nitrogen partition in freezing cast irons was examined in detail because of the serious problem of blowhole defects in castings. Based on experimentally determined values of nitrogen partition ratios and other available thermodynamic data the changes in nitrogen concentration in the phases coexisting during cast iron solidification were determined. By comparing this data to nitrogen solubility, the conditions producing nitrogen supersaturation were determined for a wide range of conditions. In general it was found that the tendency to produce nitrogen supersaturation in cast irons increased with decreasing carbon concentration. The presence of strong nitride formers in the cast iron melt reduced the tendency for nitrogen blowhole formation.

## INTRODUCTION

When cast iron solidifies, alloying elements are partitioned between solid and liquid, first on primary austenite crystallization and then on subsequent eutectic solidification. The partition coefficient of a solute element in an alloy system, defined as the ratio of the concentration of the element in the solid to that in the liquid, is a characteristic value showing the degree of microsegregation of the element. It is closely related to the formation of the solidification structure and casting defects. The coefficients for a number of iron-base binary systems have been established. Until recently, however, there has only been limited information available on the partition coefficients in ternary and multicomponent systems, especially those for eutectic solidification of cast iron.

In recent work,[1-6] the authors have experimentally and theoretically evaluated the partition coefficients of various alloying elements between austenite and liquid iron and between eutectic solid and liquid in iron-carbon base ternary and multicomponent systems. In addition, theoretical relations have been developed to connect the partition coefficients of alloying elements to their effects on graphitization of cast iron on solidification.

Among the serious foundry process problems related to partitioning of elements is blowhole formation caused by gases, such as nitrogen, hydrogen, and carbon monoxide. Several researches have been reported on the formation of nitrogen blowholes in cast iron. Davison et al.[7] indicated that nitrogen is the source of 'wormy' defects in large gray cast iron castings. They suggested that a critical nitrogen content in excess of 100 ppm produces 'wormy' defects. Caspers[8] reported the critical nitrogen content to be 110 ppm. Dawson et al.[9] showed that 'fissure' defects appear at nitrogen levels greater than 140 ppm. At the critical nitrogen content for the generation of these defects, the cast iron melts must be supersaturated with respect to nitrogen, however, the necessary degree of supersaturation is not known. The formation of blowhole defects is influenced by several factors, i.e. chemical composition, melting conditions, solidification mode, and partition behavior of solute elements. It is, therefore, important to understand the interaction of these phenomena to comprehend the causes for blowhole formation. In the present paper, the partition of nitrogen between liquid iron and austenite in freezing cast irons is computed using the partition coefficients for nitrogen and alloying elements. The degree of supersaturation of nitrogen in liquid iron and austenite is discussed in relation to the solubilities of nitrogen and the formation of nitrides in each the two phases.

## PARTITION COEFFICIENTS OF ALLOYING ELEMENTS ON PRIMARY AUSTENITE CRYSTALLIZATION AND SUBSEQUENT EUTECTIC SOLIDIFICATION

**Partition Coefficient Between Austenite and Liquid Iron**—The equilibrium condition for an alloying element $X$ partitioned between austenite and liquid iron in the Fe-C-$X$ system is represented by:

$$\overset{\circ}{\mu}_X^L + RT\ln(\gamma_X^L N_X^L) = \overset{\circ}{\mu}_X^A + RT\ln(\gamma_X^A N_X^A) \qquad [1]$$

where $\overset{\circ}{\mu}_X$ is the chemical potential of the element in a standard state; $R$ is the gas constant; $T$ is the temperature at which the phases concerned are in equilibrium, $\gamma_X$ is the activity coefficient for the element $X$, and $N_X$ is the molar fraction of the element $X$ in the phase concerned. The superscripts $L$ and $A$ denote liquid iron and austenite, respectively. From Equation 1, the following equation is obtained,

$$K_X^{A/L} = \frac{\overset{\circ}{\gamma}_X^L/\overset{\circ}{\gamma}_X^A}{F} \exp\left[\frac{\overset{\circ}{\mu}_X^L - \overset{\circ}{\mu}_X^A}{RT} + \left(\varepsilon_X^{X,L} - F\varepsilon_X^{X,A} K_X^{A/L}\right) N_X^L \right. $$
$$\left. + \left(\varepsilon_X^{C,L} - F\varepsilon_X^{C,A} K_C^{A/L}\right) N_C^L\right] \quad [2]$$

where $K_X^{A/L}$ is the equilibrium partition coefficient of the element $X$ between austenite and liquid iron, defined by the ratio of weight percent of $X$ in austenite to that in liquid iron, $\overset{\circ}{\gamma}_X$ is the activity coefficient at infinite dilution of the element $X$, $\varepsilon_X^C$ and $\varepsilon_X^X$ are the interaction coefficients of the element $X$, respectively, on carbon and $X$. $N_C$ is the molar fraction of carbon, and $F$ is a conversion factor given by:

$$F = \frac{(\%Fe)^L/M_{Fe} + (\%C)^L/M_C + (\%X)^L/M_X}{(\%Fe)^A/M_{Fe} + (\%C)^A/M_C + (\%X)^A/M_X} \quad [3]$$

where $(\%i)$ and $M_i$ denote the weight percent and the atomic weight of the element $i$, respectively.

A similar equation is obtained for carbon:

$$K_C^{A/L} = \frac{\overset{\circ}{\gamma}_C^L/\overset{\circ}{\gamma}_C^A}{F} \exp\left[\left(\varepsilon_C^{C,L} - F\varepsilon_C^{C,A} K_C^{A/L}\right) N_C^L \right.$$
$$\left. + \left(\varepsilon_C^{X,L} - F\varepsilon_C^{X,A} K_X^{A/L}\right) N_X^L\right] \quad [4]$$

The term $\overset{\circ}{\mu}_C^L - \overset{\circ}{\mu}_C^A$ is omitted in Equation 4 because graphite is taken as the standard state in both liquid and austenite. The partition coefficients of $K_X^{A/L}$ and $K_C^{A/L}$ are calculated by simultaneously solving Equations 2 and 3. For a dilute solution of the element $X$,

$$K_X^{A/L} = \left(K_X^{A/L}\right)_{Fe-X} \exp\left(\varepsilon_X^{C,L} N_C^L - \varepsilon_X^{C,A} N_C^A\right) \quad [5]$$

where $\left(K_X^{A/L}\right)_{Fe-X}$ is the equilibrium partition coefficient for the Fe-$X$ binary system. Thermochemical data pertaining to Equations 2, 4 and 5 are listed elsewhere.[5] The results of calculations using Equations 4 and 5 are shown for some alloying elements in Figure 1. For comparison the figure includes experimental values obtained by the present authors and other workers. As seen, the calculated and experimental partition coefficients agree satisfactorily over the entire range of carbon concentrations.

*References p. 161*

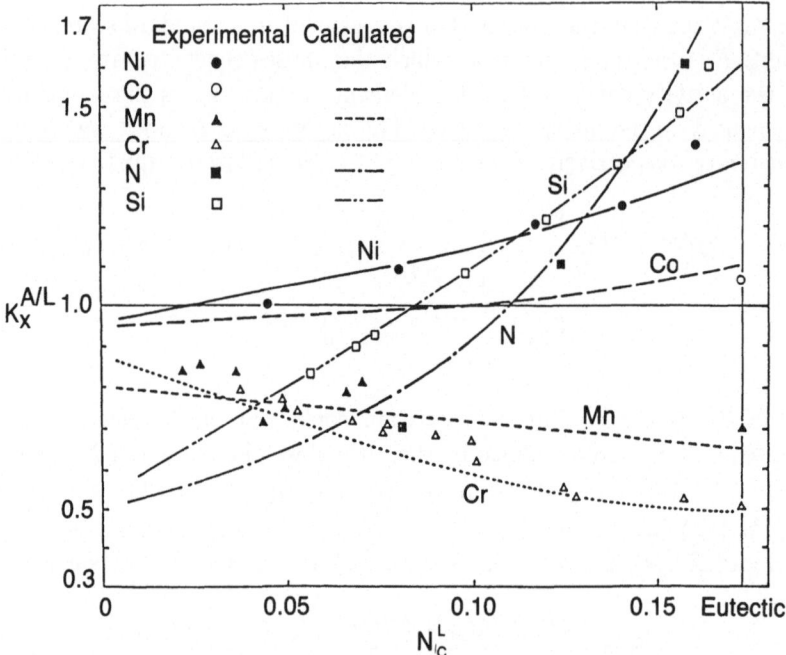

**Figure 1.** The coefficient $K_X^{A/L}$ for equilibrium partition of a third element between austenite and liquid iron in Fe − C base ternary alloys containing a small amount of the third element.

**Partition Coefficient Between Eutectic Liquid and Its Solid**—On the metastable eutectic solidification, three kinds of equilibrium for an alloying element should be established between the phases concerned, that is, between eutectic liquid and austenite, between eutectic liquid and cementite, and between cementite and austenite, and hence the three equilibrium partition coefficients are defined as $K_X^{A/L} = (\%X)^A/(\%X)^L$, $K_X^{C/L} = (\%X)^C/(\%X)^L$, and $K_X^{C/A} = (\%X)^C/(\%X)^A$, respectively, where superscripts $L$, $A$, and $C$ denote eutectic liquid, austenite, and cementite, respectively. On the stable eutectic solidification, alloying elements are in equilibrium only between eutectic liquid and austenite because of a negligibly small solubility of the elements in graphite.

The coefficients for the partition of alloy elements between eutectic liquid and solid $K^S$ and $K^M$, are defined in Equations 6 and 7 as functions of the partition coefficients, $K_X^{A/L}$, $K_X^{C/L}$ and $K_X^{C/A}$.

$$K^S = f^A K_X^{A/L} \qquad [6]$$

$$K^M = f^{A'} K_X^{A/L} + \left(1 - f^{A'}\right) K_X^{C/L}$$
$$= \left[f^{A'} + \left(1 - f^{A'}\right) K_X^{C/A}\right] K_X^{A/L} \qquad [7]$$

$K^S$ and $K^M$ are the solid/liquid concentration ratios of the elements, respectively for stable and metastable solidification. The terms $f^A$ and $f^{A'}$ are the weight fraction of austenite in the stable and metastable eutectic solids, respectively. Calculation of $K_X^{C/A}$ is described in detail by Ko et al.[10] Table 1 gives the calculated values of $K^S$ and $K^M$ as well as the calculated and observed values of $K_X^{A/L}$ at the eutectic temperature. Table 1 also includes calculated values of $K_X^{C/A}$ at the eutectic temperature as well as values determined by Ko et al.

**Partition Coefficients of Solute Elements in Iron-Carbon Base Quaternary and Multicomponent Systems**—It has been shown that the partition coefficient of an element in a multicomponent system differs from that in a binary system because of the interaction between solute elements. Thermodynamic calculations of partition coefficients in the ternary system were extended to the quaternary systems to evaluate the effect of solute interaction on the partition coefficients of the elements, and the results were compared with those from experiments.

In an Fe-C-$i$-$j$ quaternary system, the following equations are introduced. For the third element $i$,

$$FK_i^{A/L,Q} = \left[\frac{\overset{\circ}{\gamma}_i^L}{\overset{\circ}{\gamma}_i^A}\right] \exp\left[\frac{\overset{\circ}{\mu}_i^L - \overset{\circ}{\mu}_i^A}{RT} + \left(\varepsilon_i^{i,L} - F\varepsilon_i^{i,A} K_i^{A/L,Q}\right) N_i^L\right.$$

$$\left. + \left(\varepsilon_i^{C,L} - F\varepsilon_i^{C,A} K_C^{A/L,Q}\right) N_C^L + \left(\varepsilon_i^{j,L} - F\varepsilon_i^{j,A} K_j^{A/L,Q}\right) N_j^L\right] \quad [8]$$

and for carbon

$$FK_C^{A/L,Q} = \left[\frac{\overset{\circ}{\gamma}_C^L}{\overset{\circ}{\gamma}_C^A}\right]^B \exp\left[\left(\varepsilon_C^{i,L} - F\varepsilon_C^{i,A} K_i^{A/L,Q}\right) N_i^L\right.$$

$$\left. - \left(\varepsilon_C^{j,L} - F\varepsilon_C^{j,A} K_j^{A/L,Q}\right) N_j^L\right] \quad [9]$$

where superscripts Q and B denote the values in the quaternary and binary system, respectively. An equation similar to Equation 8 is obtained for the element $j$. The partition coefficients in the quaternary system can be calculated by simultaneously solving the three equations for carbon, and the elements $i$ and $j$.

The effect of the interaction between carbon and the third element $i$ in the Fe-C-$i$ ternary system on the partition coefficient of the element $i$ can be expressed by the ratio of the partition coefficient of the element $i$ in the Fe-C-$i$ ternary system, designated by $K_i^{A/L,T}$, to that in the Fe-$i$ binary system, that is, $Z_i = \left(K_i^{A/L,T}/K_i^{A/L,B}\right)$. It is given by part of the exponential term on the right-hand side of Equation 2. In the case of a dilute solution:

$$Z_i = \exp\left[\left(\varepsilon_i^{C,L} - F\varepsilon_i^{C,A} K_C^{A/L,T}\right) N_C^L\right] \quad [10]$$

*References p. 161*

## TABLE 1

### Equilibrium Partition Coefficients and Factors Influencing Graphitization.

| X | $K_X^{A/L}$ cal. | $K_X^{A/L}$ obs. | $K_X^{C/A}$ cal. | $K_X^{C/A}$ obs. | $K^S$ | $K^M$ | $\Delta K$ ($K^S - K^M$) | $\Delta T_E^S$ (K/at%) | $\Delta T_E^M$ (K/at%) | $\Delta T_E^{(X)}$ ($\Delta T_E^S - \Delta T_E^M$) | Effect on graphitization graphitizing (G) carbide stabilizing (C) | Effect on carbon activity increase (+) decrease (−) |
|---|---|---|---|---|---|---|---|---|---|---|---|---|
| Si | 1.59 | 1.6  | —    | ≃0   | 1.55 | 0.78 |  0.77 |  10.73 |  −4.14 |  14.87 | G | increase (+) |
| Ni | 1.36 | 1.5  | 0.34 | 0.32 | 1.33 | 0.90 |  0.43 |   6.44 |  −1.88 |   8.32 | G | + |
| Co | 1.10 | 1.05 | 0.58 | 0.54 | 1.07 | 0.86 |  0.21 |   1.37 |  −2.63 |   4.00 | G | + |
| Cu | 1.46 | 1.5  | —    | 0.09 | 1.43 | 0.78 |  0.65 |   8.39 |  −4.14 |  12.53 | G | + |
| Al | —    | 1.07 | —    | 0.04 | 1.05 | 0.55 |  0.50 |   0.98 |  −8.46 |   9.44 | G | + |
| Cr | 0.49 | 0.51 | 4.3  | 4    | 0.48 | 1.32 | −0.84 | −10.14 |   6.02 | −16.16 | C | decrease (−) |
| Mn | 0.65 | 0.7  | 1.7  | 1.6  | 0.64 | 0.90 | −0.26 |  −7.02 |  −1.88 |  −5.14 | C | − |
| Mo | 0.38 | 0.35 | —    | 1.7  | 0.37 | 0.52 | −0.15 | −12.29 |  −9.02 |  −3.27 | C | − |
| W  | 0.24 | 0.39 | —    | 1.9  | 0.23 | 0.35 | −0.12 | −15.02 | −12.22 |  −2.80 | C | − |
| Ti | 0.04 | —    | —    | 2.42 | 0.04 | 0.07 | −0.03 | −18.72 | −17.48 |  −1.24 | C | − |
| N  | 1.90 | 1.9  | —    | 1.2  | 1.86 | 2.09 | −0.23 |  16.77 |  20.49 |  −3.72 | C | − |
| P  | 0.14 | —    | —    | 0.63 | 0.14 | 0.11 |  0.03 | −16.77 | −16.73 |  −0.04 | C | + |
| S  | 0.06 | —    | —    | —    | 0.06 | —    |   —   |   —    |   —    |    —   | — | + |
| B  | 0.06 | —    | —    | 4    | 0.06 | 0.15 | −0.09 | −18.33 | −15.98 |  −2.35 | C | + |

In the case of the quaternary system, the ratio of the partition coefficient of the element $i$ in the quaternary system to that in the ternary system, i.e., $Z_i = \left( K_i^{A/L,Q}/K_i^{A/L,T} \right)$, is given by Equations 2 and 8 as follows:

$$\ln Z_i = \alpha_i(1 - Z_i) + \beta_i(1 - Z_C) + \gamma_i \quad [11]$$

$$\left.\begin{array}{l} \alpha_i = K_i^{A/L,T} \varepsilon_i^{i,A} N_i^L \\[4pt] \beta_i = K_C^{A/L,T} \varepsilon_i^{C,A} N_C^L \\[4pt] \gamma_i = \left( \varepsilon_i^{j,L} - \varepsilon_i^{j,A} K_j^{A/L,Q} \right) N_j^L \\[4pt] \simeq \varepsilon_i^{j,L} \left( 1 - K_j^{A/L,Q} \right) N_j^L \\[4pt] Z_C = K_C^{A/L,Q}/K_C^{A/L,T} \end{array}\right\} \quad [12]$$

where, for dilute solutions of the elements $i$ and $j$, $N_i^{L,Q} \simeq N_i^{L,T} = N_i^L$, $N_C^{L,Q} \simeq N_C^{L,T} = N_C^L$, and the values of $F$ in the ternary and quaternary systems are approximated to unity. It can be seen from Equation 12 that the value of $\alpha_i$ is characterized by the third element $i$, that of $\beta_i$ by carbon and the interaction between carbon and the element $i$, and that of $\gamma_i$ by characteristics of the fourth element $j$ and the interaction between the elements $i$ and $j$.

The partition coefficients of solute elements in Fe-C-3Cr-1Si, Fe-C-1Cr-1Ni, Fe-C-1Cr-1Ti, and Fe-C-Si-1Mn systems were calculated from Equations 8 and 9, and some results are shown in Figure 2 together with experimental data points. It is shown that noticeable effects were observed only on the partition coefficients of chromium in the Fe-C-1Cr-1Ti system. From Equation 11, the partition coefficient of the element $i$ in an Fe-C base quaternary system is determined by three factors, i.e., $P_1 = \alpha_i(1 - Z_i)$, $P_2 = \beta_1(1 - Z_C)$, and $P_3 = \gamma_i$. The factor $P_1$ depends on the characteristics of the element $i$ itself and the factor $P_2$ expresses the effects of the element $j$ on the interaction between carbon and the element $i$. The factor $P_3$ represents the interaction between the elements $i$ and $j$. The contributions of these factors to the value of $Z_i$ for chromium and manganese were evaluated for Fe-C-Cr-Si, Fe-C-Cr-Ti, and Fe-C-Si-Mn systems.[6] The results given in Figure 3 reveal that the factors $P_2$ and $P_3$ make opposite contributions to $Z_i$ at lower temperatures, and that the factor $P_1$ is negligible in the temperature range concerned. As the temperature increases, the factor $P_2$ approaches zero and the contribution of the factor $P_3$, i.e., the interaction between the elements $i$ and $j$, becomes dominant. The interaction coefficient between the elements $i$ and $j$, and the values of $Z_i$ and $Z_j$ are listed in Table 2. It is indicated from the results shown in Figure 3 and Table 2 that in Fe-C systems, alloying elements within the composition range of a few percent have only a relatively small effect on the partition coefficient of solute elements. Thus, the partition coefficients of solute elements in Fe-C base multicomponent systems differ little from those in Fe-C base ternary systems.

*References p. 161*

## TABLE 2
Effect of an Addition of a Fourth Element $j$ on the Partition Coefficient of the Third element $i$ in the Fe-C-$i$ Ternary System (3%C,1573K).

| $i$ | $j$ | $\varepsilon_i^j$ | $Z_i$ | $K_i^{A/L,Q}$ | $Z_j$ | $K_j^{A/L,T}$ | $K_j^{A/L,Q}$ |
|---|---|---|---|---|---|---|---|
| Si[†1] | B  | 11.56  | 1.064 | 1.18 | 0.936 | 0.05 | 0.05 |
|        | Cr | −2.91  | 0.987 | 1.10 | 1.011 | 0.55 | 0.56 |
|        | Cu | 4.20   | 0.977 | 1.08 | 0.989 | 1.19 | 1.18 |
|        | Mn | −2.98  | 0.982 | 1.09 | 1.010 | 0.71 | 0.72 |
|        | N  | 7.07   | 0.973 | 1.08 | 0.978 | 1.12 | 1.10 |
|        | Ni | 1.38   | 0.985 | 1.09 | 0.999 | 1.20 | 1.20 |
|        | P  | 17.24  | 1.100 | 1.22 | 0.901 | 0.10 | 0.09 |
|        | S  | 9.31   | 1.021 | 1.13 | 0.979 | 0.04 | 0.04 |
|        | V  | 6.35   | 1.035 | 1.15 | 0.958 | 0.41 | 0.39 |
| Cr[†2] | Co | −5.56  | 1.033 | 0.57 | 0.979 | 1.03 | 1.01 |
|        | Cu | −4.82  | 1.041 | 0.57 | 0.987 | 1.19 | 1.17 |
|        | Mo | −0.01  | 1.021 | 0.56 | 0.985 | 0.42 | 0.41 |
|        | N  | −12.16 | 1.027 | 0.56 | 0.955 | 1.12 | 1.07 |
|        | Ni | −1.28  | 1.022 | 0.56 | 0.999 | 1.20 | 1.20 |
|        | P  | −7.53  | 0.953 | 0.52 | 0.972 | 0.10 | 0.10 |
|        | S  | −2.63  | 1.002 | 0.55 | 1.021 | 0.04 | 0.04 |
|        | Ti | 14.03  | 1.140 | 0.63 | 1.038 | 0.15 | 0.16 |

†1 $K_{Si}^{A/L,T} = 1.11, \varepsilon_{Si}^{Si,A} = 11.25, X_{Si}^L = 0.02$
†2 $K_{Cr}^{A/L,T} = 0.55, \varepsilon_{Cr}^{Cr,A} = -6.29, X_{Cr}^L = 0.01$

## EFFECT OF ALLOYING ELEMENTS ON THE EUTECTIC TEMPERATURE AND GRAPHITIZATION OF IRON-CARBON BASE ALLOYS

**Eutectic Temperature in Fe–C Base Ternary and Multicomponent Systems**—For a theoretical approach to the solidification phenomena of cast irons, it is of fundamental importance to know the eutectic temperature and composition in multicomponent systems. From the results shown in the preceding section, the effect of interaction between solute elements except carbon are minor for a dilute solution of the elements and hence the effects of alloying elements on the eutectic temperature in Fe-C base multicomponent systems are given by the sum of the contribution of each element.

In recent work,[11] the authors have developed relationships representing the effect of an alloying element on the eutectic temperature based on the partition coefficient of the element on eutectic solidification in an Fe-C base ternary system.

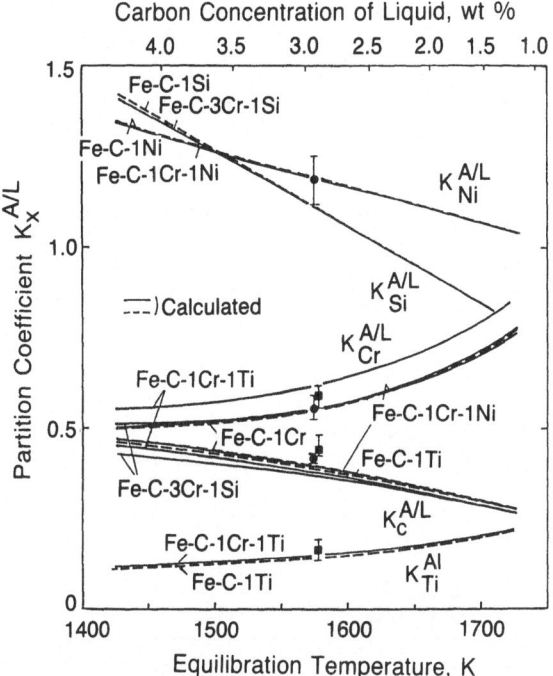

Figure 2. Calculated and measured coefficients of equilibrium partition of solute elements between austenite and liquid iron in Fe-C-Cr-j (j=Si, Ni, Ti) quaternary systems. (Carbon concentration of liquid given on the horizontal axis is for the Fe-C binary system). Measured data for Fe-C-1Cr-1Ti represented by filled squares; measured data for Fe-C-1Cr-1Ni represented by filled circles.

The relation is given by the following equations. For the stable system,

$$\Delta T_E^S = 19.8(K^S - 1) \quad [13]$$

for the metastable system,

$$\Delta T_E^M = 18.3(K^M - 1) \quad [14]$$

where $\Delta T_E^S$ and $\Delta T_E^M$ have the units, K/at %X, and represent changes in the stable and metastable eutectic temperatures, respectively, with a 1 at % addition of a third element. The sign of $\Delta T_E^S$ and $\Delta T_E^M$ indicates whether the eutectic temperature

*References p. 161*

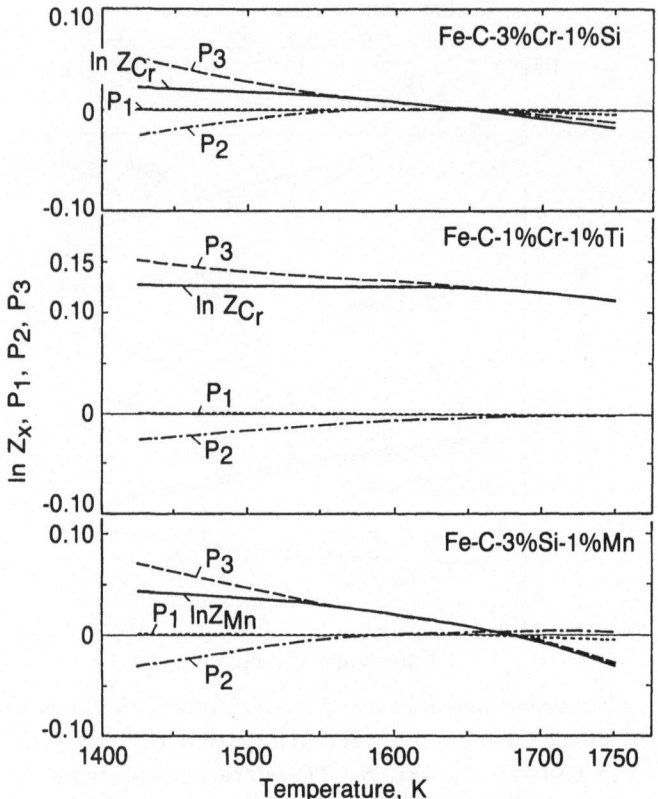

**Figure 3.** Contribution of factors $P_1$, $P_2$, and $P_3$ to $\ln Z_{Cr}$ in the Fe-C-Cr-Si and Fe-C-Cr-Ti systems and to $\ln Z_{Mn}$ in the Fe-C-Si-Mn system.

increases or decreases. The calculated results for some alloying elements are listed in Table 1. In a similar way to the introduction of $\Delta T_E^S$ and $\Delta T_E^M$,[11] equations representing the changes in the carbon concentrations of points $E$ and $A$ in Figure 4, due to a 1 at % addition of a third element, $\Delta N_E$ and $\Delta N_A$, respectively, are:

For the stable system,

$$\Delta N_E^S = 0.182(K^S - 1) - 0.067\varepsilon_C^{X,L} \qquad [15]$$

$$\Delta N_A^S = 0.167(K^S - 1) - 0.052F\varepsilon_C^{X,A} K_X^{A/L} \qquad [16]$$

for the metastable system,

$$\Delta N_E^M = 0.221(K^S - 1) + 0.508\left(K^S - K^M\right) - 0.067\varepsilon_C^{X,L} \qquad [17]$$

$$\Delta N_A^M = 0.196(K^S - 1) + 0.372\left(K^S - K^M\right)$$
$$- 0.052F\varepsilon_C^{X,A} K_X^{A/L} \qquad [18]$$

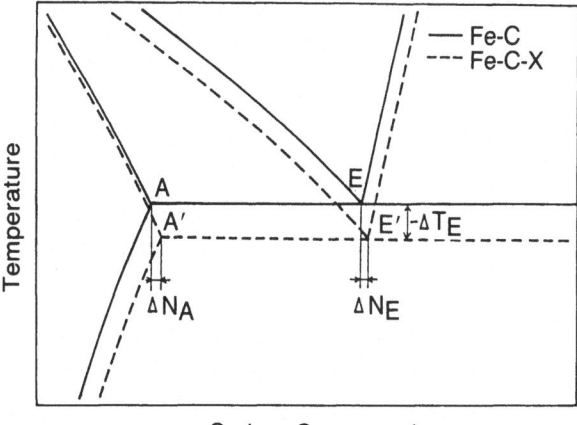

Figure 4. Schematic of the eutectic portion of the phase diagrams for the Fe-C and Fe-C-X systems.

The coefficient for partition of an alloying element between austenite and liquid iron is related to the interaction coefficient through the following equation.

$$K_X^{A/L} = \left(K_X^{A/L}\right)_{\text{Fe}-X} \exp\left[\left(1 - 0.925 K_C^{A/L}\right) N_C^L \varepsilon_X^{C,L}\right] \qquad [19]$$

Values of the term $\varepsilon_X^{C,L}$ have been shown to exhibit periodicity with respect to atomic number of $X$.[12-14] The partition coefficient in the Fe-$X$ binary system, $(K_X^{A/L})_{\text{Fe}-X}$ in Equation 19, also exhibits periodicity as shown in Figure 5. The values of $K^S$ and $\Delta T_E^S$ are periodic functions because they are related to $(K_X^{A/L})_{\text{Fe}-X}$ through Equations 6 and 13. These relationships are shown in Figures 6 and 7. The periodicity of $\Delta N_E^S$ and $\Delta N_E^M$, resulting from dependence on $K^S$, $\varepsilon_C^{X,L}$, and $K_X^{A/L}$ through Equations 15 and 16, becomes more complex as shown in Figure 8. For the metastable system, similar relationships exist with respect to atomic number for $K^M$, $\Delta T_E^M$, $\Delta N_E^M$ and $\Delta N_A^M$.[15] From the Equations 13 and 14, the stable and metastable eutectic temperatures are reduced maximally by the elements with $K^S = 0$ and $K^M = 0$, respectively. The elements having the largest effect to increase the eutectic temperatures are located in the middle of periodic rows. On the other hand, in the periodicity of $\Delta N_E^S$ and $\Delta N_A^S$ as well as $\Delta N_E^M$ and $\Delta N_A^M$, the elements giving minima belong to Groups IIIA, IVA, and VA. Finally, elements located at the beginning of the third and fourth periodic rows have a marked effect to increase $\Delta N_E^S$ and $\Delta N_E^M$.

**Effect of Alloying Elements on the Graphitization of Iron-Carbon Base Ternary Alloys**—Graphitization of cast iron on eutectic solidification has been debated in terms of the effect of alloying elements on the Fe-C binary eutectic temperature, the carbon activity of liquid iron, and the partition behavior of alloying elements between cementite and austenite. Since graphitization must be

*References p. 161*

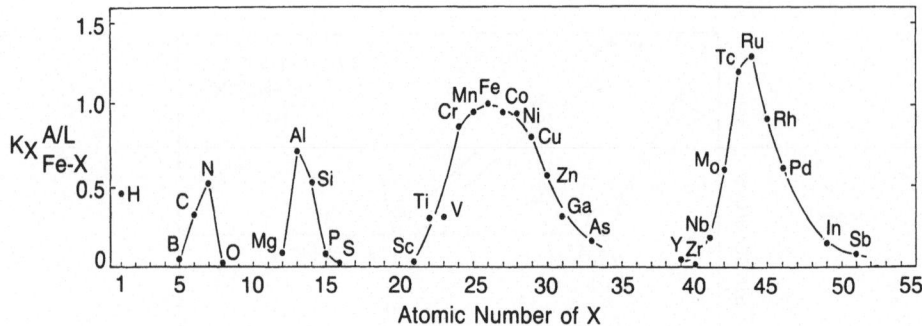

**Figure 5.** Periodicity of the partition coefficient of an alloying element between austenite and liquid iron in the Fe − X binary systems.

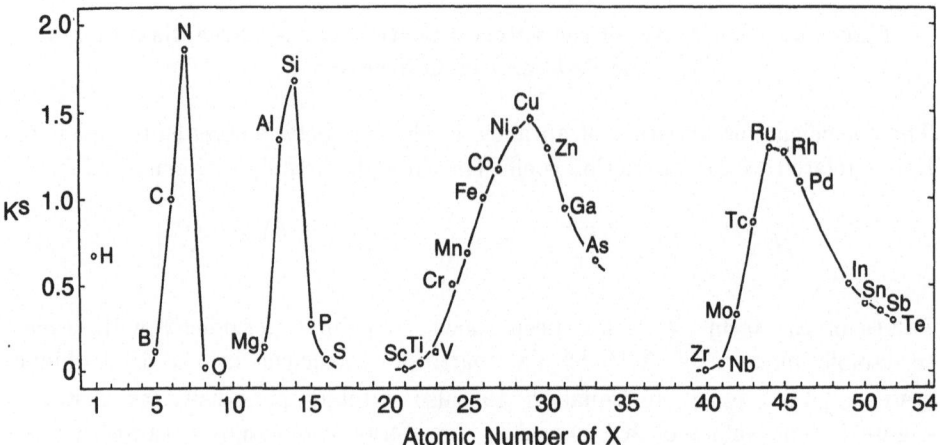

**Figure 6.** Periodicity of the partition coefficient of an alloying element during the stable eutectic solidification.

explained in terms of free energy of the phases concerned, the theoretical grounds for the points mentioned above are deduced below.

In Figure 9, the difference, $Z = (G^M - G^S)$, between the free energy of mixing, $G^S$, of the eutectic liquid equilibrating with austenite and graphite, and that, $G^M$, of the eutectic liquid equilibrating with cementite and austenite is introduced to evaluate the relative thermal stability of the phases concerned. The contribution of an alloying element $X$ to the value of $Z$, designated $Z_X$, is given by:

$$Z_X = -RT\left(K^S - K^M\right) N_X^L \ln\left(a_X^A / a_{Fe}^A\right)$$
$$= \left(K^S - K^M\right) \cdot \beta \qquad [20]$$

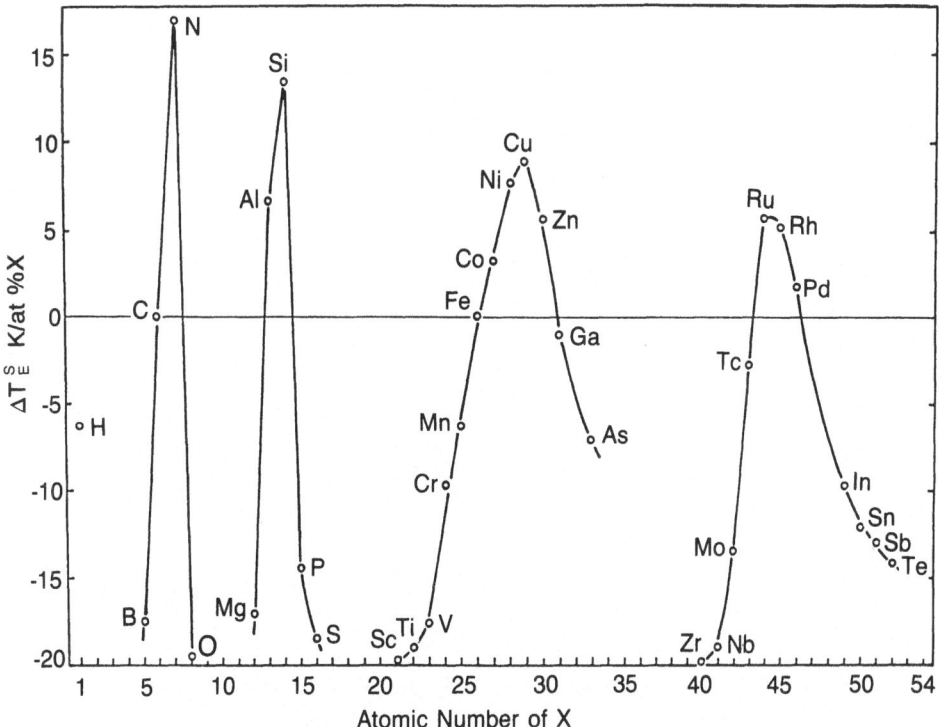

**Figure 7.** Periodicity of the influence of an alloying element on the stable eutectic temperature.

where, $\beta = -RTN_X^L \ln(a_X^A/a_{Fe}^A)$, and $a_X^A$ is the activity of the element $X$ in austenite. For dilute solutions of the element $X$, $(a_X^A/a_{Fe}^A) \ll 1$ and thus, $\beta > 0$, which means that the sign of $Z_X$ depends on the term $(K^S - K^M)$.

The values of $\Delta K = (K^S - K^M)$ for some alloying elements are shown in Table 1. A negative value of $\Delta K$ for an element, corresponding to a negative value of $Z_X$, resulting in a decrease in the free energy difference $Z$. In this case the element $X$ tends to stabilize the austenite and cementite equilibrium. On the other hand, a positive value of $\Delta K$ for an element increases the free energy difference $Z$ and the element acts as a graphitizer.

From Equations 6 and 7, the value of $\Delta K$ is given by the following equation.

$$\Delta K = \left[ \left( f^{A'} - f^A \right) - \left( 1 - f^{A'} \right) K_X^{C/A} \right] K_X^{A/L} \qquad [21]$$

The values of $f^A$ and $f^{A'}$ are approximately given from the equilibrium phase diagram. From the iron-carbon phase diagram, $f^A \simeq 1 f^{A'} \simeq 0.5$ and thus, Equation 21 becomes:

$$\Delta K \simeq -0.5 K_X^{A/L} \left( K_X^{C/A} - 1 \right) \qquad [22]$$

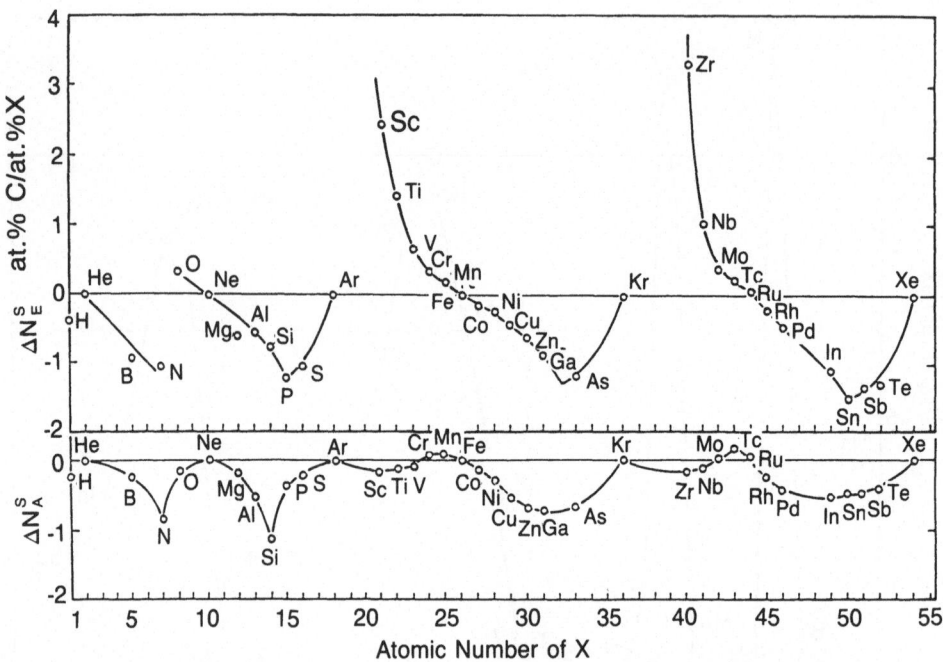

**Figure 8.** Periodicities of the influences of an alloying element on the carbon concentrations at points E and A in Figure 4 for the stable system.

Since $K_X^{A/L} > 0$, a positive value of $\left(K_X^{C/A} - 1\right)$, that is, $K_X^{C/A} > 1$, has the same meaning as a negative value of $\Delta K$ and vice versa. The relationships between the changes in the eutectic temperatures due to addition of 1 at % of an alloying element, $\Delta T_E^S$ and $\Delta T_E^M$, and the partition coefficient of the element, $K^S$ and $K^M$, are given by Equations 13 and 14. From these equations, $\Delta T_E^{(X)} = \left(\Delta T_E^S - \Delta T_E^M\right)$ is related to the value of $\Delta K$ through $\Delta T_E^{(X)} \propto \Delta K$. In turn, the parameters, $\Delta T_E^{(X)}$, $\Delta K$ and $K_X^{C/A}$ relate to the graphitization ability of alloy elements in cast iron.[16] The graphitization ability of alloying elements was measured by a chill test, and represents the normalized addition of each element required for increasing or decreasing chill by a given depth. The relation of $\Delta T_E^{(X)}$ to the graphitization ability is illustrated in Figure 10. A linear relationship exists between those quantities, although some elements such as silicon, aluminum, and titanium, which have a strong affinity for nitrogen, deviate significantly from the line. It is noted that the strong graphitizing tendency of these elements might superpose, for example, the effect of low nitrogen, due to nitride formation, on the intrinsic nature of the elements.

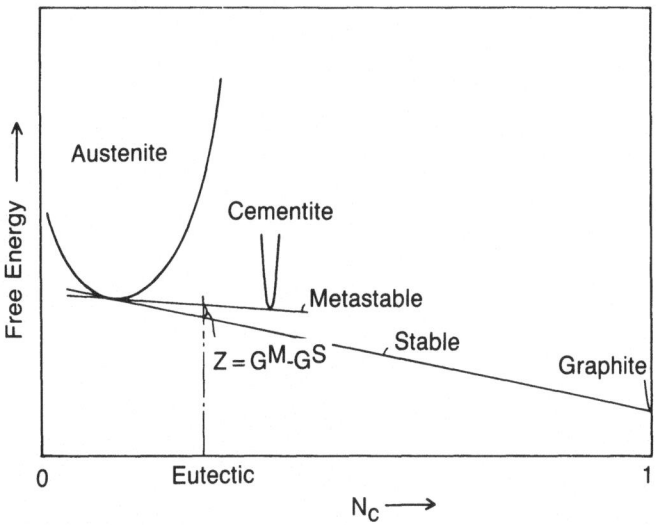

Figure 9. Schematic free energy curves for the coexisting phases in the stable or metastable eutectic solidification in the equilibrium condition.

**Partition Behavior of Nitrogen In Iron-Carbon Alloys**—From the solubility data of nitrogen in liquid iron[17,18] and austenite,[19,20] it can be deduced that nitrogen in cast iron is enriched in primary austenite during solidification and that the solubilities of nitrogen in liquid iron and austenite decreases considerably with an increase in carbon content. With the progress of austenite solidification in a hypoeutectic cast iron, the nitrogen solubility in the melt decreases more than the decrease in the nitrogen concentration in the melt. Therefore, if the melt to be solidified has a high nitrogen concentration, it can be supersaturated with nitrogen during austenite solidification and thus gas holes may be formed. This problem, however, remains obscure because the partition of nitrogen between liquid and solid during solidification is not quantitatively known in the cast iron composition range. In connection with the formation of nitrogen gas holes, the partition of nitrogen between austenite and liquid iron and between cementite and austenite was measured in the Fe-C system. Experiments conducted to determine these partition coefficients are described in Reference 4.

In Figure 11, the partition coefficient of nitrogen between cementite and austenite $K_N^{C/A}$, defined as the ratio of acid-soluble nitrogen concentration ($N_S$) of cementite to that of austenite, is shown as a function of temperature. The value of $K_N^{C/A}$ decreases from 2 to 1.5 with increasing temperature from 1073 to 1223 K.

The equilibrium partition coefficient of nitrogen between austenite and liquid iron, $K_N^{A/L}$, is shown as a function of temperature in Figure 12. Its value increases

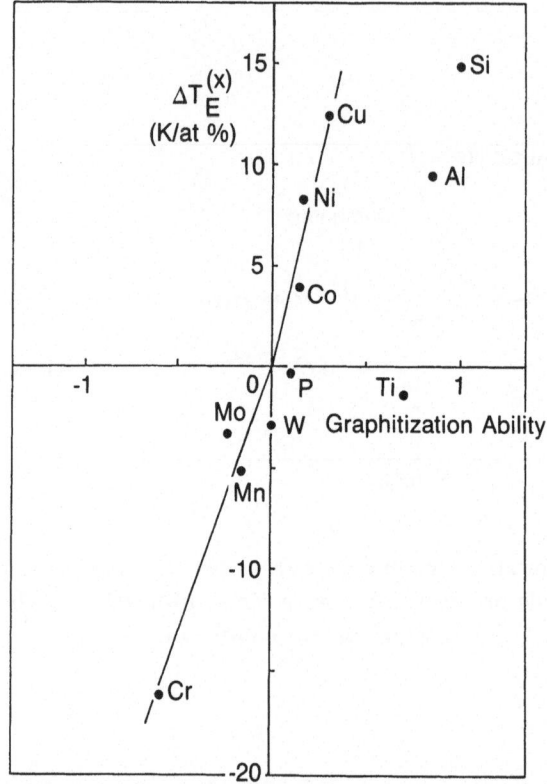

Figure 10. Relationship between graphitization ability and the effect of an alloying element on the difference between the stable and metastable eutectic temperature, $\triangle T_E^{(X)}$.

from 0.7 at 1673 K to 1.9 at 1423 K. $K_N^{A/L}$ has the value, unity, at about 1600 K. This indicates that nitrogen in a solidifying hypoeutectic iron-carbon alloy is rejected into liquid iron from austenite at temperatures > 1600 K and conversely into austenite from liquid iron at temperatures < 1600 K.

The partition coefficients of nitrogen between cementite and austenite, and between austenite and liquid iron tend to converge to a value at high temteratures. This is shown in Figures 11 and 12. The data can be expressed in Arrhenian form as follows: $K_N^{C/A} = 0.284\exp(4,130/RT)$, and $K_N^{A/L} = 0.0023\exp(19,080/RT)$.

From Equations 6 and 7, the partition coefficients of nitrogen between eutectic liquid and its solid on the stable and the metastable eutectic solidifications, $K_N^S$ and $K_N^M$, have respective values of 1.9 and 2.2. Using these values, the variation of the nitrogen concentration in each phase in solidifying iron-carbon alloys was calculated from Equations 23–25, which follow. The nitrogen concentrations in liquid iron and

*References p. 161*

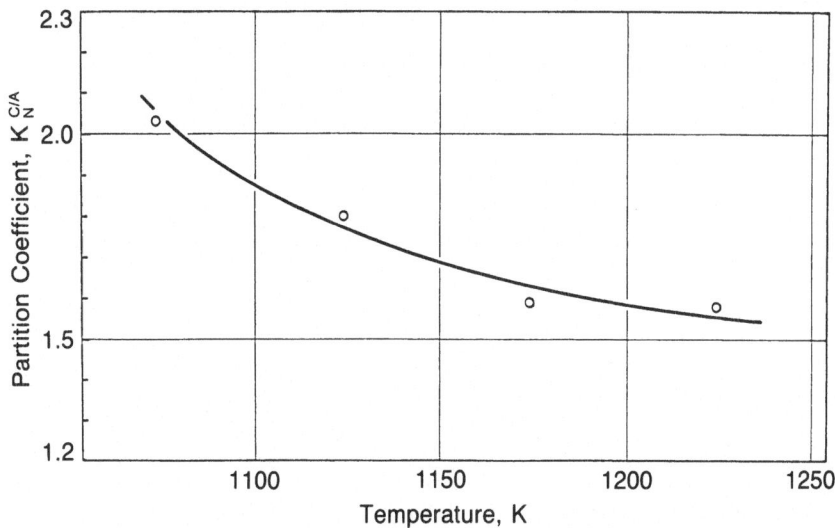

Figure 11. The partition coefficient of acid-soluble nitrogen between cementite and austenite.

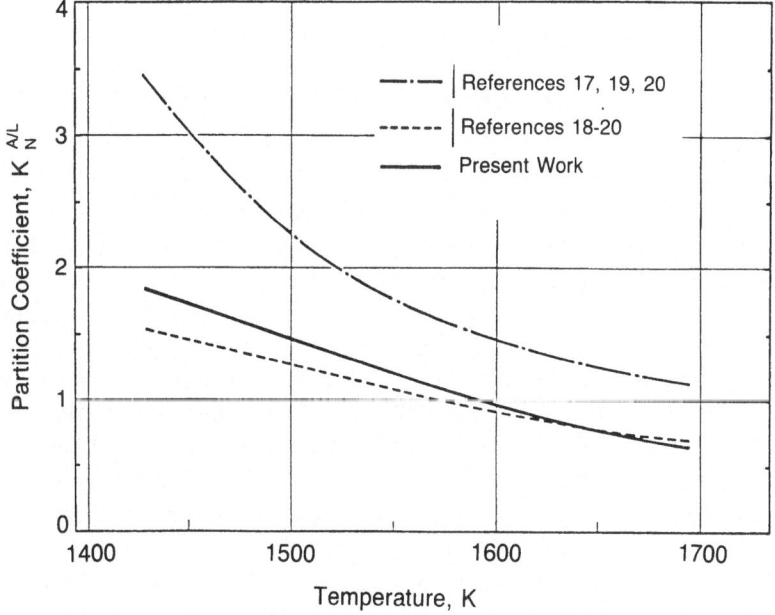

Figure 12. The partition coefficient of nitrogen between austenite and liquid iron.

austenite, $[N]^L$ and $[N]^A$, respectively, during the solidification of primary austenite are represented by:

$$\left. \begin{array}{l} [N]^L = [N_0]/\left[1 - f^A\left(1 - K_N^{A/L}\right)\right] \\ [N]^A = K_N^{A/L}[N]^L, \end{array} \right\} \quad [23]$$

where $N_0$ is the initial nitrogen concentration in liquid iron and $f^A$ the weight fraction of austenite in the stable eutectic solids. For eutectic solidification in the stable system,

$$\left. \begin{array}{l} [N]^L = [N_0]/\left[1 - f_{gr} + \left(K_N^{A/L} - 1\right) f^A\right] \\ [N]^A = K_N^{A/L}[N]^L. \end{array} \right\} \quad [24]$$

where $f_{gr}$ is the weight fraction of graphite in the stable eutectic solids.

For eutectic solidification in the metastable system, nitrogen concentration in liquid iron, austenite, and cementite, $[N]^L[N]^A$, and $[N]^C$, respectively, are

$$\left. \begin{array}{l} [N]^L = [N_0]/\left[1 - \left(K_N^{A/L} - 1\right) f^A + \left(K_N^{A/L} K_N^{C/A} - 1\right) f^C\right] \\ [N]^A = K_N^{A/L}[N]^L \\ [N]^C = K_N^{A/L} K_N^{C/A}[N]^L, \end{array} \right\} \quad [25]$$

where $f^C$ is the mass fraction of cementite in the metastable eutectic solids.

In the calculation, it was postulated that nitrogen and carbon distributed uniformly in each phase during solidification. This postulation has been generally accepted as the behavior of these elements during dendritic solidification.

On solidification of iron-carbon alloys with an initial nitrogen concentration of 100 ppm and different carbon contents, nitrogen concentrations in liquid iron and in austenite vary with temperature as shown in Figure 13. An alloy melt with 2.5% C begins to crystallize primary austenite at the liquidus temperature (point A). Nitrogen concentration in the liquid iron increases slightly with a decrease in the temperature to about 1600 K and then, decreases with decreasing temperature below 1600 K, as shown by a dotted curve. On the other hand, nitrogen concentration in the primary austenite increases with the progress of austenite solidification, as shown by a solid curve. The chain curves show the solubilities of nitrogen in liquid iron and in austenite. During the solidification of iron-carbon alloy melts with 2.5% C, 3.0% C, or 3.5% C, and an initial nitrogen concentration of 100 ppm, nitrogen concentrations in the liquid iron and in the austenite should not exceed the solubilities of nitrogen in these phases. However, in the case of iron-carbon alloy with 3.8% C, nitrogen concentrations in the liquid iron and in the austenite exceed the nitrogen solubilities in these phases below the temperature indicated by point B. Therefore, both the phases are supersaturated with nitrogen below B. The supersaturation of liquid iron with nitrogen enables gas bubbles to form. If gas bubbles do not form in the supersaturated liquid iron and no nitrides form in austenite, the nitrogen concentration in the liquid iron is given by Equation 23. In such a case, the degree of supersaturation of nitrogen in liquid iron at a temperature is given as $N^L/N^{L^*}$, the ratio of the nitrogen concentration in the liquid iron, $(N^L)$, to the solubility of nitrogen in the liquid iron equilibrated with nitrogen of 1 atm $(N^{L^*})$. In Figure 13, it is clear that the degree of supersaturation of nitrogen in liquid iron varies during solidification and becomes maximum at the eutectic temperature. The degree of supersaturation of nitrogen depends on the initial nitrogen content as well as carbon content. In Figure 14, the variations of nitrogen concentration in liquid iron and austenite are illustrated for the alloy with 3% C and two different levels of initial nitrogen concentration $N_0$. At $N_0 = 120$ ppm, nitrogen concentrations in both liquid iron and austenite reach the solubilities in these phases at the eutectic temperature. At the higher initial nitrogen concentration as $N_0 = 150$ ppm, nitrogen is supersaturated in both phases below 1515 K. Figure 15 shows the dependence of the maximum value of $N^L/N^{L^*}$ attainable during solidification, on the carbon content and the initial nitrogen concentration. It is indicated that, if the initial nitrogen concentration in the melt is below the curve of $N^L/N^{L^*}_{\max} = 1.0$, the melt remains unsaturated with respect to nitrogen during solidification. Two nitrogen solubility curves are also shown in the figure, one is for the liquid iron equilibrated with air at atmospheric pressure and the other is for nitrogen at 1 atm. When the iron-carbon alloy with 2.5% C is melted in air, the melt can contain nitrogen up to 187 ppm. On solidification of the melt with $N_0 = 187$ ppm, the value of $N^L/N^{L^*}$ reaches 1.36 at the eutectic temperature. From the figure, it is seen that the possibility of forming gas bubbles increases with decreasing carbon content. As for the critical nitrogen concentration at which gas bubbles actually form in liquid cast iron, Davison et al.[7] and Caspers[8] show 100 ppm and 110 ppm, respectively. From the composition of the cast irons used in their works, the critical degree of supersaturation of nitrogen in liquid iron was about 1.4 for Davison's work and about 2.0 for the Caspers' work.

## EFFECT OF ALLOYING ELEMENTS ON THE PARTITION BEHAVIOR OF NITROGEN AND NITROGEN BLOWHOLE FORMATION

Changes in the solubilities of nitrogen in liquid iron and austenite can be evaluated by the following equation.

$$\log[\%N]^\nu_{\text{Fe}-\text{C}-i} = \log[\%N]^\nu_{\text{Fe}-\text{C}} - e_N^{i,\nu}[\%i]^\nu \qquad [26]$$

where $[\%N]^\nu_{\text{Fe}-\text{C}-i}$ and $[N\%]^\nu_{\text{Fe}-\text{C}}$ are the solubilities of nitrogen in a $\nu$ phase (liquid iron or austenite) for the Fe - C base ternary and Fe - C binary systems, respectively. The term $e_N^{i,\nu}$ is the interaction coefficient between nitrogen and the element $i$ in a $\nu$ phase and $[\%i]^\nu$ is the weight per cent of the element $i$ in a $\nu$ phase. Employing the interaction coefficients listed in Reference 5, the nitrogen solubilities in liquid iron and austenite in equilibrium with each other are evaluated at a given temperature.

*References p. 161*

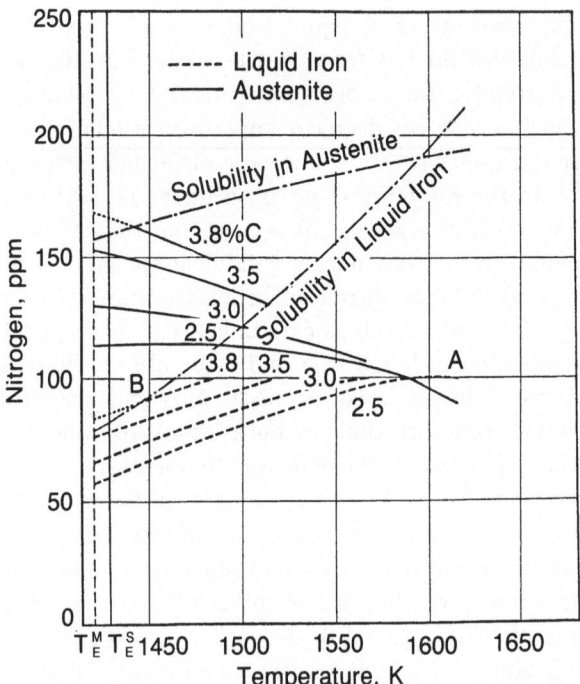

Figure 13. The variation of nitrogen concentration in liquid iron and austenite during the solidification of cast irons with initial nitrogen concentration of 100 ppm and different carbon contents.

Figure 16 shows the equilibrium nitrogen solubilities in liquid iron and austenite for Fe-C-1wt%$i$ ternary alloys at 1473 K. The data are plotted against the atomic number of the element $i$. Changes in the nitrogen solubilities of both phases, due to addition of an element $i$, show apparent periodicity at relatively low atomic numbers. The effects of alloying elements have been classified into three groups:

I. Elements that increase the solubility of nitrogen in liquid iron and/or austenite. The most effective elements are:

$$\text{Ti, V, Zr, Nb, Hf, Ta, Cr, Mn, Mo}$$

II. Elements that decrease the solubility of nitrogen in either/or both phases

$$\text{Li, Be, B, C, O, F, Na, Mg, Al, Si, P, Co, Ni}$$

III. Elements with a negligible effect on nitrogen solubility in both phases

all other elements

In group I, the first five elements have a marked effect on increasing the nitrogen solubility in liquid iron and are characterized as strong nitride formers. Silicon

PARTITION OF ALLOYING ELEMENTS 155

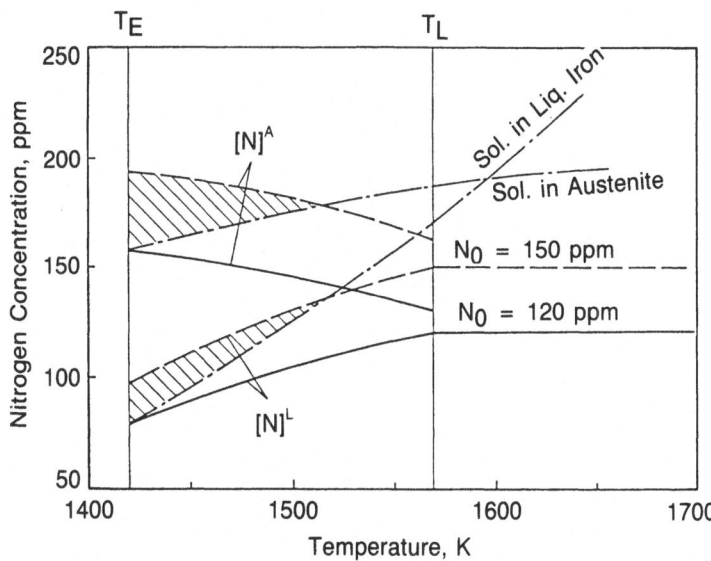

Figure 14. The variation of nitrogen concentration in liquid iron $[N]^L$ and austenite $[N]^A$ during the solidification of Fe − C alloy melts with 3%C and different levels of initial nitrogen concentration.

in group II has a large negative effect on the nitrogen solubility although it is also considered a nitride former. From the solubilities in liquid iron and austenite, the coefficients for partition of nitrogen between austenite and liquid iron at 1473 K in the Fe-C-$i$ ternary alloy were calculated and are given at the top of Figure 16. Most of the elements in group I and silicon decrease the partition coefficient of nitrogen, while the elements in group II except silicon increase it. The influence of alloying element $i$ on the partition coefficient of nitrogen during eutectic solidification is expressed as follows:

for the stable eutectic solidification,

$$K^S = \left(K^S\right)_{Fe-C} \exp\left[2.303[\%i]^L \left(e_N^{i,L} - e_N^{i,A} K_i^{A/L}\right)\right] \quad [27]$$

for the metastable eutectic solidification,

$$K^M = \left[f^{A'} + \left(1 - f^{A'}\right) K_N^{C/A}\right] K^S/f^A \quad [28]$$

where

$$K_N^{C/A} = \left(K_N^{C/A}\right)_{Fe-C} \exp\left[2.303 \left(e_N^{i,A}[\%i]^A - e_N^{i,C}[\%i]^C\right)\right] \quad [29]$$

For silicon as the element $i$, the solubility of silicon in cementite is negligible,[21] $(\%Si)^C = 0$, and thus, Equation 28 is simplified for silicon and the value of $K_N^{C/A}$ is calculated using only the data for $e_N^{Si,A}$. Other elements are soluble in

*References p. 161*

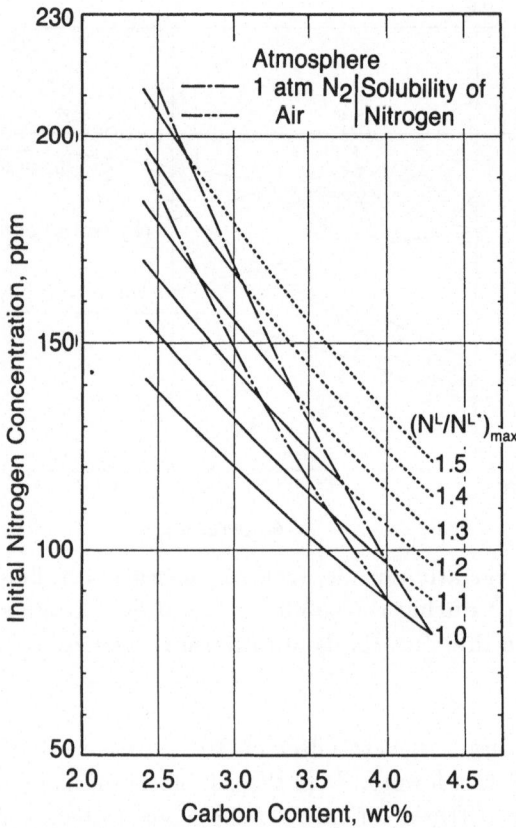

**Figure 15.** Effects of initial nitrogen concentration and carbon content in iron-carbon alloy melts on the maximum degree of nitrogen saturation $(N_L/N_L^*)_{max}$ attained during solidification.

cementite, however, data for $e_N^{i,C}$ are not available in the literature. Using assumed values for $e_N^{i,C}$ and $e_N^{i,A}$ for chromium as the element $i$, the partition coefficients of nitrogen, $K_N^S$ and $K_N^M$ were calculated. The values for chromium and silicon bearing alloys are shown as functions of the concentration of the respective elements in Figure 17. The partition coefficients, $K_N^S$ and $K_N^M$, are larger than unity. It is known from Equation 27 that $K^S$ possibly decreases for the elements having the partition coefficient, $K_i^{A/L}$, much larger than unity. From Figure 6 it is seen that all alloying elements except nitrogen have the coefficient, $K_i^{A/L}$ smaller than that of silicon. Since $K^S$ is greater than unity for all values of silicon in Figure 17, it is indicated that the coefficient $K^S$ for nitrogen in all Fe-C-$i$ ternary systems remains larger than unity. As for the partition coefficient $K^M$, it is known from the Equation 28 that alloying elements which decrease the coefficient $K^M$ for nitrogen should have an affect to make the partition coefficient $K_N^{C/A}$ less than unity. From Equation 29, it is suggested tht carbide forming elements such as Cr, Mn, and

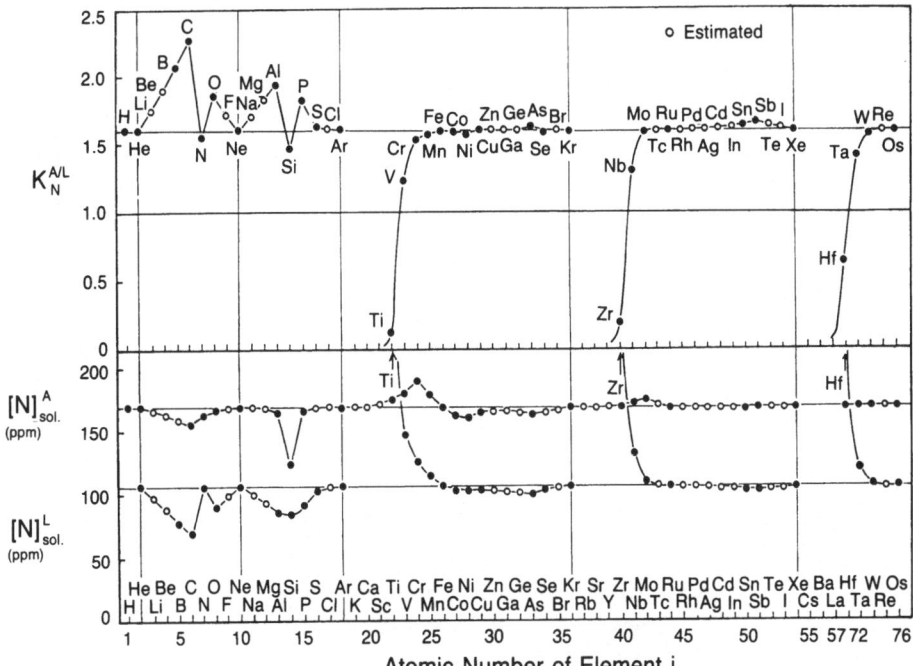

Figure 16. Effect of an element i on the solubilities of nitrogen in liquid iron $[N]^L_{sol.}$ and austenite $[N]^A_{sol.}$, and the coefficient for partition of nitrogen between austenite and liquid iron, $K_N^{A/L}$ in the in the Fe-C-i ternary alloys.

Mo tend to condense into cementite ($[\%i]^C > [\%i]^A$) and have negative values of $e_N^{i,A} (\simeq e_N^{i,C})$, resulting in an increase in the coefficient $K^M$ for nitrogen. Graphitizing elements such as Si, Ni, and Co, on the other hand, are enriched in austenite rather than in cementite ($[\%i]^C < [\%i]^A$) and have positive values of $e_N^{i,A}$ which result in an increase in the coefficient $K^M$ for nitrogen. It is, therefore, concluded that the partition coefficients of nitrogen for stable and metastable eutectic solidification are larger than unity no matter whether the additive alloying elements is a carbide stabilizer or graphitizer. This means that nitrogen contained in eutectic liquid is captured entirely into solidifying eutectic solid. The problem of the formation of a nitrogen bubble is, therefore, focused on primary austenite crystallization.

For iron-carbon-silicon, the basic alloy of cast iron, the maximum values of the initial nitrogen concentration not to supersaturate nitrogen in liquid iron during primary austenite crystallization were evaluated in a similar manner to the earlier analysis of the Fe-C binary alloy (see Figure 15). The results of the analysis for the Fe-C-Si system are given in Figure 18. The solubilities of nitrogen in Fe-C-Si alloys at the liquidus and the solidus temperatures were calculated using the results of nitrogen solubilities in liquid iron and austenite and the data of the interaction

*References p. 161*

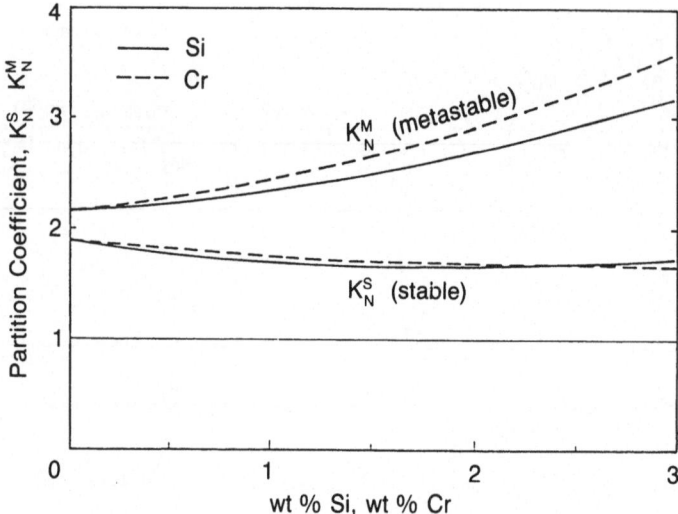

Figure 17.  Effects of silicon and chromium on the coefficients, $K_N^S$ and $K_N^M$, for partition of nitrogen on the stable and metastable eutectic solidification.

coefficients between silicon and nitrogen; $e_N^{Si,L} = 171/T - 0.031 + 0.0037(\%C)$ for liquid iron[18] and $e_N^{Si,A} = -360/T + 0.349$ for austenite.[19]

In the calculation of the degree of supersaturation of nitrogen, silicon distribution during solidification was assumed to be given by the Scheil equation. It is indicated from the thermodynamic data on silicon nitride[21] that $Si_3N_4$ is not formed in both liquid iron and austenite for an alloy containing nitrogen of 100 ppm and silicon less than 3%. The solid curves ($N^L/N^{L^*} = 1.0$) in Figure 18 show the critical values of $N_0$, above which nitrogen is supersaturated in liquid iron. Attention must be paid to the fact that there is greater possibility of gas hole formation in cast irons with lower carbon content.

Figure 19 shows the effect of a small addition of aluminum into a hypoeutectic cast iron melt with the composition of 3% C, 2% Si, and $N_0 = 100$ ppm. The calculation of changes in soluble nitrogen [N], soluble aluminum [Al] and the amount of AlN formed with lowering temperature are based on solubility product data for AlN given in Reference 22. For cast iron, free of aluminum, the soluble nitrogen in the melt reaches the solubility curve of liquid iron at 1544K then decreases along the curve with lowering temperature. Below 1498K, nitrogen is partitioned between austenite and liquid iron. However, a supersaturated condition exists in both liquid iron and austenite. The effect of 0.01% Al addition is shown by broken curves in the figure. In this case, the formation of aluminum nitride starts at 1498 K and primary austenite crystallization occurs at the same temperature. It is seen that a small addition of aluminum reduces effectively the soluble nitrogen concentrations in liquid iron and austenite. The addition of strong nitride formers such as titanium,

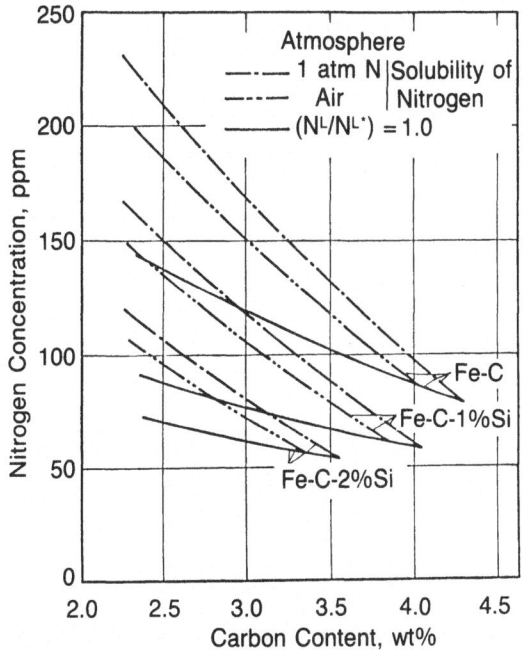

Figure 18. Relation between carbon content of cast iron and initial nitrogen concentration in the melt with which the degree of nitrogen saturation ($N_L/N_L^*$) becomes unity during solidification.

zirconium, or hafnium is also effective in reducing the soluble nitrogen concentration in cast iron melts. It is, however, noted that available amounts of these elements is reduced by forming carbides.

## CONCLUSION

Partition of alloying elements and nitrogen in iron-carbon base ternary and quaternary alloys were evaluated experimentally and theoretically and, from the results obtained, partition behavior of nitrogen and the condition for nitrogen blowhole formation in freezing cast irons were investigated. On primary austenite crystallization, the coefficients for partition of alloying elements between austenite and liquid iron, $K_X^{A/L}$, evaluated thermodynamically show a satisfactory agreement with experimental values over a wide range of carbon concentrations up to the eutectic composition. The coefficients for partition of alloying elements between eutectic liquid and solid, for stable and the metastable solidification, $K^S$ and $K^M$, respectively, were calculated from the partition coefficient $K_X^{A/L}$ and the coefficient for partition of the element between cementite and austenite, $K_X^{C/A}$. The contribution of solute interaction to the partition coefficient $K_X^{A/L}$ was examined for iron-carbon base quaternary systems and it was found that the effect was minor except that

*References p. 161*

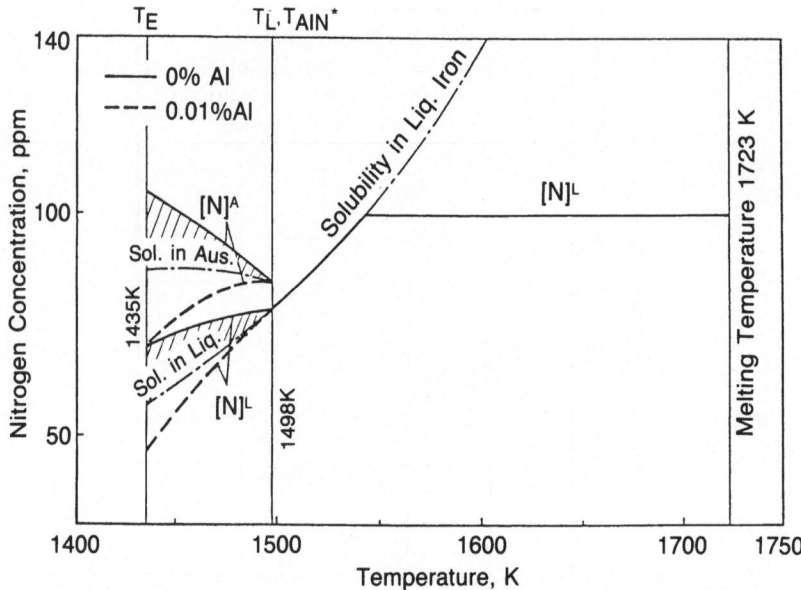

Figure 19. Effect of aluminum on the variation of nitrogen concentration in liquid iron $[N]^L$ and austenite $[N]^A$ during the solidification of a hypoeutectic cast iron with the composition of 3%C, 2%Si, and initial nitrogen concentration of 100 ppm.

between carbon and an alloying element. On this basis the partition coefficients of solute elements in Fe-C base multicomponent systems differ little from those in Fe-C base ternary systems. A linear relationship was thermodynamically developed between the effect of an alloying element on the eutectic temperature and the partition coefficients, $K^S$ and $K^M$. Periodicity of the effect of an alloying elements on the eutectic temperature was observed, which results from the periodicity of the partition coefficients of the element on primary austenite crystallization and eutectic solidification. From the thermodynamic stability of the related phases on eutectic solidification, the difference in the partition coefficients $K^S$ and $K^M$ for an alloying element, i.e., $\Delta K = (K^S - K^M)$ relates to the partition coefficient $K^{C/A}$.

When primary austenite crystalizes, nitrogen is rejected from austenite to liquid iron at temperatures higher than about 1600 K and the partition behavior is reversed at temperatures lower than 1600 K. On the partition of nitrogen between cementite and austenite, nitrogen is more concentrated in cementite than in austenite and the partition coefficient reduces with increasing temperature. Using partition coefficient data, nitrogen concentrations in liquid iron and austenite on primary austenite crystallization were calculated for iron-carbon and iron-carbon-silicon alloys and the effect of a small addition of aluminum as a strong nitride former was evaluated. From the results of calculations, it was indicated that, when nitrogen in a cast iron melt is in equilibrium with atmospheric air, there is a higher possibility of forming nitrogen bubbles as the carbon content of the cast iron is reduced. An addition of aluminum of more than 0.01% significantly decreases the soluble nitrogen concentrations in both liquid iron and austenite by forming aluminum nitride and is effective in suppressing nitrogen blowhole formation.

## REFERENCES

1. A. Kagawa and T. Okamoto: *Met. Sci.*, 1980, *14*, p. 519.
2. A. Kagawa and T. Okamoto: *Trans. Jpn. Inst. Met.*, 1981, *22*, pp. 137.
3. A. Kagawa, S. Moriyama, and T. Okamoto: *J. Mater. Sci.*, 1982, *17*, pp. 135.
4. A. Kagawa and T. Okamoto: *Trans. Jpn. Foundrym. Soc.*, 1983, *2*, pp. 12.
5. A. Kagawa and T. Okamoto: *J. Mater. Sci.*, 1984, *19*, p. 2306.
6. A. Kagawa, K. Iwata, A. A. Nofal, and T. Okamoto: *Mater. Sci. Techn.*, 1985, vol. 1, p. 678.
7. M. H. Davison, F. P. H. Chen, and J. Keverian: *Modern Castings*, 1963, *85*, p. 528.
8. K. H. Caspers: *AFS International Cast Metals J.*, 1977, *2*, p. 29.
9. J. V. Dawson, J. A. Kilshaw, and A. D. Morgan: *Trans. AFS*, 1965, *73*, p. 224.
10. M. Ko, T. Sakuma, and T. Nishizawa: *Nippon Kinzoku-gakkai-shi*, 1976, *40*, p. 593 (in Japanese).
11. A. Kagawa and T. Okamoto: *The Physical Metallurgy of Cast Iron*, H. Fredriksson and M. Hillert, Elsevier Science Pub. Co., Amsterdam, 1985, p. 201.
12. F. Neumann and H. Schenck: *Arch. Eishenhüttenwes.*, 1959, *30*, p. 477.
13. K. Sanbongi, M. Ohtani, and K. Toita: *Tohoku Daigaku Senkoseiren Kenkyusho Iho*, 1957, *12*, p. 97 (in Japanese).
14. T. Fuwa and J. Chipman: *Trans. Met. Soc. AIME*, 1959, *215*, p. 708.
15. A. Kagawa and T. Okamoto: to be published in *J. Mater. Sci.*
16. K. Taniguchi: *Jitsuyo-kinzoku-zairyo-koza (Lectures on Applied Metallic Materials)* Vol. 5, *Chilled castings*, Kyoritsu-sha, Tokyo, 1936, p. 74 (in Japanese).
17. D. W. Gomersall, A. McLean, and R. G. Ward: *Trans. Met. Soc. AIME*, 1968, *242*, p. 1309.
18. M. Uda and R. P. Pehlke: *Cast Metals Res. J.*, 1974, *10*, p. 30.
19. T. Mori, E. Ichise, Y. Niwa, and K. Kuga: *Nihon Kinzoku-gakkai-shi*, 1967, *31*, p. 887.
20. H. Schenck, M. G. Fronberg, and F. Reinders: *Stahl u. Eisen*, 1963, *83*, p. 93.
21. T. Okamoto and A. Kagawa: *Met. Sci.*, 1977, *11*, p. 471.
22. H. Sawamura and T. Mori: *Tetsu-to-Hagane*, 1955, *41*, p. 1082.

# DISCUSSION

**S. Katz** *(General Motors Research Laboratories.)*

Are the Equations 13 and 14 that you use, respectively, to calculate the effect of an element on the stable and metastable eutectic temperature empirically obtained or are they theoretically derived?

**A.** These equations are derived theoretically. The derivation of the equations is given in our recent work to be published in *J. Mater. Sci.* (1986).

**R. Singh** *(General Motors Advanced Engineering Staff)*

You have presented arguments for the use of aluminum for the reduction of gas defects. Dr. Turkdogan, earlier this morning, presented rationale for the role of aluminum to increase gas defects. Can you rectify these opposite points of view?

**A.** It may depend on the species of gaseous elements whether aluminum is harmful or not for the blowhole formation in cast iron. It is well known that aluminum in cast iron has a detrimental effect to enhance the formation of hydrogen blowholes. In our presentation, the effect of aluminum on the partition of nitrogen during austenite crystallization was discussed and our conclusion was that aluminum effectively reduced the supersaturation of nitrogen by forming aluminum nitride. The problems of the absorption of gaseous elements due to metal-mold reaction and nucleation of gas bubbles in the melt were not considered. As Dr. Turkdogan pointed out, aluminum may enhance the nucleation of gas bubbles by lowering the surface tension of the cast iron melt. It is, however, noted that an excess supersaturation should be necessary as a driving force for the nucleation and growth of nitrogen gas bubbles. We stressed that sufficient amounts of aluminum in cast iron effectively diminish the supersaturation of nitrogen and thus the driving force for the nucleation of nitrogen gas bubbles in the melt.

# CHEMICAL PROCESSES AND HEAT LOSS IN CUPOLAS

### CRAIG LANDEFELD
*General Motors Research Laboratories*
*Warren, Michigan*

## ABSTRACT

The first slag to form consists of coke ash and metal oxidation products. Limestone is incorporated later. Lower in the cupola, 0.4 m below the tuyeres, slag composition is much nearer to that of the tapped slag. However, it has less $SiO_2$ and more FeO and MnO than tapped slag. Final slag composition is attained in the slag layer of the cupola as metallic Si is oxidized by reaction with FeO.

The observed C pickup in a cupola agrees well with predictions of a revised model. The revision considers that iron and steel charge components melt and descend without mixing until leaving the cupola. Two rate parameters, one for each charge component, are derived from tests in a small cupola. With those parameters, the model accurately predicts C pickup in a large cupola.

Heat losses through uninsulated cupola shells were measured previously. One way to reduce those losses is insulating the shell. Insulating the shell is estimated to increase iron melting rate 11% and decrease the coke/iron ratio 10%. Another way is using a reverse taper shell. Heat flux measurements on cupola shells suggest that convective heat transfer is less with the reverse taper.

Previous modelling of the kinetics of the endothermic reaction of $CO_2$ with coke is extended to include the lower temperature range in which reaction rate is controlled jointly by gaseous diffusion within the porous coke and by chemical reaction. The principal variable studied is the intrinsic reaction rate constant, $k_v$. Normal variation of $k_v$ is estimated to produce small changes in required coke.

## INTRODUCTION

Numerous chemical reactions occur within a cupola, and many do not attain equilibrium. Furthermore, the extents of reaction are quite dependent on the heat transfer and fluid flow processes. Consequently, the connection between inputs and outputs consists mostly of empirical relations valid only for specific conditions. Therefore, an accurate numeric model encompassing the important cupola processes is needed to speed development of cupola melting practices.

Each of these simultaneous processes is amenable to modelling, so an overall numeric model of the cupola is theoretically feasible. The practical difficulty is that many of the individual processes can not yet be described accurately. Consequently, a model, although formally correct, would give wrong answers. The needed information is gradually being acquired, however. With continued progress, a useful cupola model could soon be developed.

In the following sections, information of the type needed for developing a quantitative cupola model is presented. Most is unpublished. Much of it could usefully be extended. Some of the topics appear unconnected, but they all are pieces to the cupola puzzle.

## SLAG-METAL REACTIONS

**Slag Formation**—Decreasing $CO/CO_2$ ratio in the cupola stack increases coke efficiency at the price of greater oxidation of the metal charge[1]. For cold blast air cupolas, unity ratio was considered ideal[1], although oxidizing to iron above 550°C. Present cupolas using hot blast and oxygen enrichment have greater $CO/CO_2$ ratios, as shown by extensive gas measurements on two General Motors cupolas. For gas sampled 3 m above the tuyeres of a 2.64 m diameter cupola using basic slag, hot blast, $O_2$ enrichment and large coke percentages, the ratio $CO/CO_2$, exclusive of $CO_2$ from limestone, averaged almost 2. At that ratio, iron oxidizes to FeO above 850°C. For a similar cupola using acid slag and smaller coke percentages, the $CO/CO_2$ ratios ranged from 1 to 2 at the gas offtake. So present cupolas are oxidizing, but apparently to a smaller degree than older cupolas.

As a result, the slag in the coke bed above the tuyeres consists mostly of metal oxidation products and coke ash, along with molding sand. Compositions of two slag samples extracted through a tuyere demonstrate this (Table 1). Sample 1 was black and porous. The predominance of $SiO_2$, FeO, and MnO, along with small concentrations of CaO, MgO and $Al_2O_3$, shows it to be comprised of metal oxidation products and sand. The three particles of sample 2 are evidently derived from coke ash, which is typically 50% $SiO_2$ and 30% $Al_2O_3$, with small amounts of CaO and iron oxide. These findings agree with those of Meister[2].

The slags in Table 1 contain very little CaO relative to tapped slags. Rambush and Taylor[3], in a quenched cupola, found calcined limestone at the tuyere level that had not combined with the other slag-forming oxides, and Morris and Woolf[4], in a study of a quenched blast furnace, noted that lime was the last slag constituent

## TABLE 1

### Compositions of Two Slag Samples from the Tuyere Level

Sample 1. Bulk Chemical Analysis

| $SiO_2$ | FeO | MnO | CaO | MgO | $Al_2O_3$ | C | S |
|---|---|---|---|---|---|---|---|
| 50.2 | 20.7 | 8.86 | 6.55 | 2.20 | 0.54 | 2.8 | 0.2 |

Sample 2. Electron Microprobe Analysis of Slag Particles

|  | $SiO_2$ | $Al_2O_3$ | CaO | MgO | FeO |
|---|---|---|---|---|---|
| Particle 1 | 62 | 17 | 3.8 | 9 | 0.2 |
| Particle 2 | 64 | 27 | 9.4 | trace | trace |
| Particle 3 | 56 | 49 | 10 | trace | trace |
| Average | 61 | 30 | 8 | 3 | 0.1 |

incorporated. Based on these observations, it is clear that, above the tuyere level, the lime and other fluxes are largely uncombined with the other slag constituents.

We can speculate on the means by which coke ash and the limestone flux combine. One can watch coke being consumed in front of a tuyere and see very few slag drops washing its surface, yet no accumulation of ash is apparent. This observation was confirmed by examination of coke lumps recovered from the coke bed near a tuyere. A very porous layer of reddish-brown ash only a fraction of a millimeter thick covered the surface. Under that was a layer of partially oxidized coke 2 mm thick, and the interior of the coke lump was visibly unaltered. Evidently the ash is blown off by the force of the air blast and strikes other surfaces in the coke bed. Some of those surfaces are CaO; these encounters produce liquid slag which subsequently reacts with more ash.

Meister[2] states that insufficient limestone causes sintering together of coke lumps, and also viscous slags. Those undesirable things must normally occur to some extent, because of the time needed for lime to combine. So earlier slag formation is probably desirable; for example, it would be expected to promote carbon pickup[5]. The preceeding observations suggest that increasing the specific surface area of CaO would cause earlier slag formation. That could be achieved by charging smaller limestone lumps or using limestone that decrepitates.

Further evidence regarding the progress of slag formation was provided by a sample of metal and slag obtained by inserting a water-cooled sampling trough through the safety tuyere of a General Motors cupola using basic slag, hot blast

*References pp. 186–189*

## TABLE 2
## Compositions of Tapped Slag and Slag 0.4m below the Tuyeres (mass %)

(A) Standard Chemical Analyses of Bulk Slag Samples

|  | CaO | SiO$_2$ | MgO | Al$_2$O$_3$ | FeO | MnO | F | S |
|---|---|---|---|---|---|---|---|---|
| **Tapped Slag:** | | | | | | | | |
| 5 min after | 46.3 | 25.8 | 13.1 | 4.90 | 1.08 | 1.03 | 5.38 | 0.88 |
| 10 min after | 45.2 | 24.1 | 13.4 | 4.65 | 1.06 | 1.78 | 5.45 | 0.78 |
| Average | 45.8 | 25 | 13.3 | 4.73 | 1.07 | 1.4 | 5.42 | 0.83 |
| **Sampler:** | | | | | | | | |
| point | 44.2 | 9.48 | 8.69 | 3.95 | 25.1 | 1.85 | 4.49 | 0.44 |
| composite | 27.8 | 11.6 | 14.0 | 5.70 | 21.5 | 2.60 | 3.14 | 0.69 |

(B) Equivalent Compounds (normalized to 100%)

|  | CaO | CaF$_2$ | CaS | SiO$_2$ | MgO | Al$_2$O$_3$ | FeO | MnO |
|---|---|---|---|---|---|---|---|---|
| Tapped slag (avg.) | 38.4 | 11.7 | 1.97 | 26.4 | 14.0 | 4.99 | 1.13 | 1.48 |
| Sampler point | 38.4 | 9.5 | 1.03 | 9.87 | 9.05 | 4.1 | 26.1 | 1.93 |
| Sampler composite | 25.7 | 7.6 | 1.8 | 13.6 | 16.4 | 6.7 | 25.2 | 3.1 |

(C) Electron Microprobe Analyses of Two Slag Particles

|  | Ca as CaO | SiO$_2$ | FeO | Al$_2$O$_3$ | MgO | S |
|---|---|---|---|---|---|---|
|  | 47 | 7.1 | 18 | 5.7 | 2.6 | 0.40 |
|  | 51 | 6.6 | 18 | 3.8 | 3.9 | 0.45 |
| Average | 49 | 6.9 | 18 | 4.8 | 3.3 | 0.43 |

and O$_2$ enrichment. The sampling height was 0.4 m below the tuyere centerline, a small but unknown distance above the slag layer.

In Table 2 the composition of the sampled slag is compared with that of slag tapped 5 and 10 minutes after the safety tuyere sample. The composite sample in Table 2(A) represents the average slag composition in the sampler. Three small, physically separate, samples were also analyzed for an indication of compositional uniformity. One of these was on the point of the sampler (2(A)), and the other two were separate particles; see Table 2(C). Table 2(B) simply restates 2(A) in terms of probable compounds.

Table 2 shows that slag formation is well advanced 0.4 m below the tuyere. The slag contains materials from all the major sources and $CaF_2$, another flux charged to this cupola. However, slag composition still varies from point to point, as indicated by the differences in CaO and MgO. Also, the overall slag composition is quite different from that of the tapped slag. The slag 0.4 m below the tuyeres contains much greater concentrations of FeO and smaller concentrations of $SiO_2$ and S than does the tapped slag. The implications of these differences are discussed in the next section.

### Reactions in the Slag Layer

**Si Oxidation**—The FeO and $SiO_2$ concentrations in Table 2 indicate that the reaction

$$2FeO + \underline{Si} = 2Fe + SiO_2 \tag{1}$$

occurs in the slag layer. The ratio $\Delta(\%FeO)/\Delta(\%SiO_2)$ between the composite sample and tapped slags is $(25.2-1.1)/(13.6-26.4) = -1.9$, which is 80% of the -2.4 stoichiometric value. MnO should react similarly to FeO, and the ratio $\Delta(\%FeO+\%MnO)/\Delta(\%SiO_2) = -2.2$. Thus Equation 1 appears qualitatively correct.

Only one sample was taken 0.4 m below the tuyeres, due to experimental difficulties. A test of its representativeness is therefore desirable. A mass balance for metallic Si provides it. For that calculation the data in Table 3 are needed to calculate the slag/metal mass ratio (0.088, based on CaO) and mean %Si in the cupola charge (1.23%). Use again 26.4% $SiO_2$ in the tapped slag and 13.6% at the top of the slag layer inside the cupola. The expected Si loss is therefore 0.53%, so the tapped iron should contain 0.70% Si. Typical iron tapped from that cupola contains 0.5% Si, which is more than the mean 0.27% Si for the three samples of Table 3. (Tapped %Si is quite erratic.) Comparison of the calculated and typical values of tapped %Si suggests the typical $SiO_2$ concentration 0.4 m below the tuyeres may be less than in Table 2, but that the Table 2 values are qualitatively correct.

Si loss is reputedly greater with basic slags than with acid slags[6]. That observation is consistent with the differences in FeO contents of tapped slags: 1% for basic cupolas and up to 5% for acid cupolas. Also, the activity coefficients of FeO and $SiO_2$ in basic slags favor Si oxidation (Equation 1): larger for FeO and smaller for $SiO_2$ than in acid slags. However, a mass balance predicts little difference in Si loss. Assuming the same amount of iron oxide is produced with either acid or basic slag, the difference in absolute Si loss by Equation 1 is 0.2 times the difference in absolute amount of FeO remaining in the two kinds of slags. The slag/metal ratio is ~ 0.05 for acid cupolas and ≤ 0.09 for basic cupolas. Using these values and the 1% and 5% FeO contents mentioned above, the Si loss with basic slag is expected to be $(5 \cdot 0.05 - 9 \cdot 0.01)0.2 = 0.03$ percentage points greater than with acid slag. This question needs further study.

The doubt about a significant effect of slag composition on Si loss is partially supported by an earlier analysis of Si loss in normal operation of commercial

## TABLE 3
### Cupola Charge and Composition of Tapped Iron (Mass %)

| Cupola Charge | | Tapped Iron Composition | | | |
|---|---|---|---|---|---|
| Charge constituent | Mass (kg) | Element | Minutes after safety tuyere sample | | |
| | | | -5 | 5 | 10 |
| Steel | 1818 | C | 3.88 | 3.85 | 3.72 |
| Nodular iron remelt* | 1790 | Si | 0.39 | 0.24 | 0.18 |
| FeMn briquettes | 27 | Mn | 0.72 | 0.72 | 0.70 |
| Coke | 590 | S | 0.020 | 0.018 | 0.021 |
| Limestone | 173 | P | 0.004 | 0.008 | 0.012 |
| Dolomite | 164 | | | | |
| Fluorspar | 59 | | | | |

*Nominally 3.8% C, 2.50% Si

cupolas[7]. The supporting evidence was that the average loss was 0.6% for both acid and basic cupolas, even though nearly twice as much Si was charged to the acid cupolas. However, closer examination revealed the absolute loss was linearly proportional to the amount of Si charged. This reasonably suggests additional Si oxidation in the cupola stack above the slag layer. This explanation is supported by the observation that the fraction oxidized in an incremental Si addition to the charge was the same for acid and basic cupolas.

**Emulsified Metal**—Tapped cupola slags contain numerous suspended metallic drops, spherical to nearly spherical in shape and 1-20 $\mu$m diameter. Compositions of eight such drops in two basic slags were measured by electron microprobe (Table 4) to gain information on reactions in the slag layer.

The Ca contents in Table 4 are due to the slag surrounding and underlying the drops; much of the Si and S should also be attributed to the slag. Table 5 gives compositions of the corresponding tapped slags and bulk iron. Based on previous analyses of such slags[8], the first slag in Table 5 is heterogeneous, consisting of liquid and $2CaO \cdot SiO_2$ phases.

Several features of Tables 4 and 5 are noteworthy. One is that the contents of C, Mn and Si are smaller in the drops than in the bulk tapped iron. This indicates that oxygen potential of the drops is greater than in the bulk iron. Another feature is that, for the drops in a given slag, %C varies more than %Mn does. This indicates that Mn/MnO equilibrium is easier to establish than C/CO equilibrium.

## TABLE 4

### Electron Microprobe Analyses of Emulsified Metal in Two Basic Cupola Slags (mass %)

|       | Slag 1 | | | | | Slag 2 | | | | |
| --- | --- | --- | --- | --- | --- | --- | --- | --- | --- | --- |
| Drop: | 1 | 2 | 3 | 4 | Aver. | 1 | 2 | 3 | 4 | Aver. |
| Fe | 98.9 | 98.8 | 102.3 | 101.4 | 100.4 | 95.9 | 98.9 | 94.3 | 96.2 | 96.3 |
| Si | 0.02 | 0.03 | 0.02 | 0.01 | 0.02 | 0.06 | 0.09 | 0.06 | 0.06 | 0.07 |
| Mn | 1.07 | 0.18 | 0.38 | 0.28 | 0.48 | 0.23 | 0.24 | 0.25 | 0.30 | 0.26 |
| C | 2.32 | 2.88 | 1.11 | 0.97 | 1.82 | 2.27 | 0.65 | 3.14 | 2.75 | 2.20 |
| S | 0.02 | 0.05 | 0.09 | 0.11 | 0.07 | 0.05 | 0.01 | 0.03 | 0.02 | 0.03 |
| Ca | 0.44 | 0.56 | 0.40 | 0.38 | 0.45 | 0.76 | 0.66 | 0.58 | 0.76 | 0.67 |
| Sum: | 102.7 | 102.5 | 104.2 | 103.1 | | 99.1 | 100.6 | 98.8 | 100.1 | |

## TABLE 5

### Compositions of Tapped Slag and Iron Corresponding to Emulsified Metal Compositions in Table 4

| | Tapped Slag | | | | | | | | Tapped Iron* | | | | |
| --- | --- | --- | --- | --- | --- | --- | --- | --- | --- | --- | --- | --- | --- |
| | CaO | CaF$_2$ | CaS | SiO$_2$ | MgO | FeO | MnO | Al$_2$O$_3$ | C | Si | Mn | S | T(°C) |
| Slag 1 | 38.2 | 12.0 | 1.88 | 25.7 | 14.3 | 1.13 | 1.90 | 4.96 | 3.72 | 0.18 | 0.70 | 0.021 | 1432 |
| Slag 2 | 47.5 | 0.0 | 2.74 | 38.7 | 2.4 | 0.31 | 0.46 | 7.7 | 3.90 | 1.50 | 0.44 | 0.052 | 1436 |

*Iron composition for slag 2 is an average for the period.

To quantify these points, oxygen potentials were calculated for C/CO and Mn/MnO equilibria with slag 2 (Table 6). Similar calculations for slag 1 were not feasible because of the uncertain MnO activity in heterogeneous slag. Si/SiO$_2$ equilibrium was also calculated for bulk iron and slag 2, but not for the drops, because relative error in the small %Si values is probably large. Fe/FeO equilibrium was not calculated because the FeO percentage in slag containing metal is uncertain. The estimate $\gamma_{MnO} = 1.35$ was based on reference 9; other data were from reference 8[a].

---

a. Erratum: $\ln(\gamma_C) = (5.275 - 0.00545\ T + 1.266 \cdot 10^{-6}\ T^2) + (12.43 - 0.00138\ T)N_C + (5.53 + 27.1\ N_C)N_{Si}$

## TABLE 6
### Equilibrium log $P_{O_2}$ Values for Iron and Slag 2

|  | C/CO (1 atm) | Mn/MnO | Si/SiO$_2$ |
|---|---|---|---|
| Drop 1 | -14.56 | -15.02 | - |
| 2 | -12.93 | -15.25 | - |
| 3 | -15.12 | -14.97 | - |
| 4 | -14.88 | -15.18 | - |
| Average drop | -14.52 | -15.12 | - |
| Tapped iron | -15.71 | -15.37 | -16.15 |

Table 6 shows that the drops are more oxidized than the tapped iron and that the oxygen potential for C/CO is greater than for Mn/MnO in the drops, while the order is reversed in the bulk metal. The above evidence is insufficient to determine the origin of the drops. Brief descriptions of two reasonable mechanisms, however, may provide useful insights.

The first mechanism is reduction of Fe and Mn from slag in contact with coke. We find coke in the slag layer is not penetrated by slag, suggesting the overall reaction would consist of coupled two-phase reactions: $C+CO_2=2CO$ at the coke-gas interface, and $CO+FeO=Fe+CO_2$, for example, at the gas-slag interface. Carburization would follow, presumably by direct contact between iron and coke. Carburization by $2CO=CO_2+\underline{C}$ is unlikely because a more consistent relation than found between C/CO and Mn/MnO equilibria would be expected from the coupled reaction mechanism.

The second mechanism is disintegration of a large drop, high in C, into the 1-20 µm drops. This emulsification has been shown to occur when metal and slag are initially far from equilibrium[10,11]. Interfacial energy then approaches zero, allowing nucleation of CO bubbles and, as a consequence of agitation, separation of drops as small as 1µm[11]. Variable, but extensive, decarburization of iron was found when using slags containing FeO$_x$, along with oxidation of other metallic elements[11].

This second mechanism accounts for observations adequately. When an iron drop enters the slag layer, vigorous reactions lead to partial emulsification. The emulsified metal reacts quickly with the FeO-rich slag in the upper part of the slag layer. These small drops do not equilibrate with the more reducing slag nearer the taphole because of their limited contact with coke. The larger unemulsified portion

### TABLE 7

Conventional Chemical Analyses of Metal from 4 Areas in a Sample from 0.4 m below the Tuyere Centerline (mass %)

|         | C    | S     | Si   | Mn   |
|---------|------|-------|------|------|
|         | 0.88 | 0.023 | —    | —    |
|         | 0.93 | 0.026 | 0.01 | 0.04 |
|         | 1.47 | 0.035 | 0.01 | 0.04 |
|         | 1.30 | 0.030 | —    | —    |
| Average | 1.15 | 0.029 | 0.01 | 0.04 |

of the initial iron drop is less extensively oxidized in the upper part of the slag layer, and it can recover C, Mn and Si while in contact with coke during its descent.

**Nonmixing of Metallic Charge Materials**—Composition of the metal recovered from the safety tuyere was measured by conventional chemical analysis (Table 7) and by electron microprobe. Microprobe determination of Si and Mn at 11 places in the sample found <0.1% Si for all and $0.05 \leq \%Mn \leq 0.09$, in essential agreement with Table 7. The Si, Mn and C contents show the sampled metal was derived from the steel half of the charge. This implies that the individual charge constitutents melt and descend independently to the slag layer. Mixing occurs in the slag layer or in the pool of liquid metal under the slag layer.

Slag sample 1 of Table 1, taken from the coke bed, provides another example of nonmixing. It contained several metallic spheres analyzing 40% Si, evidently 50% ferrosilicon alloy.

This finding of nonmixing obviously complicates analysis of metal reactions with slag, coke and gas. The extent of metal mixing in the slag layer is unknown. Some mixing could be expected in the slag layer because slag reduces the metal drops' velocity, leading to a greater volume fraction of metal and greater probability of collisions between drops. The nonmixing finding influenced the model of carbon pickup described in the next section.

## CARBON PICKUP

**Description of the Model**—An earlier model of carbon pickup from coke by the metal[12] has been refined. The earlier model will first be described, for most

*References pp. 186–189*

of its features are retained in the present model.

The models assume the rate of carbon pickup, d%C/dt, is first order in the quantity (1-$a_C$), where $a_C$ is carbon activity relative to graphite. The term carbon equivalent, which is familiar to foundrymen and is proportional[12] to $a_C$, was substituted for carbon activity. The most common expression for carbon equivalent, CE, is

$$CE = \%C + 0.30 \cdot \%Si \tag{2}$$

where the coefficient 0.30 accounts for the effect of Si on C solubility. To use CE as a substitute for activity in unsaturated melts, it was necessary to replace this coefficient with the derivative -d%C/d%Si at constant carbon activity. An expression for this derivative was obtained from thermochemical data[12].

The earlier model assumed the iron and steel charge materials mix immediately upon melting, thus forming a homogeneous metallic phase. Since the time-temperature history of the metal within the cupola is not accurately known, the cupola was considered isothermal (temperature of tapped iron), and a single rate parameter P was substituted for the product of time and the conventional rate constant. According to the model, P is defined as

$$e^{-P} = \frac{\%C_{sat} - (CE)_f}{\%C_{sat} - (CE)_i} \tag{3}$$

where the subscripts sat, f and i mean saturated, final and initial, and

$$\%C_{sat} = 1.3 + 0.00257 \cdot T(°C) \tag{4}$$

for binary Fe-C[13]. (From the definition of CE, $\%C_{sat}=(CE)_{sat}$.)

The refinement to the model, suggested by the finding that the melted iron and steel streams do not mix appreciably in the cupola, was to apply the model separately to the streams of iron and steel. This required determining individual P values for iron and steel. These were derived from analysis of data from various cupola studies by the British Cast Iron Research Association (BCIRA), using a 0.76 m diameter cupola.

Figure 1 shows the dependence of $P_{iron}$ on temperature. The data was from cupola runs in which the charge was pig iron and, in some cases, ferrosilicon. Runs with both iron and steel in the charge were also analyzed, using the previously determined $P_{iron}$ values, to determine $P_{steel}$ values as a function of temperature (Figure 2).

For a given temperature, $P_{steel} \simeq 2P_{iron}$. Two possible explanations are suggested. One is that steel is known to reach a higher temperature because it remains solid to a lower height in the cupola[14]. Thus a molten steel drop is considered to dissolve more coke because it is hotter, even though it does not fall as far through the coke bed as a cast iron drop. The other explanation is that the steel portion of the charge contains less sulfur than cast iron or pig iron. Sulfur concentrates on the liquid surface and has been found to retard carbon pickup[15].

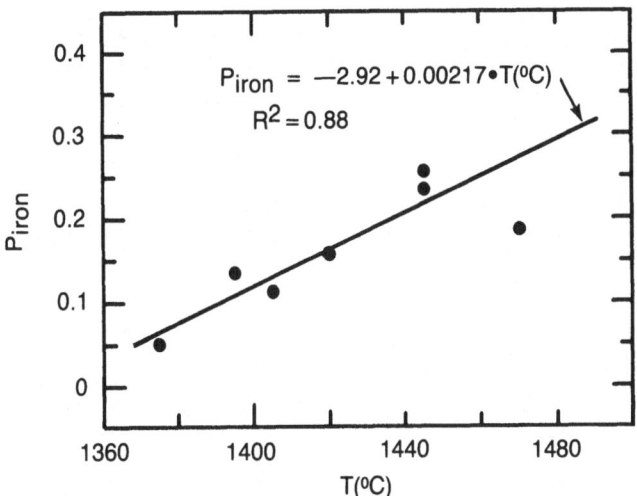

Figure 1. $P_{iron}$ values derived from BCIRA data using pig iron charges.

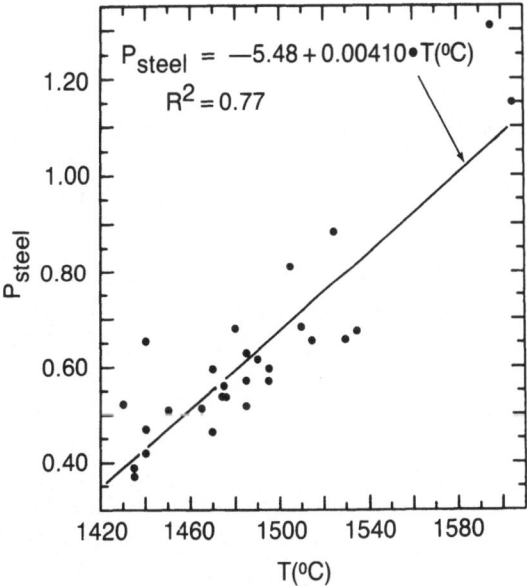

Figure 2. $P_{steel}$ values derived from BCIRA data with pig iron + steel charges and the $P_{iron}$ values determined previously.

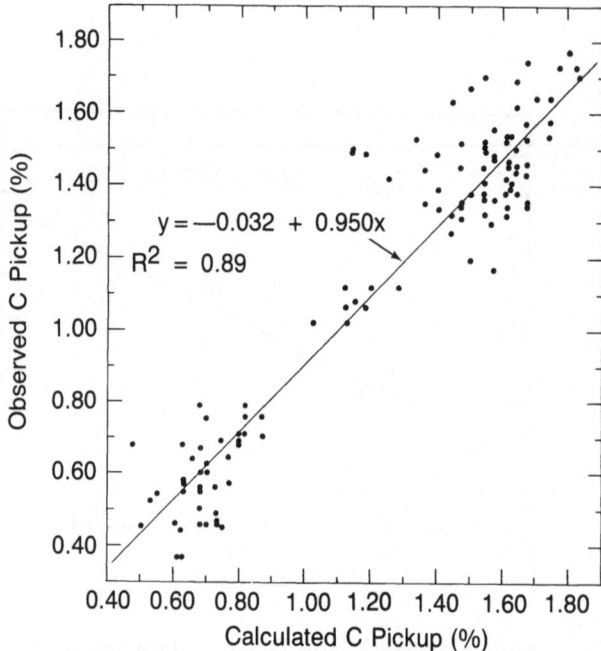

**Figure 3.** Comparison between carbon pickup observed and predicted by the model.

The inverse calculation of %C tapped was then made using the determined P values. The agreement was very good. The correlation coefficient between actual and calculated %C values was 0.92 for operation with pig iron only and 0.98 for charges of iron + steel.

**Testing the Model**—The BCIRA cupola was small, 0.76 m diameter, and the experimental conditions were very carefully controlled. Because of our relative ignorance of the factors governing carbon pickup in a cupola, the application of the P values determined from BCIRA data to large commercial cupolas cannot be assumed valid. To assess this applicability, the linear regression expressions for $P_{iron}$ and $P_{steel}$ as functions of temperature were applied directly to 128 observations on a General Motors 2.9 m diameter gray iron cupola. It, like the BCIRA cupola, used acid slags. The agreement between actual and calculated carbon pickup was again quite good (Figure 3), the correlation coefficient R being 0.94. The good agreement tends to confirm the validity of the assumptions, particularly that the tap temperature is a useful measure of the unknown time-temperature history within the cupola.

**A Limitation of the Model**—The parallel-stream model is potentially useful for predicting carbon pickup and for studying the effects of variables such as slag composition, sulfur content of the charged metal, cupola well depth, etc.

The principal limitation of the model is that tapped iron temperature is used in the calculation. For use in predicting carbon pickup, another model is needed to predict tapped iron temperature. However, the need for tapping temperature is no disadvantage when analyzing data to determine the effects of variables on carbon pickup.

## HEAT LOSSES AND THEIR EFFECT ON EFFICIENCY

**Introduction**—Energy losses are of three kinds: 1) conduction through the walls, 2) sensible heat of the offgas, and 3) latent heat of CO combustion of the offgas. An approximate heat balance can be calculated from extensive data on a 2.9 m diameter General Motors cupola without refractory except in the well (Table 8). Heats of metal oxidation and decomposition of limestone are omitted. "Other losses" in Table 8 are equivalent to 10 MW, and represent mainly heat conduction to the tuyeres and walls of the cupola, especially the bare steel water-cooled shell.

Table 8 is useful for identifying methods for increasing cupola efficiency. Increasing the $CO_2/CO$ ratio in the offgas would greatly increase combustion energy. Increasing the hydraulic radius of the coke bed, by increasing coke size, was shown to increase that ratio by reducing the surface available for the Boudouard reaction[16]. Any measure that allows shortening the height of the coke bed would help for the same reason. Increasing blast temperature is economically effective when the blast heating is done recuperatively. On the output side, the sensible heat in the offgas is small. Some shorter cupolas have 500°C offgas temperatures, however, and in that case the sensible heat would be an appreciable 15050 J. Conduction losses, included in "other losses," will be discussed in some detail.

**Conduction Losses and Their Effect**—Losses to the tuyeres and shell of cupolas similar to that of Table 8 were extensively measured[17]. From those measurements the total losses to the shell and tuyeres of the example cupola are about 5.5 MW, of which 80-85% are losses to the shell. That is about half the 10 MW equivalent of "other losses."

Losses to the shell could be greatly reduced by lining the shell. The upper limit on savings resulting from lining the shell is estimated by assuming the shell losses are entirely eliminated and that the energy saved is directed entirely into the iron. For our example cupola, these assumptions lead to a prediction that 23% more iron would be melted with the original amount of coke, or the coke/iron ratio could be reduced 19%.

Data from a 2.13 m diameter cupola before and after lining the shell[18] suggests significant benefits can be attained. Heat losses to the cooling water were initially 58% of the iron enthalpy. After installing a refractory shell lining, these heat losses decreased 73%. Of the energy saved, 64% was absorbed by the iron. (Iron melting rate increased 29% and the coke/iron ratio decreased 27%.) Applying the same 73% and 64% factors to the example cupola, lining the shell would increase melting rate 11% and decrease the coke/iron ratio 10%.

*References pp. 186–189*

## TABLE 8
## Cupola Heat Balance

*Operating Conditions*

| | |
|---|---|
| Air blast temperature | 400°C |
| $O_2$ in enriched blast | 23 mole % |
| Blast rate | 525 m³/min at STP |
| Offgas temperature | 175-250°C |
| Distance, tuyeres to gas offtake | 8 m |
| Offgas molar composition | 19% CO |
| | 11.1% $CO_2$ (calculated) |
| | 1% $H_2$ |

*Enthalpies per gmole Offgas, Relative to 25° C (J)*

Inputs
| | |
|---|---|
| Combustion to 19% CO, 11.1% $CO_2$ | 66260 |
| Blast | <u>10160</u> |
| | 76420 |

Outputs
| | |
|---|---|
| Iron | 44380 |
| Slag | 4300 |
| Offgas sensible heat at 175°C | 4550 |
| Other losses, by difference | 23190 |

**Heat Flux to the Shell**—The loss in cupola efficiency due to conduction losses leads to consideration of methods for reducing heat loss to the cupola shell. Durability of refractory is a concern, for temperatures are high immediately above the tuyeres, and these areas require weekly maintenance. Extending the tuyeres into the cupola bed is a standard practice which moves the region of greatest temperature away from the shell. The "reverse taper" design in which the shell tapers 5° from vertical (Figure 4), similar to a blast furnace bosh, has the same purpose as extending tuyeres. For all these methods, information on heat flux to the cupola shell would be useful. Such data was measured by a calorimetric method.

The water flow rate down the shell, per unit circumference, was first measured using a special container. Then the variation of water temperature with height on the cupola was measured at 100 mm intervals with a copper-Constantan thermocouple embedded in a copper slab. Heat flux was calculated from the

# CHEMICAL PROCESSES AND HEAT LOSS IN CUPOLAS

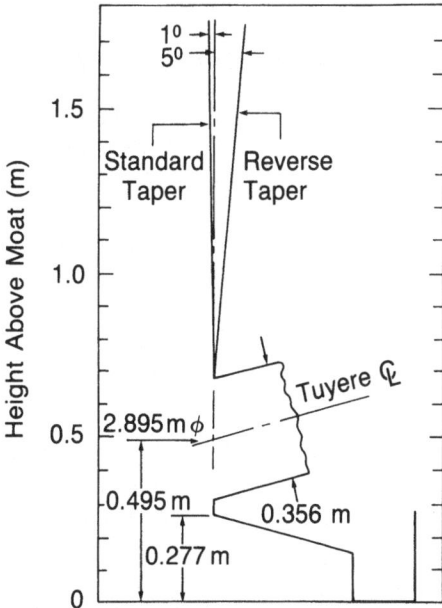

Figure 4. Dimensions of cupolas with standard and reverse tapers.

derivative of temperature versus height. Least square interpolation by the cubic spline and incremental polynomial methods gave nearly the same derivatives.

Figures 5 and 6 compare heat fluxes on reverse taper and standard taper shells (Figure 4), respectively. The generally smaller fluxes with the reverse taper shell indicates the effect of increasing the distance between the shell and the region of maximum temperature. Both figures show that maximum heat flux occurs at or a short distance above the tuyeres.

Figure 6 shows a second maximum flux occurs at a point beyond 1.3 m above the moat, or 0.8 m above the tuyere center. If the first maximum is attributed to radiant heat transfer to the shell, a reasonable estimate of the coke bed temperature is obtained. Assuming emissivities of coke and steel shell ($\epsilon_s$) are 1 and 0.5, respectively, and neglecting the shell temperature, the relation

$$q = \sigma \epsilon_s T^4 \tag{5}$$

where $\sigma$ is the radiation constant, gives an estimate of coke bed temperature, T. For heat fluxes q of 300-500 kW, coke bed temperatures of 1530-1870°C result. The average of those temperatures is the 1700°C value measured by many investigators, e.g.[14]. The second maximum is attributed to convective transfer from the gas, which is injected radially through the tuyere and evidently returns to the shell $\geq$ 0.8 m higher. If a second maximum occurs in the reverse taper shell, it is at a higher level, perhaps because of the enlarging shell diameter. At the higher level, gas temperature is less, and the rate of convective heat transfer would be correspondingly reduced.

*References pp. 186–189*

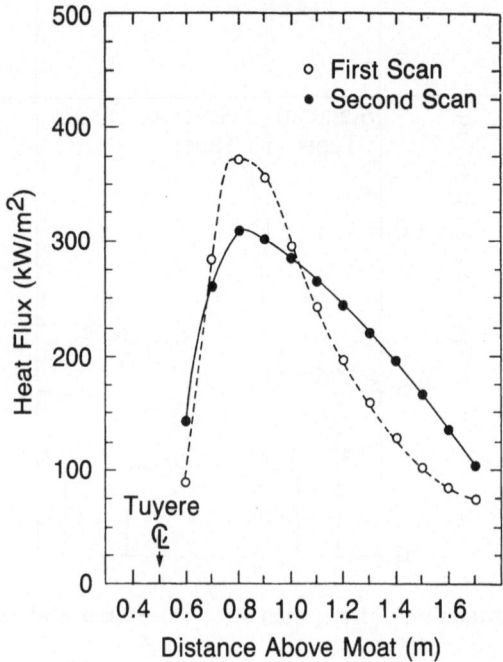

**Figure 5.** Heat fluxes through a reverse taper cupola shell.

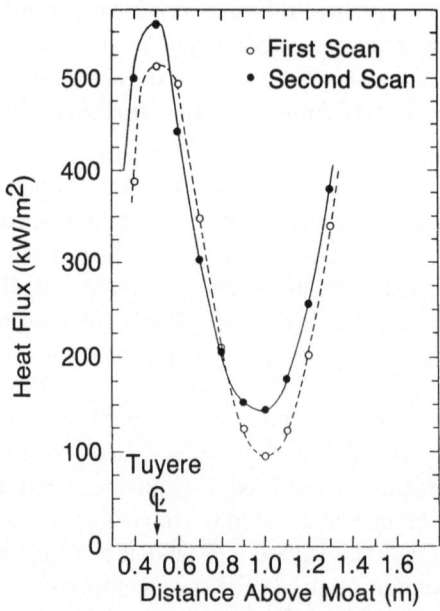

**Figure 6.** Heat fluxes through a standard taper cupola shell.

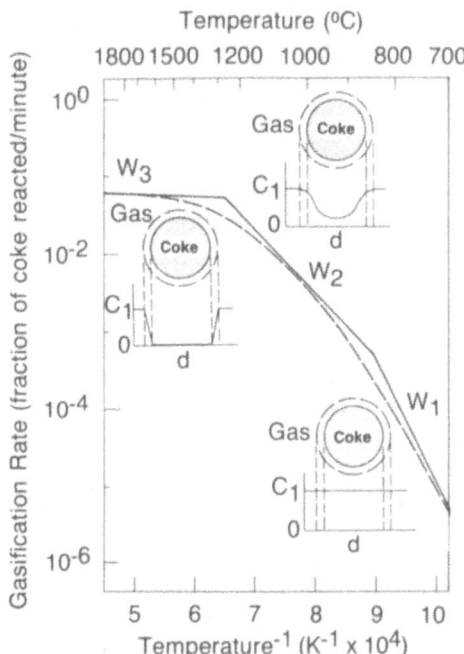

Figure 7. Temperature ranges for gasification rate control by three processes.

## COKE GASIFICATION

**Introduction**—The second topic relating to thermal efficiency is modelling the gasification of coke, $CO_2 + C = 2CO$, which is alternatively called the Boudouard or solution loss reaction. The reaction consumes carbon and is endothermic, and therefore is undesirable. The $CO_2$ is the main combustion product until all the $O_2$ reacts, and then gasification begins[19-22].

The effect of temperature on the reaction kinetics is illustrated in Figure 7. At low temperatures, $CO_2$ diffuses throughout the coke lump and reacts slowly, i.e. pore diffusion is faster than reaction. Reaction rate, $W_1$, is determined by the intrinsic rate constant, $k_v$, which increases rapidly with temperature. At somewhat higher temperatures the $CO_2$ reacts before penetrating to the center of the lump, and the reaction rate $W_2$ depends on both $k_v$ and effective diffusivity through the porous coke. In this pore diffusion regime, reaction rate is still quite temperature sensitive. At still higher temperatures, $CO_2$ reacts instantly upon reaching the coke surface, and reaction rate is determined by mass transfer through a boundary layer between the bulk gas and the coke surface. In this interphase mass transfer regime the reaction rate, $W_3$, is insensitive to temperature.

In previous modelling of gasification rate[16], Katz considered only the interphase mass transfer regime and found it accounted for the literature data. Particularly, the use of large coke was shown to be effective, for that reduces the external surface area available for reaction. Gasification appeared to cease at 1400°C, corresponding to the onset of pore diffusion control[16].

The previous modelling work[16] will be extended here to reaction rates, $W_2$, controlled jointly by pore diffusion and chemical reaction. The object is to determine more precisely whether significant gasification occurs by that mechanism. If not, then the various coke reactivity tests are useless for cupola coke.

An indication that gasification in the pore diffusion regime might be significant came from unpublished General Motors data from a cupola for which sodium aluminate ($Na_2O \cdot Al_2O_3$) was a charge constituent. Soda ($Na_2O$) is known to catalyze the gasification reaction, thereby increasing the intrinsic rate constant $k_v$. Soda content of the charge was 0.75-1.14% of the metal weight.

Figure 8 shows, for comparison, the normal coke percentage, %C of tapped iron, and moles of C consumed per mole of $O_2$ added, determined from gas composition 3 m above the tuyeres (corrected for limestone decomposition). Figure 9 shows one effect of adding soda. Carbon content of the iron decreased gradually, indicating decreasing coke bed height due to coke being consumed faster than being replenished from above. Figure 10 shows an instance in which extra coke was added to counter that effect. The decline of %C was less, but the $C/O_2$ ratio increased dramatically. In both cases soda increased gasification, presumably by increasing the intrinsic rate constant.

The soda effect showed only that gasification under mixed control (rate $W_2$) can be significant if $k_v$ is large enough. But the $k_v$ values with soda in the charge are unknown, so they cannot be compared with $k_v$ values for coke under normal cupola conditions. In a later section the importance of the intrinsic rate constant is estimated by varying $k_v$ over the extreme range of values likely to be found and calculating the corresponding coke consumptions by gasification.

**Gasification Rate Model**—The present model considers the series processes of diffusion of $CO_2$ to the outer coke surface, diffusion into the pores, and reaction within the pores to produce CO. Gasification by water vapor, $C+H_2O=CO+H_2$, is not considered. The principal quantity calculated is W ($=W_1+W_2+W_3$), the fraction of carbon reacted per minute; the expression for W is derived in reference 16. The variables used in the model are defined in Table 9.

The coke shape factor $\alpha$ was determined from measurements of surface areas and volumes[23]. $D_b$ and $\mu$ were based on Chapman-Enskog kinetic theory[24]. The values given by the expressions

$$\mu_{CO_2} = \exp(-12.48 + 0.6667 \ln(T)) \tag{16}$$

$$\mu_{N_2} = \exp(-12.45 + 0.6697 \ln(T)) \tag{17}$$

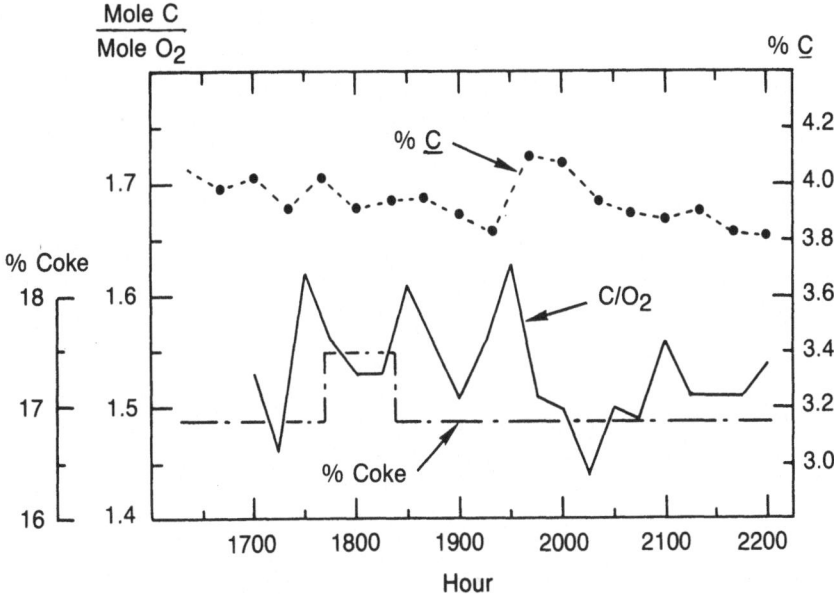

Figure 8. Normal cupola operating characteristics.

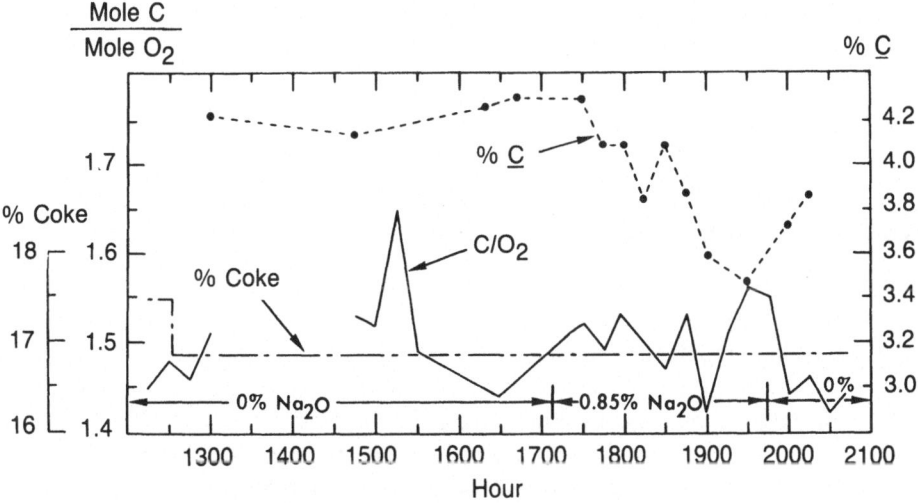

Figure 9. Cupola operating characteristics with $Na_2O$ in the charge and normal coke percentage.

*References pp. 186–189*

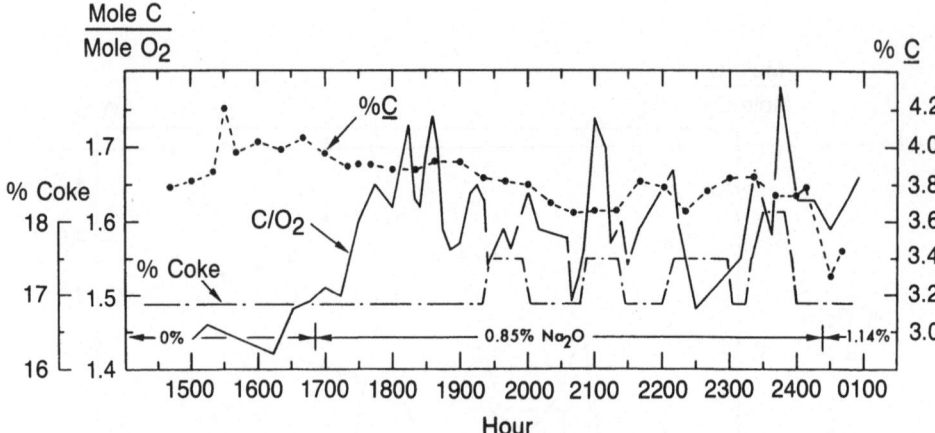

Figure 10. Cupola operating characteristics with $Na_2O$ and extra coke in the charge.

equalled those from Chapman-Enskog theory, and gas viscosity was approximated by the law of mixtures, Equation 18, using mole fraction concentrations.

$$\mu = X_{CO_2}\mu_{CO_2} + (1 - X_{CO_2})\mu_{N_2} \qquad (18)$$

The results of Aderibigbe[25] for blast furnace coke were used for the intrinsic rate constant. At the high temperatures of interest the Langmuir-Hinshelwood formulation reduces to only one kinetic constant. After converting $CO_2$ concentration from pressure to mole/volume units, $k_v$ is

$$k_v = 4.31 \cdot 10^{15} \exp(-36382/T). \qquad (19)$$

## Application of the Model—

**Testing the Model**—In Equation 6 (Table 9) the gasification rate at a point depends primarily on gas velocity, geometrical characteristics of the coke bed, $CO_2$ concentration, temperature, and $k_v$ and $D_e$ of the coke. To test the model, data from a 40 cm diameter cupola[26] were used for gas composition (Figure 11), temperature (Table 10) and gas velocity (91.6 m/min for STP and empty cupola). Temperatures shown in Table 10 are not for conditions identical to those for Figure 11. The gas compositions shown by solid lines in Figure 11 were obtained with 18% coke sized 5.0-6.5 cm and cold air blast containing 26 g $H_2O/Nm^3$ (3.2 mole %). Note that % CO reaches its final value at ~100 cm, where T≃1120°C.

The extent of reaction by a given amount of isobaric gas was calculated at each of successively greater height increments, along with the revised mass balance

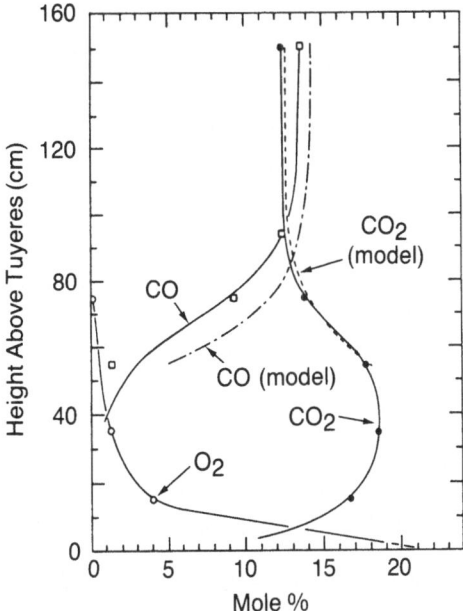

Figure 11. Gas composition versus distance above the tuyeres for a 40 cm diameter cold blast cupola[26].

to set the initial condition for the next higher increment. The model assumes that net CO formation begins only after all the $O_2$ reacts, but that is only approximately true. In order to avoid the region of coexisting $O_2$, the calculation started at 55 cm height, where %$CO_2$=18 and %CO=5. That is no disadvantage, because reaction rate at the estimated 1580°C temperature is controlled entirely by interphase mass transfer, so intrinsic reactivity is unimportant. The chosen 6 cm harmonic mean coke size fairly represents the actual coke size range. The broken curves in Figure 11 show the model represents the measured gas profile fairly well. Thus the model is likely to adequately represent the effect of a change in coke reactivity.

**Effect of Varying $k_v$**—The effect of coke reactivity on coke consumption was estimated by calculations identical to that for Figure 11 except for varying $k_v$ and coke size. Only the preexponential factor in Equation 19 was varied. Coke sizes of 5, 6 and 7 cm were used to compare the importance of $k_v$ for varying coke sizes.

The appropriate range of $k_v$ values is uncertain. Little reactivity data for foundry coke is available[27,28], so reactivity data for the more reactive blast furnace cokes was also consulted[29-37]. The conventional reactivity test, measuring weight loss in flowing $CO_2$, has numerous shortcomings. Apparatus design and test conditions vary greatly, so results of different experimenters cannot be usefully compared. Experimental scatter is sometimes a problem because of coke inhomogeneity[31]; and the results are affected significantly by pore structure and fraction of sample reacted[38-41], so the test does not measure a unique value of $k_v$ alone.

*References pp. 186-189*

## TABLE 9
## Variables and Equations in the Gasification Model

$$W = \left[\frac{k_v \eta}{1 + k_v \eta C_s D/(6h_m)}\right] C_1 : \frac{gC}{gC} \frac{1}{\min} \qquad (6)$$

$k_v$ intrinsic rate constant : $\dfrac{\text{mole C reacted}}{\text{mole C} \cdot \min} \dfrac{\text{cm}^3 \text{ gas}}{\text{mole CO}_2}$

$C_s$ C concentration in coke ($= 0.0775$) : $\dfrac{\text{mole C}}{\text{cm}^3 \text{ coke}}$

$C_1$ CO$_2$ concentration in gas : $\dfrac{\text{mole CO}_2}{\text{cm}^3 \text{ gas}}$

$$D = \text{harmonic mean coke diameter} = 1/\Sigma(\overline{w}_i/d_i) \quad : \text{cm} \qquad (7)$$

$\overline{w}_i$ = weight fraction of i-th size coke :(−)
$d_i$ = characteristic dimension of i-th size coke :cm

$$\eta = \text{pore diffusion effectiveness factor} = \frac{3}{\phi^2}[\phi \coth(\phi) - 1] \quad :(-) \qquad (8)$$

$$\phi = \frac{D}{2}\left[\frac{k_v C_s}{60 D_e}\right]^{\frac{1}{2}} \quad :(-) \qquad (9)$$

$$D_e = \text{effective diffusivity} = D_b \epsilon_p^3 \quad : \frac{\text{cm}^2}{\text{s}} \qquad (10)$$

$D_b$ binary gas interdiffusivity : $\dfrac{\text{cm}^2}{\text{s}}$

$\epsilon_p$ pore fraction in coke $= 0.5$ : $(-)$

$$h_m = \frac{\text{interphase mass}}{\text{transfer coefficient}} = \frac{D_b N_{Sh}}{D_h} : \frac{\text{mole C reacted}}{\min \bullet \text{cm}^2 \text{ coke}} \frac{\text{cm}^3 \text{ gas}}{\text{mole CO}_2} \qquad (11)$$

$$D_h = \text{hydraulic diameter of coke bed} = \frac{4\epsilon_b D\alpha}{6(1 - \epsilon_b)} \quad : \text{cm} \qquad (12)$$

$\alpha$ coke shape factor ($= 0.66$) : $(-)$
$\epsilon_b$ interparticle void fraction of bed ($= 0.5$) : $(-)$

$$N_{Sh} = j_d N_{Re} N_{Sc}^{1/3} \quad :(-) \qquad (13)$$

$$N_{Re} = \frac{D_h G}{\psi \mu \epsilon_b} \quad :(-) \qquad (14)$$

G mass flux : $\dfrac{\text{g gas}}{\text{s} \bullet \text{cm}^2 \text{ cupola}}$

$\psi = 0.95$ : $(-)$
$\mu$ gas viscosity : g/(cm $\bullet$ s)
N Schmidt number ($= 0.86$) : $(-)$

$$j_d = 0.23 N_{Re}^{-0.3} \psi \quad :(-) \qquad (15)$$

## TABLE 10

### Temperatures at Five Distances above Tuyere Level[26]

| Height (mm) | 150 | 350 | 550 | 750 | 950 | 1500 |
|---|---|---|---|---|---|---|
| T (°C) | 1780 | 1800 | 1580 | 1300 | 1190 | 800 |

## TABLE 11

### Effects of Coke Intrinsic Reactivity and Size on Gasification
$0.79N_2 + 0.21CO_2 + xC = 0.79N_2 + 2xCO + (0.21-x)CO_2$

| Harmonic mean coke $\phi$ (cm) | Intrinsic rate constant | Moles C reacted with $CO_2$ ($= x_f$) | Mole C reacted with $CO_2$ above 55 cm ($=x_f-x_i^*$) | $\dfrac{x_f - x_i}{0.21 + (x_f)_{std}^{**}}$ |
|---|---|---|---|---|
| 5 | 0.5 $k_v^{***}$ | 0.0774 | 0.0520 | 0.180 |
| 5 | 1.0 " | 0.0889 | 0.0635 | 0.220 |
| 5 | 1.5 " | 0.0959 | 0.0705 | 0.244 |
| 6 | 0.5 $k_v$ | 0.0694 | 0.0440 | 0.152 |
| 6 | 1.0 " | 0.0793 | 0.0539 | 0.186 |
| 6 | 1.5 " | 0.0854 | 0.0600 | 0.207 |
| 7 | 0.5 $k_v$ | 0.0634 | 0.0380 | 0.131 |
| 7 | 1.0 " | 0.0721 | 0.0467 | 0.161 |
| 7 | 1.5 " | 0.0774 | 0.0520 | 0.180 |

\* $x_i = 0.0254$
\*\* $(x_f)_{std} = 0.0793$, row 5 of table.
\*\*\* $k_v$ from Equation 19

Nevertheless, reactivity test data provides the only estimate of the variation of $k_v$ among cokes. After calculating first order rate constants from weight losses, the ratio of maximum/minimum rate constants was calculated for the range of cokes tested by each investigator. For the blast furnace cokes[29-37] the highest and lowest ratio was deleted, and the ratio for the rest averaged 2.3. The ratio for foundry cokes[27,28] was essentially the same. Accordingly, model calculations were made using the $k_v$ expression of Equation 19 multiplied by factors of 0.5, 1.0 and 1.5. The results are in Table 11.

As expected, the amount of C reacted with $CO_2$ over the range of calculation,

$x_f$-$x_i$, increases with increasing reactivity and decreasing coke size. To measure the significance, the $x_f$-$x_i$ values are divided by total coke consumption, $0.21+(x_f)_{std}$, for the standard condition represented by row 5 of Table 11. For 6 cm coke, the large reactivity change from $0.5k_v$ to $1.5k_v$ increases coke consumption 5.5%. Roughly 4% extra coke would have to be burned to replace the heat absorbed by the endothermic reaction, so the total coke penalty would approach 10%. The smaller change from $1.0k_v$ to $1.5k_v$ increases direct coke consumption only 2%, and the total increase would be less than 4%.

Two tests of the effects of coke reactivity in small cupolas led to only tentative conclusions, as could be expected from the the small changes in coke consumption estimated above. Draper et al[27] found higher reactivity coke gave 4% greater C/O combustion ratio and 4% greater melt rate, but the expected decrease in tapping temperature was not found. Boyer and Durand[28] found a convincing relation between high reactivity and low thermal yield (iron temperature), but the relationship was obscured by coke size effects in a later test series.

Table 11 also indicates that a given change in coke reactivity causes a greater increase in coke consumption with smaller coke sizes. That is another reason to prefer large coke in cupolas, but reducing the interphase mass transfer rate is a more important reason[16].

This analysis suggests that differences in coke reactivity are likely to produce only small changes in required coke percentage in cupolas. It also provides a basis for estimating the price premium that is justified for coke of low reactivity.

This model gives the most complete known description of gasification in a cupola, but its accuracy and utility could be improved in two ways. One is determining accurate temperature profiles for industrial-size cupolas, which would permit absolute calculations for them. The temperature profiles could be measured or derived from a model. The other advance is to include gasification by water vapor. That would be a major step toward quantitative description of the well-known detrimental effect of water vapor on cupola thermal efficiency.

## ACKNOWLEDGEMENT

The work described was done jointly with Dr. S. Katz, also of the GMR Metallurgy Department. Compositions were measured by members of the GMR Analytical Chemistry Dept: R.L. Passeno, H.E. Vergosen and M.P. Balogh (iron); D.W. Walton (slag); R.A. Waldo (electron microprobe).

## REFERENCES

1. *Cupola Handbook*, 4th ed, American Foundrymens' Soc., Des Plaines, IL, 1975, 290.
2. F. Meister, "Zur Schlackenbildung im Kupolofen" ("Slag Formation in Cupola Furnaces"), *Giessereitechnik*, *25*, No. 10 (1979), 296–300.

3. N.E. Rambush and G. B. Taylor, "A New Method of Investigating the Behaviour of Charge material in an Iron-Foundry Cupola and Some Results Obtained," *Foundry Trade Journal*, Nov. 8, 1945, 197–212.

4. J.P. Morris and P. L. Woolf, "Examination of an Experimental Iron Blast Furnace After Quenching with Nitrogen," *Report of Investigations 6217*, Bureau of Mines, United States Dept. of the Interior, 1963.

5. *Cupola Handbook*, 4th ed, American Foundrymens' Soc., Des Plaines, IL (1975), 474,478,489.

6. ibid, 487.

7. C.F. Landefeld and W.J. Peck, "The Relation between Silicon Loss and Metallic Silicon in the Cupola Charge," *American Foundrymens' Soc. Trans.*, *91* (1983), 1–6.

8. S.Katz and H.C. Rezeau, "The Cupola Desulfurization Process," *American Foundrymen's Society Transactions*, *87* (1979), 367–76.

9. E.W. Filer and L.S. Darken, "Equilibrium between Blast-Furnace Metal and Slag as Determined by Remelting," *Trans. AIME, Journal of Metals*, March 1952, 253–57.

10. H. Gaye and P.V. Riboud, "Oxidation Kinetics of Iron Alloy Drops in Oxidizing Slags," *Metallurgical Transactions*, *8B* (Sept. 1977), 409–15.

11. P.V. Riboud and L.D. Lucas, "Influence of Mass Transfer upon Surface Phenomena in Iron and Steelmaking," *Canadian Metallurgical Quarterly*, *20*, No. 2 (1981), 199–208.

12. S. Katz and C. F. Landefeld, "A Kinetic Model for Carbon Pickup in the Cupola: A Step beyond the Levi Equation," *American Foundrymen's Soc. Transactions*, *93* (1985), 209–14.

13. F. von Neumann, H. Schenck, and W. Patterson, *Giesserei*, *47* (1960), 25–32.

14. Hideo Nakae and Nobutaro Kayama, "Effect of Charging Manner of Metal Charge on Tapping Temperature in Cupola Melting," *Imono* (Castings), *41*, No. 7 (1969), 1–9. Translated by Berkeley Scientific Translation Service.

15. P.M. Coon and T.N. Blackman, "The Effect of Sulphur on Carbon Pick-up in an Acid-lined Cold-blast Cupola," *BCIRA Report 1560* (January 1984), 29–33.

16. S. Katz, "The Properties of Coke Affecting Cupola Performance," *American Foundrymen's Soc. Transactions*, *90* (1982), 825–33.

17. C.F. Landefeld, "Thermal Power Losses to the Cupola Shell and Tuyeres," *American Foundrymen's Society Transactions*, *93* (1985), 385–88.

18. F.M. Degner and F.T. Kaiser, "Increasing Cupola Energy Efficiency," *American Foundrymen's Soc. Transactions*, *88* (1980), 609–14.

19. S.P. Kinney, P.H. Royster and T.L. Joseph, U.S. Bureau of Mines Technical Paper 391, "Iron Blast-Furnace Reactions," U.S. Dept. of Commerce, 1927, 13.

20. J. Hiles and R.A. Mott, "The Mode of Combustion of Coke," *Fuel in Science and Practice*, *23* No. 6 (1944), 154–171.

21. W. Ruhenbeck and R. Gunther, "A Mathematical Model of the Cupola Operating Process," *Giessereiforschung*, *25* No. 2 (1973), 47–59.

22. M.W. Thring and R.H. Essenhigh, *Chemistry of Coal Utilization, Supplementary Volume*, ed H.H. Lowry, John Wiley & Sons, N.Y. (1963), 769.

23. C.F. Landefeld, "A Flow Method for Measuring Hydraulic Radius of a Coke Bed," *American Foundrymen's Soc. Transactions*, *91* (1983), 541–48.

24. R.B. Bird, W.E. Stewart and E.N. Lightfoot, *Transport Phenomena*, Wiley, NY, 1960, 22–26, 508–513.

25. D.A. Aderibigbe and J. Szekely, "Studies in Coke Reactivity: Part I–Reaction of Conventionally Produced Coke with $CO$-$CO_2$ Mixtures over Temperature Range 850-1000°C," *Ironmaking and Steelmaking*, 1981, No. 1, 11–19.

26. H. Nakae, E. Kato and N. Kayama, "Some Considerations on the Chemical Reactions in a Cupola Based on the Results of the Gas Analysis Obtained by a Mass Spectrometer," *Imono*, *41*, No. 2 (1969), 1–9.

27. A.B. Draper, T. DebRoy, K. Nyamekye, and M. Alam, "Cupola Output as a Function of Coke Reactivity," *American Foundrymen's Soc. Transactions*, *91*, (1983), 585–92.

28. A.F. Boyer and G. Durand, "Reactivity of Cokes. II.–Study of Foundry Cokes and Comparison with Their Practical Value," (in French) *Chimie & Industrie*, *82* No. 3 (Sept 1959), 309–28.

29. G. Heynert, W. Zischkale and E. Schurman, *Stahl und Eisen*, *80* (1960), 981–90.

30. H. Fujita, M. Hijiriyama and S. Nishida, "Gasification Reactivities of Optical Textures of Metallurgical Cokes," *FUEL*, *62* (August 1983), 875–79.

31. J.W. Patrick and H.C. Wilkinson, "Coke Reactivity," *The Coke Oven Manager's Yearbook*, (1980), 191–220.

32. K. Koba, K. Sakata and S. Ida, "Gasification Studies of Cokes from Coals. The effects of carbonization pressure on optical texture and porosity," *FUEL*, *60* (June 1981), 499–506.

33. D. Hunter, "A Summary of Factors Affecting Coke Reactivity," British Steel Corp. report.

34. M. Alam and T. DebRoy, "Reaction of High and Low Reactivity Coke with $CO$-$CO_2$ Mixtures," *Iron & Steel Soc. Transactions*, *5* (1984), 7–12.

35. R. Mutso and W. DuBroff, "Correlation of Reactivity and Electrical Resistivity of Coke," *FUEL*, *61* (March 1982), 305–6.

36. R.J. Gray, "Comparison of Coke Reactivity to $CO_2$ as Determined at U.S. Steel and Sumitomo," U.S. Steel Corp. report, October 1980.

37. K. Koba and S. Ida, "Gasification Reactivities of Metallurgical Cokes with Carbon Dioxide, Steam and Their Mixtures," *FUEL, 59* (January 1980), 59–63.

38. N.J. Desai and R.T. Yang, "Kinetics of High-Temperature Carbon Gasification Reaction," *AIChE Journal, 28* No. 2 (March 1982), 237–44.

39. D.A. Aderibigbe and J. Szekely, "Structural Changes and Reactivity of Conventional Coke and Form Coke during Reaction with $CO$-$CO_2$ Gas Mixtures at 1000°C," *Ironmaking and Steelmaking, 9* No. 3 (1982), 130–35.

40. W.G. May, R.H. Mueller and S.B. Sweetser, "Carbon-Steam Reaction Kinetics from Pilot Plant Data," *Industrial and Engineering Chemistry, 50* No. 9 (Sept 1958), 1289–96.

41. M. Sakawa, Y. Sakurai and Y. Hara, "Influence of Coal Characteristics on $CO_2$ Gasification," *FUEL, 61* (Aug 1982), 717–20.

## DISCUSSION

**Ned Rehder** *(University of Toronto)*

How was the heat flux through the cupola shell measured? Was the water temperature simply measured at various levels on the shell, or were you measuring the shell temperature?

**A.** First we put a special container against the shell to measure the water flux down the shell. Then we measured the mean water temperature as a function of height, using a thermocouple embedded in a slab of copper. The heat fluxes were calculated from that data.

**R. J. Fruehan** *(Carnegie-Mellon)*

I'm intrigued with the parallel stream model for carbon pickup. Does the simple first order equation used imply that liquid phase mass transfer is rate-controlling?

**A.** In laboratory studies by others of carbon pickup by iron, rates were first order in carbon content and the rate constant measured mainly liquid phase mass transfer. Other rate processes in series can still result in first order behavior, however.

**Fruehan**

The P value for steel being twice that for iron is intriguing because the steel doesn't melt until further down the cupola. So, the actual kinetic term, k, in the k T product could be even more than twice as great for steel than for iron.

**A.** The steel does melt lower in the cupola, but therefore it is hotter when it melts.

**Fruehan**

In that rate equation is the carbon equivalent calculated using the final silicon content? Initial silicon content of the steel must be quite small. What is it for iron?

**A.** Nominal silicon initial concentrations are 2.3 to 2.7% for iron and roughly 0.1% for steel. For calculating tapped carbon equivalent, the initial model used tapped silicon concentration, but the parallel stream modification fits data better when initial silicon concentration is used. That is to be expected from our finding that silicon concentration changes in the slag layer, but carbon concentration does not. The distinction between initial and final contents is important because the coefficient of (% Si) in the modified carbon equivalent definition is about 0.3 at carbon saturation, but gets very small as carbon concentration falls.

**Ned Rehder** *(University of Toronto)*

I will comment about bed height. The top of the bed must be the point at which the metal becomes liquid; the metal drops then fall very rapidly. Most cupolas operate with mixed charges of cast iron and steel. The final liquefaction temperature of steel is greater than that of iron. Since they descend toward greater temperatures, the cast iron melts, by definition and by demonstration, a considerable distance above the steel. In other words, there is not a definite height of bed.

**A.** Yes, the top of the bed is not clearly defined because all the charge materials don't melt at the same temperature. That is why I wasn't more definite when I said we may find some feature in those curves that would be identified as the top of the bed. If we were melting 100% steel for instance, or 100% pig iron, I would be pretty confident that we could.

**E. Kato** *(Waseda University)*

We have a small experimental cupola in our institute, and in the summertime its operation is unstable. How do you think humidity affects heat loss and ease of operation?

**A.** It is well known that water or water vapor entering the cupola are decomposed to hydrogen and CO, and the reaction is extremely fast. The reaction is endothermic, so that would of course hurt the fuel efficiency. We have not studied that reaction ourselves, so I'm describing what we've read in the literature.

**Norm Lillybeck** *(Deere & Co.)*

I'd like to make an observation on the heat losses and ask a question. The double peak that you showed in the heat flux on the shell suggests, I think correctly, that a strong increase in the convective heat transfer has occurred, and it's interesting to note that the location of this is just about where you'd expect the upper limit of the fully liquid zone to exist. There's much evidence that at the transition from a zone containing only liquid metal to a zone with much crumbling solid about to melt, especially steel scrap, there is quite a throttling effect; and the gas velocity near the shell undoubtedly increases greatly. That could explain the upper peak, which is a very interesting measurement. My question concerns oxygen enrichment of the blast. In the most recent measurements I've made on shell losses, I seem to have noticed a strong effect of oxygen enrichment, and I wondered whether you made any measurements with high or low oxygen enrichment. By high I mean something like 3%, enough to significantly change the flame temperature. In the specific case that I have in mind, we had about 4%, and I think the overall shell losses increased by about 20%.

**A.** The oxygen enrichment varied between 0 and 2% in these measurements. It did affect heat loss, but not strongly. I was a little surprised at that. I did elaborate analyses of the heat loss data without gaining much quantitative understanding of the effects of operating variables.

### R. J. Warrick *(Lynchburg Foundry)*

Norm, did you decrease air velocity when using 4% oxygen?

**A.** (Lillybeck) It was something lower than normal. One of the reasons that I asked the question is that I'm trying to learn more definitely the effect of oxygen. The data that I have for that cupola is predominately with the oxygen on, so I lack a good on and off comparison. But the heat loss with the oxygen on is much higher than would be expected for operation without oxygen, based on much data from other cupolas.

### Ned Rehder

I was going to comment on that question about operation of the cupola in the summer humidity. I've been using a card I constructed some years ago that relates the humidity of the air to the percentage of coke needed to maintain constant iron tapping temperature. For some cupolas in Sarnia, Ontario. the humidity is measured twice a day in summer, and the coke is adjusted accordingly to get a constant metal temperature. I developed it purely on the basis of the amount of heat absorbed by water decomposition. And in fact, it just about works that way in practice.

### Carl Loper *(University of Wisconsin)*

Craig, I wonder if you might mention how your Boudouard reaction results might be affected by blast pre-heat temperature and particularly how high pre-heat temperatures might affect the equilibrium that you recorded.

**A.** (Landefeld) Rate of initial $CO_2$ formation is controlled by interphase mass transfer, and so would not be affected by blast temperature. And the following gasification, from the results so far, is also controlled mostly by interphase mass transfer, and is so relatively unaffected by temperature. But we've seen that a certain fraction of gasification is controlled by pore diffusion, and that fraction would increase with blast temperature. So I think the affect on coke percentage would be small, something like the 5% change that we calculated by doubling the reactivity.

## Carl Loper

I think it's going to be quite complicated, as you try to say, because changing the blast temperature will also change the point of melting or bed height, and the entire bottle will be affected geometrically.

**A.** Yes. This brings up the point that we assume a temperature profile for the cupola. To really understand gasification, one needs a model for other processes in the cupola, for they affect temperature, and thus gasification.

## Ned Rehder

I'll just make a short comment that. About 1913 the Bureau of Mines showed quantitative differences with the temperature of the air in the maximum $CO_2$ development zone. Moderate air temperature changes produce quite a strong effect.

# THE MODELLING OF FLUID FLOW PHENOMENA IN FOUNDRY OPERATIONS

## JULIAN SZEKELY

*Department of Materials Science and Engineering*
*Massachusetts Institute of Technology*
*Cambridge, Massachusetts 02139*

### ABSTRACT

An overview is given of fluid flow phenomena in foundry operations. It is shown that fluid flow phenomena play a key role in affecting the kinetics of many physicochemical processes in foundry systems associated with the production, refining and casting of foundry iron.

A general methodology is described for formulating problems of this type, with emphasis on the numerous problems that may now be quite accurately addressed with the aid of mathematical modeling.

The examples cited include the behavior of gas bubble stirred ladles, thermal stratification of melts held in ladles, induction furnaces and a simplified case of mold filling. In all these cases the theoretical predictions were found to be in good agreement with experimental measurements.

## INTRODUCTION

The physicochemical phenomena associated with the production and purification of foundry iron involve fluids, principally gases, slag and metal phases—the motion and agitation of which may play a profound role in affecting the overall process kinetics. Several papers at this conference touch on the fluid flow and mass transfer aspects of these problems.

The purpose of the present work is to provide an overview of fluid flow aspects of foundry operations, with emphasis on what may be predicted on the basis of fluid flow fundamentals. Such an exercise is worthwhile now, because the ready availability of computational hardware and software is greatly expanding the scope of modelling these phenomena.

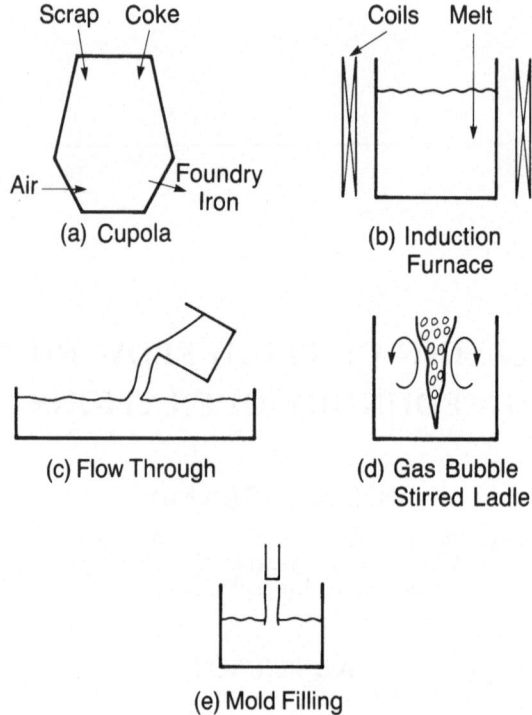

**Figure 1.** Schematic sketch of some physical situations in foundry practice that involve fluid flow.

This overview will be complemented by a series of examples illustrating the utility of these approaches.

Figure 1 shows sketches of typical examples of fluid flow phenomena of interest within the foundry context; these include cupolas, induction furnaces, flow-through troughs and open channels, gas bubble agitated ladles and finally mold filling. Fluid flow is a key component of all these processing operations and may have a profound effect on the associated physicochemical phenomena.

In cupolas, the oxygen in the air injected through tuyeres reacts with the coke, and the resultant hot combustion gases melt the scrap charge. Here, gas flow or gas flow maldistribution may be quite important in affecting the thermal efficiency of the process.

The alternative scrap melting process—the use of induction furnaces—is attractive, but here the principal remaining problem is to devise a coil configuration and refractory wall arrangement such that the wall erosion is minimized.

Flow through troughs is a relatively straightforward problem from the standpoint of fluid mechanics, but the use of this type of information has yet been utilized only to a very limited extent in foundry practice for exploring metal purification (e.g. by filtration), alloying additions, temperature control and the like.

Gas bubble agitated vessels are frequently employed in metallurgical practice, good examples being ladle metallurgy in steel processing (see papers by Fruehan, Irons and Asai in this Conference), and the secondary treatment of aluminum and copper; thus there has evolved a good methodology for tackling problems of this type, which may be readily adapted to foundry practice.

Finally, mold filling, the last step in foundry operations, is an important component of the operating sequence, which may be tackled for the first time by using recently developed computational algorithms.

## FORMULATION OF FLUID FLOW PROBLEMS

The general statement of the transport equations is readily accomplished by writing down the equations of continuity, motion, thermal energy balance and convective mass transfer. These take the following form, using vector notation (1,2):

$$\underline{\nabla} \cdot \rho \underline{U} = 0 \tag{1}$$

continuity

$$\rho \frac{D\underline{U}}{Dt} = -\underline{\nabla} p + \underline{\nabla} \mu_{eff} \cdot \underline{\nabla} \underline{U} + \underline{F}_b \tag{2}$$

motion

$$\rho C_p \frac{DT}{Dt} = \underline{\nabla} k_{eff} \underline{\nabla} T + q''' \tag{3}$$

thermal energy balance

$$\frac{DC}{Dt} = \underline{\nabla} D_{eff} \underline{\nabla} C + r''' \tag{4}$$

convective mass transfer, where:

$\underline{U}$ = velocity vector;

$\mu_{eff}$ = effective viscosity;

p = pressure;

$\underline{F}_b$ = body force;

$\rho$ = density;

T = temperature;

$k_{eff}$ = effective thermal conductivity;

$q'''$ = rate of heat generation;

$C_p$ = specific heat;

$D_{eff}$ = effective diffusivity;

C = concentration;

$r'''$ = rate of reaction.

*References pp. 213-214*

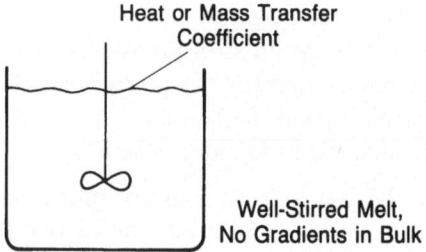

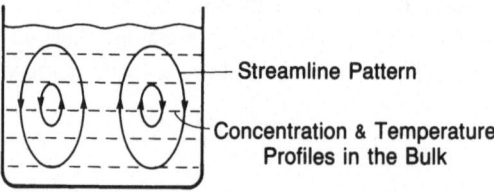

**Figure 2.** Lumped and distributed parameter representation.

These are, of course, quite standard relationships, which have been part of the textbook literature for some time. What will make the statement of the problem unique from a metals processing point of view are the subsidiary relationships that have to be employed, viz.:

—Maxwell's equations to describe the electromagnetic force field, e.g. in case of induction furnace calculations (3,4);

—Expressions for the turbulence models, in case of turbulent flow (5,6);

—Expressions for describing the gas bubble or void fraction distribution in gas bubble stirred systems;

—Relationships involving a balance between surface forces and inertial forces to determine the shape of free surfaces in teeming or mold filling problems;

—Kinetic expressions and thermodynamic equilibrium relationships;

—The composition and temperature dependence of the property values.

Furthermore, one will need to invoke the appropriate boundary conditions, specifying the velocities, temperatures and concentrations at the bounding surfaces.

Within this general framework we may consider lumped parameter approximations and distributed parameter representations, as shown in Figure 2.

# MODELLING OF FLUID FLOW PHENOMENA

In the former case, we consider the fluid phase well mixed and postulate that the key resistance to the transfer process resides at the phase boundary. This resistance may be characterized in terms of a transfer coefficient which must, however, be determined empirically or experimentally.

In the latter case, we may, in principle, obtain solutions on a totally fundamental basis, with the minimum of empiricism; such solutions can provide rather more detailed information on the velocity, temperature and concentration fields within the system.

Until quite recently the detailed fundamental transport-based solution of these problems was not possible, because of the limited experience in computational algorithm development and the high cost of computer time. However, during the past few years there have been dramatic developments regarding major cost reductions in computing, paralleled by greater availability of experience and software packages for solving two and three dimensional transport problems.

Ultimately the choice that has to be made between these two approaches will depend on the particular circumstances and the background and experience of the investigators involved. However, the current trend is clearly toward the more fundamental, distributed parameter computational approaches.

In the following we shall present some examples of recent work in this area—but before we do so, let us address a fundamental question of mixing and agitation in metallurgical vessels.

## MIXING AND AGITATION IN METALLURGICAL VESSELS

Figure 3 shows a schematic diagram of a vessel agitated by a rising stream of gas bubbles, together with the ensuing gross circulation pattern—that is, upward in the middle and downward near the walls. While this particular system will be used for illustrating these concepts, similar considerations can be applied to the more complex flow fields in other metallurgical systems, such as the AOD Process, induction stirring and the like.

In these situations mixing will take place by two mechanisms—that is, by bulk motion, and by eddy diffusion due to turbulence. It is of interest to note that, as shown in Figure 4, the mixing time in a broad range (and size) of metallurgical equipment may be related to the rate of energy dissipation[7,8]. Since the first publication of this idea as a purely empirical relationship, numerous variations on this theme have appeared; indeed, examples of this type of a correlation are also being presented at this conference. While these relationships are approximate, they provide an excellent guide regarding expected performance.

It is of interest to note that on postulating that mixing takes place by eddy diffusion and that turbulence is isotropic, one may develop on a fundamental basis the following expression for the mixing time:[9]

$$\tau_m = \frac{80}{\eta^{1/3}} (L_v^2 / P_t)^{1/3} \tag{5}$$

*References pp. 213–214*

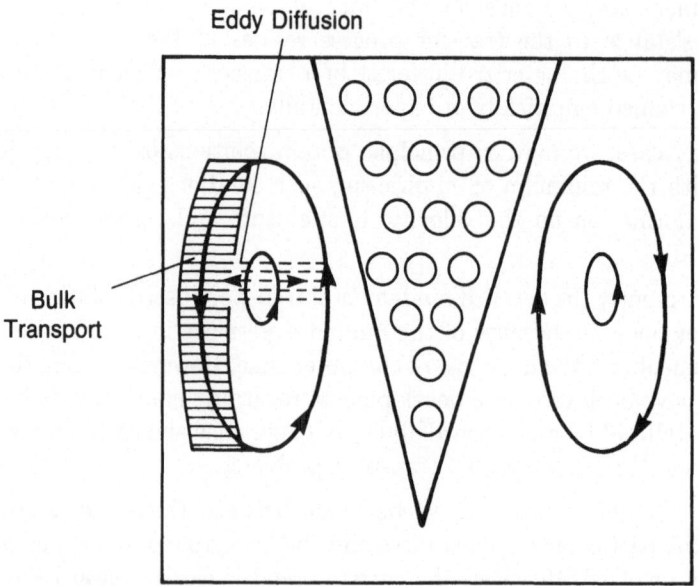

Figure 3. Schematic sketch of a gas bubble stirred ladle, also showing the mixing mechanisms.

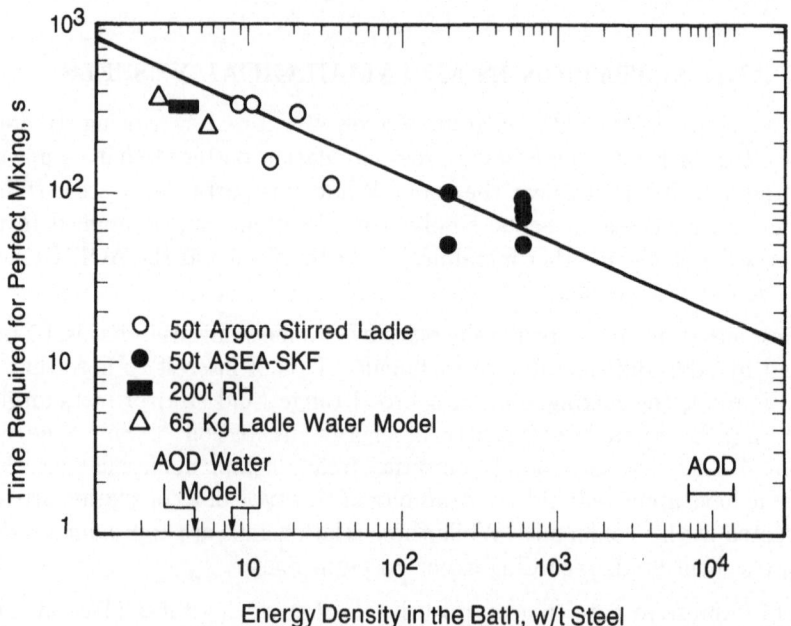

Figure 4. Plot of the time required for complete mixing against the energy density in the bath, after Nakanishi et al. (7).

where $P_t$ is the rate of energy dissipation per unit mass of fluid

$L_v$ is the characteristic vessel dimension

$\eta$ is the efficiency factor which should be of the order of 0.5; and

$\tau_m$ is the mixing time.

Equation 5 is of very similar form to the empirical relationships cited earlier, and is in quantitative agreement with them. The important point is that, having established a fundamental basis for these expressions, one may use them with greater confidence.

It should be stressed that this "bulk" or "lumped parameter" approach will provide only a rough preliminary guide and in general will be inadequate to perform the design of specific processing equipment, such as a unit for alloying additions or desulfurization.

In the following we shall present some examples pertaining to foundry and metals processing practice, where the detailed knowledge of the fluid flow field may be readily developed and where this information has a significant bearing on the system performance. In presenting the calculations we should stress that when dealing with one phase systems, such as ladle stratification or induction stirring, calculations may be performed very readily and accurately. In contrast, the understanding of two phase systems, e.g. gas bubble stirred ladles, is less well developed, and here some empiricism must be used.

## STRATIFICATION IN LADLES[10]

Whenever molten metal is held in ladles, heat losses will occur through the walls and also through the top slag cover, unless a fairly thick cover is being employed. As discussed in earlier publications[11,12,13], the contact between the melt and the cold vertical walls will give rise to thermal natural convection (downward flow near the walls and upward in the middle) which will produce some mixing. The question is, however, whether this circulation can overcome the significant cooling that takes place through the ladle bottom. If stratification occurs, especially just before the casting step, this can cause serious problems because of the critical dependence of cast structures and their integrity on the superheat.

This problem is readily formulated by writing down

— The transient turbulent Navier-Stokes equations for an axi-symmetric cylindrical system;

The corresponding convective heat flow equation and the usual boundary conditions, including one specifying transient conduction into the ladle walls.

The property values used in the calculations are summarized in Table 1; the calculations, which involved using the $k - \varepsilon$ turbulence model and the PHOENICS computational package required about 3 to 5 hours on an Apollo DN 330 digital

## TABLE 1
### Principal Input Parameters in the Ladle Stratification Calculation

| Property | 80-ton ladle | 7.5-ton ladle |
|---|---|---|
| Melt | steel | steel |
| Melt temp. | 1600°C | 1640°C |
| Gr | $2.0 \times 10^{11}$ | $1.2 \times 10^{11}$ |
| k (refractory) | 30.3 W/Km | 1.39 W/Km |
| $\alpha$ (refractory) | $5.6 \times 10^{-6} m^2/s$ | $5.5 \times 10^{-7} m^2/s$ |
| I (cases with stirring) | 300A | — |
| f | 5Hz | — |
| n | 30 turns | — |

computer. Some typical computed results are given in Figures 5, 6 and 7 for an 80 ton ladle.

It is seen in Figure 5 that quite significant circulation is taking place due to natural convection, with maximum velocities of the order of 5 cm/s. Figure 6 shows marked stratification notwithstanding these flows, with a maximum temperature difference of the order of 20°C. Finally, Figure 7 shows the temperature distribution with induction stirring and it is seen that this will greatly minimize the extent of stratification.

Finally, Figure 8 shows a comparison between the theoretical predictions and the experimental measurements obtained by Hlinka[11,12] in a 7.5 ton ladle, holding molten steel. The agreement between measurements and predictions is seen to be very good.

At this point it is important to pause for a moment and to examine the physics of the situation. In unstirred ladles, stratification will occur, because the bouyancy driven flow will not be able to provide adequate agitation to overcome the imbalances in temperature, brought about by heat losses from the walls, particularly as far as the bottom region is concerned. Our ability to predict these effects quantitatively, including possible remedial actions, should be helpful in process design.

## FLOW PHENOMENA IN GAS-BUBBLE STIRRED LADLES

Agitation by means of injected gases is perhaps the least expensive and easiest way of homogenizing a melt. This is a wide-spread practice in ladle metallurgy in steel processing operations. In mathematically modelling these systems, one can make the assumption that for an axi-symmetric situation (which is rather infrequently realized in practice) the central core, where the bubbles are located, is much more bouyant (see Figure 1a); thus, the overall flow may be represented as being bouyancy driven.

By making some assumptions regarding bubble behavior, such as slip flow, it

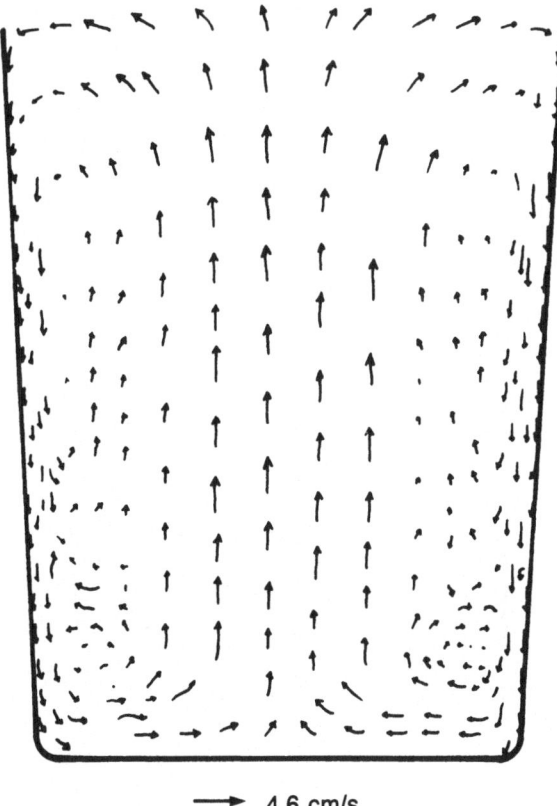

→ 4.6 cm/s

Figure 5. The computed map of the velocity vectors after the passage of 5 minutes in the 80 ton ladle example.

was possible to predict reasonably well the experimentally measured velocity fields both in a water model of an argon stirred ladle and velocities measured in a real molten steel system.[14]

However, measurements and predictions agreed rather less well as far as the turbulent kinetic energy was concerned. Knowledge of the turbulent kinetic energy distribution is, of course, a key factor–both in mixing and in the predictions of reaction rates between the melt and injected powders, as in the case of desulfurization.[15]

It was not clear whether the discrepancy between the model predictions was due to inherent shortcomings of the model, or attributable to experimental uncertainties. In recent work, Brimacombe and Castillejos[16] reported on extensive experimental measurements describing the void fraction distribution in a water model of an argon-stirred ladle. Figure 9 and 10 show comparisons between the experimentally measured velocities and turbulent kinetic energy distribution by Grevet et al.[17] and theoretical predictions,[18] based on the experimentally measured void fraction distribution.[16] It is seen that by using the experimentally measured

*References pp. 213-214*

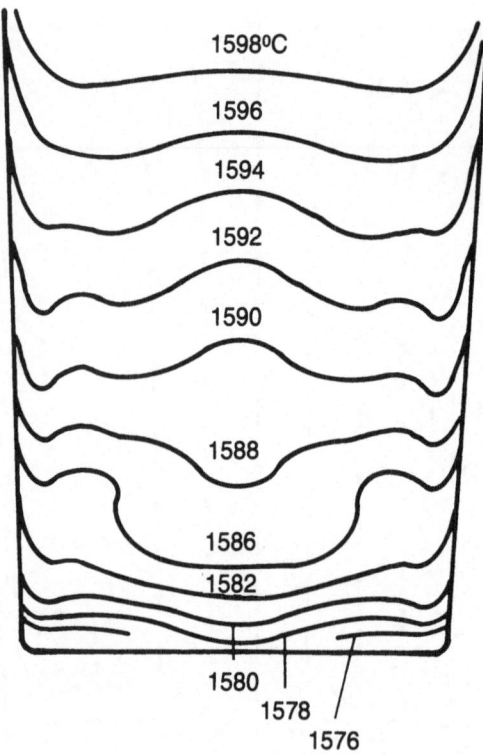

Figure 6. The computed map of the temperatures after the passage of 5 minutes in the 80 ton ladle example.

void fraction distribution—which constitutes the driving force for the flow—we can obtain very good agreement between measurements and predictions, regarding both the velocity fields and the turbulent kinetic energy distribution.

This improved understanding of the flow phenomena should provide a sound basis for further theoretical developments involving process kinetics, such as desulfurization and the dissolution of alloying additions.

In closing this section, we should note very recent and interesting efforts aimed at representing gas bubble-melt interactions from first principles, employing the PHOENICS code and two-phase flow considerations.[19] While far from proven at this stage, this represents a very promising and novel approach, if combined with appropriate experimental work.

## INDUCTION FURNACES

A schematic sketch of induction furnace operations has been given previously in Figure 1b. In essence, these systems consist of a crucible, surrounded by one or more sets of induction coils, through which an alternating current is passed. The passage of the current through the coils will generate a magnetic field; at the same

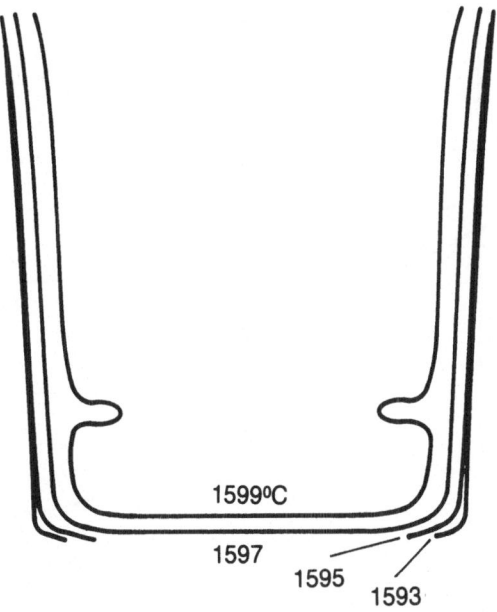

Temperature stratification in an inductively-stirred
80-ton ladle after 5 mins. $I_o$ = 300A, f = 5Hz

**Figure 7.** The computed temperature map for the
80 ton ladle after 5 minutes, if induction stirring is applied.

time, current will be induced in the melt. This induced current will provide "joule heating" (often the key objective), but at the same time the interaction between the induced current and the magnetic field provides an electromagnetic or Lorentz force, which in turn will lead to melt circulation.

These problems are readily formulated by invoking the Navier-Stokes equations, i.e. Equations (1-2), where $F_b$ the body force term has to be evaluated by solving Maxwell's equations. This procedure is well documented[4,6,20].

Before presenting some computed results, it may be important to cite some order of magnitude expressions, which can provide useful insight into the general behavior of the system.

The extent of the field penetration into the metal (or conducting phase) may be described in terms of the "skin depth," given by:

$$\delta_s = \left[\frac{1}{\pi f \mu_0 \sigma}\right]^{1/2} \tag{6}$$

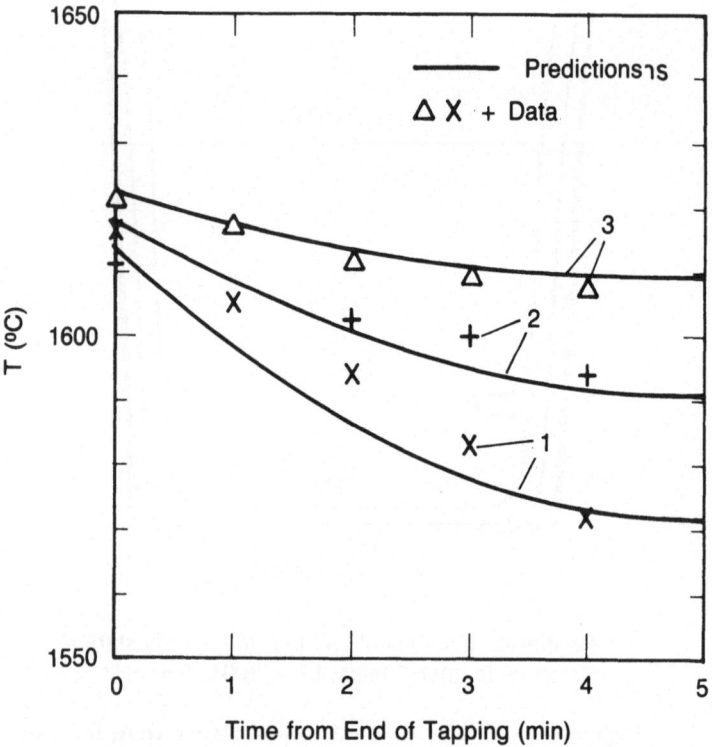

Figure 8. Comparison of the experimentally measured values of the time-dependent temperatures at three different locations with the theoretical predictions for a 7.5 ton ladle.

where f = frequency

$\sigma$ = electrical conductivity

$\mu_o$ = magnetic permittivity of free space

which is seen to be inversely proportional to the square root of the frequency.

The ratio of the energy used for driving the flow and for heating is given by the following expression:

$$\frac{[\text{Stirring}]}{[\text{Heating}]} \sim \frac{U_o}{fL} \tag{7}$$

where $U_o$ is characteristic liquid velocity from which it readily follows that high frequencies will promote heating, while low frequencies will promote stirring.

Finally, the order of magnitude of the velocity may be estimated by equating the inertial term in the Navier-Stokes equations with the electromagnetic force term, which will give:

$$U_o \sim J_o \frac{\sqrt{\sigma f}}{\rho} L^2 \mu_o \tag{8}$$

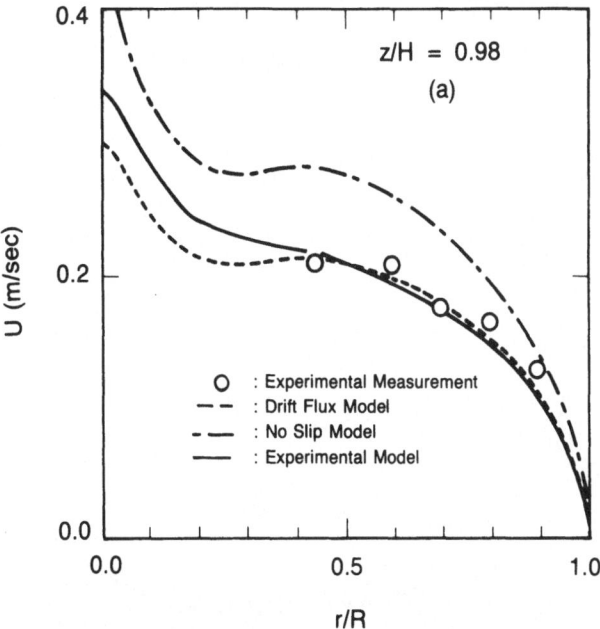

Figure 9. Comparison of the radial distribution of the mean axial velocity with predictions. (R=0.3 m, H=0.6 m, gas flow rate=13 liter/min)

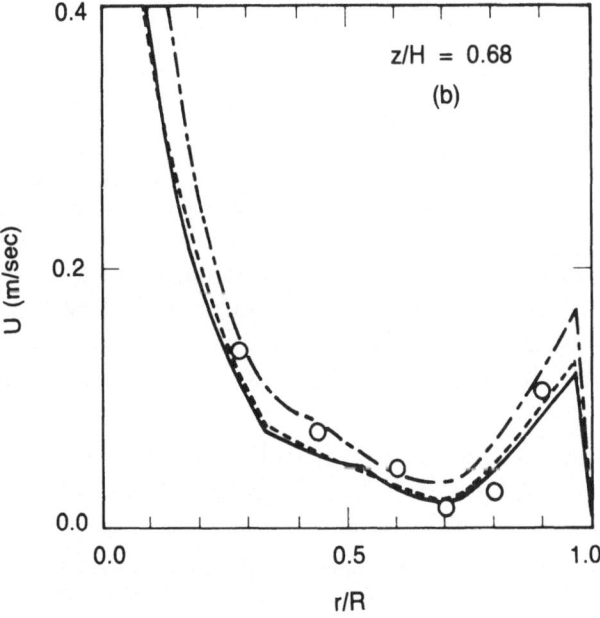

Figure 9b.

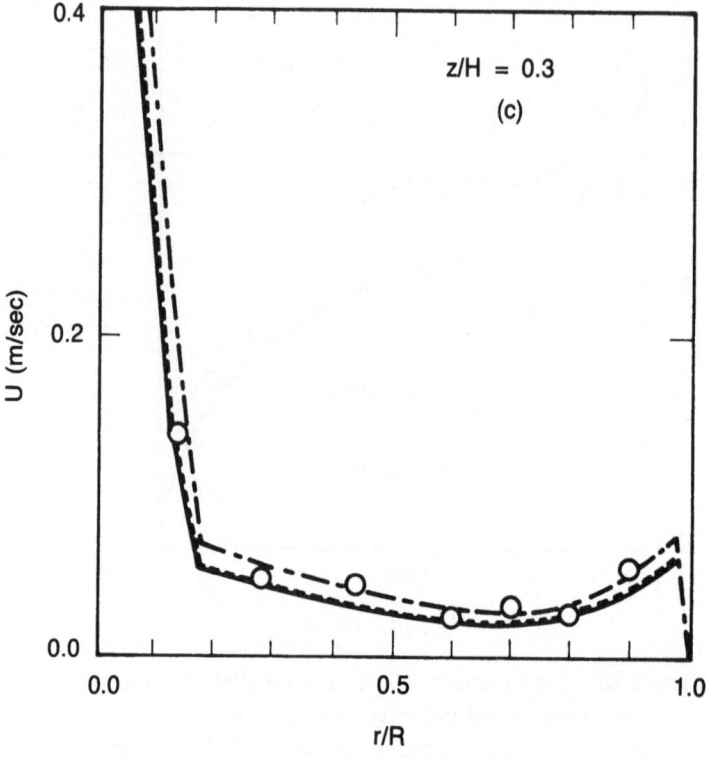

Figure 9c.

where $U_o$ = characteristic velocity

$L$ = characteristic length

$J_o$ = rms coil current.

In presenting computed results for the velocity fields in inductively stirred systems we should note that as result of recent work this is a well established field, where predictions may be made with considerable accuracy and certainty. The reason for this is simple. The calculation of the electromagnetic force field and that of the induced current involves straightforward manipulation of the linear Maxwell's equations, which is readily done on a computer. Similarly, the computation of the fluid flow field is also quite readily done, because we are dealing with one-phase systems.

Figure 11 (a) and (b) shows a comparison between an experimentally derived profile of the velocity and the turbulent kinetic energy and the corresponding predictions; the good agreement is readily apparent.[20]

Figure 12 (a,b,c) shows a similar comparison, but now for the whole circulation pattern, in the absence and in the presence of magnetic shields,[21] again with excellent agreement between measurements and predictions. The model correctly

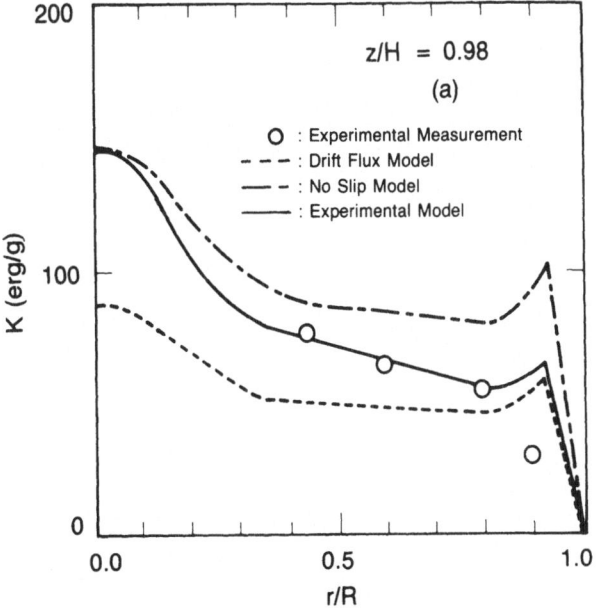

Figure 10. Comparison of the radial profiles of the experimentally measured turbulent kinetic energy with the theoretical predictions at various vertical locations. (R=0.3 m, H=0.6 m, gas flow rate=13 liter/min)

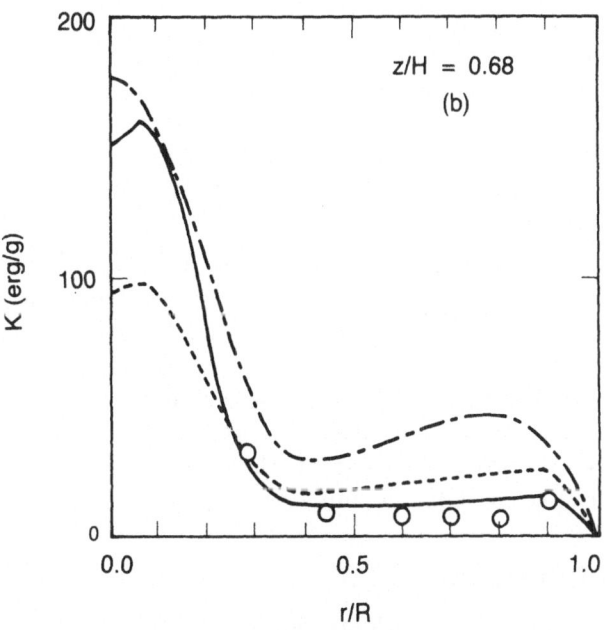

Figure 10b.

*References pp. 213–214*

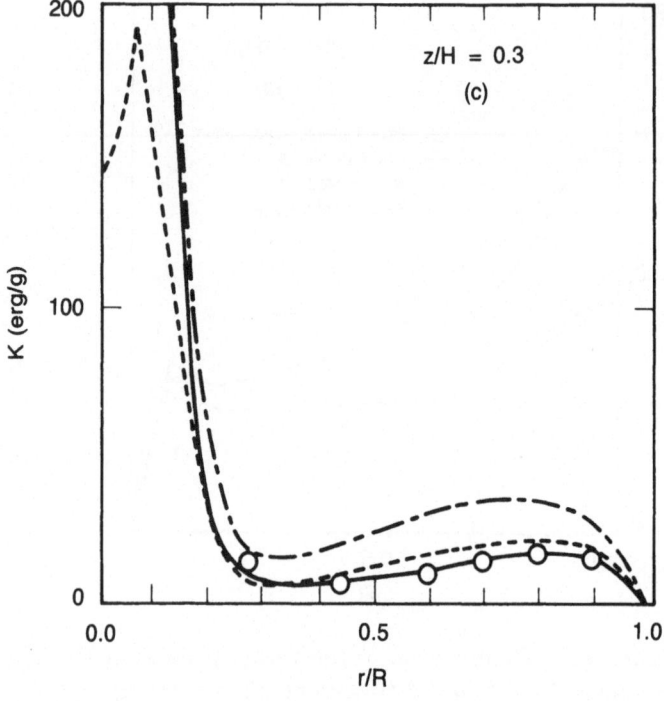

**Figure 10c.**

predicts the marked change in the circulation pattern brought about by the insertion of the shield.

With respect to operation of larger-scale, real metallurgical systems, Figure 13 shows a comparison between the experimentally measured and the theoretically predicted melt velocities in a four-ton inductively stirred steel melt, as a function of the coil current.[22] The excellent agreement is again readily apparent; furthermore, the results show the same linear dependence of the velocity on the coil current as was predicted by Equation 8.

It should be noted that, while the calculation of velocities, temperatures and melting rates in induction furnaces may now be performed in a more or less routine manner using established procedures, some interesting problems still remain. In particular, we would need a better basic understanding of melt-crucible (furnace) wall interactions, which in turn will require a more precise knowledge of the boundary layer regions in the vicinity of the walls. This is a very important problem area, because excessive wall erosion is perhaps the most serious impediment to the more widespread use of induction furnaces.

## MOLD FILLING PROBLEMS

It is probably appropriate to conclude this discussion of fluid flow phenomena in foundry processes with a brief reference to mold filling problems. Ultimately,

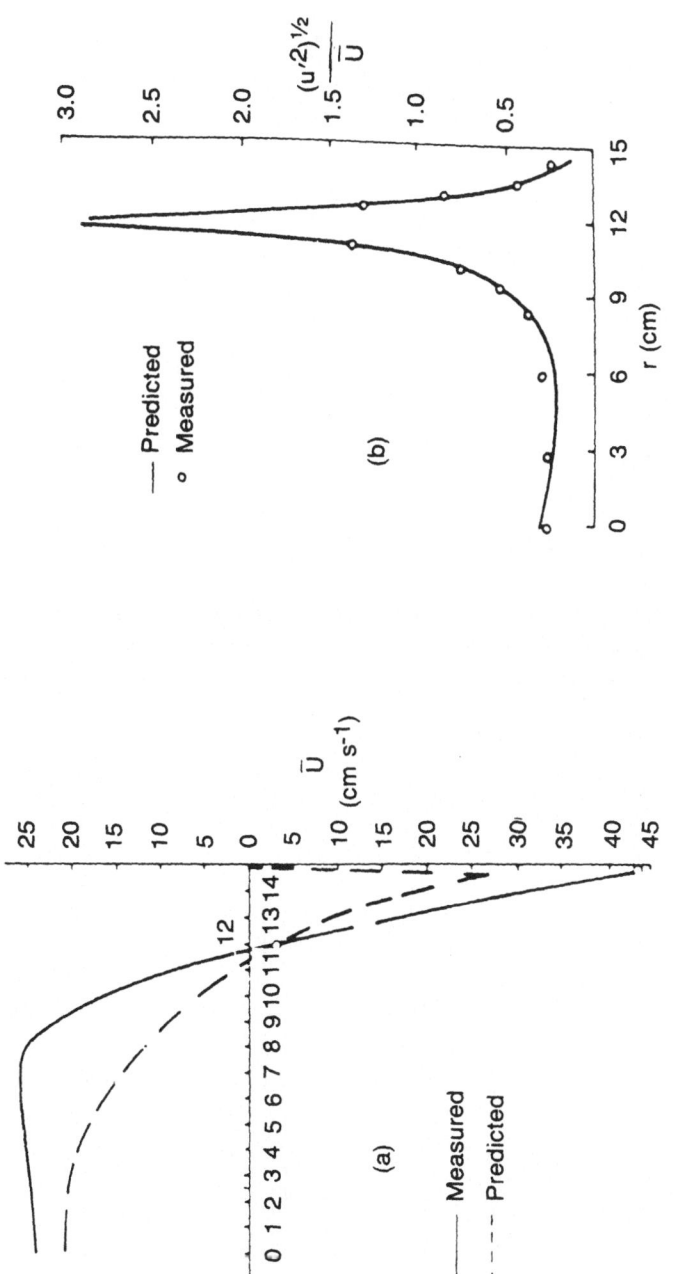

Figure 11. Comparison of the experimentally measured and the theoretically predicted velocity and turbulent kinetic energy profiles in an inductively stirred metal.
(a) The velocity profiles.
(b) The turbulent kinetic energy profiles.

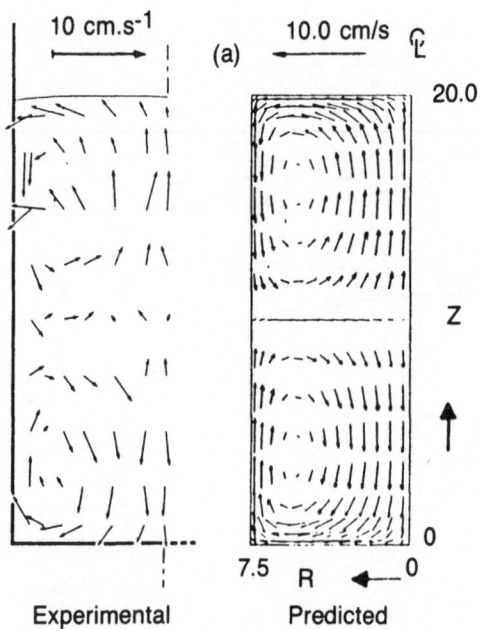

Figure 12. A comparison between the theoretically predicted and the experimentally measured velocity fields in an inductively stirred melt. (a) In the absence of magnetic shields. (b) and (c) In the presence of shields.

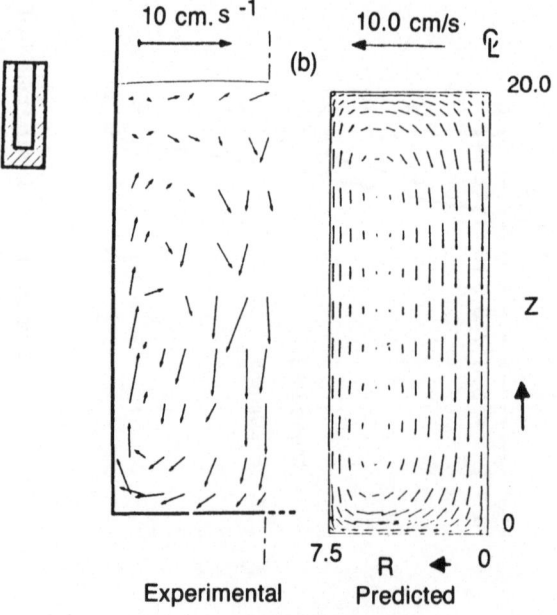

Figure 12b.

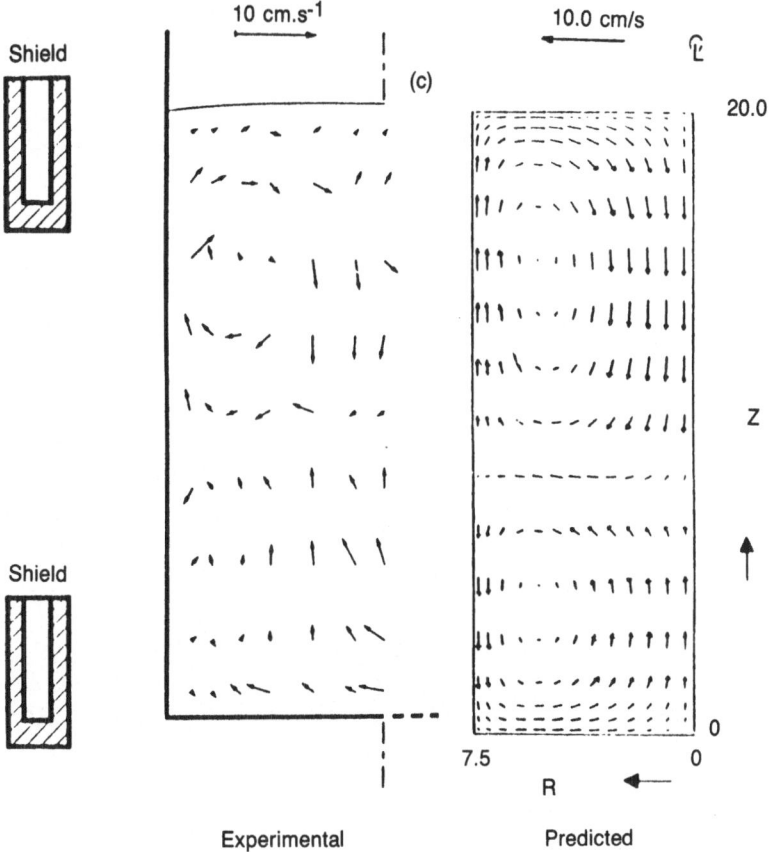

Figure 12c.

the hot metal produced and refined in foundry operations is poured into molds, where solidification takes place. At present, mold design is a highly empirical process requiring several trial-and-error steps. If such mold design operations could be automated through the use of CAD-CAM techniques, this could have major economic consequences.

In a generic sense, mold filling problems are very different from the systems discussed in the preceeding, because they involve a free surface and the possibility of surface waves.

In recent years, a number of computational techniques have been proposed for tackling free surface problems, particularly with potential applications in the ocean engineering field (27). The first application of these ideas to metals processing operations is due to Stoehr,[23] and rapid growth of this field is to be expected.

As a simple illustration, Figure 14 shows the sequence of filling a cavity after the removal of a barrier; the SOLA code was used for doing the calculations, and this very simple problem required several hours of CPU time on an IBM PC AT computer.[24] Clearly the tackling of complex geometries will be quite time-

*References pp. 213–214*

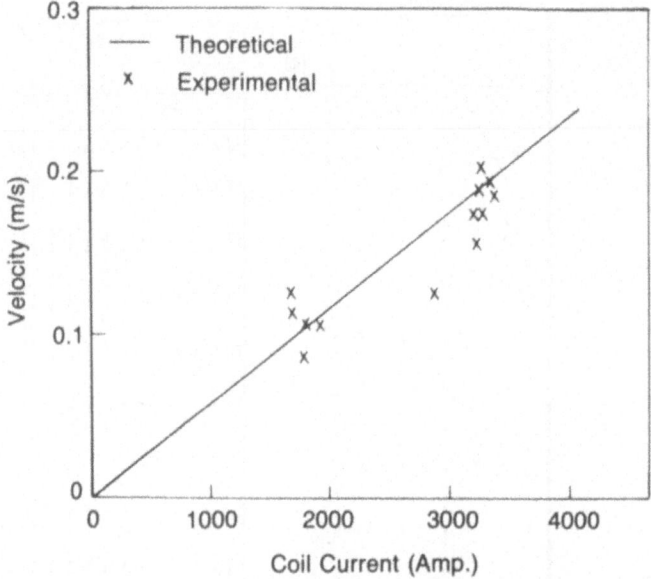

Figure 13. Comparison of the theoretically predicted and the experimentally measured velocities as a function of the coil current in a 4-ton induction furnace holding molten steel.

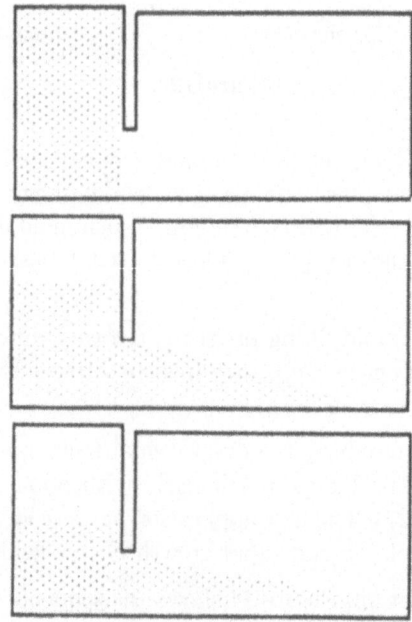

Figure 14. Computer simulation of a mold filling by the removal of a barrier.

consuming, but with the great advances in computing speed and software packages, one can confidently expect significant results in the near future, relating not only to flow, but also to the associated heat transfer and solidification phenomena.

## CONCLUDING REMARKS

In the paper an overview was given of fluid flow phenomena in foundry operations. It was shown that fluid flow plays a key role in affecting the kinetics of many foundry processes, and it is now generally realized that a better understanding of flow phenomena may provide the key to process improvement in many instances.

It has been traditional in metallurgical practice to treat most transport processes using a "lumped parameter" approach, representing fluid flow phenomena indirectly through semi-empirically obtained heat or mass transfer coefficients. This is an important step forward and entails relatively simple mathematics; however, in many instances the technique lacks generality and the empirical transfer coefficients obtained in one system may not apply, without modifications, to other operations.

The approach outlined in this paper—that is, computation of the fluid flow field in the system as an intermediate step toward addressing the chemical and thermal process kinetics, is now being adopted in ladle metallurgy practice and has a great deal to offer to the foundry community. The examples cited are relatively simple, but nevertheless have important practical implications. The important conclusion that has to be drawn from them is that through the greater availability of computer hardware and software, many of our kinetic problems may now be tackled by performing transport calculations using first principles. In many cases, this option will represent the most cost-effective approach, and thus will require serious consideration.

## REFERENCES

1. R. B. Bird, W. E. Stewart and E. N. Lightfoot, *Transport Phenomena*, John Wiley, New York (1960).
2. J. Szekely and N. J. Themelis, *Rate Phenomena in Process Metallurgy*, John Wiley, New York (1971).
3. W. F. Hughes and F. J. Young, *The Electromagnetodynamics of Fluids*, John Wiley, New York (1966).
4. J. Szekely and C. W. Chang, *Ironmaking and Steelmaking 4*, 190, 196 (1977).
5. B. E. Launder and D. B. Spalding, *Lectures on the Mathematical Modelling of Turbulence*, Academic Press, London (1972).
6. J. Szekely, *Fluid Flow Phenomena in Metals Processing*, Academic Press, New York (1979).
7. K. Nakanishi, T. Fujii and J. Szekely, *Ironmaking and Steelmaking 3*, 193 (1975).
8. M. Sano and K. Mori, *Proc. SCANINJECT III*, Lulea, Sweden (1983).
9. A. Murthy and J. Szekely, *Met. Trans. 17B*, 487 (1986).

10. O. J. Ilegbusi and J. Szekely, "The Stratification of Ladles," to be submitted to *Trans. ISIJ*.
11. J. W. Klinka, in *Mathematical Process Models in Iron and Steelmaking*, The Metals Society (1973).
12. J. W. Hlinka and T. W. Miller, *Iron and Steel Engineer* (Aug. 1970), p. 123.
13. J. Szekely and J. H. Chen, *Met. Trans. 2* (1), 189 (1971).
14. N. El-Kaddah and J. Szekely, *Proc. SCANINJECT III*, Lulea Sweden (1983).
15. N. El-Kaddah and J. Szekely, *Proc. 3rd Process Technology Conference*, Pittsburgh, The Iron and Steel Society (1982).
16. A. H. Castillejos and J. K. Brimacombe, *Proc. SCANINJECT IV*, Lulea, Sweden (1986).
17. J. H. Grevet, J. Szekely and N. El-Kaddah, *Int. J. Heat and Mass Transfer*, 25, 487 (1982).
18. J. S. Woo and J. Szekely, forthcoming publication.
19. M. P. Schwartz and P. T. L. Koh, *Proc. SCANINJECT IV*, Lulea, Sweden (1986).
20. N. El-Kaddah and J. Szekely, *J. Fluid Mech. 147*, 53 (1984).
21. J-L. Meyer, N. El-Kaddah, J. Szekely, C. Vives and R. Ricou, *Met. Trans.* 18B (1987), in press.
22. N. El-Kaddah, J. Szekely and G. Carlsson, *Met. Trans. 15B*, 633 (1984).
23. R. Stoehr, *Journal of Metals* (1984).
24. J. Szekely and F. Foreman, *Journal of Metals* (October 1986).

# DISCUSSIONS

**Norman Lillybeck** *(Deere & Co.)*

Is the purpose of suppressing the current in induction melters to retard erosion of the lining?

**A.** No. We try to devise a critical test of our model by changing the electromagnetic force field configuration and testing whether we can accurately predict the change in stirring. Incidentally, the use of shields is practiced in the electromagnetic casting of aluminum to generate an adequate force to restrain the metal without promoting stirring. With respect to lining erosion. I have no immediate recipes to prevent it. However, now we have the tools to precisely calculate the casual relationship between the coil arrangement, coil configuration and velocity fields in the vicinity of the lining, and I suspect that if we use these tools in an optimal fashion, we could reduce lining wear.

**N. Lillybeck**

If we assume that success could be obtained in reducing lining wear in an induction furnace, would it be possible to then answer how this would affect the transfer of heat to the cold scrap, in other words, the melting rate of the vessel?

**A.** These two things can be neatly uncoupled. As far as the melting rate is concerned, it will depend on the rate of heat input into the melt and also on how much turbulence is generated. If you operate in such a way that the highest forces are not in the vicinity of the wall, this might be a benefit. The issues are complex, and I don't think one can come up with an immediate solution. But, it certainly can't hurt to precisely understand the forces and velocities. I think the two considerations of minimizing wear on the lining and still having an adequate melting rate are not orthogonal.

**R. J. Warrick** *(Lynchburg Foundry)*

When you show your eighty-ton ladles cooling in five minutes, you have a temperature contour that came down, spiked in, came back out and curved. What causes that? Why do you get that pattern, and that spike coming in and disappearing?

**A.** This is a good question. Whenever we carry out calculations it is important to check whether the results are consistent with the physical picture of the process. In the present case we set up a circulation pattern due to natural convection, which will lead to a relatively quiescent region at the bottom of the ladle. Initially this will result in a "cold finger" type behavior. As the cold front moves upwards, primarily due to conduction, this "cold finger" will disappear. To my mind this seems reasonable on physical grounds.

**Klaus Lange** *(Tech. Univ. of Aachen)*

You showed gas-stirred ladles can be well described. Now, can you also include a layer of slag on top of the liquid metal and then calculate motions without changing the slag? My other question is whether one could also calculate when emulsification starts, bringing warm slag droplets back into the liquid metal.

**A.** The answer is no to both of them. The models have two parts. One is the governing equations, and there's little doubt about these. The second part is the boundary conditions. If you are dealing with a single-phase system these can be stated in a fairly unequivocal way; you define velocity and temperatures. Now, when dealing with a two-phase or a multi-phase system, I don't think I have at this point good enough assumptions about boundary conditions.

**K. Lange**

What about the whole cell? You have two different density liquids and a pretty good circulation pattern.

**A.** Yes, but you are not emulsifying the bath.

**K. Lange**

Okay, that makes a difference. But you can take two phases.

**A.** Yes. But when it comes to the interaction between these phases, the model is only as good as the boundary conditions. In our current state of knowledge, the three-phase situation is a lot more difficult, and Professor Asai will be talking about emulsification. A very important philosophical point is that models are very good at describing what happens in the bulk. But, describing what happens at the interfaces will depend on how clever you are specifying those boundary conditions. So, for understanding the slag metal phase, the breakthrough will not come from the model, but will come from maybe the work that Professor Asai and others are doing on emulsification. Now, how well the slag phase itself is mixed due to the presence of the bas bubbles, this is something that one can address.

**Norman Lillybeck** *(Deere & Co.)*

I think the models that you showed or maybe have done haven't showed convective movement of the metals. Does your model allow you to introduce a heat flux on all surfaces including the top surface? Can you manipulate that well?

**A.** There's no problem.

**N. Lillybeck**

So you could use heating and cooling on the surface?

**A.** There's no problem at all. Now, I would say that if you are heating the surface, that's a very easy thing to do. If you are really rapidly cooling the top surface, at that point you may set up a natural convection pattern having absorbable cells, and that is not terribly well represented by models of this type. So you'd have to exercise some caution, as you always must when you handle these models. But, as far as the heat fluxes are concerned, there's no problem. We put in the actual transient conduction type that takes place into the walls, so that's pretty reasonable, and heating is no difficulty at all. For example, if a surface heater were above the melt, one could readily calculate how hard to stir to dissipate that thermal energy.

**Gino Sovran** *(General Motors)*

No computational code has the resolution normally to compute all the way to the wall, to the boundary of the vessel. So, you'd have to use models called wall functions to get close to the wall. So you are subject to the degree of accuracy of the existing wall models.

**A.** That is correct.

**Gino Sovran**

Now, in multiphase flows, to define boundary conditions you need some phenomenological model to connect the phases.

**A.** That is true.

**Gino Sovran**

You have the Navier-Stokes equation for each phase, and it is the connection between them that's the difficult part. For multi-phase flows, the capability to utilize computational techniques is not as great as it is in single-phase.

**A.** This is precisely the same point I was making.

**Gino Sovran**

In some recent interactions I've had at the multi-phase flow conventions, it appears that the single, most difficult problem that they have is scaling a system for transferal of laboratory experiments to full scale. Now, to what degree that might be true in the foundry business, I don't know, but if there are multi-phase flows it's a potential area of difficulty.

**A.** You have gas bubbles to assist and you have just a cone right at the two-phase region that's the engine for driving the flow, but the principal region of interest may well be what is outside the plume—what sort of circulation you get. That is a question that one can reasonably well address.

# ELECTROCHEMICAL SENSING OF CARBON, OXYGEN AND SILICON IN IRON MELTS

## ALBERTO R. ROMERO, KIYOSHI ICHIHARA, HANS-JÜRGEN ENGELL and DIETER JANKE

### ABSTRACT

Activities of carbon and oxygen in Fe-O-C-$X_n$ melts ($X_n$ = Cr, Mn, Si) and of silicon in Fe-C-Si melts were determined in-situ by emf measurements. Electrochemical probes were developed and tested, based on $ZrO_2$ and $ThO_2$ solid electrolytes for C and O sensing, and based on molten silicate electrolytes for Si sensing. The probes can be immediately immersed into the metal bath at a response time of 20 to 30 s and 30 to 60 s, respectively. Accurate C and O activities were obtained from the emf measurements with $ZrO_2$ and $ThO_2$ electrolyte probes up to carbon saturation when a controlled $p_{CO}$ is maintained. A reproducible emf vs silicon activity relationship resulted from measurements with a silicate electrolyte probe in carbon-saturated Fe-C-Si melts up to 1.8 wt% Si. The present results are based on laboratory-scale experiments. Further efforts are thought to be promising to develop these probes to an industrial standard.

### INTRODUCTION

High-temperature electrochemical cells based on solid oxide electrolytes are widely used to determine activity of oxygen dissolved in liquid metals[1]. Commercial oxygen probes are particularly applied in the processes of iron and steel making[2].

High-temperature electrochemical cells based on molten silicate, aluminate or phosphate electrolytes have been used for decades in laboratory tests to measure thermodynamic properties of liquid metal alloys[1]. But so far cells of this kind have not been developed to an industrial standard to measure the activity of constituents in Fe-based melts.

The present experimental work refers to the *in-situ* measurement of the activities of carbon, oxygen and silicon dissolved in Fe-C-$X_n$ ($X_n$ = Cr, Mn, Si) and Fe-C-Si melts which represent fundamental systems in the production of blast furnace iron, carbon steels, ferroalloys and cast iron. Solid electrolyte cells have been used to determine carbon and oxygen in iron melts, while tests on liquid electrolyte cells were made to measure silicon in iron melts.

## CARBON AND OXYGEN SENSING

**Basic considerations**—Up to the present, there is a lack of reliable emf measurements in Fe-O-C and Fe-O-C-$X_n$ melts at elevated carbon contents. When oxygen activity is measured in carbon-containing iron, accurate performance of the solid oxide electrolyte probes at extremely low oxygen activities is a special requirement. Two main problems exist:

a. Partial electronic conductivity of the $ZrO_2$ electrolyte which lowers the measured emf by partially short-circuiting the cell and causes oxygen ion transfer across the electrolyte, resulting in polarization effects at the electrolyte interface;

b. Chemical reduction of the $ZrO_2$ electrolyte by the carbon present in the iron melt, resulting in increased oxygen activities at the interface.

In view of the fluctuating $h_o$ values obtained in Fe-O-C melts by earlier inverstigators[3-6], it was obvious that further experimental work had to be done to study the fundamental relationships and to improve the reliability of the EMF measurements.

The probes used in the present study represent oxygen concentration cells of the type

$$(Ia) \ominus \text{Fe melt} \| ZrO_2(CaO) \| Cr, Cr_2O_3 \oplus$$

and

$$(Ib) \ominus \text{Fe melt} \| ThO_2(Y_2O_3) \| Cr, Cr_2O_3 \oplus$$

According to Schmalzried's analysis[7], the emf for oxide electrolytes with mixed ionic and electronic conduction at low oxygen partial pressures

$(p_h \cdot \gg p_{O_2}(Cr, Cr_2O_3) > p_{O_2}(Fe) > p_{e'})$ is expressed as

$$E = \frac{RT}{F} \ln \frac{p_{e'}^{1/4} + p_{O_2}^{1/4}(Cr, Cr_2O_3)}{p_{e'}^{1/4} + p_{O_2}^{1/4}(Fe)} \tag{1}$$

where the parameter $p_{e'}$ is defined as the oxygen partial pressure at equal ionic and excess electron conductivity ($t_{ion}$ = 0.5), depending on temperature and the type and composition of the electrolyte. Explanation of the symbols is given in Table 1 and more detailed background information is found in[8]. Using the known parameters $p_{e'}(t_{ion} = 0.5)$ and taking $p_{O_2}(Fe) = (h_O/K_O)^2$ according to the

## TABLE 1

### Explanation of symbols and units

| | |
|---|---|
| $a_i$ | activity of dissolved component i ($= \gamma_i \cdot x_i$) |
| E | electromotive force, V |
| $e_i^j$ | first-order interaction parameter, indicating influence of j on i (dilute solution, wt.%) |
| $f_i$ | activity coefficient of dissolved component i ($\lim f_i = 1$) wt.% i→0 |
| F | Faraday's constant (96570 As · mol$^{-1}$) |
| $\Delta G_O^\circ$ | free energy change of the reaction $\frac{1}{2} O_2$ (1 bar) $\rightleftharpoons [O]_{Fe,1wt.\%}$ ($-137120+7.79T$ J·mol$^{-1}$) [16] |
| $\Delta G_{CO}^\circ$ | free energy change of the reaction $[C]_{Fe,1wt.\%}+[O]_{Fe,1wt.\%} \rightleftharpoons CO$ (g) ($-22400 - 39.69$ T J · mol$^{-1}$) [11] |
| $h_i$ | activity of dissolved component i ($=f_i \cdot$ wt% i), (dilute solution, 1 wt.%) |
| K | equilibrium constant ($= \exp(-\frac{\Delta G^\circ}{RT})$) |
| $p_{O_2}$ (met) | oxygen partial pressure in liquid metal, bar |
| $p_{O_2}$ (ref) | oxygen partial pressure of the reference, bar |
| $p_{e'}$ ($t_{ion} = 0.5$) | oxygen partial pressure at $t_{ion} = 0.5$ for solid oxide electrolyte, bar |
| $r_i^j$ | second-order interaction parameter, indicating influence of j on i (dilute solution, wt.%) |
| R | gas constant ($= 8.317$ J · K$^{-1}$· mol$^{-1}$) |
| T | temperature, K |
| $t_{ion}$ | transference number of ions |

reaction $\frac{1}{2}O_2$ (1 bar)$\rightleftharpoons \underline{O}_{Fe}$, 1 wt.%, $h_O$–EMF functions can be calculated from Equation 1.

$$E = \frac{RT}{F} \ln \frac{p_{e'}^{1/4} + p_{O_2}^{1/4}(Cr, Cr_2O_3)}{p_{e'}^{1/4} + (h_O/K_O)^{1/2}} \tag{2}$$

*References pp. 235-236*

$$h_O = \exp\left(-\frac{\Delta G_O^\circ}{RT}\right)\left[(p_{e'}^{1/4} + p_{O_2}(Cr, Cr_2O_3)^{1/4}) \cdot \exp\left(-\frac{EF}{RT}\right) - p_{e'}^{1/4}\right]^2 \quad (3)$$

Plots of Equation 2 as emf (V) vs. log $h_O$ are shown for a $ZrO_2(CaO)$ electrolyte in Figure 1a and for a $ThO_2(Y_2O_3)$ electrolyte in Figure 1b in a wide range of oxygen activities. It is evident that an idealized relationship according to

$$E = \frac{RT}{4F}\ln\left[p_{O_2}(Cr, Cr_2O_3) \cdot \left(\frac{K_O}{h_O}\right)^2\right] \quad (4)$$

exists only at higher oxygen activities and lower temperatures. For the evaluation of $h_o$ values from measurements in Fe-C melts with elevated carbon contents it is therefore necessary to apply Equations 2 and 3 instead of Equation 4. At higher carbon contents in the iron melt, it is also recommended to use $ThO_2$-based electrolytes which not only exhibit a lower partial electronic conductivity but also a higher chemical stability as compared with $ZrO_2$-based electrolytes.

At controlled $p_{CO}$, oxygen activity in Fe-O-C melts is closely related to carbon activity. Experimental results for the reaction

$$\underline{C}_{Fe}, 1wt\% + \underline{O}_{Fe}, 1wt.\% \rightleftharpoons CO(g) \quad (5)$$

investigated by the gas equilibration technique are fairly well in accordance[9-12]. Based on Equation 3 for the cell of Figure (1b) and the free energy change of reaction 5 (according to ref. 11), a nomograph is shown in Figure 2 correlating the liquid iron at temperatures from 1300 to 1700°C and $p_{CO} = 1$ bar. In the plot $h_o$ vs. wt% $\underline{C}$, oxygen activity is expressed as

$$\log h_O = -\log wt\%\underline{C} - e_{\underline{C}}^{\underline{C}} \cdot wt\%\underline{C} - r_{\underline{C}}^{\underline{C}} \cdot wt\%\underline{C}^2 - \log K_{CO} \quad (6)$$

with $e_{\underline{C}}^{\underline{C}} = 0.14$ and $r_{\underline{C}}^{\underline{C}} = 0.0074$[13]. The $\underline{O} - \underline{C}$ equilibrium relationship is only slightly temperature-dependent. At 4 wt%$\underline{C}$, e.g., $a_O$ values around $10^{-4}$ are expected, corresponding to oxygen partial pressures in the Fe-C melts in the order of $10^{-15}$ bar.

**Test performance of sensors in Fe-O-C and Fe-O-C-$X_n$ melts—** The two types of sensors used in the laboratory experiments are schematically sketched in Figure 3[8]. Materials and dimensions of the sensor parts are also given in Ref. 8. The experiments were carried out in a Tammann furnace on 500 g of liquid iron contained in alumina crucibles (Figure 4). The gas-tight furnace was flushed with CO at $p_{CO} = 1$ bar. An externally attached molybdenum contact lead was immersed into the melt from the top together with the oxygen sensor. In this arrangement, one emf reading was obtained for one individual carbon content from each run.

Measurements in Fe-O-C melts were made at C contents from 0.05 to 4 wt% and temperatures of 1400, 1500 and 1600°C using sensors of type A and B in Figure 3[8]. In Figure 5, emf and oxygen activity are given as a function of carbon

# ELECTROCHEMICAL SENSING OF CARBON

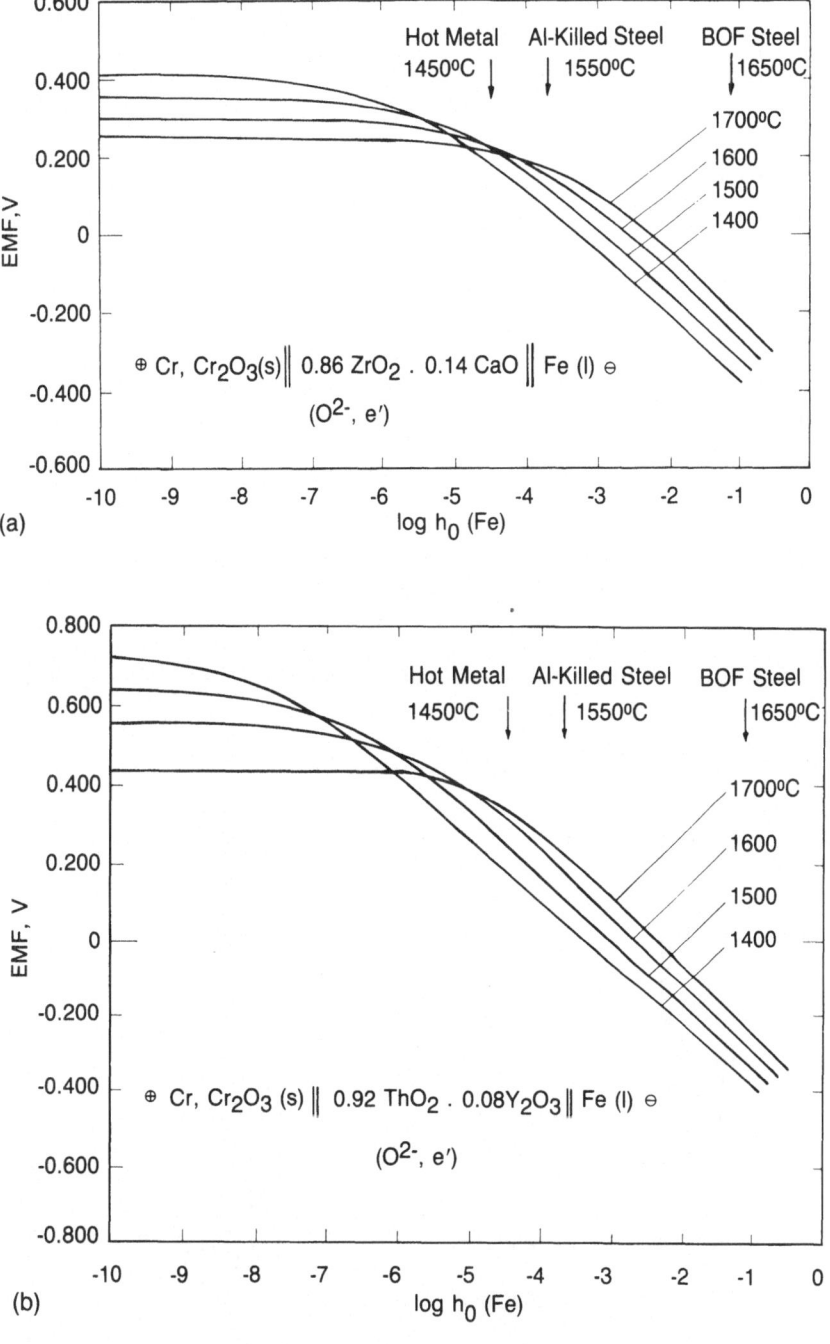

Figure 1. EMF-log $h_O$(Fe) functions for oxygen sensors with Cr-Cr$_2$O$_3$ reference using (a) ZrO$_2$ (CaO) and (b) ThO$_2$ (Y$_2$O$_3$) electrolytes.

*References pp. 235–236*

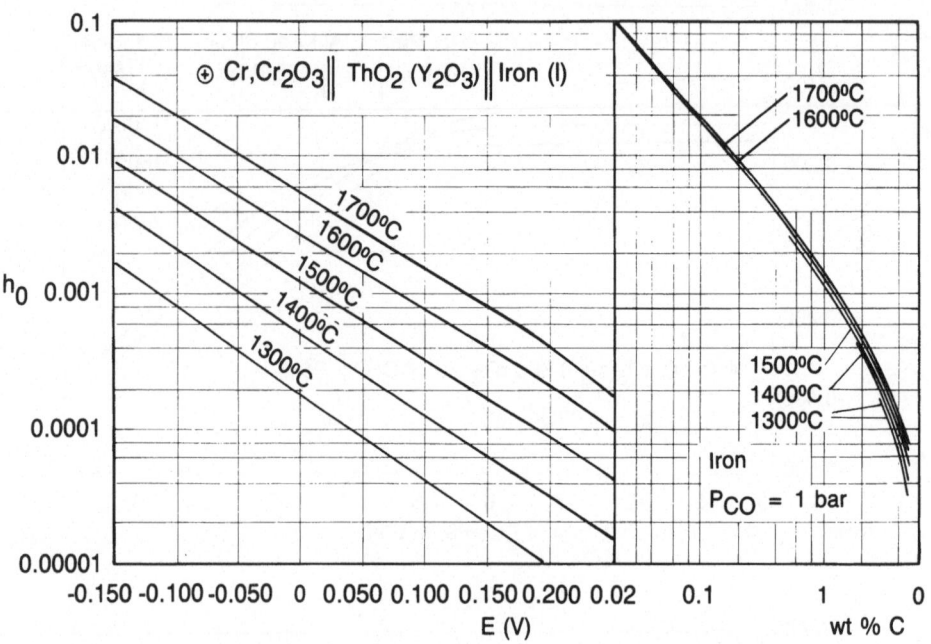

Figure 2. Nomograph correlating measured EMF with $h_O$ and wt. % $\underline{C}$ in iron melts.

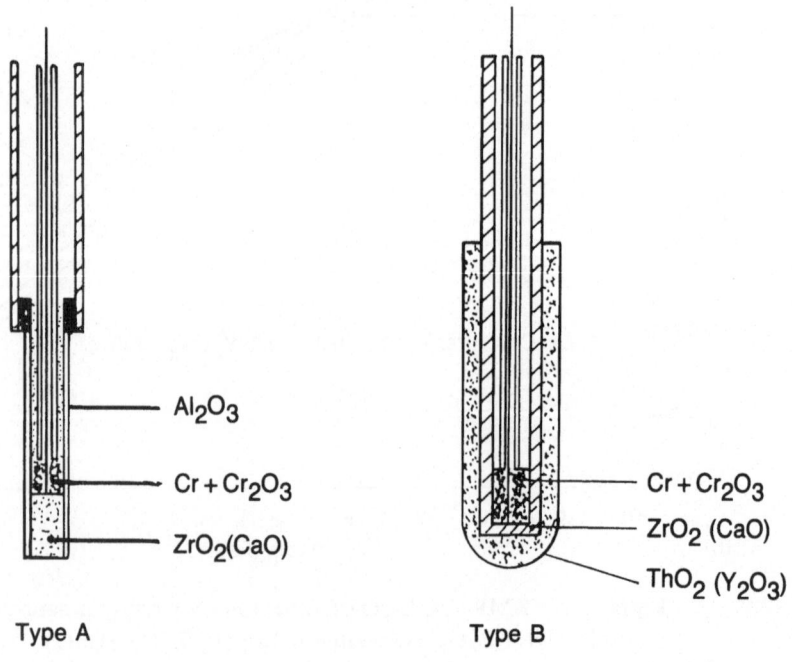

Figure 3. Types of sensors.

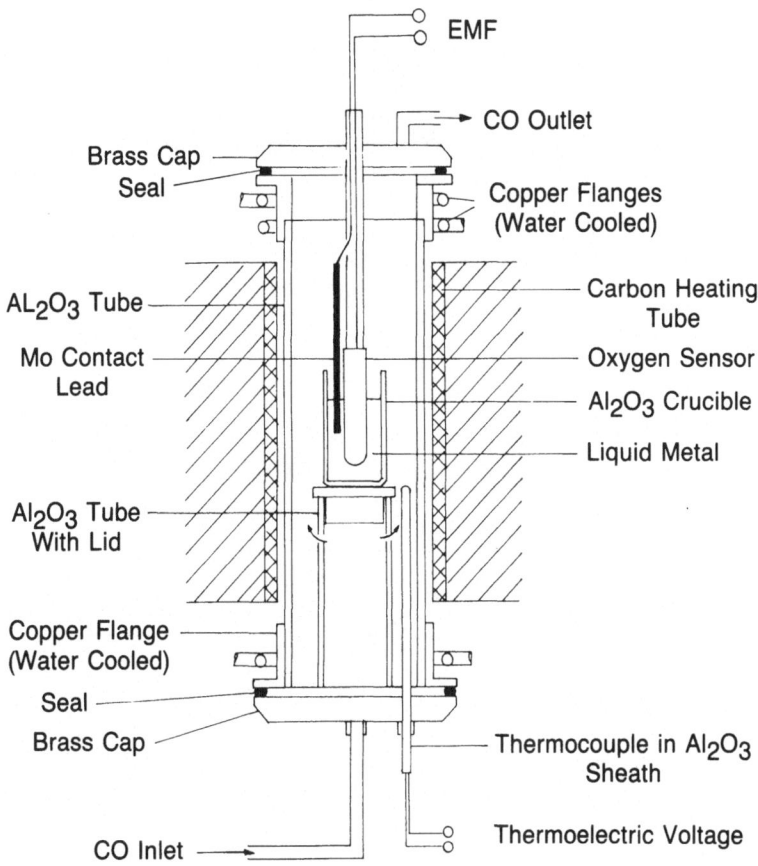

Figure 4. Experimental set-up for EMF measurements under pure CO gas.

content at 1600°C, and Figure 6 is a double-logarithmic plot of $h_O$ vs. wt% $\underline{C}$ at 1600, 1500 and 1400°C. It becomes clear that the experimental data satisfactorily represents the well known O-C equilibrium relationship[11]. Using the $h_O$ data from the emf measurements and the results of O and C analysis from quenched samples, the activity coefficients $f_O$ and $f_C$ can be calculated as follows:

$$f_O = h_O(\text{emf}) \cdot \text{wt}\%\underline{O}^{-1} \tag{7}$$

$$f_C = \frac{p_{CO}}{h_O(\text{emf}) \cdot \text{wt}\%\underline{C}} \cdot \exp\left(\frac{\Delta G°_{CO}}{RT}\right)(h_O(\text{emf}) \text{ see Eq. [3]}) \tag{8}$$

In Figure 7 log $f_O$ and log $f_C$ are graphically represented as a function of carbon content. From these functions, interaction parameters of first and second order were evaluated as shown in Table 2 in comparison with the data compiled by Sigworth and Elliott[13].

*References pp. 235-236*

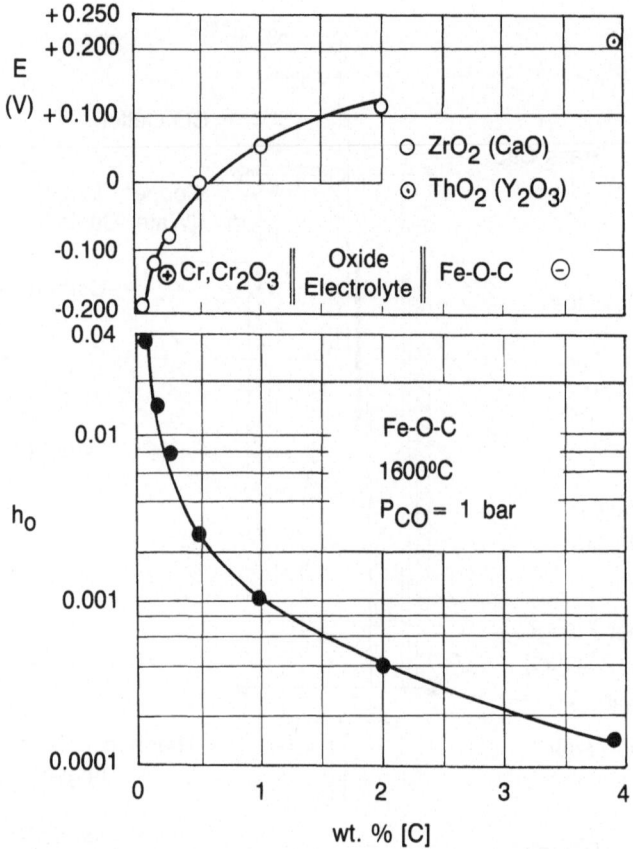

Figure 5.  EMF measurements in Fe-O-C melts at 1600°C and $p_{CO} = 1$ bar.

Oxygen activities were also measured in Fe-O-C-$X_n$ melts at a fixed carbon content of 4 wt.% and varying contents of the alloying elements $X_n$=Cr, Mn or Si under pure CO gas. It was shown that $h_O$ is decreased by additions of silicon and increased by additions of chromium or manganese. The variation of the activity coefficients $f_O$ (Equation 7) and $f_C$ (Equation 8) with % $X_n$ is given in Figure 8 at 1400°C.

**Further development**—It is obvious from the results of this fundamental study that reliable oxygen and carbon sensing in Fe-O-C and Fe-O-C-$X_n$ melts is achieved at a well-defined carbon monoxide pressure. For potential use under industrial conditions, an alternative sensor was therefore designed which permits maintenance of a known carbon monoxide pressure at the tip of the immersed sensor where the solid electrolyte is in contact with the liquid metal. The sensor is mainly based on the earlier described plug type sensor developed for long-term measurements in steel melts[14,15]. Essentially two features are added (Figure 9):
a. An $Al_2O_3$ capillary tube is inserted through a bore in the solid electrolyte plug serving as a channel for CO gas.

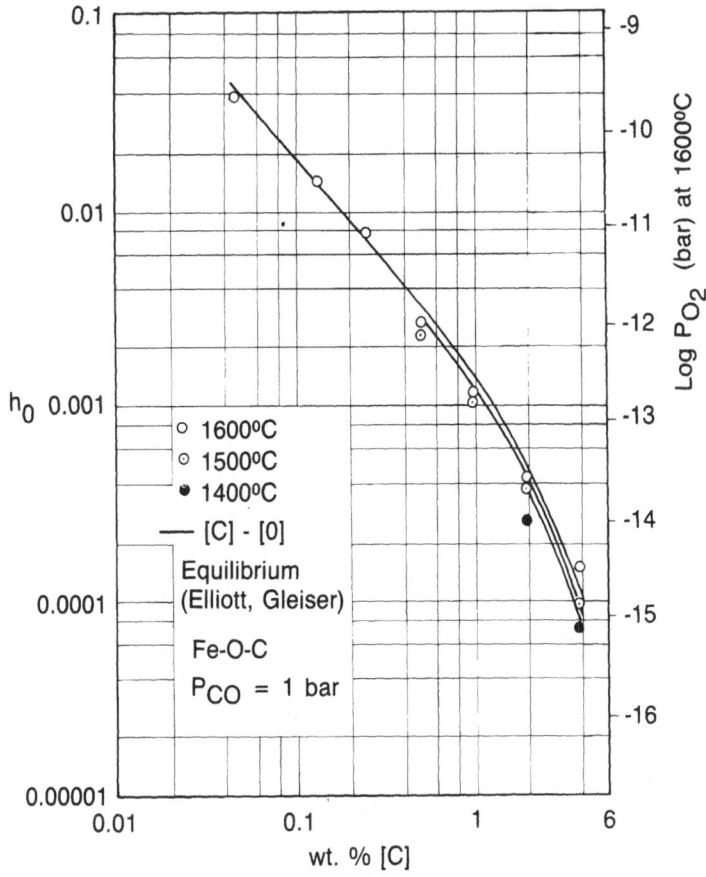

Figure 6.  $h_O$–wt.% $\underline{C}$ relationship in iron melts.

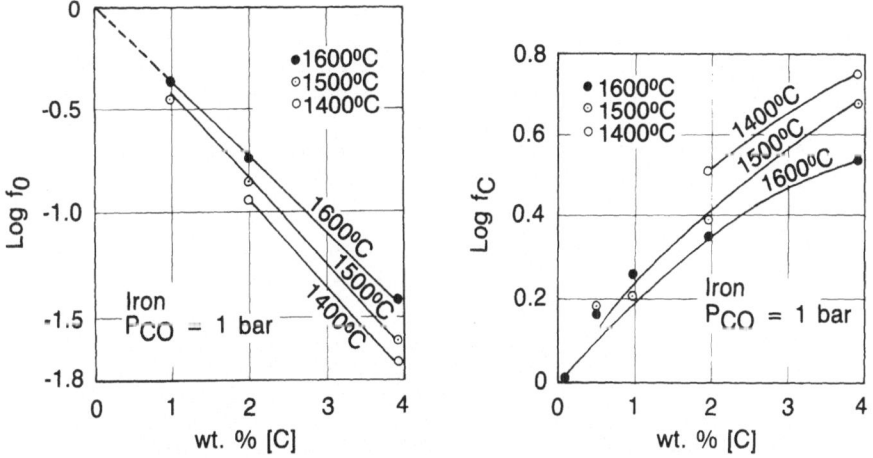

Figure 7.  Activity coefficients $f_O$ and $f_C$ in Fe-O-C melts.

## TABLE 2

### Interaction parameters in liquid Fe-O-C alloys

|  |  | present work | G. K. Sigworth, J. F. Elliott[13] |
|---|---|---|---|
| $e_O^C$ | 1400°C | −0.49 | — |
|  | 1500°C | −0.45 | — |
|  | 1600°C | −0.43 | −0.45 |
|  |  | $-\frac{1000}{T} + 0.11$ | — |
| $r_O^C$ |  | 0.01 | 0 |
| $e_C^C$ | 1500°C | 0.19 | 0.15 |
|  | 1600°C | 0.17 | 0.14 |
|  |  | $\frac{800}{T} - 0.26$ | $\frac{158}{T} + 0.06$ |
| $r_C^C$ |  | 0.0002 | 0 |

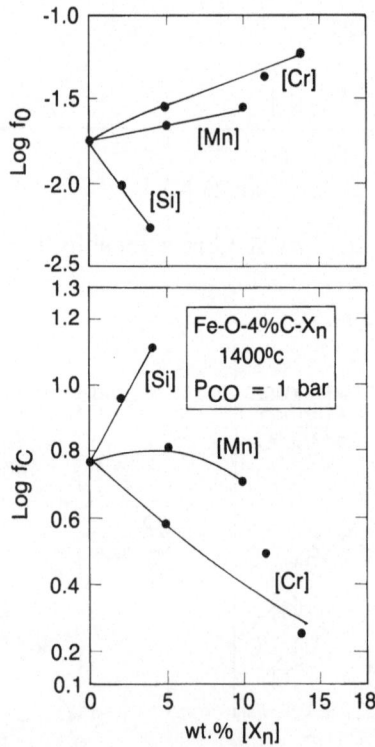

Figure 8. Activity coefficients $f_O$ and $f_C$ in Fe-O- 4 wt.% $\underline{C}$-$\underline{X}_n$ melts at 1400° C ($X_n$ = Cr, Mn, Si).

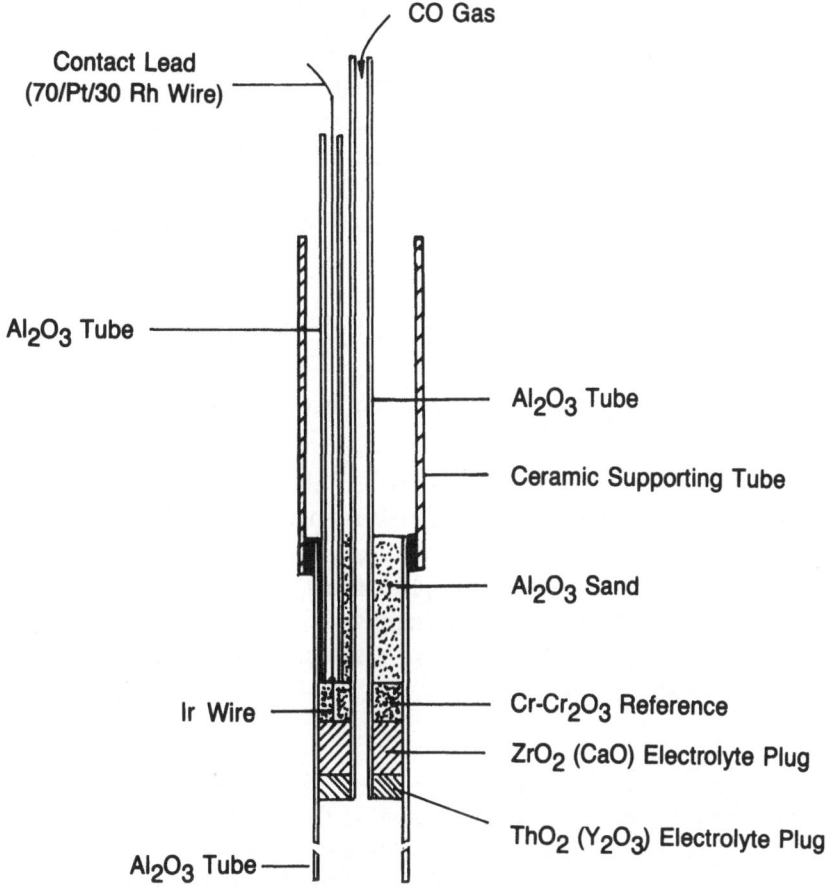

Figure 9. Schematic sketch of the carbon - oxygen sensor.

b. A bi-electrolyte plug is used consisting of a $ThO_2$ $(Y_2O_3)$ tablet at the bottom and a $ZrO_2$ (CaO) tablet on top of it.

As shown in Figure 9, the solid electrolyte plug is fixed in an elevated position 2 to 3 cm above the lower end of the tube so that the liquid metal filling the small chamber is saturated with the CO gas flowing through the capillary tube. The emf of the sensor is correlated with $h_O$ and $h_C$ of the melt in equilibrium with the CO gas as follows:

$$E = \frac{RT}{F} \ln \frac{p_{e'}^{1/4} + p_{O_2}^{1/4}(Cr, Cr_2O_3)}{p_{e'}^{1/4} + \left(\dfrac{p_{CO}}{f_C \cdot K_{CO} \cdot K_O \cdot wt.\%C}\right)^{1/2}} \qquad (9)$$

$$h_C = a_O^{-1}(emf) \cdot \exp(\frac{\Delta G°_{CO}}{RT}) \cdot p_{CO} \qquad (10)$$

*References pp. 235-236*

$$\text{wt.\%C} = h_O^{-1}(\text{emf}) \cdot \exp(-\frac{\Delta G^\circ_{CO}}{RT}) \cdot p_{CO} \cdot f_C^{-1} \qquad (11)$$

To obtain actual $h_O$ and $h_C$ data in the metal bulk two approaches may be followed. Measurements using one sensor, as described, can be performed when the $p_{CO}$ of the gas flowing through the capillary tube is adjusted to the $p_{CO}$ in the melt. If the two carbon monoxide pressures deviate markedly from each other, a usual sensor as shown in Figure 3 can be used in addition and the actual carbon activity will be obtained from the relationship

$$h_C = K_{CO}^{-1} \cdot \frac{p_{CO,\,eq.}}{h_{O,\,eq.}} + (h_O - h_{O,\,eq.}) \qquad (12)$$

where the subscript eq. denotes the equilibrium with the CO gas supplied through the capillary tube.

The C-O equilibrium sensor, as described above, was tested on 500 g iron melts. The data given in Figure 6 at 1600°C were confirmed by these measurements. Additional experimental work is needed to develop the sensor to an industrial standard and to check its reliability in plant trials. Further development includes the possibility of replacing the CO gas supply through the capillary tube with a small reservoir of CO released, e.g., from a metal carbonyl and the need for a suitable iron bath contact lead.

## SILICON SENSING

**Basic considerations**—The objective of the present study was to develop an electrochemical cell for the direct measurement of silicon activity and silicon content in Fe-C-Si melts. The concept of the sensor is based on a galvanic cell

$$\text{Me}_I - \text{Si(l)} \| \text{silicate electrolyte (l)} \| \text{Me}_{II} - \text{Si(l)}$$

where one of the metal alloys represents the silicon reference and the other one the Fe-based alloy with a variable Si content. The reliable performance of such "concentration cells" has been confirmed[17,18] with special emphasis on the fact that the transference number of ions, $t_{ion}$, is approximately unity and that the measured emf is thermodynamically interpreted in terms of the activities $a_{Si}$ of the two molten alloys. It has also been demonstrated in previous work[19,20] that cells of the type

$$\text{Pt}, O_2(g) \| \text{ silicate electrolyte (l)} \| \text{Me} - \text{Si(l)}$$

can be used to determine silicon activities in molten Me-Si alloys.

The probe designed and used in the present work is represented by the "silicon concentration cell"

$$(II) \text{Ni} - \text{Si(l)} \| \text{CaO} - \text{Al}_2\text{O}_3 - \text{SiO}_2(l) \| \text{Fe} - \text{C} - \text{Si(l)}$$

**Test performance of the sensor in Fe-C-Si melts**—A drawing of the probe is shown in Figure 10[21]. A liquid Ca-Al silicate saturated with tridymite and contained in a quartz glass tube (Figure 11a) represents the ionic conductor. A Ni-Si melt saturated with solid Ni and connected with a nickel lead wire (Figure 11b) serves as a reference electrode. The dimensions of the sensor were made as small as possible to achieve rapid thermal equilibrium upon dipping into the iron melt. Special care was taken in its design and manufacture to provide intimate contact between the liquid iron, the liquid electrolyte and the liquid-solid reference alloy. A graphite rod (+) was used as a contact lead to the metal bath. All measurements were carried out on Fe-C-Si melts saturated with graphite in a Tammann furnace using carbon crucibles at a temperature of 1400°C. As shown in Figure 12, stable emf readings were obtained after 30 to 60 s. Since the thermoelectric voltage, $E_{th}$, between nickel and graphite is involved in the measured emf, a correction had to be made according to E (galvanic) = E (measured) $- E_{th}$. $E_{th}$ was found to be 0.03 v at 1400°C for the thermocouple graphite plus / nickel minus. The experimental data of galvanic emf vs. log wt.%Si are given in Figure 13[21] and expressed as

$$E = -0.096 \log \text{wt.\%Si} - 0.123 \text{ (Volt)} \tag{13}$$

in the range between 0.05 and 1.8 wt.%Si. The pre-logarithmic factor 0.096 is equivalent to a charge transfer number of n= 3.5. Assuming n= 4 and $t_{ion} = 1$, the emf of the cell should follow the relationship

$$E = -\frac{RT}{4F} \ln \frac{a_{Si}(Fe)}{a_{Si}(Ni)} \tag{14}$$

The Si activity, $h_{Si}$, in the Fe-C-Si melts as based on Henry's law and 1 wt% in the dilute solution range is obtained from the existing thermodynamic data[13]

$$\log h_{Si} = e_{Si}^{Si} \text{wt\%\underline{Si}} + r_{Si}^{Si} \text{wt\%\underline{Si}}^2 + e_{Si}^{C} \text{wt\%\underline{C}} + \log \text{wt\%\underline{Si}} \tag{15}$$

$h_{Si}$ is converted to $a_{Si}$ (relative to Raoult's law and mole fraction) using the equations

$$[Si] \rightleftharpoons [Si]_{Fe, 1wt\%} \tag{16}$$

$$\Delta G_{Si}^{\circ} = -31430 - 4.12T \text{ [reference 22]}$$
$$= RT \ln \frac{\gamma_{Si}^{\circ} \cdot M_{Fe}}{100 \cdot M_{Si}} = RT \ln \frac{a_{Si}(Fe)}{h_{Si}(Ni)} \tag{17}$$

$$a_{Si}(Fe) = 0.983 \cdot 10^{-5} \cdot h_{Si} \tag{18}$$

The activity of Si in the Ni-Si melt is defined as

$$a_{Si}(Ni) = \gamma_{Si}(Ni) \cdot x_{Si}(Ni) \tag{19}$$

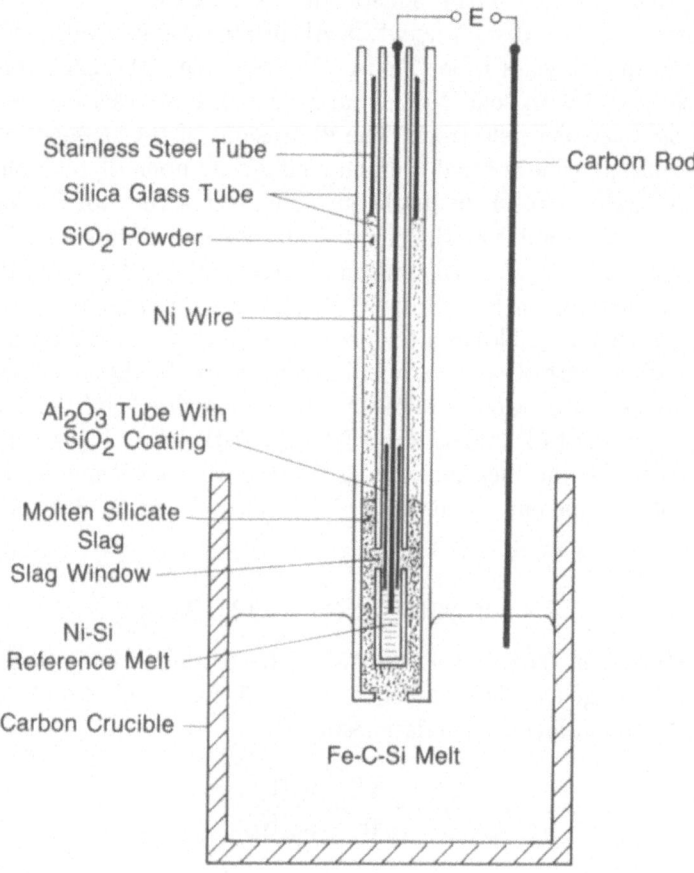

Figure 10. Schematic sketch of the silicon sensor.

The datum, $\log \gamma_{Si}(Ni) = -4.2$, determined at 1480°C and $x_{Si} = 0.07$[20] has been taken as an approximation in the present evaluation. $x_{Si} = 0.07$ (3.5 wt% Si) represents the composition of the liquidus line in the Ni-Si system at 1400°C, and $a_{Si}(Ni) = 4.45 \cdot 10^{-6}$ is obtained for this composition. The results of the aforementioned calculation are given in Table 3 as $h_{Si}$ and E depending on the melt composition. The calculated emf − wt% Si function is included in Figure 13 and agrees satisfactorily with the experimental data. A larger deviation was found in the study published by Egami et al where a pure liquid silicon reference was used with a graphite contact rod[23].

**Further development**—For a potential industrial application, some additions and alterations have to be considered. To further reduce the response time of the sensor, its dimensions should be diminished, at least in the case of short-term measurements. It is also necessary to provide a reliable contact lead to the

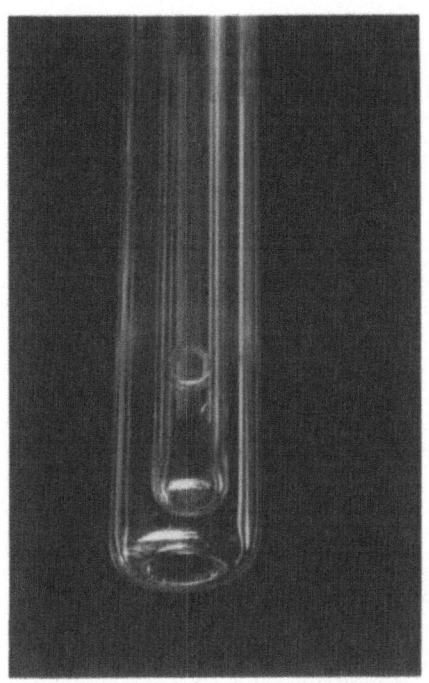
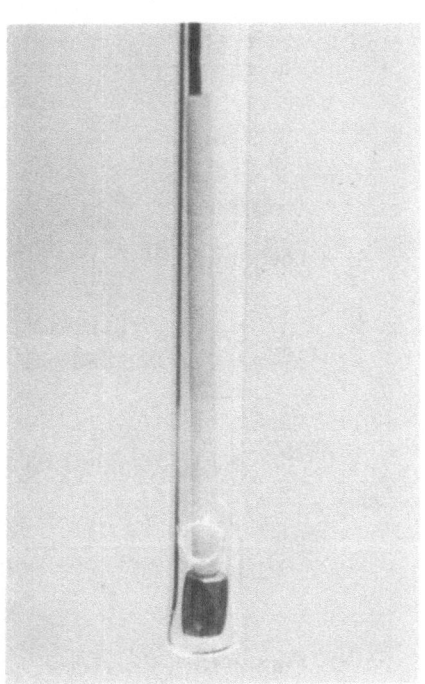

Figure 11.  Quartz glass parts (a) and Ni-Si reference electrode (b)
of the silicon sensor.

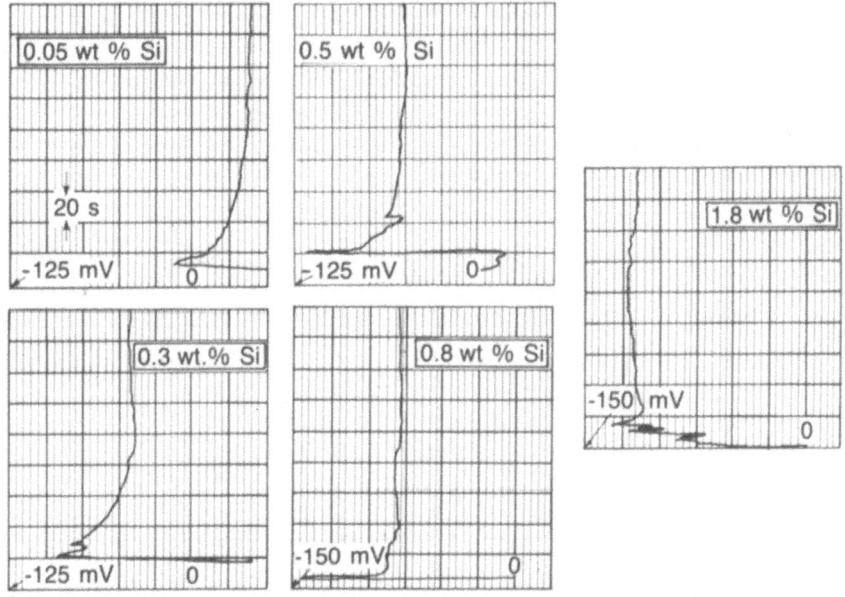

Figure 12.  EMF readings from sensors at various Si contents
in C-saturated Fe-C-Si melts.

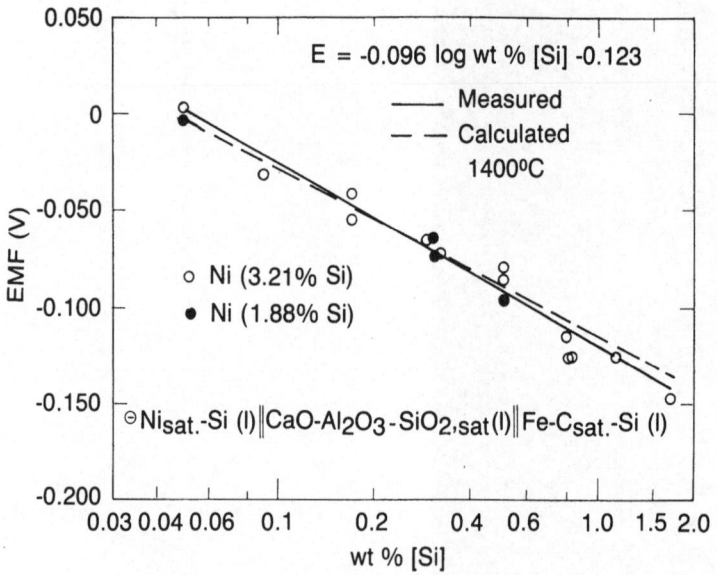

Figure 13. Corrected EMF vs. wt.% $\underline{Si}$ in C-saturated Fe-C-Si melts at 1400°C.

### TABLE 3

Calculated data for theoretical EMF's at various silicon contents and activities

| Fe-C-Si alloy | | | | Ni-Si alloy | | E (V) |
| Si wt.% | C wt.% | $h_{Si}$ | $a_{Si}$ | Si wt.% | $a_{Si}$ | Ni-Si⊖ |
| --- | --- | --- | --- | --- | --- | --- |
| 0.1 | 4.82 | 0.988 | $9.7 \cdot 10^{-6}$ | 3.5 | $4.45 \cdot 10^{-6}$ | −0.030 |
| 0.3 | 4.76 | 3.03 | $2.98 \cdot 10^{-5}$ | ″ | ″ | −0.070 |
| 0.5 | 4.70 | 5.16 | $5.07 \cdot 10^{-5}$ | ″ | ″ | −0.090 |
| 0.8 | 4.61 | 8.52 | $8.52 \cdot 10^{-5}$ | ″ | ″ | −0.108 |
| 1.2 | 4.49 | 13.3 | $1.31 \cdot 10^{-4}$ | ″ | ″ | −0.124 |

Fe-C-Si melt. In carbon-saturated melts and for one-reading probes, a graphite rod attached to the sensor can be used. For longer measuring intervals, an iron rod inserted through the bottom or the wall of the furnace or ladle is preferable. When Si measurements in Fe-C-Si melts at temperatures remarkably above 1400°C are required, metal-silicon reference alloys with a higher melting point of the base metal, e.g. Fe or Pt, should replace the Ni-Si alloy. Reliable emf readings for periods up to 5 minutes were obtained from the probes described in this study. Further experimental work has to be carried out to investigate the probe behavior at extended measuring intervals.

## REFERENCES

1. W. A. Fischer and D. Janke: *Metallurgische Elektrochemie*, Verlage Springer, Berlin, Heidelberg, New York, and Stahleisen mbH, Düsseldorf, 1975.

2. *Elektrochemische Sauerstoffmessung in der Metallurgie*, Stahleisen-Berichte, Verein Deutscher Eisenhüttenleute (VDEh), ed., Verlag Stahleisen mbH, Düsseldorf, 1985.

3. K.-H. Ulrich and K. Borowski, *Archiv Eisenhüttenwesen*, 1968, *39*, 259–63.

4. W. A. Fischer and D. Janke, *Archiv Eisenhüttenwesen* 1971, *42*, 249–55.

5. H. U. Lindenberg and P. Meierling, in: *Preprints 3rd Germany-Japan Seminar*, Verein Deutscher Eisenhüttenleute (VDEh), 1978, 197–230.

6. K. Yamada, K. Iwasaki, H. Nakamura, O. Yamase, S. Kuriyama, and E. Ogura, *Trans. Iron Steel Inst. Japan*, 1983, *23*, B-87.

7. H. Schmalzried, *Ber. Bunsengesellschaft Phys. Chem.*, 1962, *66*, 572–576.

8. A. R. Romero, J. Härkki, and D. Janke, *Steel Research*, 1986, in press.

9. S. Banya and S. Matoba, *Physical Chemistry of Process Metallurgy*, Part 1, G. R. St. Pierre, ed., New York, 1959, 373–401.

10. T. Fuwa and J. Chipman, *Trans. Metallurg. Soc. AIME*, 1960, *218*, 887–91.

11. J. Elliott, M. Gleiser, and V. Ramakrishna, *Thermochemistry for Steelmaking*, *2*, Reading, Mass., Palo Alto, London, 1963.

12. H. Schenck and H. Hinze, *Arch. Eisenhüttenwesen*, 1966, *37*, 545–50.

13. G. K. Sigworth and J. F. Elliott, *Metal Science*, 1974, *371*, 298–310.

14. D. Janke, *Stahl und Eisen*, 1983, *103*, 29–30.

15. D. Janke, *Archiv Eisenhüttenwesen*, 1983, *54*, 259–66.

16. D. Janke and W. A. Fischer, *Archiv Eisenhüttenwesen*, 1975, *46*, 755–760.

17. O. A. Esin and L. K. Gavrilov, *Izv. Akad. Nauk SSR*, OTN, 1951, *8*, 1234–42.

18. K. Sanbongi and M. Ohtani, *Sci. Rep. Tohoku Univ.*, 1953, *5*, 350.

19. K. Schwerdtfeger and H.-J. Engell, *Arch. Eisenhüttenwesen*, 1964, *35*, 533–40.

20. K. Schwerdtfeger and H.-J. Engell, *Trans. Metallurg. Soc. AIME*, 1965, *233*, 1327–32.

21. K. Ichihara, D. Janke, and H.-J. Engell, *Steel Research*, 1986, *57*, 166–71.

22. F. Wooley and J. F. Elliott, *Trans. Metallurg. Soc. AIME*, 1967, *239*, 1872.

23. T. Onoye, A. Egami, S. Nishi and K. Narita, *Trans. Iron Steel Inst. Japan*, 1984, *24*, B-45.

## DISCUSSIONS

**K. S. Goto** *(Tokyo Institute of Technology)*

Can you be sure of C-CO equilibrium, especially when Si is in the metal?

**A.** The results obtained from the EMF measurements with $ZrO_2$- and $ThO_2$-based oxygen probes in Fe-O-C melts under pure CO gas and from chemical analysis of quenched samples indicate that a situation close to the C-O equilibrium is attained. The experimental data reflect the earlier results[9,10,12] achieved by a gas equilibration technique. Figure 8 indicates that Si, Cr and Mn did not interfere with the C-O equilibrium.

**K. S. Goto**

In Japan results from oxygen sensors are scattered when %O<50 ppm in steel. In cast irons the %O≃1 ppm, so much scatter would be expected. Can you comment?

**A.** I agree that scatter of data may occur when $a_O$ is measured in rimming steels due to the fluctuating release of CO gas. On the other hand, reliable and reproducible $a_O$ data are obtained from probe measurements in Al-killed steels at levels of 2 to 4 ppm O. Moreover, I have shown that reliable $a_O$ values around 1 ppm can also be measured in Fe-C melts under controlled $p_{CO}$.

**R. J. Fruehan** *(Carnegie-Mellon University)*

As you know, we have extensively evaluated commercial oxygen tips in laboratory conditions. Dr. Niedringhaus did that and found that several different commercial brands of oxygen sensors do not give the real oxygen activity in a well-controlled laboratory experiment in which CO is over the melt and the commercial sensor tip is immersed. The potential gradient across your plug sensor is less since it is much thicker, and therefore the amount of oxygen ions transmitted because of electrical conductivity is smaller. So you could get less polarization, and the erroneous readings we got could be due to polarization on the liquid metal side. Would you comment?

**A.** The advantage of the plug type sensor at a greater thickness of the solid electrolyte ceramic body over a thin-walled electrolyte tube is supposed to be the smaller gradient of oxygen potential across the solid electrolyte which reduces polarization effects at the electrolyte interfaces.

**R. J. Fruehan**

What was your experience with thin-wall commercial sensors?

**A.** Under those same conditions, one will have the plateau for 30 to 60 seconds, and then the emf will decay.

**R. J. Fruehan**

Will they give the same top reading?

**A.** One will achieve practically the same top reading.

**R. J. Fruehan**

In cast iron?

**A.** This was in cast iron.

**David Robertson** *(University of Missouri, Rolla)*

Your results showed that carbon decreases the activity coefficient of oxygen in molten iron. But our results [N. H. El-Kaddah and D. G. C. Robertson, "Equilibria in Reactions of CO and $CO_2$ with Dissolved Oxygen and Carbon in Liquid Iron," *Metallurgical Transactions, 8B* (1977), pp 569–579] showed that carbon *increases* the activity of coefficient of oxygen. This has been confirmed by another of my students [Guzman and Robertson, *Met. Trans. B*, in press]. In my opinion, high-carbon iron melts pick up oxygen from the crucible, and this leads to the measurement of too high an oxygen in the melt. Containerless melting and high total CO pressures must be used to measure the very low oxygen concentrations properly. Would you like to reply to these comments?

**A.** The interaction coefficients of oxygen, $e_O^C$, in Fe-O-C melts determined in our work are satisfactorily in agreement with more than 20 previous studies which all indicate a negative effect of carbon on oxygen activity (see references 9, 10 and 12 in our paper). Reduction of the $Al_2O_3$ containers used in our work by the carbon in the melt is fairly small. Al contents of < 0.003 wt.% at 1 wt.% $\underline{C}$ and 0.014 wt.% at 4 wt.%$\underline{C}$ were found at 1600°C. Some further experiments in extremely stable crucibles made of CaO and with inductive stirring of the melt would be helpful to support the present results.

**Joyce Niedringhaus** *(ARMCO)*

Were your cells commercial or made in your laboratory?

**A.** We make the plug-type probes ourselves. The commercial PSZ tubes we used were from Corning Glass.

**Joyce Niedringhaus**

Did you look into the possibility that impurities in the electrolyte may contribute to electronic conductivity of the cell which may not be adequately compensated for by the $P_{e'}$ by Schmalzried?

**A.** In former experiments we have investigated $ZrO_2$-based electrolytes at different levels of impurities such as FeO, $Fe_2O_3$, $Al_2O_3$, $SiO_2$. No significant deviations in the EMF plateaus were observed. Attention should be drawn to the proportions of the semi-conducting tetrahedral $ZrO_2$ phase and the ionic conducting cubic $ZrO_2$ (CaO) or $ZrO_2$ (MgO) phase in the two-phase electrolyte materials.

# A Comment to Electrochemical Sensing of Carbon, Oxygen and Silicon in Iron Melts

### A. R. Romero, K. Ichihara, H.-J. Engell and D. Janke

K. S. Goto
*Tokyo Institute of Technology*

This paper is very impressive, because it has been shown that galvanic cells can give the contents of carbon and silicon dissolved in iron. However to estimate the carbon and silicon contents, thermodynamic equilibrium should prevail in liquid iron. The present author and his coworkers[1] have measured the oxygen pressures both in slag and in pig iron, and also in cast iron just after flowing out from the blast furnaces. The latter work was done in 1979, 1980 and 1981.

Figure 1, from top, shows temperatures of slags and pig iron, oxygen pressures in slag, in pig iron and in equilibrium with carbon, and at bottom, the silicon content in pig iron. Figure 2 shows the affinities of the following reactions, respectively (minus Gibbs free energy change);

$$\underline{Si} \text{ (in iron)} + 2\underline{O} \text{ (in iron)} = SiO_2 \text{ (in slag)} \tag{1}$$

$$\underline{Mn} \text{ (in iron)} + \underline{O} \text{ (in iron)} = MnO \text{ (in slag)} \tag{2}$$

$$\underline{S} \text{ (in iron)} + CaO \text{ (in slag)} = CaS \text{ (in slag)} + \underline{O} \text{ (in iron)} \tag{3}$$

The results shown in Figure 2 suggest that only the second reaction is in equilibrium and the other reactions are not.

This implies that the oxygen in pig iron is supersaturated with respect to silicon content.

The small horizontal lines in the center part of Figure 1 are generally lower than the measured oxygen pressure in pig iron. Therefore, the oxygen in pig iron is also supersaturated with respect to carbon.

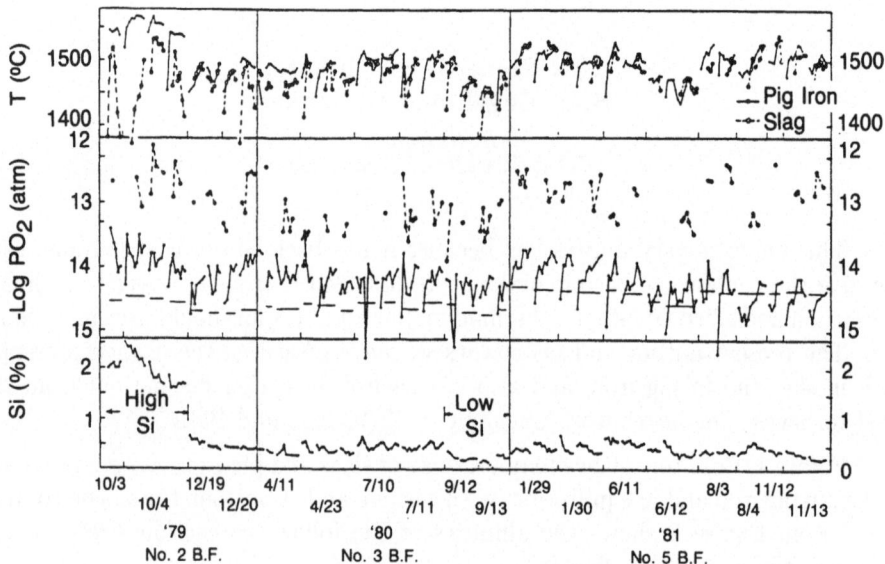

Figure 1. Change of Po₂ and temperatures in pig iron and slag, Po₂ in equilibrium with C and CO (horizontal line), and Si content in three blast furnace during long period.

# ELECTROCHEMICAL SENSING OF CARBON

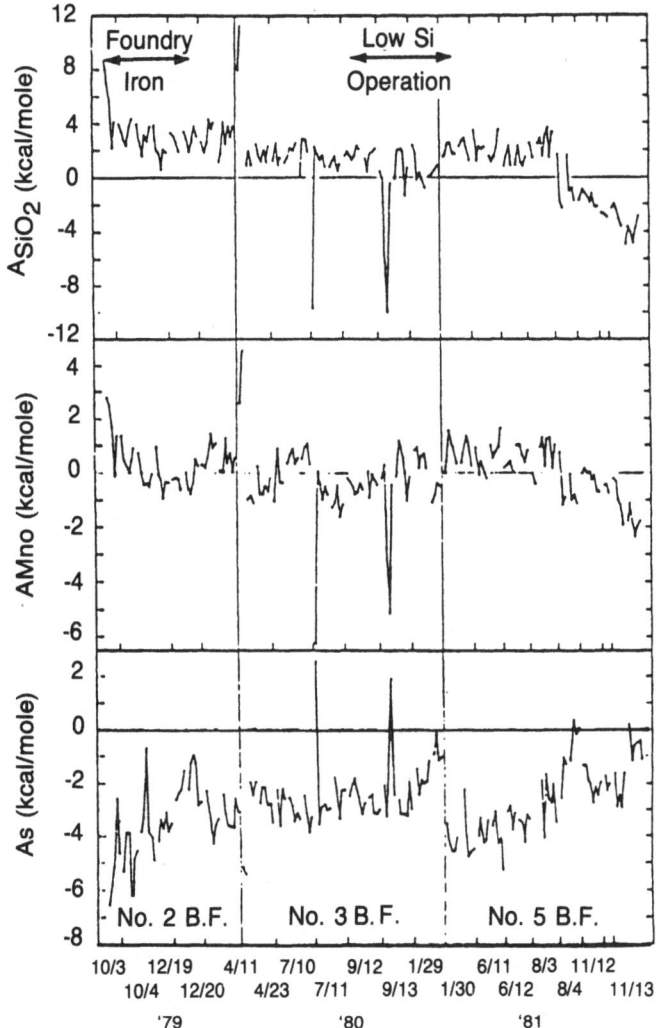

Figure 2. Chemical affinities of the transfer reactions of Si, Mn, and S between slag and pig iron calculated from $P_{O_2}$ and temperature in pig iron.[5]

In conclusion, the carbon sensor invented by Dr. Janke and his coworkers can be used for industrial measurements but with the corrections depending upon the magnitude of the supersaturation.

## REFERENCE

1. K. S. Goto, K. Nagata and S. Yamaguchi, "Review on New Uses of Oxygen Sensors for Iron and Steelmaking Slags" *Iron and Steelmaker*, November 1983, 43–52.

# POSSIBLE USES OF SENSORS IN THE ALUMINUM FOUNDRY INDUSTRY

## DEREK J. FRAY

*Department of Metallurgy and Materials Science*
*University of Cambridge*
*Cambridge, CB2 3QZ, UK*

## ABSTRACT

The need for on-line monitoring of elements in molten aluminum is discussed. One possible method is an electrochemical technique in which a voltage is produced which is related to the concentration of the element in the aluminum. The principle of electrochemical sensors is outlined and examples are given for hydrogen, lithium, and sodium determination in molten aluminum using sensors based upon calcium hydride, lithium silicate-phosphate and sodium $\beta$-alumina, respectively.

## INTRODUCTION

Although modern analytical techniques are fast, time is spent taking a sample and sending it to the analytical facility and, during this time period, the composition of the melt may be changing either due to a refining reaction or interaction with the environment. As a consequence, the analysis may only be of retrospective use and an on-line monitoring technique would be of considerable advantage. However, in order to perform satisfactorily in a foundry environment the technique would have to be rugged, reliable and inexpensive, have a quick response, and also be implemented by a relatively unskilled workforce. These criteria are particularly demanding, and no one technique is likely to satisfy all the requirements. A method which offers the possibility of meeting some of them is an electrochemical one based upon a simple thermodynamic measurement. The electrochemical cell can be represented by

$$W, M_{(ref)}/M^+ \text{conductor}/M_{(metal)}, W$$

where W represents the leads, $M_{(ref)}$ is the reference material of known activity or concentration, $M^+$ conductor is an ionic conductor of $M^+$ ions and M is the metal

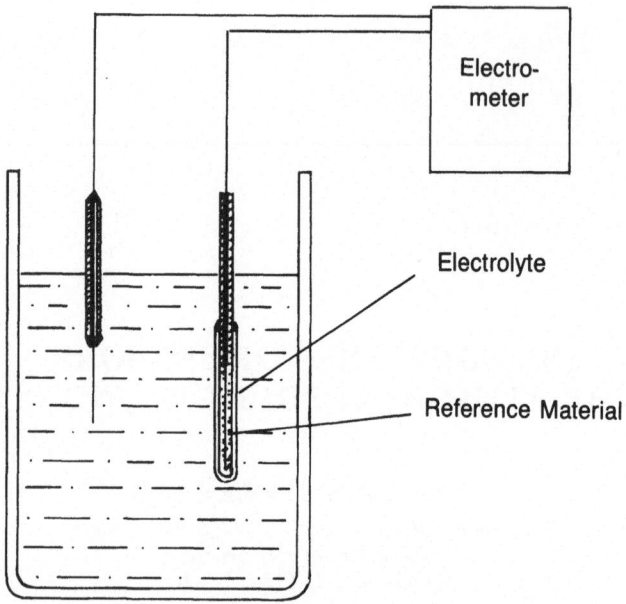

**Figure 1.** Schematic diagram of equipment required to measure the concentration of an element in solution using an electrochemical sensor.

in solution whose activity or concentration is to be determined. Under equilibrium conditions, the potential is given by the Nernst equation

$$-ZEF = RT \ln \frac{a_M}{a_{M\,(\text{ref})}} \qquad (1)$$

where $Z$ is the number of charge equivalents per g mole reacted, $E$ is the potential measured by a high impedance voltmeter, $F$ is Faraday's constant, $R$ is the gas constant, $T$ is the temperature, $a_M$ is the activity of M in the metallic solution and $a_{M(\text{ref})}$ is the activity of M in the reference. The activities can be related to the atom fraction via the relationship

$$a_M = \gamma_m \chi_m \qquad (2)$$

where $\gamma_m$ is the activity coefficient and $\chi_m$ is the atom fraction, which can simply be converted into any other unit of concentration. The main problem is, therefore, selecting suitable electrolytes and reference materials as the experimental setup is remarkably simple: a solid electrolyte tube about 5 cm long and 0.5 cm in diameter, containing the reference, two leads and a high impedance voltmeter. A schematic diagram is shown in Figure 1.

## SELECTION OF ELECTROLYTE

To obtain a stable reproducible potential, the following are requirements:

1. The electrolyte must have a high ionic conductivity, preferably greater than $10^{-5} \text{ohm}^{-1} \text{cm}^{-1}$.
2. The electrolyte/reference and electrolyte/metal interfaces must be reversible to $M^+$ ions.
3. The electronic contribution to the overall conductivity must be negligible. Otherwise the measure potential will be less than given by the Nernst expression (Equation 1) and, secondly, the active species will transfer through the electrolyte.
4. The electrolyte must not conduct other species present in the melt.

Worrell[1] has characterised four main classes of solid electrolyte and some of these may find application in the aluminum foundry industry. The four classes are:
1. Compounds containing intrinsic defects and a large band gap. Sodium chloride, potassium chloride and calcium flouride are examples. The latter is the most useful, although it does suffer from the disadvantage that it responds to a variety of species e.g. flourine[2] calcium[3] and oxygen[4].
2. Compounds which undergo an order-disorder transition below the melting point often yield a disordered solid with a high conductivity. Some examples are silver iodide, silver rubidium iodide[5] and lithium sulphate[6]; unfortunately, many of these compounds are relatively unstable.
3. Many compounds with layered or skeleton structure exhibit high ionic conductivities. The $\beta$-alumina[7] and Nasicon compounds[8] fall into this category.
4. The last group are solid solution electrolytes in which the number of defects is increased by doping. Stabilized zirconia, $CaO\text{-}ZrO_2$,[9] and thoria yttria, $ThO_2Y_2O_3$[10] are examples.

Although there might not be a suitable electrolyte with a specific mobile ion, it is still possible to utilize other electrolytes, providing equilibrium is reached between the melt and the electrolyte. For example, Wilder and Galin[11] developed a zinc sensor for molten brass using stabilized zirconia, where only the oxygen ion is mobile.

## SELECTION OF REFERENCE MATERIAL

Performance of a solid electrolyte sensor is strongly influenced by the reference material. It needs to come to equilibrium quickly and should be both an ionic and electronic conductor. For sensors measuring oxygen, sulphur, hydrogen and phosphorous it is possible to use a mixture of a metal and its oxide[10], sulfide[11], hydride[12] or phosphide[13] respectively. Examination of the phase rule

$$P + F = C + 2 \qquad (3)$$

where P is the number of phases, F is the number of degrees of freedom and C is the number of components, shows that at constant temperature and total pressure, the

*References pp. 252–253*

number of degrees of freedom is zero and, therefore, the pressure of oxygen, sulphur, hydrogen or phosphorus is fixed. For systems involving sodium and lithium, the selection of the reference is more difficult as the pure elements are reactive and most of the intermetallic compounds are unstable at elevated temperatures and subject to oxidation. However, it is possible to establish a reference $Na_2O$ activity using a mixture of sodium $\beta$-alumina and alumina or a mixture of sodium ferrite and iron oxide[15]. The $Na_2O$-$Al_2O_3$ phase diagram[16] has two regions where two phases coexist. According to the phase rule, the activities of alumina and sodium oxide are fixed when two phases coexist. To maintain a constant sodium activity via the reaction

$$Na_2O = 2Na(v) + \frac{1}{2}O_2 \tag{4}$$

it is necessary to maintain a constant oxygen partial pressure, and this can either be achieved by a metal/metal oxide mixture or by exposing the reference to atmospheric oxygen. Similar concepts can be applied to the sodium ferrite/sodium oxide system and to the development of a lithium reference. Using a slightly different approach, Dubreuil and Pelton[17] have used the equilibria between

$$Na_3AlF_6 + 3Na = 6NaF + Al \tag{5}$$

and

$$Li_3AlF_6 + 3Li = 6LiF + Al \tag{6}$$

to give constant sodium and lithium activities.

## HYDROGEN SENSORS

Hydrogen has a very deleterious effect on the properties of aluminum castings. This can be attributed to the formation of porosity upon solidification, as hydrogen solubility decreases dramatically at the solidification point. The hydrogen arises from the interaction between the aluminum and water vapour in the air, from furnace refractories and from water occluded on the initial charge. Most high quality alloys require degassing before casting.

In order to monitor the degassing process, an accurate and reliable method for measuring the dissolved hydrogen is necessary. At the present time, three methods are commonly used to determine hydrogen in molten aluminum. Vacuum extraction[18], in which a solid sample is heated under high vacuum and the quantity of evolved hydrogen is measured, is the most accurate but, also, the most time consuming. A quicker method, taking a few minutes, is to equilibrate nitrogen bubbles with the molten metal or to extract the hydrogen from a molten aluminum sample using a flow of nitrogen and to determine the hydrogen content using a katharometer or gas chromatograph[19]. A much more qualitative approach is the Straube-Pfeiffer test, in which a visual estimate is made of the formation of hydrogen bubbles as the sample is subjected to a decreasing pressure. However, as shown by Brondyke and Hess[20], the results are somewhat dependent on the inclusion content of metal.

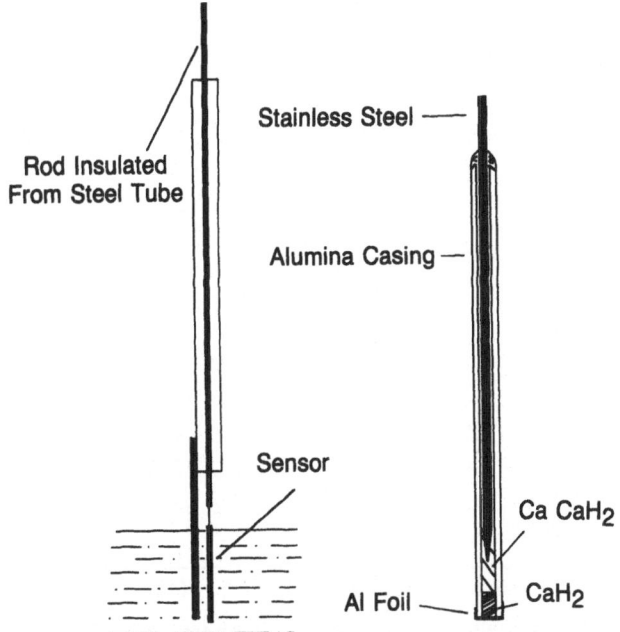

Figure 2. Schematic diagram of hydrogen probe and lance assembly. Taken from Gee and Fray.[12]

The advantages of a very rapid hydrogen analysis are obvious but none of the above methods are entirely suitable for use in the foundry. In order to mitigate this problem, Gee and Fray[12] investigated the use of calcium hydride as an electrolyte with a mixture of $Ca/CaH_2$ as the reference. To protect the electrolyte and reference from reaction with the environment, the sensor, shown in Figure 2, was sealed with cement at one end and an aluminum foil at the other. The reproducible results, shown in Figures 3, were in good agreement with those of the standard subfusion technique which takes several hours to complete, compared with 20 seconds for the sensor to reach equilibrium. Williams and McGeehin[21] report the use of a hydrogen β-alumina electrolyte in a similar sensor, but it was found that the device also responded to the presence of sodium. One way of improving the performance of both devices might be to incorporate a hydrogen permeable membrane, such as iron, between the aluminum and the electrolyte.

Recently a very sensitive room temperature hydrogen sensor based on uranyl phosphate has been developed, and it is now used industrially to monitor hydrogen concentration in steel pipe carrying sour gas streams. The hydrogen produced by corrosion at the inside pipe surface causes hydrogen absorption into the steel, and, in some cases, the hydrogen pressure or activity can reach thousands of atmospheres. When the sensor is applied to the outer surface, the activity of hydrogen

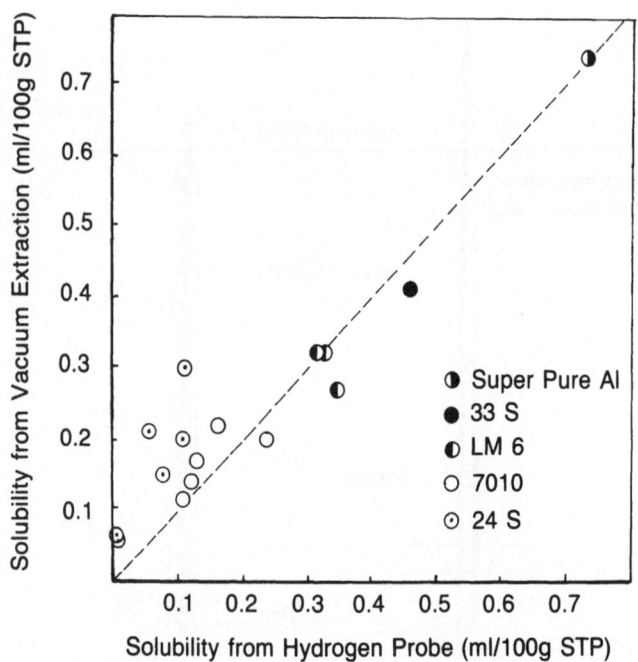

Figure 3. Comparison of solubilities, calculated from hydrogen probe, versus vacuum extraction results. Taken from Gee and Fray.[12]

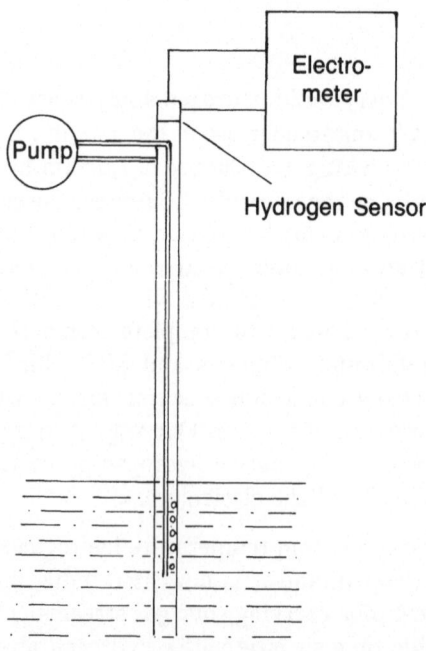

Figure 4. Schematic diagram of apparatus to measure hydrogen content of aluminum using room temperature hydrogen sensor.

at the surface is measured but as the rate of combination of hydrogen atoms, dissolved in the steel, to form hydrogen molecules in the gas is relatively slow, it can be assumed that the surface activity or concentration is equal or close to the bulk concentration. Opportunities exist for the device to be incorporated into an instrument, Figure 4, for measuring the hydrogen content of molten aluminum. Argon would be repeatedly bubbled through the melt to establish in the argon an $H_2$ partial pressure in equilibrium with H in the melt. The sensor would measure $H_2$ pressure at the room temperature part of the circuit.

## LITHIUM SENSORS

Lithium is introduced into aluminum from two sources. In the Hall Heroult cell, lithium salts are added to the cryolite electrolyte in order to raise the conductivity of the electrolyte. According to Dewing and Gilbert[23], cryolite containing 2.5% lithium flouride is in equilibrium with aluminum containing 25 ppm lithium and 80 to 120 ppm sodium. The second source of lithium is likely to arise from scrap aluminum-lithium alloys which become mixed with non-lithium containing aluminum scrap; here an on-line method for lithium analysis would be an advantage. A considerable number of lithium conductors have been developed for lithium batteries[24] but, as mentioned above, sodium is also likely to be present in the aluminum; so it is important to select an electrolyte which does not respond to sodium. Dubreuil and Pelton[17] have used the sodium $\beta$-alumina electrolyte with a reference of $Li_3AlF_6$, LiF and aluminum to measure lithium activity in molten aluminum. The presence of sodium in the aluminum would probably influence the probe potential.

A search of the properties of sodium and lithium conductors shows that some silicate electrolytes which have very much better lithium ion conduction than sodium ion conduction; and it is likely that, on insertion into liquid aluminum, these electrolytes will come to equilibrium with the lithium rather than the sodium.

One possible electrolyte, used by Yao and Fray[25], is a two-phase mixture of $Li_3PO_4$ and $Li_4SiO_4$, with an overall composition of $Li_{3.6}Si_{0.6}P_{0.4}O_4$. Slight attack of the electrolyte was observed in pure liquid lithium, but no attack was detected in aluminum-lithium melts after 56 hours of continuous immersion. The reference was a mixture of $Li_4Ti_5O_{12}$ and $TiO_2$, which was formed eutectoidly from $Li_2Ti_3O_7$. More recently, a mixture of lithium ferrite and $Fe_2O_3$ has been used[26]. As in the case of the sodium reference, both these mixtures need to be exposed to a constant oxygen partial pressure to fix the lithium activity. The typical response of two probes while adding aluminum-lithium master alloy to aluminum is shown in Figure 5. The activities, calculated from the potentials, agree closely with other determinations[27,28] and these probes may offer a way of instantaneously determining the lithium content of molten aluminum alloys.

*References pp. 252-253*

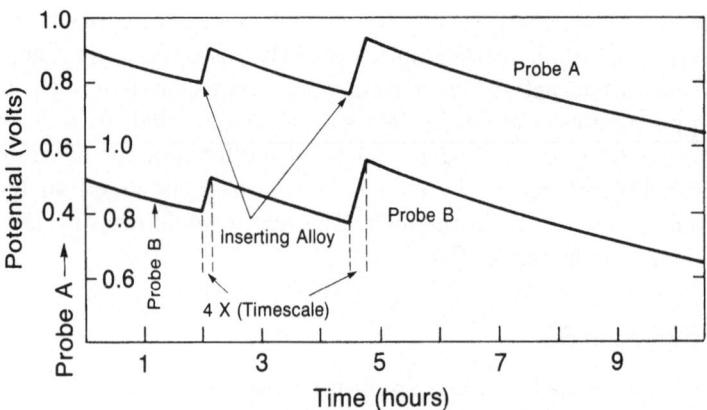

Figure 5. Response of two lithium probes tested simultaneously in the same melt. Taken from Yao and Fray[25].

## SODIUM SENSORS

The occurrence of sodium in aluminum is similar to lithium. It is co-deposited with aluminum in the Hall-Heroult cell and is also added to aluminum-silicon eutectic alloys to improve the mechanical properties. However for some wrought alloys, notably aluminum-magnesium alloys, it is very important to keep sodium below 2 ppm to prevent hot shortness occurring during rolling at elevated temperatures. Sodium is a particularly volatile element. This leads to problems when attempting to make a known sodium addition and in sampling when the analytical results may prove only to be of retrospective use. In order to develop an on-line sensor, Brisley and Fray[29] used sensors based upon the sodium $\beta$-alumina electrolyte and a reference of $\alpha$ and $\beta$-alumina. The potential vs concentration plot for super purity aluminum is shown in Figure 6 and these results were in good agreement with the data of Dewing[30] and Ransley and Neufeld[31] who both used a different experimental approach and, also, the data of Dubreuil and Pelton[17] who used a sodium $\beta$-alumina tube with a $Na_3AlF_6$,NaF reference. For the aluminum-silicon alloys it was found that the potential versus concentration curve did not follow the Nernst equation and that the activity coefficient was a strong function of the sodium concentration, indicating a strong interaction between the sodium and silicon in the aluminum melt. In all cases it was found that the sensors gave reproducible readings for about 8 hours in the melt before failing.

## GENERAL DISCUSSION AND CONCLUSIONS

Solid electrolyte sensors can either be used once, as with oxygen sensors in the steel industry, or left immersed for extended periods in the melt as in oxygen measurement in copper. The sensor lifetimes described in this paper vary. The calcium hydride sensor lasts about 20 minutes and can be used only once, whereas the hydrogen uranyl phosphate sensor has worked for months at room temperature. The sodium and lithium sensors lasted for several hours in the melts and both

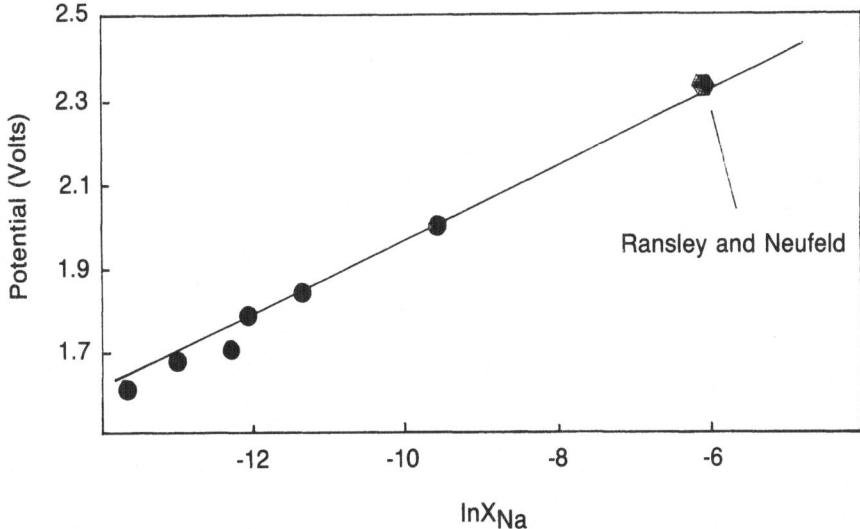

Figure 6. Plot of potential versus concentration of sodium. Taken from Brisley and Fray.[29]

survived several immersions into the molten aluminum provided care was taken to preheat the device to prevent thermal shock. These lives could probably be extended with optimization of electrolyte preparation.

This paper has described solid state sensors which have been developed using solid electrolytes in which the species to be measured is mobile in the electrolyte. This obviously restricts the elements to the available solid electrolytes. For some elements, these are no suitable solid electrolytes but it may be possible to utilize a fused salt electrolyte, such as calcium or strontium chloride contained in a ceramic membrane, for elements such as calcium and strontium. A similar approach has been used to measure zinc in molten lead using a fused zinc chloride electrolyte[32]. Other elements of interest in aluminum include iron, phosphorus and silicon. Usually these elements are likely to be present as immobile anion species in the solid electrolytes and, therefore, the direct measurements of the type described in this paper are not likely to be appreciable. However, it may be possible to measure these elements indirectly using other electrolytes by adopting the approach mentioned in the introduction for zinc in copper using stabilised zirconia[11].

One of the major problems of the development of these sensors is the transfer of the ideas from the research laboratory to the industrial process. In the steel and copper industries, oxygen sensors, based on stabilized zirconia, are widely used. The Japanese steel industry used several hundred thousand such devices in 1983[33]. Similarly, the Toyota Motor Corporation uses about 2 million oxygen sensors annually[33] for control of their engines. In other industries, sensors with an identical performance to those described in this paper, are widely and successfully used. This has resulted from a concerted effort by these industries to develop the research devices into industrial sensors. For the foundry industry to take advantage of this method of on-line analysis, a similar commitment will probably be required.

# REFERENCES

1. W. L. Worrell: *Metal-Slag-Gas Reactions and Processes*, Z. A. Foroulis and W. W. Smeltzer, eds. The Electro Chemical Society, 1975, Princeton, 822–33.
2. T. L. Markin: *Emf Measurements in High Temperature Systems*. C. B. Alcock, ed. IMM, 1968, London, 91–97.
3. J. Delcet and J. J. Egan: *J. Less Common Metals*, 1978, *59*, 229–36.
4. T. A. Ramanarayaman, M. L. Narula and W. L. Worrell: *J. Electrochem. Soc.*, 1979, *126*, 1360–63.
5. J. N. Bradley and P. O. Greene: *Trans. Farad. Soc.*, 1966 *62*, 2069–75.
6. A. Kvist and A. Bengtzelius: *Fast Ion Transport in Solids*, W. van Gool, ed. North Holland, 1973, Amsterdam, 193–99.
7. J. T. Kummer: *Progress in Solid State Chemistry*, H. Reiss and J. O. McCaldin, Vol. 7, Pergamon, 1972, Oxford, 141–75.
8. H.Y-P. Hong: *Mat. Res. Bull.* 1976, *11*, 173–82.
9. T. H. Etsell, S. Zador and C. B. Alcock: *Metal-Slag-Gas Reactions and Processes*, Z. A. Foroulis and W. W. Smeltzer, eds. The Electrochemical Society, 1975, Princeton, 834–50.
10. E. T. Turkdogan and R. J. Fruehen: *Can. Met. Quart.*, 1972, *11*, 371–82.
11. T. C. Wilder and W. E. Galin: *Trans. TMS-AIME*, 1969, *245*, 1287–90.
12. R. Gee and D. J. Fray: *Met. Trans.*, 1978, *9B*, 427–30.
13. V. W. A. Fischer and D. Janke: *Arch. Eisenhüttw*, 1966, *37*, 853–62.
15. D. J. Fray: *Metall. Trans.*, 1977, *8B*, 152–6.
16. R. C. DeVries and W. L. Roth: *J. Amer. Ceram. Soc.*, 1969, *52*, 364–69.
17. A. A. Dubreuil and A. D. Pelton: *Light Metals '85*, H. O. Bohner, ed., Met.Soc., AIME 1985, Warrendale, PA, 1197–1205.
18. C. E. Ransley and D. E. H. Talbot: *J. Inst. Metals 1955-58*, *84*, 445–52.
19. C. E. Ransley, D. E. J. Talbot and H. C. Barlow: *J. Inst. Metals*, 1957–58, *86*, 212–19.
20. K. J. Brondyke and P. D. Hess: *Trans TMS-AIME*, 1964, *230*, 1542–46.
21. D. E. Williams and P. McGeehin: *Roy.Soc.Chem. Specialist Periodical Reports. Electrochem*, 1984, *9*, 246–90.
22. S. B. Lyon and D. J. Fray: *Brit. Corr. J*: 1984, *19*, 23–29.
23. E. W. Dewing and M. J. Gilbert: *Light Metals 1980*, C. J. McMinn, ed., Met.Soc., AIME, 1980, Warrendale PA, 221–26.
24. W. Weppner: *Solid State Ionics*, 1981, *5*, 3–8.
25. P. C. Yao and D. J. Fray: *Met. Trans.*, 1985, *16B*, 41–46.
26. R. V. Kumar and D. J. Fray: Unpublished data.
27. S. P. Yatsenko and E. A. Saltykova: *Russ. J. of Physical Chemistry*, 1974, *48*. 1402–03.
28. M. L. Saboungi and M. Blander: *J. Electrochem. Soc.*, 1977, *124*, 6–13.

29. R. J. Brisley and D. J. Fray: *Metall. Trans.*, 1983, *14B*, 435–40.
30. E. W. Dewing: *Metall. Trans.*, 1972, *3*, 495–501.
31. C. E. Ransley and H. Neufeld: *J. Inst. Metals*, 1950, *78*, 25–46.
32. D. J. Fray: *Extraction Metallurgy '81*, IMM, 1981, London, 321–29.
33. K. S. Goto: *Proceedings First International Meeting on Chemical Sensors*, 1983, Fukuoka, Japan.

## DISCUSSION

**E. F. Ryntz,** *(General Motors)*

Dr. Fray, you showed a slide of the difference in voltage with different degrees of modification and sodium levels. How would these voltages be influenced by variations in composition in typical commercial melts? I am thinking of Si, Cu, Mg and Zn.

A. Since the slide was not included in the paper, it is reproduced below (Figure 7). It shows how the modification varies with sodium content and the voltages given by a sodium probe using $\beta$-alumina exposed to the ambient oxygen partial pressure.

The sodium sensor measures the sodium activity. Given the normal variation of Si, Cu, Mg and Zn in a typical melt, I would only expect the voltage to vary by millivolts, and this is likely to be within the scatter of the results. I believe it should be emphasized that the sensor does not measure modification directly but only the sodium content. If you were to cause modification by fast cooling or by additions of other elements, the sodium sensor would not give an accurate prediction of the degree of modification.

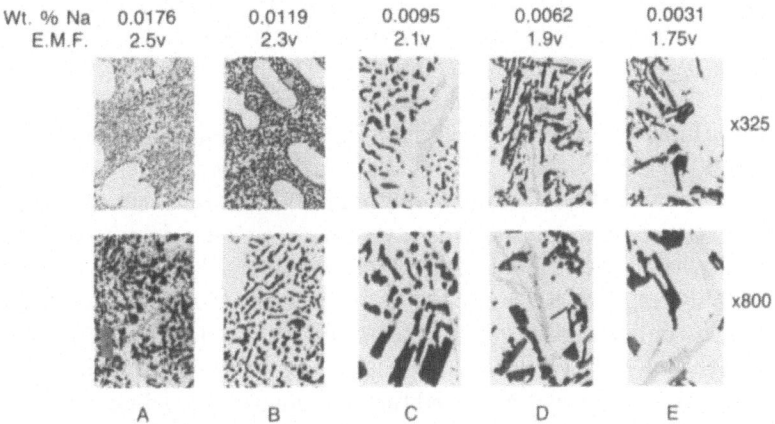

Figure 7. Microstructures of five aluminum-12.4% silicon alloy sand castings, exhibiting increasing degrees of modification as a result of increasing sodium content.

**B. L. Tiwari** *(General Motors)*

In aluminum foundries, the hydrogen concentration varies from 0.01 to 0.2 ml hydrogen per 100 gram aluminum, and the temperature range is anywhere from 700 to 800°C. My question is what is the lower limit for measuring the hydrogen concentration using your device? Secondly, isn't the temperature range too high to detect the required hydrogen concentration without decomposing the $CaH_2$?

A. At about 700°C we measured the hydrogen content down 0.01 ml per 100 gram aluminum using the sensor, whereas vacuum extraction gave values of 0.06 ml per 100 gram aluminum. It seems that at 700°C, the sensor will respond to the desired range of hydrogen concentrations. The calcium hydride is in fact unstable at these low values but the rate of decomposition is slower than the time taken to reach equilibrium with the melt. Therefore, it is possible to obtain a stable reading for several minutes before the sensor eventually fails. At 800°C, this is likely to be a greater problem as the stability of calcium hydride decreases with increase in temperature.

**B. L. Tiwari** *(General Motors)*

Was the hydrogen concentration varied by bubbling hydrogen through the melt?

A. No; hydrogen generating reagents were added.

**K. S. Goto** *(Tokyo Institute of Technology)*

There are several different proton, sodium and lithium conductors. How did you select the particular electrolytes discussed in your paper.

A. In the case of electrolytes for the sodium sensor. sodium $\beta$-alumina is the best characterized sodium conducting electrolyte. It is easy to handle and sinter, whereas Nasicon is difficult to sinter. For the lithium sensor, an electrolyte is needed which does not respond to other alkali metals, especially sodium. As many lithium electrolytes, such as lithium $\beta$-alumina, are the exact analogue of a sodium electrolyte, it is probable that a sensor made from this material would respond to both sodium and lithium. In the case of the mixed phosphate-silicate, the sodium equivalent is a very poor ionic conductor compared to the conductivity of the lithium compounds. Therefore, it is likely that the probes would only respond to changes in lithium content. A different problem exists for hydrogen conductors in that all known conductors decompose at about 700°C and in the case of hydrogen $\beta$-alumina, it also responds to changes in sodium content. At the time we performed the measurements, calcium hydride appeared to be the best electrolyte.

**K. S. Goto** *(Tokyo Institute of Technology)*

Have you considered the electrolyte developed by Iwahara and coworkers (H. Iwahara, T. Esaka, H. Vchida and N. Maeda, *J. of Solid State Ionics*, 3/4 (1981), 359)?

**A.** From the paper in *Solid State Ionics*, it appears that the electrolyte will only respond to hydrogen in the presence of water vapor. As water vapor is unstable in the presence of liquid aluminum, the application of this electrolyte to the aluminum foundry industry is perhaps unlikely.

**John Puckett** *(Nelson Products)*

Derek, I wondered if you might speculate on whether phosphorus in aluminum can be measured by this technique.

**A.** We are presently developing sensors for arsenic in zinc and phosphorus in copper in the part per million range. The sensors certainly work for arsenic in zinc and as the same principle is applied in the phosphorous case, the prospects for a sensor to measure phosphorus in aluminum are reasonably promising.

# A SUPPLEMENT TO "POSSIBLE USES OF SENSORS IN THE ALUMINUM FOUNDRY INDUSTRY" (by D.J. Fray)

## K. S. GOTO
*Tokyo Institute of Technology*

I would like to offer the following to supplement Dr. Fray's discussion of possible uses of the sensors with solid electrolytes in the aluminum foundry industry. For a proton conductor, Iwahara and his coworkers[1] have invented a solid solution of SrO and $CeO_2$ doped with $Yb_2O_3$ in 1981. Fig. 1 is the relation between the electrolyte and the partial pressure of water vapor in sample gases. The results in this figure were measured by the present author and his coworkers[2]. The results suggest that this proton conductor can be used for a hydrogen sensor at about 600°C.

Table 1 contains information about conduction mechanism of oxides and other compounds. This table was compiled by the present author from references (3) (4) and (5). From this table one can select a suitable solid electrolyte for a specific sensor.

## REFERENCES

(1) H. Iwahara, T. Esaka, H. Uchida and N. Maeda, *J. of Solid State Ionics*, 3/4 (1981), 359.

(2) K. Nagata, M. Nishino and K. S. Goto, submitted to *J. of Electrochemical Soc.*, (July 1986).

(3) K. S. Goto, "A Review on Chemical Sensors with Solid Electrolytes at High Temperature" *Proceedings of International Meeting on Chemical Sensors*, (1983) edited by T. Seiyama et al, Elsevier, Tokyo, 338–347.

(4) *Proceedings of 4th International Conference on Solid State Ionics*, (1983) Grenoble, France, North-Holland, Amsterdam.

(5) *Proceedings of 5th International Conference on Solid State Ionics*, (1985) Lake Tahoe, USA, North-Holland Publ. Co., Amsterdam.

SENSORS IN THE ALUMINUM FOUNDRY INDUSTRY 257

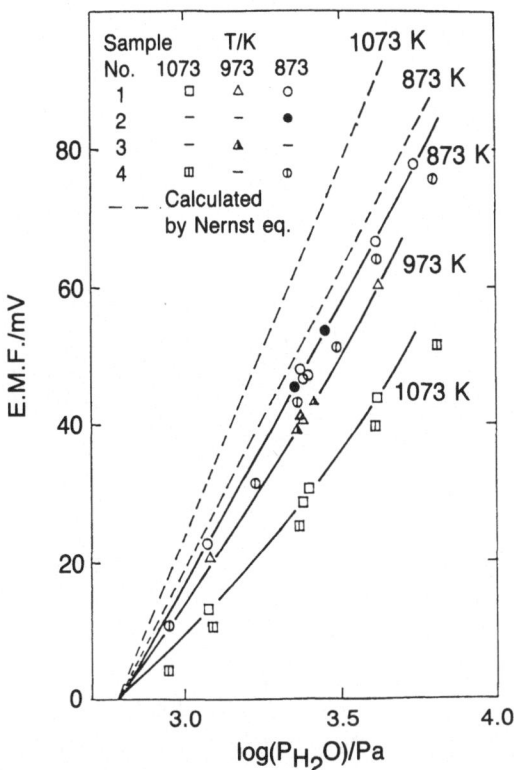

Figure 1. Relation between the elecromotive force of the following galvanic cell and the partial pressure of water vapor; $H_2O$, $H_2$ and $O_2$ | $SrCe_{0.95}Yb_{0.05}O_3$ | air with $H_2O$ in sample gas solid electrolyte as reference gas

## TABLE 1

### Classification of Oxides and Other Compounds according to Conduction Mechanism at High Temperature

| Classification | Examples of Conductors | Main Charge Carrier |
| --- | --- | --- |
| p-type Semiconductor | $Fe_{1-y}O$, NiO, CoO, $Cu_2O$, $MnO_2$, $UO_{2+x}$ $Cr_2O_3$, etc. | Positive holes |
| n-type Semiconductor | ZnO, CdO, $SnO_2$, $TiO_2$ $CeO_2$, $Fe_3O_4$, $WO_3$ $V_2O_5$, $Nb_2O_5$, $\beta$-$Ta_2O_5$ | Excess electrons |
| Amphoteric Semiconductor with n- and p-type conduction | MnO, $Y_2O_3$, $Bi_2O_3$ PbO, $Sc_2O_3$ | holes or electrons |
| Ionic Conductor | MgO | $Mg^{2+}$(?) |
| | $Al_2O_3$ | $Al^{3+}$(?) |
| | $SiO_2$ | ? |
| | Spinel $MgAl_2O_4$ | ? |
| | Mullite $3Al_2O_3 \cdot 2SiO_2$ | ? |
| | Forsterite $Mg_2SiO_4$ | ? |
| | Liquid $PbO$-$SiO_2$ | $Pb^{2+}$ |
| | Liquid $Li_2O$-$SiO_2$ | $Li^+$ |
| | Liquid $Na_2O$-$SiO_2$ | $Na^+$ |
| | Liquid $K_2O$-$SiO_2$ | $K^+$ |
| | Liquid $CaO$-$SiO_2$-$Al_2O_3$ | $Ca^{2+}$ |

(Table 1 continues on next page)

## TABLE 1, continued

| Classification | Examples of Conductors | Main Charge Carrier |
|---|---|---|
| Ionic Conductor | CaS and CaS-TiS$_2$ | Ca$^{2+}$ |
| | SrS and SrS-Ce$_2$S$_3$ | Sr$^{2+}$(?) |
| | LiH | ? |
| | SrO-CeO$_2$-Yb$_2$O$_3$(-H$_2$O) | H$^+$(?) |
| | K$_2$SO$_4$ | K$^+$(?) |
| | Na$_2$SO$_4$ | Na$^+$(?) |
| | AlN-Y$_2$O$_3$, AlN-Al$_2$O$_3$ | Al$^{3+}$(?) |
| | Si$_3$N$_4$-Al$_2$O$_3$ (called Sialon) | Si$^{4+}$ or Al$^{3+}$ |
| Superionic Conductor | ZrO$_2$-15mole%CaO | O$^{2-}$ |
| | ZrO$_2$-15mole%MgO | O$^{2-}$ |
| | Na$_2$O-11Al$_2$O$_3$ ($\beta$-Alumina) | Na$^+$ |
| | Na$_3$Zr$_2$Si$_2$PO$_{12}$ (called NASICON) | Na$^+$ |
| | Li$_{14}$Zn(GeO$_4$)$_4$ (called LISICON) | Li$^+$ |
| | Glassy Na$_2$O-B$_2$O$_3$-NaCl-SiO$_2$ | Na$^+$ |
| | Glassy AgI-Ag$_2$MoO$_4$ | Ag$^+$ |
| | Glassy AgI-Ag$_2$O-B$_2$O$_3$ | Ag$^+$ |
| | AgI | Ag$^+$ |
| | Ag$_{19}$P$_2$O$_7$I$_{15}$ | Ag$^+$ |
| | Rb$_3$Cu$_7$Cl$_{10}$ | Rb$^+$ |
| | Li$_3$N | Li$^+$ |

# FLUID FLOW AND MASS TRANSFER IN GAS STIRRED LADLES

### SHIGEO ASAI, IWAO MUCHI

*Department of Iron & Steel Engineering,*
*Nagoya University*
*Nagoya 464, Japan*

### and MASAYUKI KAWACHI

*Aichi Steel Ltd.*
*Thokai 476, Japan*

## INTRODUCTION

The interest in ladle refining is increasing due to a growing demand for high grade products and the spreading of hot metal pretreatment. Theoretical and experimental approaches have been evaluated to improve refining operations and to provide a better understanding of the complex phenomena involved in the field of ladle metallurgy. Particularly, it is necessary to understand the rate processes because they substantially govern the efficiency of the operations and the quality of the products.

To evaluate the degree of mixing of the melt, Nakanishi and Fujii[1] first proposed a relation between mixing time and mixing power density. Afterwards, its theoretical background has been much clarified to enable scale-up based on water model experiments and theoretical analyses.[2,3] On the other hand, with respect to the effect of agitation on mass transfer rate, only empirical relations have been presented. However, theoretical study is not sufficient to appraise mass transfer rate for given operating conditions or to understand the influence of fluid flow on it.

The procedure for determining whether the mass transfer in a given batch reactor is controlled by mixing in the bulk liquid or interphase mass transfer is discussed first. Then, the relationships between mixing time and mixing power density are derived by applying dimensional analysis to the Navier-Stokes and mass balance equations. Based on these relationships, a procedure for scale-up is proposed.

Finally, the effects of gas flow rate on mass transfer rate in three different systems, i.e., gas-metal, slag-metal and solid-metal/slag are reviewed, and the significant points applicable to ladle metallurgy are taken from the chemical engineering literature. In this work, focusing on the connection between mass transfer and the behavior of liquid-liquid interface, a model experiment was conducted to measure the mass transfer rate and to observe the interface behavior in an agitated vessel. A critical metal velocity at which a part of the slag layer is entrapped in the metal phase in the form of particles is derived from the energy balance for a slag particle.

## PROCEDURE FOR DETERMINING THE CONTROLLING STEP

In order to determine whether the mass transfer in a given batch reactor is controlled by mixing in the bulk liquid or interface mass transfer, the evaluation of a circulation Stanton number is crucial. The circulation Stanton number[4] ($St = k \cdot a \cdot t_c$, where $k \cdot a$ is the capacity coefficient and $t_c$ is the circulation time in liquid phase) indicates the interaction between interface mass transfer and convective transport.

In general, mass transfer rate can be described by Equation 1.

$$V(dC/dt) = k \cdot A \cdot (C_e - C) \tag{1}$$

Equation 1 is integrated under the condition of $C = C_i$ at $t = 0$.

$$C = (C_i - C_e) \exp(-k \cdot A \cdot t/V) + C_e$$
$$= (C_i - C_e) \exp(-k \cdot a \cdot t) + C_e \tag{2}$$

Using data on the variation of concentration obtained in a given reactor, the capacity coefficient $k \cdot a$ can be evaluated from Equation 2.

The circulation time $t_c$ is determined by a number of experimental techniques, such as impulse response[5,6] or the relation between $t_c$ and mixing time, $\tau$, such as $t_c \cdot \tau =$ const.[7] Then we can evaluate the value of $St$.

When $St > 2$, promotion of mixing is essential since convective mass transport is slower than interphase mass transport. A procedure for scale-up based on the relation between $\tau$ and $\varepsilon$, which will be mentioned later, is useful.

When $St < 2$, interface mass transfer is crucial. A detailed discussion of mass transfer between phases will also be given later.

## RELATION BETWEEN MIXING TIME AND MIXING POWER DENSITY

**Relation between Flow Velocity and Mixing Power Density**—The flow of molten metal or slag can be described by the Navier-Stokes equation:

$$\rho(\partial \bar{v}/\partial t + \bar{v} \cdot \nabla \bar{v}) = -\nabla P + \mu_e \nabla^2 \bar{v} + \overline{F} \tag{3}$$

where, $\overline{F}$ denotes the body force generated by gas blowing or electromagnetic force, etc. Depending on each of the terms in Equation 3 balanced with the body force term, fluid motion is classified as follows:

(1) Laminar Fluid Motion Predominated by Viscous Force

The following balance is obtained from Equation 3.

$$\mu \nabla^2 \overline{v} = -\overline{F} \tag{4}$$

If a similar pattern of velocity distribution is assumed in a different size apparatus, the differential operator $\nabla$ in Equation 4 can be replaced by $L^{-1}$. Then, Equation 5 is derived through dimensional analysis.

$$v \propto (FL^2/\mu) \tag{5}$$

(2) Laminar Fluid Motion Predominated by Inertia Force

Equation 3 is simplified as Equation 6:

$$\rho \overline{v} \cdot \nabla \overline{v} = \overline{F} \tag{6}$$

Following the derivation of Equation 5, Equation 7 is obtained as:

$$v \propto (FL/\rho)^{1/2} \tag{7}$$

(3) Fluid Motion Predominated by Turbulent Viscous Force

The body force term $\overline{F}$ is equally balanced with the viscosity term as Equation 5, but molecular viscosity, $\mu$, must be replaced by turbulent viscosity, $\mu_t$.

$$\mu_t \nabla^2 \overline{v} = -\overline{F} \tag{8}$$

Based on Boussinesq and Prandtl's hypothesis, turbulent viscosity can be expressed by Equation 9.

$$\mu_t = \rho \ell^2 | \text{grad } \overline{v} | \tag{9}$$

Substituting Equation 9 into Equation 8 and applying dimensional analysis yields Equation 10.

$$v \propto (FL^3/\rho \ell^2)^{1/2} \tag{10}$$

Based on the work-energy theorem, the work $W$ done by a force $F$ acting during the displacement $ds$ is defined as

$$W = \int F ds \tag{11}$$

Differentiating Equation 11 with respect to time, $t$, Equation 12 is obtained

$$dW/dt = F \cdot (ds/dt) \tag{12}$$

*References pp. 288–289*

Mixing power density, $\varepsilon[(N \cdot m)/s \cdot m^3] = W/m^3$], is defined as $\varepsilon = dW/dt$, and $v$ (m/s) is proportional to $ds/dt$. By substituting Equation 12 into Equation 5, 7 and 10 to eliminate $F$, the following relations between characteristic velocity, $v$, and mixing power density, $\varepsilon$, are finally obtained.

(1) Laminar Fluid Motion Predominated by Viscous Force

$$v \propto (L^2 \varepsilon / \mu)^{1/2} \tag{13}$$

(2) Laminar Fluid Motion Predominated by Inertia Force

$$v \propto (L\varepsilon/\mu)^{1/3} \tag{14}$$

(3) Fluid Motion Predominated by Turbulent Viscous Force

$$v \propto (L^3 \varepsilon / \rho \ell^2)^{1/2} \tag{15}$$

**Relation between Mixing Time and Flow Velocity**—Equation 16 is a statement of the law of conservation of mass.

$$\partial C / \partial t + \bar{v} \cdot \nabla C = D_e \nabla^2 C \tag{16}$$

On the basis of Equation 16, the decaying process of tracer can be classified as follows:

**Molecular Diffusion Control**—Equation 16 is simplified as follows:

$$\partial C / \partial t = D \nabla^2 C \tag{17}$$

If the same profile of the tracer concentration is kept regardless of vessel size, the differential operator $\nabla$ is in inverse proportion to the characteristic length L. Thus, Equation 17 can be expressed as Equation 18. Equation 19 is obtained by integrating Equation 18.

$$dC/dt = -kDC/L^2 \tag{18}$$

$$C \propto \exp(-kDt/L^2) \tag{19}$$

where, $k$ is a proportional constant.

Mixing time is defined as the time it takes the concentration of a tracer to decay to within a narrow band around the final concentration. As the width of the band is arbitrarily determined, it is meaningless to compare the absolute values of mixing time reported by different researchers. However, the mixing time determined through this procedure is so bound as to make the index number of the exponential function in Equation 19 have a constant value. Thus, the expression of Equation 20 for mixing time is obtained from Equation 19

$$\tau \propto L^2/D \tag{20}$$

**Convection Control**—The following equation is derived from Equation 16.

$$\partial C/\partial t = -\bar{v} \cdot \nabla C \tag{21}$$

After replacing $\nabla$ and $\bar{v}$ with $(1/L)$ and the characteristic velocity, $v$, respectively, Equation 21 is rewritten in an ordinary differential form.

$$dC/dt = -kvC/L \tag{22}$$

Equation 23 is obtained by integrating Equation 22.

$$C \propto \exp(-kvt/L) \tag{23}$$

From Equation 23, the following relation is obtained as mixing time.

$$\tau \propto L/v \tag{24}$$

**Turbulent Diffusion Control**—A formula equivalent to Equation 17 can be obtained. However, in this case molecular diffusivity, $D$, should be replaced by turbulent diffusivity, $D_t$. As the value of turbulent diffusivity, $D_t$, can be regarded as large as eddy kinematic viscosity[8], $\nu_t (= \mu_t/\rho)$, $D_t$ is readily obtained from Equation 9.

$$D_t = \ell^2 |\text{grad}\,\bar{v}| \tag{25}$$

Substituting Equation 25 into Equation 20 and arranging the resultant equation, Equation 26 is obtained.

$$\tau \propto L^3/(\ell^2 v) \tag{26}$$

## EFFECT OF MIXING POWER DENSITY ON MIXING TIME

Combining the specific types of fluid motion (1,2,3) and those of the decaying process of tracer, the dependence of $\tau$ on physical properties and operational variables can be tabulated as in Table 1 (the unrealistic combinations were eliminated). As shown in Table 1, the value of $n$ showing the exponents of $\varepsilon$ changes as $n = 0 \sim 1/2 \sim 1/3$ and $\gamma$, that is, the exponent of characteristic length, varies within $\gamma = 0 \sim 8/3$.

In this analysis the whole effect of turbulence is stuffed into a mixing length $\ell$, the value of which substantially depends on the structure of turbulence. As the flow of an agitated fluid in a vessel generally indicates complex turbulence, the evaluation of $\ell$ is difficult. On the basis of the theoretical results shown in Table 1, $\gamma = 0$ and $n = 1/2$ hold in the case of molecular diffusion control and type 1 fluid motion and $\gamma + \xi = 2/3$ and $n = 1/3$ hold in the cases of convection control and turbulent diffusion control with fluid motion types 2 and 3.

*References pp. 288–289*

## TABLE 1
## The Effect of Physical Properties and Operational Variables on Mixing Time.

$$\tau \propto \varepsilon^{-n} L^\gamma \ell^\xi \rho^\alpha \mu^\beta D^\kappa$$

| Decaying process of tracer | Fluid flow predominated by | | |
|---|---|---|---|
| | (1) Viscous force | (2) Inertia force | (3) Turbulent viscous force |
| (A) Molecular diffusion | $n=0, \gamma=2$ $\xi=0, \alpha=0$ $\beta=0, \kappa=-1$ | — | — |
| (B) Convection | $n=0.5, \gamma=0$ $\xi=0, \alpha=0$ $\beta=0.5, \kappa=0$ | $n=\frac{1}{3}, \gamma=\frac{2}{3}$ $\xi=0, \alpha=\frac{1}{3}$ $\beta=0, \kappa=0$ | $n=\frac{1}{3}, \gamma=0$ $\xi=\frac{2}{3}, \alpha=0$ $\beta=0, \kappa=0$ |
| (C) Turbulent diffusion | — | $n=\frac{1}{3}, \gamma=2\frac{2}{3}$ $\xi=-2, \alpha=\frac{1}{3}$ $\beta=0, \kappa=0$ | $n=\frac{1}{3}, \gamma=2$ $\xi=-1\frac{1}{3}, \alpha=\frac{1}{3}$ $\beta=0, \kappa=0$ |

## EXPERIMENTAL RESULTS

Experimental results in a vessel 405mm⌀×400mm high are shown in Figure 1. The fluid velocity $v$ increases with increase of mixing power density, $\varepsilon$. At first, the relation between $v$ and $\varepsilon$ is $v \propto \varepsilon^{0.48}$ and then shifts $v \propto \varepsilon^{0.24}$ after the deflection point around $\varepsilon = 5kg/ms^2$. Observed values of the exponents of $\varepsilon$, 0.48 and 0.24 are nearly equal to the theoretical values of 0.5 and 1/3, which are expected from Equations 13, 14 and 15, respectively.

Mixing time, $\tau$, decreases with increase of mixing power density, $\varepsilon$. The relation for small $\varepsilon$ is $\tau \propto \varepsilon^{-0.68}$ and is $\tau \propto \varepsilon^{-0.32}$ for large $\varepsilon$. It is noted that the value of $\varepsilon$ at the deflecting point for mixing time is nearly equal to that for fluid velocity. The observed values of the exponents of $\varepsilon$, 0.68 and 0.32, correspond to the values of 1/2 in (1) and that of 1/3 in (2) and (3), as shown in Table 1. Thus, this fact confirms that such transitional behaviors observed in the case of fluid velocity and mixing time have to be closely related to each other. Similar tendencies were observed for both mixing time and fluid velocity in the experiment by use of a half size vessel of similar geometrical shape. As apprehended from Figure 1, examination of the effect of mixing power density on mixing time enables one to classify fluid motion into

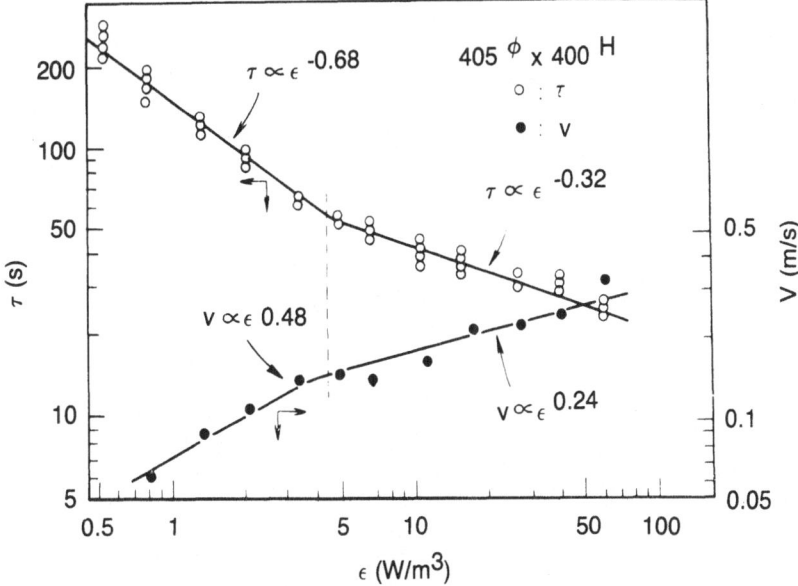

**Figure 1.** Effects of mixing power density on mixing time and fluid velocity

two major flows, i.e. the flow predominated by viscous force and the flow caused by inertia force or turbulent viscous force.

Figure 2 shows the surface velocities of molten steel observed by Lehner et al. in a 60 $t$ ladle agitated by argon gas. The surface velocities at different radial positions appear to be in proportion to the volumetric gas flow rate to one-third power, $v \propto Q^{1/3}$. Since $Q$ is approximately proportional to $\varepsilon$, the result shown in Figure 2 indicates the existance of the relation $v \propto \varepsilon^{1/3}$. It has been confirmed that the theoretical result developed in this paper applies to not only water model experiments but also large industrial vessels and that the fluid motion shown in Figure 2 corresponds to that described by Equation 14 or 15.

In Figure 3, the variations of $\tau$ observed in three different size vessels with similar geometries are lumped together. According to the theoretical results listed in Table 1, under the flow predominated by viscous force, the exponent of mixing power density, $\varepsilon$, is $n = 1/2$ and the exponent of characteristic length, i.e. diameter of vessel, $L$, is $\gamma = 0$, which means that $\gamma$ should be independent of vessel size. The experimental result shown in Figure 3 demonstrates this theoretical prediction for viscous flow. In the case of flow dominated by inertia force or turbulent viscous force, the observed relation $\tau \propto L^{0.36} \varepsilon^{-0.32}$ was obtained. It indicates that there should be an effect of $L$ on $\tau$. As described in the theoretical analysis, the exponent of characteristic length should have been $2/3$ from $\gamma + \xi = 2/3$, if the condition of $\ell \propto L$ was presumed. However, this experimental result reveals that the exponent of characteristic length is not $2/3$, but $0.36$.

*References pp. 288-289*

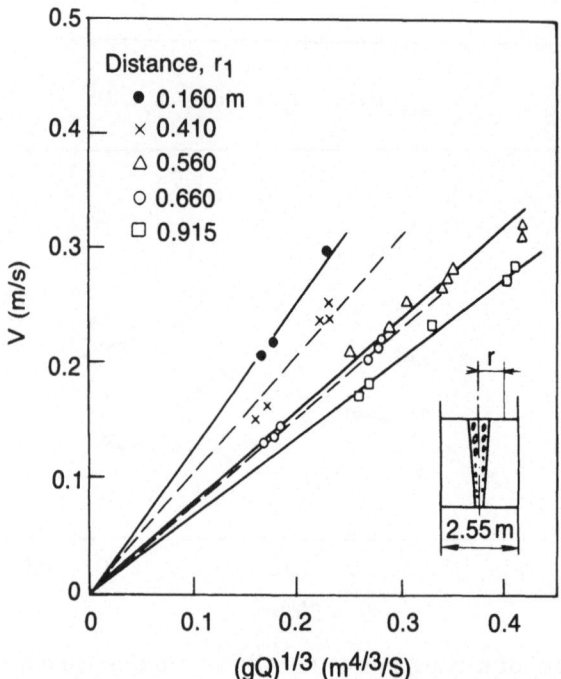

Figure 2. The correlations between surface velocity and gas volumetric flow rate in the 60 ton ladle[9].

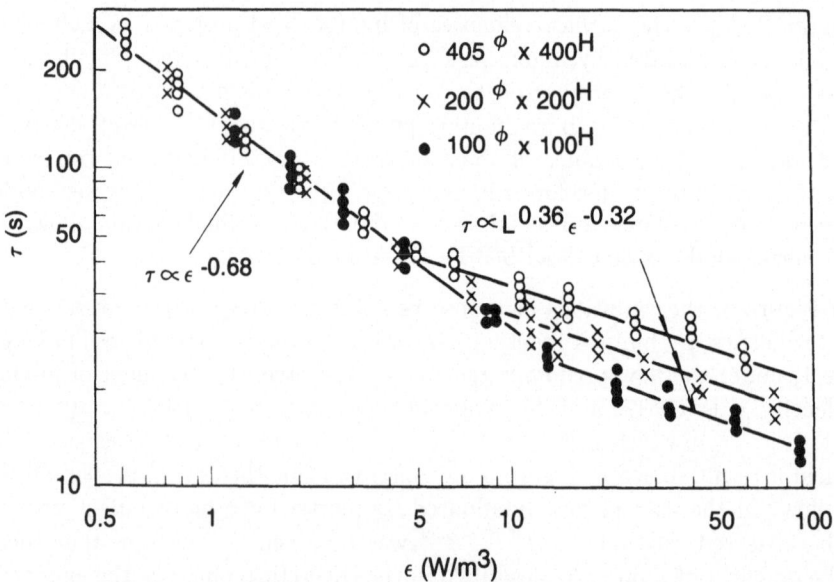

Figure 3. Effect of vessel size on the relation between mixing time and mixing power density.

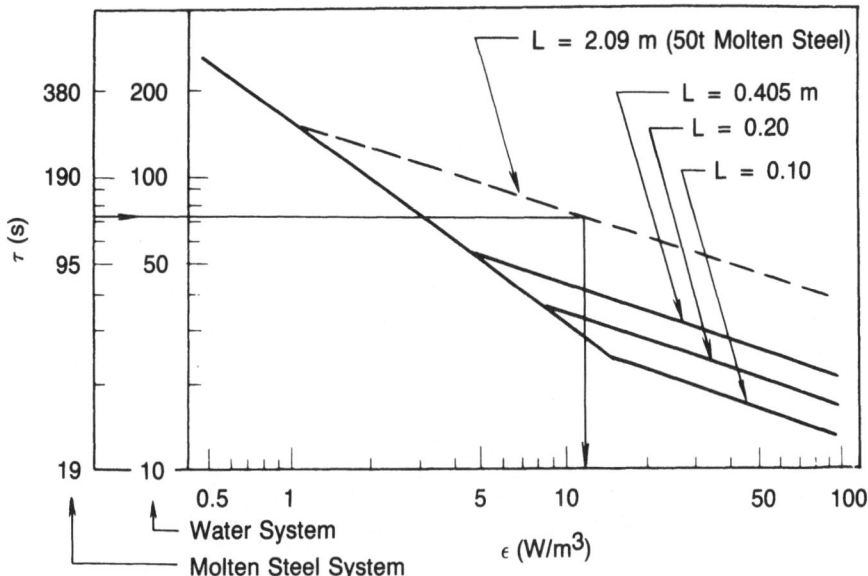

Figure 4. The relation between mixing time and mixing power density, which shows the procedure for scale-up.

## PROCEDURE FOR SCALE-UP

Some of the proposed relationships between mixing time and mixing power density were based on the data including equipment with non-similar geometry or completely different ones. However, it can not be guaranteed from theory that the data from vessels of differing design might be formulated by one relation. Thus, the following procedure would be required to carry out scale-up.

(1) In order to estimate the value of the exponent of characteristic length, $L$, water model experiments have to be carried out in equipment of a geometry similar to that of the objective one and a figure corresponding to Figure 3 has to be drawn (the value is 0.36 in the case of Figure 3).

(2) By multiplying 1.9 $\{(\rho_{Fe}/\rho_W)^{1.3} = 7^{1.3} = 1.9\}$ by the ordinate value of the figure for water model, the ordinate can be transferred into that of a molten iron system. Figure 4 shows an example where gas is injected from a nozzle placed at the bottom center of a cylindrical vessel. The capacity of molten steel and the ratio of liquid depth to inner diameter are as 50 tons and 1, respectively. Taking into account scale effect, the relation between $\tau$ and $\varepsilon$ in the objective apparatus was added on the figure by a broken line. Advancing in the direction of the arrow, the value of mixing power density is determined from a given value of mixing time.

(3) By using the relation between gas volumetric flow rate $q(\text{Nm}^3/\text{m}^3\text{s})$ and input power density $\varepsilon(\text{W/m}^3)$ [$\varepsilon = 371qT_\ell\{\ln(1+9.8\rho_\ell \text{ h}/P_a) + \eta \ (1-T_n/T_\ell)\}$, or $\varepsilon'(\text{W/kg})$, $\varepsilon' = \varepsilon/\rho_\ell$, where, $T_n$: blowing gas temperature (K), $T_\ell$:metal tempera-

*References pp. 288–289*

ture (K) and $P_a$:atmospheric pressure (Pa)][21], the required gas flow rate can be evaluated.

## MASS TRANSFER BETWEEN PHASES

Mass transfer in process metallurgy can be classified into three groups; the mass transfer taking place between gas and liquid (gas absorption or gas desorption), between liquid and liquid (slag-metal reaction), and between solid and liquid (melting of particle and refractory in slag or metal phase). Let us consider the effect of gas agitation on mass transfer between phases and discuss the obtained experimental results.

**Mass Transfer Between Gas and Liquid Phases**—Decarburization, degassing and absorption of hydrogen and nitrogen, etc. are important examples of mass transfer between gas and liquid in process metallurgy. The effect of gas flow rate on the capacity coefficient of mass transfer $K$, was studied in model experiments[10-18], and the results are listed in Table 2. The exponents of gas flow rate are scattered from 0.65 to 1.5. Since the capacity coefficient consists of the product of specific interfacial area and mass transfer coefficient, it is interesting to separately consider the influence of gas flow rate on interfacial area and on mass transfer coefficient.

Yoshimatsu and Fukuzawa[13] attempted to evaluate the gas-liquid interfacial area and the mass transfer coefficient in a bottom blowing vessel, separately, by using the method of gas-absorption with the pseudo-first order reaction, in which mass transfer coefficient with chemical reaction is described as $k'_\ell = \sqrt{k_r C_b D}$. Thus, $k'_\ell$ can be evaluated if the reaction rate constant $k_r$ is known. On the other hand, the capacity coefficient $k'_\ell \cdot a$ can be obtained from experiments. Then, the specific interfacial area $a$ is available. The method to evaluate $a$ is well known to chemical engineers[19]. Figure 5 shows the capacity coefficients with and without chemical reaction against superficial velocity. By using both of the capacity coefficients in Figure 5, the gas-liquid interfacial area $A = a \times V$ was evaluated as shown in Figure 6, where the symbols denote the position of gas inlet. Based on the fact that both $k \cdot a$ and $a$ are nearly in proportion to gas flow velocity, it is understood that mass transfer coefficient does not strongly depend on gas flow rate and the increase in mass transfer rate is mainly due to the increase in interfacial area.

The weak dependence of mass transfer coefficient on gas flow rate has been recognized in the chemical engineering field where Calderbank and Moo-Young[20] correlated mass transfer coefficient in aerated mixing vessels and gas bubble columns as shown in Figure 7. For gas bubbles of average diameter larger than 2.5 mm, corresponding to the range CD in Figure 7, the correlation of mass transfer coefficient is given as follows:

$$Sh = 0.42 Gr^{1.3} \cdot Sc^{1/2} \qquad (27)$$

where, $Gr \equiv d_p^3 g \rho \Delta \rho / \mu^2$: Grashof number, $Sh \equiv k_\ell d_p / D_f$: Sherwood number, $Sc \equiv \mu / \rho D_f$: Schmidt number.

## TABLE 2
### Correlation of Mass Transfer in Gas-Liquid System.

| System | Correlation | Remarks | Ref. |
|---|---|---|---|
| Water-$CO_2$ | $K \propto q^{1.5}$ <br> $K \propto q^{0.7}$ | Side blown <br> $We \cdot Fr^{-0.4} > 1500$ <br> $We \cdot Fr^{-0.4} < 1500$ | 10 |
| NaOH solution -$CO_2$ | $K \propto q^{0.7}$ | Gas absorption with chemical reaction | 11 |
| Water-$CO_2$ | $K \propto q^{0.64}$ | | 12 |
| NaOH solution -$CO_2$ | $K \propto q$ | $A \propto q$ | 13 |
| Fe-C(melt)-$N_2$ | $K \propto q^{0.65}$ | Bottom blown | 14 |
| Fe-C(melt)-Ar | $K \propto q^{0.33}$ | Bottom blown | 15 |
| Water-Air | $K \propto q^{0.8}$ <br> $K \propto q^{3.3}$ | Bottom blown <br> Top blown | 16 |
| Ag-$O_2$ <br> Cu-$O_2$ | $K \propto q$ | Orifice | 17 |
| NaOH solution -$CO_2$ | $K \propto q^{1.22}$ | Side blown | 18 |

For bubbles with an average diameter of less than 0.06 mm, corresponding to the range AB in Figure 7, the correlation is given as follows:

$$Sh = 0.31 Gr^{1/3} \cdot Sc^{1.3} \qquad (28)$$

Themelis et al.[21] suggested that Equations 27 and 28, which had been derived from chemical engineering practice, were applicable to the estimation of mass transfer coefficient in a gas-molten metal system.

In gas-liquid agitation, such as degassing and absorption of gas through bubble interfaces, increase of gas-liquid interfacial area by making swarms of tiny bubbles is considered to be essential due to the weak dependence of mass transfer coefficient on gas flow rate. Szekely[22] attempted to make swarms of bubbles in

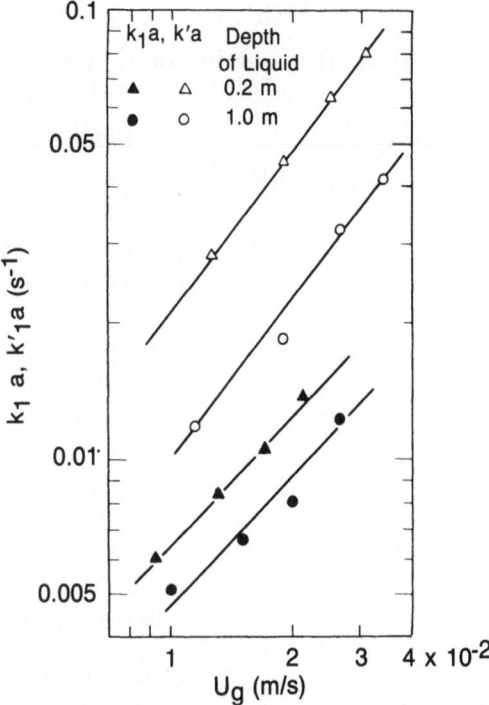

Figure 5. The relations between the superficial velocity $U_g$ and the capacity coefficient, $k'_\ell \cdot a$ and $k_\ell \cdot a$ which are accompanied with and without chemical reaction, respectively[13].

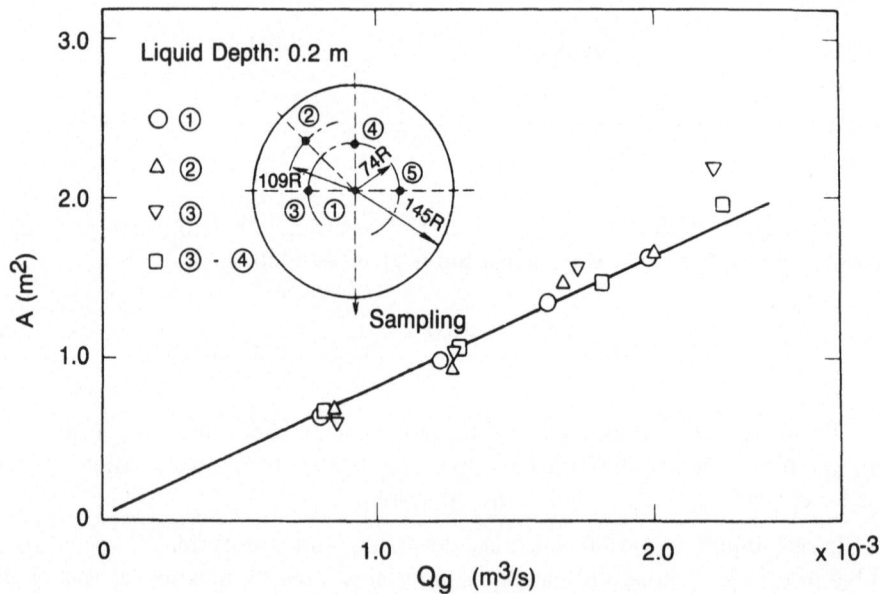

Figure 6. The effect of gas flow rate on the gas-liquid interfacial area[13].

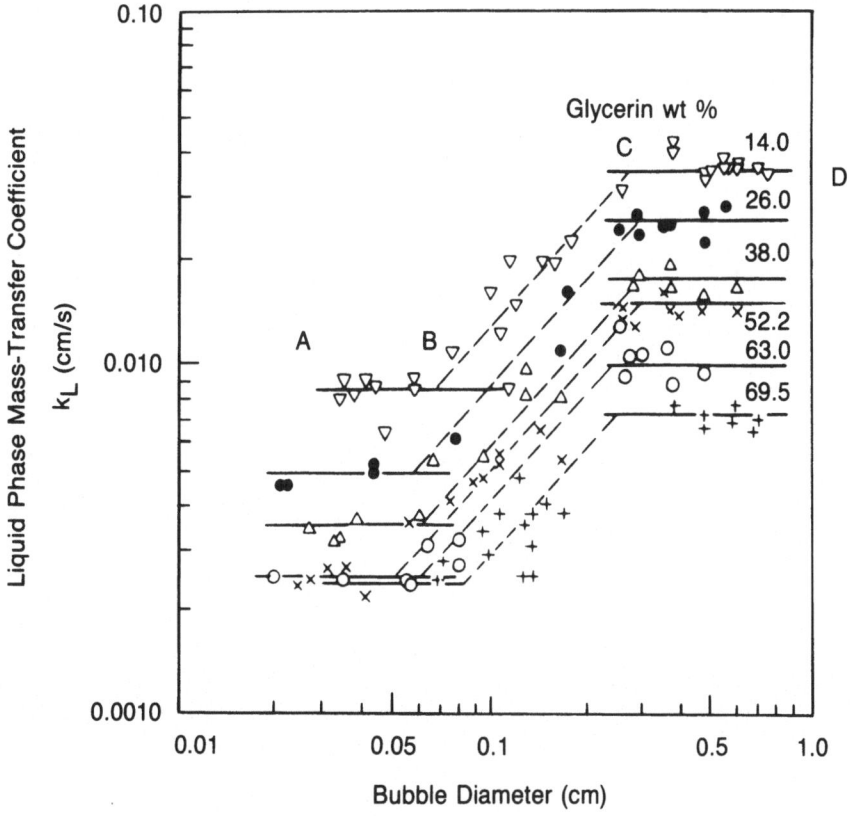

Figure 7. Dependence of bubble diameter on mass transfer coefficient in agitated vessels[20].

molten aluminum mechanically by using a spinning nozzle. Though this method increases mass transfer rate considerably, the bubble size produced by mechanical methods is limited due to high surface tension of molten metal. Thus, for future work we should consider chemical rather than mechanical methods. There are many materials which produce gas in reaction with chemical components in molten metal, such as injection of oxide powder which can produce fine bubbles due to nucleation on the surface of a particle. Actually, in the steelmaking field it is reported[23] that the injection of oxide materials under a reduced atmosphere can produce an ultra low, as low as 10 ppm, carbon steel. New proposals for producing swarms of fine bubbles by making use of chemical reactions are eagerly waited.

## Mass Transfer Between Liquid and Liquid Phases

### Mass Transfer Behavior

Liquid-liquid reactions, such as desulfurization and dephosphorization, are important in process metallurgy. Considerable work[9,24-39] has been done to evaluate the effect of gas flow rate on mass trans-

*References pp. 288–289*

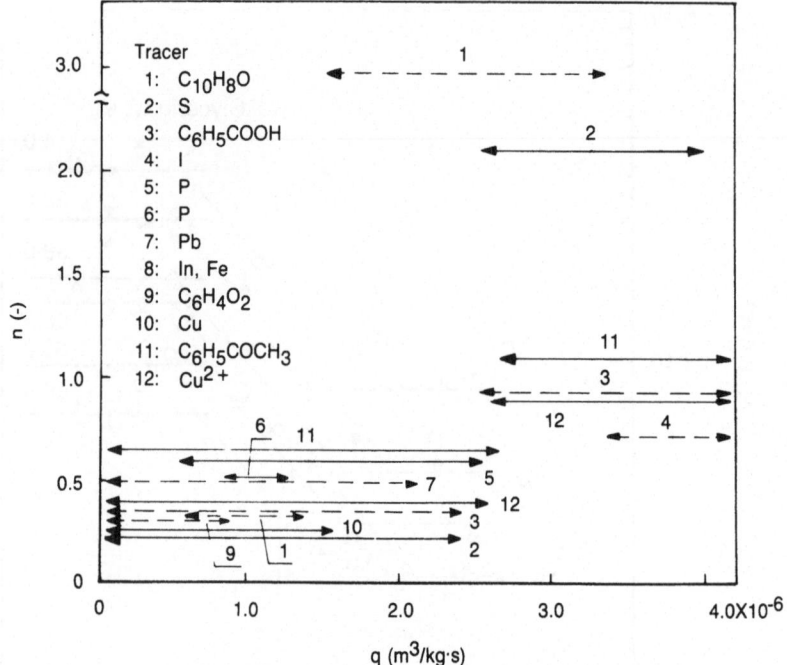

**Figure 8.** The power number, $n$, in the relation of $K \propto q^n$, against gas flow rate per unit mass of liquid, q.

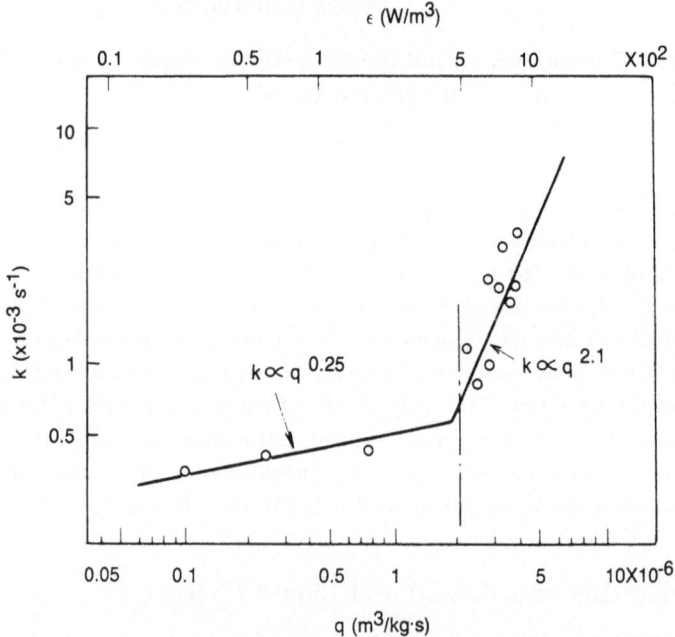

**Figure 9.** The effect of gas flow rate on the capacity coefficient of desulphurization reaction[24].

fer rate between slag and metal. Results of this work, including high temperature experiments and model studies, are listed in Table 3. As an illustration, the exponent $n$ in the relation of $K \propto q^n$ is shown against gas flow rate per unit mass of liquid, $q$, in Figure 8. The solid line indicates the high temperature experiment and the dotted line implies the low temperature model study. The noticeable feature is the substantial increase in the value of $n$ in the high ranges of $q$. This tendency is obvious in the experimental results shown in Figure 9. This work was done to study the effect of gas flow rate on desulfurization reaction rate in a 2.5 ton converter by Ishida and co-workers[24]. It is noted that the dependence of capacity coefficient on gas flow rate abruptly changes around $\varepsilon = 450 W/m^3$. These relations can be expressed by $K \propto \varepsilon^{0.25}$, $\varepsilon < 450 W/m^3$ and $K \propto \varepsilon^{2.1}$, $\varepsilon > 450 W/m^3$. The exponent 0.25 for weakly agitated regions may be explained as follows:

Based on Higbie's model, mass transfer coefficient is written as:

$$k = 2\sqrt{d_f/\pi t_e} \tag{29}$$

where, the exposure time $t_e$ is the time required for a liquid element to travel along the liquid-liquid interface from center to wall. Thus, it is considered that $t_e$ is inversely related to characteristic velocity. On the other hand, the interfacial area $A$ is regarded to be nearly constant under a rather quiet agitation, since the solid skin usually covers all the slag surface except plume and the whole slag layer may be viscous. Thus, the capacity coefficient can be expressed as:

$$K \propto A\sqrt{v/L}/V \tag{30}$$

Substituting the characteristic velocity in laminar flow given by Equation 13 into Equation 30, the following relation is obtained.

$$K \propto (\varepsilon/\mu)^{1/4} \propto (q/\mu)^{1/4} \tag{31}$$

The exponent on gas flow rate, 1/4, coincides with the value seen in weakly agitated regions in Figure 9.

On the other hand, in the vigorously agitated region of $\varepsilon > 450 W/m^3$ in Figure 9, the exponent on $q$ appears as high as 2. This high value and the abrupt change of the dependence on gas flow rate could account for the significant increase of interfacial area and the change of physical properties and thermodynamic conditions such as viscosity and sulfur capacity of slag, due to the increased average temperature of agitated slag.

By measuring the electrical conductivity in the liquid phase, the authors observed the capacity coefficient of mass transfer of benzoic acid between tetraline (slag phase) and water (metal phase). Figure 10 shows the relation between this capacity coefficient and gas flow rate. It is noted that the tendency shown in Figure 10 is similar to that in Figure 9, that is, an abrupt change appears in the dependence of gas flow rate on the capacity coefficient though the exponent on $q$ is different. It should be emphasized here that the entrapment of a few drops of tetraline into

*References pp. 288–289*

## TABLE 3
### Correlation of Mass Transfer in Liquid-Liquid System.

| System | Stirring | Reaction | Correlation | Remarks |
|---|---|---|---|---|
| Slag-steel | Ar gas | Desulfurization | $K \propto \varepsilon^{0.25}$ ; $\varepsilon < 60W/ton$ $K \propto \varepsilon^{2.1}$ ; $\varepsilon > 60W/ton$ | 2.5 ton converter $q < 150\ell/min \cdot ton$ $150 < q < 240$ |
| Water-Hg | $N_2$ gas | Reduction of quinone | $K \propto \varepsilon^{0.3-0.4}$ | $q < 58$ $\phi = 0.22$ |
| Slag-Steel | Ar gas mechanical stirring | Dephosphorization | $K \propto \varepsilon^{0.60}$ | $30 < q < 160$ |
| Slag-Steel | Ar gas | $\underline{Cu} \to (Cu)$ | $K \propto \varepsilon^{0.27}$ | $q < 100$ |
| Slag-Steel | | Desulfurization | $K \propto \varepsilon^{1.0}$ | |
| Oil-water | | | $K \propto \varepsilon^{0.33}$ | |
| Amalgams-aqueous sol. | | $[In] + 3(Fe^{3+})$ $= (In^{3+}) + 3(Fe^{2+})$ | $K_m \propto \varepsilon^{0.33}$ $K_s \propto \varepsilon^{0.42}$ | $q < 10$ $\phi = 0.5$ |
| n-hexane-aqueous sol. | $N_2$ gas | $I_2 + 2OH^- = IO^- + I^- + H_2O$ $3IO^- = IO_3^- + 2I^-$ | $K \propto \varepsilon^{0.72}$ | $199 < q < 994$ $\phi = 0.5$ |

## TABLE 3
### Correlation of Mass Transfer in Liquid-Liquid System (Continued).

| System | Stirring | Reaction | Correlation | Remarks |
|---|---|---|---|---|
| Amalgams-aqueous sol. Lead-molten salt | | $2[Tl] + Pb^{2+} = 2(Tl^+) + [Pb]$ | $K \propto \varepsilon^{0.5}$ | $q < 130$ $\phi = 0.5$ |
| Slag-Steel | $O_2$ gas | Dephosphorization | $K \propto \varepsilon^{0.54}$ | $50 < q < 80$ |
| Liquid paraffin-water | | Mass transfer of $C_{10}H_7OH$ | $K \propto \varepsilon^{0.36}$ $K \propto \varepsilon^{3.0}$ | $30 < q < 80$ $80 < q < 200$ $\phi = 0.17$ |
| Slag-Steel | $Ar, N_2$ gas | Desulfurization | $K \propto \varepsilon^{0.93}$ | LF |
| Tetraline-aqueous sol. | air | Mass transfer of $C_6H_5COOH$ | $K \propto \varepsilon^{0.36}$ $K \propto \varepsilon^{1.0}$ | $q < 150$ $150 < q < 650$ $\phi = 0.1$ |
| Benzene-water | $N_2$ gas | Mass transfer of $C_6H_5COCH_3$ | $K \propto \varepsilon^{0.66}$ $K \propto \varepsilon^{1.1}$ | $q < 160$ $160 < q < 1300$ $\phi = 0.1$ |
| Benzene-water | Ar gas | Mass transfer of $Cu^{2+}$ | $K \propto \varepsilon^{0.4}$ $K \propto \varepsilon^{0.9}$ | $q < 158$ $158 < q < 816$ $\phi = 0.1$ |

*References pp. 288–289*

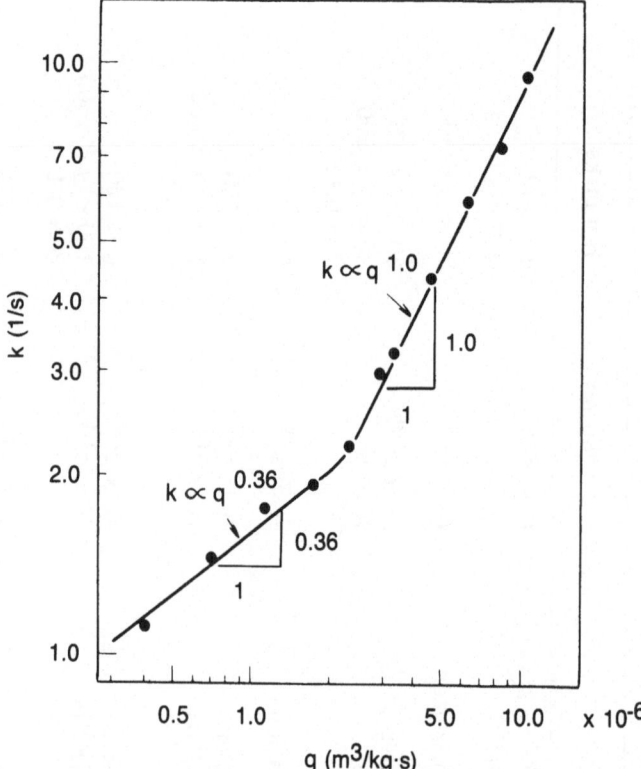

Figure 10. The effect of gas flow rate on the capacity coefficient of the benzoic acid transfer between tetraline and water.

water phase was observed at nearly the same $q$ at which the abrupt change of mass transfer rate appears. Similar phenomenon was observed in model experiments by R. Fruehan[38] and K. Nakanishi et al[33].

T. Yamada et al[34] reported the noticeable results of desulfurization reaction in LF (ladle furnace) as shown in Figure 11, where the effects of mixing power density on capacity coefficient is indicated under the conditions with one and two gas blowing plugs. In such a case that slag-metal reaction is considered as the mass transfer controlling step, it is noticed that capacity coefficient can not be adjusted only by mixing energy density, but depends on the number of plugs. This evidence could explain that the amount of local mass transfer through a region of slag-metal interface disturbed by bubbles is much larger than that through other regions due to the entrapment of slag particles. This explanation may be endorsed by the fact that the capacity coefficient in the case with two plugs has a value nearly two times that of the case with single plug. A similar result was also reported for the desiliconization reaction[39]. From the above considerations and evidence it can be understood that the quantitative evaluation of the condition of slag entrapment is important.

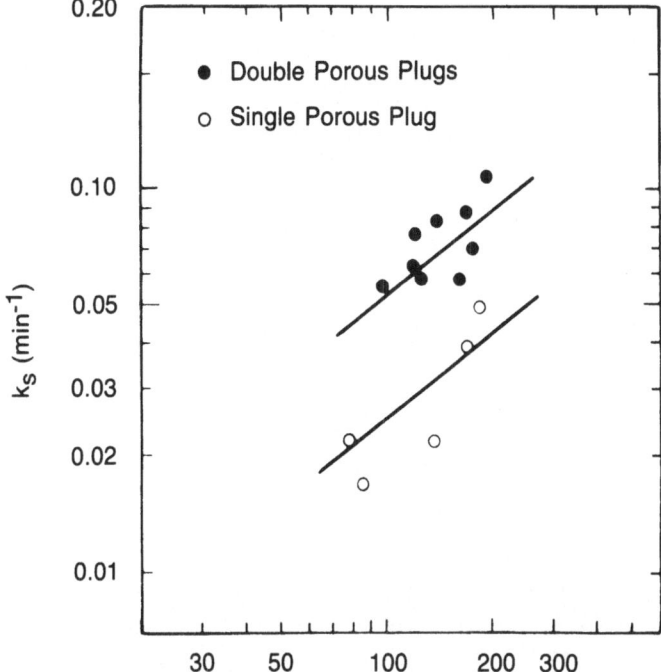

Figure 11. The effects of mixing power density on the capacity coefficient of desulphurization reaction under the conditions with single porous plug and double ones[34].

## Derivation of the Condition for Slag Entrapment

The gas flow rate at which a part of slag layer is entrapped into metal layer is called "critical gas flow rate." At this flow rate the dependence of the capacity coefficient of mass transfer on gas flow rate drastically changes as is shown in Figure 10. Thus, evaluation of critical gas flow rate from given physical properties of metal and slag and the geometry of vessel is important. As the first step in evaluating the critical gas flow rate, we derive a "critical metal velocity" at which a slag particle can be entrapped into metal phase. Taking the energy balance for a slag particle with a velocity $v$ to penetrate into metal phase through slag-metal interphase, Equation 32 is derived as the penetrability condition.

$$(1/2) \cdot (\rho_s 4\pi R_p^3/3)v^2 \geq 4\pi R_p^2 \sigma + g(\rho_m - \rho_o)4\pi R_p^3 \cdot R_p/3 \qquad (32)$$

(Kinetic energy)　　(Surface energy)　　(Energy due to buoyancy)

In equation 32, the drag term proportional to a relative velocity of slag and metal has been omitted since the velocity of a slag particle to be entrapped was observed to be nearly equal to the velocity of metal layer in the model experiment and the relative velocity is considered to be zero.

*References pp. 288–289*

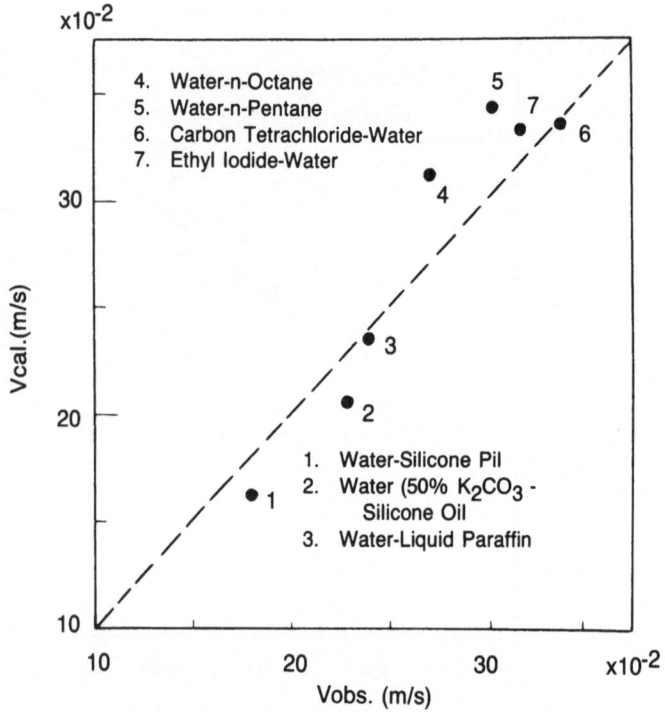

Figure 12. Comparisons between observed velocity in the model experiment and calculated velocity from the theoretical formula of Equation 35.

Rewriting Equation 32 into Equation 33, a quadratic equation in respect of the particle radius $R_p$ is obtained.

$$g(\rho_m - \rho_s)4\pi R_p^2/3 - (1/2) \cdot (\rho_s 4\pi v^2/3)R_p + 4\pi\sigma \leq 0 \qquad (33)$$

Based on the condition where Equation 33 holds under a real value of $R_p$, the discriminant $D$ of Equation 33 should be positive.

$$D \equiv (1/4)(\rho_s 4\pi/3)^2 v^4 - 4g(\rho_m - \rho_s)(4\pi/3) \cdot (4\pi\sigma) \geq 0 \qquad (34)$$

From Equation 34, the minimum velocity for the entrapment of a slag particle, which is corresponding to the critical metal velocity, is given in Equation 35.

$$v \geq \{48g(\rho_m - \rho_s)\sigma/\rho_s^2\}^{1/4} \qquad (35)$$

By substituting Equation 35 into Equation 33 the corresponding slag particle radius is found: $R_p = \{3\sigma/[g(\varrho m - \varrho s)]\}^{1/2}$.

Equation 35 was confirmed, as shown in Figure 12, by the two-dimensional model experiment in which the critical velocities were measured in several slag-metal model systems by use of a video tape recorder. The critical velocity turns out to be the actual value. The physical properties of the adopted slag-metal systems are listed in Table 4.

## TABLE 4
## Physical Properties of Slag and Metal Phases in the Model Systems.

| No. | Slag phase | Density, $\rho \times 10^{-3}$ (kg/m³) | Metal Phase | Density, $\rho \times 10^{-3}$ (kg/m³) | Interface Tension, $\sigma \times 10^3$ (N/m) |
|---|---|---|---|---|---|
| 1 | Silicone oil | 0.96 | water | 1 | 39.1 |
| 2 | Silicone oil | 0.96 | water (50%$K_2CO_3$) | 1.54 | 6.1 |
| 3 | Liquid paraffin | 0.85 | water | 1 | 39.0 |
| 4 | n-octane | 0.70 | water | 1 | 50.8 |
| 5 | n-pentane | 0.63 | water | 1 | 50.2 |
| 6 | Water | 1 | carbon tetrachloride | 1.59 | 45.0 |
| 7 | Water | 1 | ethyl iodide | 1.93 | 28.8 |

## Mass Transfer Between Solid and Liquid Phases

In ladle metallurgy, the study of agitation in solid-liquid system is required not only for reduction of diffusion resistance around solid particles and protection of refractories from wear, but also for promotion of the agglomeration of non-metallic inclusions.

When a solid bar is dissolved in liquid under the controlling step of mass transfer, the correlation is given as Equation 36.

$$Sh = cRe^{\alpha} Sc^{\beta} \qquad (36)$$

Among the water model experiments and also the high temperature ones, the correlations obtained for mass transfer are similar to each other as can be seen in the references[9,40-45] in Table 5.

In such cases as the dissolution of sponge iron in electric furnace and the reduction of solid $Cr_2O_3$ by C in AOD, the evolution of CO gas on solid surfaces plays an accelerative role on the dissolution rate. Sakuraya and Mori[45] studied the dissolution rates of oxygen-bearing iron bars in carbon-saturated iron melts, and found that the highest dissolution rate at 1673 K for iron containing 0.96% oxygen

*References pp. 288-289*

## TABLE 5
### Correlation of Mass Transfer in Solid-Liquid System.

| System | Correlation | Remarks | Ref. |
| --- | --- | --- | --- |
| C(bar)-Fe-C(melt) | $Sh=0.05Re^{0.74}Sc^{1/3}$ | dissolution | 9 |
| Fe(bar)-Fe-C(melt) | $Sh=0.064Re^{0.75}Sc^{1/3}$ | " | 40 |
| Fe(bar)-Fe-C(melt) | $Sh=0.112Re^{0.67}Sc^{0.356}$ | " | 41 |
| Benzoic acid -water | $Sh=0.079Re^{0.70}Sc^{0.356}$ | " | 42 |
| $Cr_2O_3$(bar)-Fe-C(melt) | $Sh=0.104Re^{0.70}Sc^{1/3}$ | " | 43 |
| Benzoic acid -water | $k \propto \varepsilon^{1.37}, \varepsilon < 60W/\text{ton}$<br>$k \propto \varepsilon^{0.35}, \varepsilon > 60W/\text{ton}$ | Suspension of particle | 44 |
| Fe(bar)-Fe-C(melt) | $k \propto V_{co}^{0.35\sim0.45}$ | Dissolution due to CO-bubbling | 45 |

was 590 times larger than the lowest one observed at 1473 K for aluminum-killed iron.

Dissolution of solid suspended in liquid follows a slightly different tendency from the dissolution of a bar. Nagata and coworkers[49] studied the rate of dissolution of suspended solid particles in liquid and found that with increasing agitator speed $N$ the dissolution rate has two distinct bending points at $N_f$ and $N_a$ as shown in Figure 13. $N_f$ refers to the critical agitating speed at which all the solid particles are suspended and $N_a$ is the speed at which air is sucked into the liquid. The rate of mass transfer from solid particles is correlated in each range by the same formula as Equation 36. Let $\alpha_1$ denote the slope in the range of agitating speeds less than $N_f$, $\alpha_2$ denote the one in between $N_f$ and $N_a$, and $\alpha_3$ that in the range of speeds larger than $N_a$. They are in the order of $\alpha_1 > \alpha_2 > \alpha_3$. A similar behavior was observed by Nakao et. al.[44] in the dissolution of benzoic acid in water agitated by bottom blowing, impeller and top blowing as shown in Figure 14 where the apparent mass transfer coefficient is well correlated with mixing power density.

Figure 15 shows the influence of ultra fine particle with the diameter $d_p$ on the mass transfer coefficient $k$ in which case agitator speed is used as a parameter. Mass

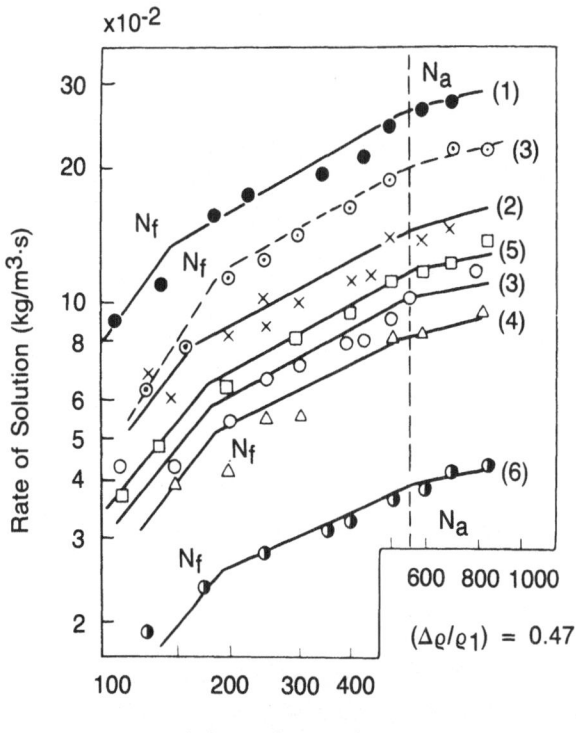

Figure 13. Rate of solution vs agitator speed ($H_3BO_3$-water system).[49]
solid dissolved: (1)-(6) 2.00 g. (3') 4.00 g
Sphere: (1) 60-100#, (2) 45-60#, (3), (3') 28-45#($d_p$ = 0.519mm),
(4) 16-28# ($d_p$ = 0.737mm) rhombic crystal (5) 28-46#($d_p$ = 0.557mm),
(6) 16-28#($d_p$ =0.858mm) temp. 298 K

transfer coefficient increases with decreasing particle size and reaches the maximum at 40 ~ 50μm. Thereafter $k$ decreases again, and the effect of agitator speed becomes negligible for particles smaller than 10μm.

According to the theory presented by Razc and Marshall[47], the limiting value of Sherwood number converges to 2 at $Re \to 0$. This theoretical prediction is shown by the chain line in Figure 15, which conflicts with the experimental results.

Nagata and coworkers[46] explained this phenomenon with regard to mass transfer from tiny particles as follows: The tiny particle with a size smaller than Kolmogoroff's smallest eddy size, $\eta = (\mu^3/\rho^2\varepsilon)^{1/4}$, moves with the motion of the eddies or entraps in the smallest eddies in which viscous effect is prevailing in spite of a high turbulence. For instance, the values of $\eta$ are calculated in molten steel as $\eta = 90\mu m$ at $\varepsilon' = 6 \times 10^{-3}$W/kg, $\eta = 50\mu m$ at $\varepsilon' = 6 \times 10^{-2}$W/kg. If we consider the segregation of $\varepsilon'$ in the plume zone up to 100 times of the mean value in vessel[48], $\eta = 15 \sim 30\mu m$ can be obtained. These sizes of $\eta$ coincide with the non-

*References pp. 288-289*

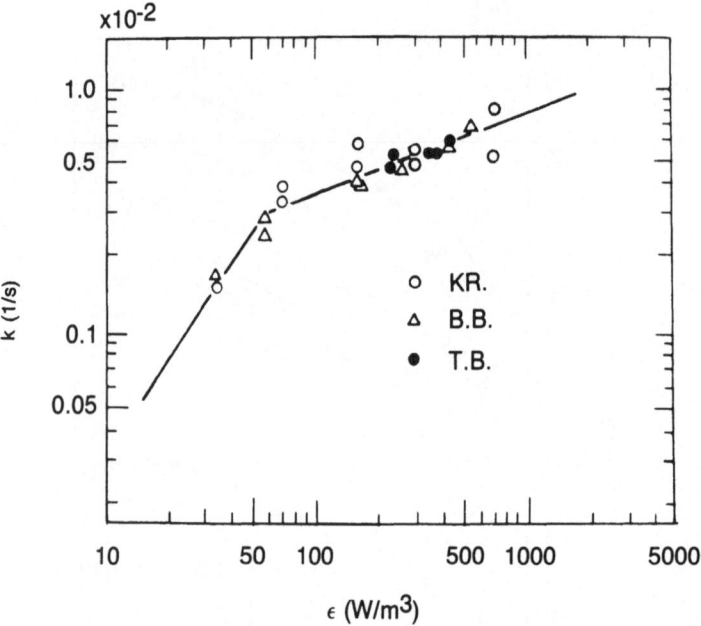

Figure 14. Effect of mixing power density on the capacity coefficient of dissolution of benzoic acid[44].
KR: paddle agitator, BB: bottom blowing, BT: top blowing

metallic inclusion size in which we are interested and with the powder size used in injection metallurgy. This phenomenon regarding mass transfer of tiny particles could be helpful to understand the deoxidation mechanism in molten metal and is substantial in the injection process with fine powder.

## CONCLUSION

The strategy to improve the efficiency of a given batch reactor such as a ladle was indicated by use of the circulation Stanton number which is a criterion determining the controlling step of mass transfer. Then, the dependences of fluid velocity and mixing time on mixing power density were clarified by dimensional analysis on the basis of the Navier-Stokes and the mass-balance equations. These theoretical results were confirmed by water model experiments.

Subsequently, the effects of gas flow rate on mass transfer rate in three different systems, namely, gas-liquid, liquid-liquid and solid-liquid were reviewed from the viewpoint of transport phenomena and some interesting information from the chemical engineering field was explored. Particularly, the connection between

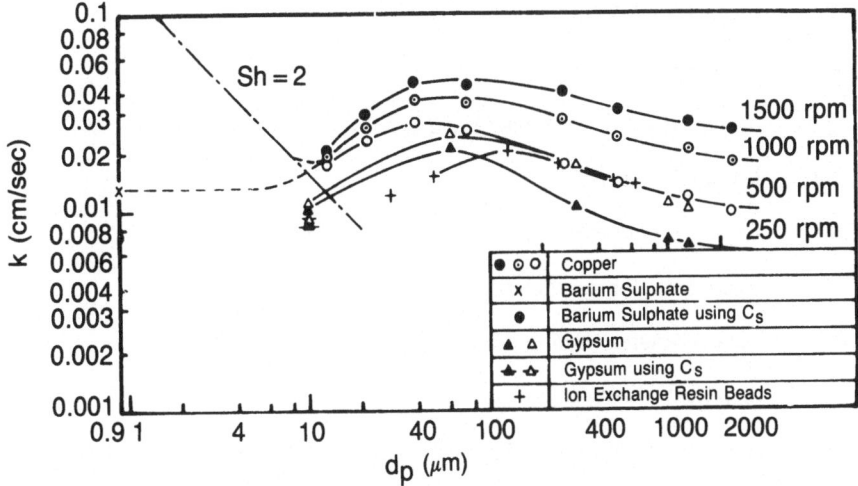

Figure 15. Effect of diameter of small particle on mass transfer coefficient.[46]

mass transfer rate and the behavior of liquid-liquid interphase were studied in model experiments. It was found that an abrupt change occurs at nearly the same gas flow rate as that at which a part of slag layer begins to be entrapped in metal layer. The critical metal velocity for entrapment of a slag particle was theoretically derived and confirmed by model experiments.

## NOMENCLATURE

$A$: interfacial area, m

$a$: specific contact area, 1/m

$C$: concentration, kg/m$^3$

$C_e$: equilibrium concentration, kg/m$^3$

$C_i$: initial concentration, kg/m$^3$

$C_b$: bulk concentration, kg/m$^3$

$D$: discriminant, kg$^2$/m$^2$s$^4$

$D, D_f$: molecular diffusivity, m$^2$/s

$D_e$: effective diffusivity, m$^2$/s

$D_t$: turbulent diffusivity, m$^2$/s

$d_p$: diameter of solid particles, m, μm, mm

*References pp. 288–289*

| | |
|---|---|
| $F, \overline{F}$ : | body force, N/m³ |
| $Gr$ : | Grashof number, – |
| $g$ : | acceleration due to gravity, m/s² |
| $h$ : | liquid depth, m |
| $K$ : | capacity coefficient of mass transfer, 1/m |
| $k$ : | mass transfer coefficient, m/s, proportional constant, – |
| $k, k_\ell$ : | mass transfer coefficient without chemical reaction in liquid phase, m/s |
| $k'_\ell$ : | mass trasnfer coefficient with chemical reaction in liquid phase, m/s |
| $k_r$ : | reaction rate constant, 1/s |
| $L$ : | characteristic length, m |
| $\ell$ : | mixing length, m |
| $N$ : | agitator speed, 1/s, rpm |
| $n$ : | exponent of gas flow rate, – |
| $P$ : | pressure, Pa |
| $P_a$ : | atmospheric pressure, Pa |
| $Q$ : | gas flow rate, Nm³/s |
| $q$ : | gas flow rate per unit mass of liquid, m³/kg·s |
| $R_p$ : | radius of slag particle, m |
| $r$ : | distance from the center of vessel, m |
| $S$ : | distance, m |
| $Sc$ : | Schmidt number, – |
| $Sh$ : | Sherwood number, – |
| $St$ : | circulation stanton number, – |
| $s$ : | displacement, m |
| $T$ : | temperature, K |
| $T_g$ : | blowing gas temperature, K |
| $t$ : | time, s |

| | |
|---|---|
| $t_c$ : | circulation time, s |
| $t_e$ : | exposure time, s |
| $U_g$ : | superficial gas velocity, m/s |
| $V$ : | liquid volume, m³ |
| $v, \bar{v}$ : | characteristic velocity, velocity, m/s |
| $W$ : | work per unit volume, J/m³ |
| $\alpha$ : | exponent of Reynolds number, – |
| $\beta$ : | exponent of Schmidt number, – |
| $\gamma$ : | exponent of characteristic length, – |
| $\varepsilon$ : | mixing power density, W/m³ |
| $\varepsilon'$ : | mixing power density, W/kg |
| $\eta$ : | size of Kolmogoroff's smallest eddy, m |
| $\mu$ : | moleculer viscosity, Pa· s |
| $\mu_e$ : | effective viscosity, Pa· s |
| $\mu_t$ : | turbulent viscosity, $Pa \cdot s$ |
| $\nu_t$ : | turbulent kinematic viscosity, m²/s |
| $\xi$ : | exponent of mixing length, – |
| $\rho$ : | density, kg/m³ |
| $\sigma$ : | surface tension, Pa· m |
| $\tau$ : | mixing time, s |
| (*Subscripts*) | cal., obs.: calculated value and observed one, respectively |
| | $\ell, m, s, Fe, w$ : liquid, metal, slag, iron and water, respectively |
| | a: atmospheric pressure |

*References pp. 288–289*

## REFERENCES

1. N. Nakanishi and T. Fujii: *Tetsu-to-Hagane*, 59 (1973), S460.
2. M. Sano and K. Mori: *Tetsu-to-Hagane*, 68 (1982), 2451.
3. S. Asai, T. Okamoto, J. C. He and I. Muchi: *Trans. ISIJ*, 23 (1983), 43.
4. G. Andre, C. W. Robinson and Y. M. Young: *Chem. Eng. Soc.*, 38 (1983), 1854.
5. O. Levenspiel: Chemical Reaction Engineering, 2nd ed. 1972, 304 [John Wiley & Sons].
6. T. Maruyama, N. Kamishima and T. Mizushina: *J. of C.E.J.*, 17 (1984), 120.
7. C. Koen: *The Chemical Engineer*, Feb. (1975), 91.
8. Y. Katto:Den-netsugairon, 1966, 111 [Yoken-do].
9. T. Lehner, G. Carlsson and T. C. Hsiao:SCANINJECT II, 2nd Int'l Conf. on Injection Metallurgy, Mefos and Jernkontoret, June 1980, Lulea, Sweden, 22.
10. M. Shimada, M. Ishibashi, K. Shiraishi and H. Morise: *Tetsu-to-Hagane*, 61 (1975), S447.
11. S. Inada and T. Watanabe: *Tetsu-to-Hagane*, 61 (1975), S449.
12. N. Bessyo, N. Taniguchi and J. Kikuchi: *Tetsu-to-Hagane*, 68 (1982), S125.
13. S. Yoshimatsu and A. Fukuzawa: Report presented on the Meeting of Steel Refining Reaction Committee of ISIJ, Jan. 1983.
14. M. Kawakami, K. Ito, M. Okuyama, T. Kikuchi and S. Sakase: Paper presented at the 19th Committee of Japan Society for the Promotion of Science, Feb. 1983, No. 19-10468.
15. H. Takeuchi, N. Harada, H. Nakamura, T. Fujii and Y. Habu: Paper presented at the 19th Committee of Japan Society for the Promotion of Science, 1984, No. 19-10565.
16. Y. Kato, T. Fujii, T. Sakuraya and Y. Habu: *Tetsu-to-Hagane*, 69 (1983), S1011.
17. R. J. Fruehan: *Metal Technol.* (1980), 95.
18. T. Touge, Y. Fujita and K. Chikazawa: *Tetsu-to-Hagane*, 69 (1983), S873.
19. F. Yoshida and H. Miura: *A.I.Ch.E.J.*, 9 (1963), 331.
20. P. H. Calderbank and M. B. Moo-Young: *Chem. Eng. Sci.*, 16 (1961), 39.
21. N. J. Themelis and P. Goyal: *Canadian Metallurgical Quarterly*, 22 (1983), 313.
22. A. G. Szekely: *Met. Trans.*, 17B (1976), 259.
23. T. Aoki, T. Matsuo and K. Shin-me: *Tetsu-to-Hagane*, 69 (1983), S178.
24. J. Ishida, K. Yamaguchi, S. Sugiura, K. Yamano, S. Hayakawa, and N. Demukai: Denki-Seiko (Electric Furnace Steel), 52 (1981), 2.
25. S. D. Clinton and J. J. Perona: Ind. Eng. Chem. Fundam., 21 (1982), 269.

26. K. Umezawa, S. Nishugi, R. Arima and H. Matsunaga: *Tetsu-to-Hagane*, 67 (1981), S182.

27. Paper presented by Nippon Kokan K. K. 60th Meeting of Special Steel Committee of ISIJ, 21–22, Nov. 1979.

28. S. Linder: KTH Jernets Metallurgy Report R15172.

29. N. Subramanian and F. D. Richardson: *J. of ISI*, June (1968), 576.

30. P. Patel, M. G. Frohberg and D. Papamantellos: *Trans. Met. Soc. of AIME*, 245, (1969), 855.

31. D. G. C. Robertson and B. B. Staples: *Process Engineering of Pyrometallurgy*, Inst. of Mining and Metallurgy, 1974, 51.

32. K. Kawakami, K. Takahashi, Y. Kikuchi, T. Usui, T. Ebizawa and H. Tanabe: *Tetsu-to-Hagane*, 69 (1983), A33.

33. K. Nakanishi, Y. Kato, T. Nozaki and T. Emi: *Tetsu-to-Hagane*, 66 (1980), 1307.

34. T. Yamada and N. Futamura: Technical Report of Aichi steel, (1985) [private communication].

35. S. Asai, M. Kawachi and I. Muchi: SCANINJECT III, 12 (1983), [Jernkontoret].

36. Y. Ooga, N. Taniguchi and J. Kikuchi: *Tetsu-to-Hagane*, 71 (1985), S897.

37. S. Endo and M. Hasegawa: *Tetsu-to-Hagane*, 71 (1985), S899.

38. R. Fruehan: this conference.

39. S. Narita, K. Arima, T. Takahashi, M. Iwamoto, J. Sakane and Ibaragi: *Tetsu-to-Hagane*, 70 (1984) S125.

40. M. Kosaka and S. Minowa: *Tetsu-to-Hagane*, 53 (1967), 983.

41. Y. U. Kim and R. D. Pehlke: *Met. Trans.*, 5 (1974), 2527.

42. M Eisenberg, C. W. Tobias and C. R. Wilke: *A.I.Ch.E.*, Symp. Ser., 51 (1955), 1.

43. K. Suzuki, K. Mori and T. Ito: *Tetsu-to-Hagane*, 65 (1979), 1131.

44. Y. Nakao, T. Ohno, H. Horiuchi, M. Mineyuki, K. Umezawa and H. Matsunaga: *Tetsu-to-Hagane*, 67 (1981), S867.

45. T. Sakuraya and K. Mori: *Tetsu-to-Hagane*, 69 (1983), 60.

46. S. Nagata, M. Nishikawa, K. Ozaki and K. Hashimoto: *Asahigarasu-kougyogizyutsu-syoureiken-kyu-houkoku*, 27 (1975), 405.

47. W. E. Ranz and W. R. Marshall: *Chem. Eng. Prog.*, 48 (1952), 141.

48. Y. Watanabe, J. C. He, S. Asai, and I. Muchi: *Tetsu-to-Hagane*, 69 (1983), 596.

49. S. Nagata: *Mixing* [Kodansha, John Wiley & Sons], (1975), 259.

## DISCUSSIONS

**Y. Sahai** *(Ohio State University)*

I have two questions for Dr. Asai and one comment before that. The comment regards mixing time with more than one gas injector. About twelve years ago Dr. Szekely and Dr. Nakanishi put that equation for mixing time versus energy rate. They came up with some power; it was a very nice correlation and was used for seven years. Later in the Scaninject second conference about five or six years ago there was a Japanese paper describing actual plant trials in which they showed mixing was fastest with one injector, and the rate of mixing went down as the number of injectors increased. The reason was that each injector created its own cell, as probably you have seen in your experiments. Dye at one place mixes very well in one cell, but it does not go to the other cells. That has also been seen in previous work with channel reactors; work which has been done at Missouri Rolla and at Ohio State University; and even earlier work has indicated that with more than one injector there are local cells. The mixing within the cell is very good, but from one cell to another cell it's not that good. My first question for Dr. Asai concerns the use of the parameter L to describe size of the vessel. I would expect the depth of liquid and the vessel diameter should both affect mixing, but in separate ways—one should be directly proportional and one should be inversely proportional. You have only one power on L. Would you clarify that.

**A.** Strictly speaking when we measure mixing time the geometry of vessels should be the same. We can not put a correlation for mixing time versus energy rate in the vessels with different geometry. That means that the ratio of depth to diameter should be kept constant. So we can use one parameter.

**Y. Sahai**

Some of our previous work has shown that the depth to diameter ratio plays a very important role in mixing, so that's my point. The second equation concerns the two extreme cases for which you reduced the Navier-Stokes equation. In one case the dominant force is viscous when the mixing is bulk flow dominated and the other one is viscously dominated. You form two separate powers on epsilon. I would expect there should be one more region where there is a mixed mode, where the transition is not drastic, and even your experiment showed the point where the curve changed. I would like to have a comment on that.

**A.** I think that perhaps torpedo type vessels would be the transition region. But in my experiment the depth and diameter is the same, so the extreme condition very easily shifts abruptly from the laminar region to the turbulent region. But if you use a long vessel like torpedo type, maybe in the center the flow is turbulent but in the side part it is still laminar flow, so maybe the transition is not drastic and shows the rather combined shape.

**J. Szekely** *(MIT)*

I'd like to congratulate Dr. Asai on very nice work. I'd like to comment that mixing time is a bit like a thermometer is in medical practice. It gives a rough idea as to what has happened. It tells whether the patient is okay or has a high fever. It will cetainly not diagnose pneumonia in the left lower lung. Mixing time will give you a good general idea classifying how the systems will behave. But when you start comparing apples and oranges, then it tends to break down, and I would caution against using mixing time or mixing time correlations for precise design. It's very useful for talking broadly about these systems. I agree with Dick Fruehan that bulk mixing time has very little relation to emulsification, because they are entirely different phenomena. To understand the really violent reactions between slags and metal, such as in a basic oxygen furnace where you do get slag-metal emulsions, one must get into the fundamentals of how emulsions form and the criteria for generating these two phase dispersions. As an interim step, the correlations are very good, but one should be very, very cautious in applying them for processes.

**D. Apelian** *(Drexel University)*

My question for Dr. Asai concerns the formal relation you presented to calculate the critical velocity $V_c$ at which the slag particles mix with metal. You fixed the R, the size of the particle of the slag. Isn't that correct? In other words, you get a value for $V_c$ for a given diameter slag particle. It sounds like a backwards way of doing it.

**A.** No. We can calculate the slag particle size at the critical velocity.

**D. Apelian**

You want to calculate when the slag particles get entrained but in so calculating, you're fixing the particle size. That seems to totally ignore the surface tension forces. So I'm wondering whether I missed something or there's a flaw in the logic? You're fixing the size of the particle that's getting entrained.

**A.** By taking the discriminant of the equation of the energy balance for a slag particle we can determine both the critical velocity and diameter of the particles together.

**D. Apelian**

Okay, not for a given R.

**A.** Never given.

**R. Guthrie** *(McGill University)*

I'd like to make one comment to Dr. Asai about the mixing time in ladles. We have done much work solving the stirring equations. We find that in the typical ladle stirring systems it's a mixed control process, that the stirring is a function both of convection and eddy diffusion. So we find and we believe that, for typical industrial processes, in the middle region convection is very important for mixing in the bulk of the steel, and so also is eddy diffusion. We witnessed that.

**Y. Sahai** *(Ohio State University)*

Dr. Asai talked about laminar flow and turbulent flow. I don't think that in any of the systems you can say that the flow is laminar. Rod said we have a combination of these two in the earlier stages, so that transition is definitely not from laminar to turblent.

**A.** Strictly speaking the combination of laminar and turbulent flows exists.

**J. Szekely** *(MIT)*

The actual hallmark of turbulence is that inertial forces dominate. So you can't say that its either inertial or turbulent. Those two things really mean the same thing.

# TWO PHASE MASS TRANSFER IN GAS STIRRED REACTORS

## R. J. FRUEHAN and S-H KIM

*Carnegie Mellon University*
*Pittsburgh, Pennsylvania 15213*

### ABSTRACT

Gas stirred reactors are used in several metallurgical processes, such as the desulfurization of liquid cast iron, to increase the rate of the chemical reaction and homogenize the melt. Mixing times and two phase mass transfer rates were measured in a model system of a liquid steel ladle. Mixing times were increased by the presence of a second phase and were faster for off-center tuyeres. However, two phase mass transfer is faster for a center tuyere as compared to an off-center tuyere. The rate of mass transfer was independent of tuyere size, lance versus tuyere injection and only slightly dependent on the viscosity of the second phase. The dramatic increase in mass transfer at a critical flow rate is explained in terms of an increase in the interfacial area between the two phases.

### INTRODUCTION

Several metallurgical processes employ gas-stirred reactors to increase the rate of refining reactions and homogenize the melt. For example, in the production of liquid cast iron, the iron is desulfurized in a reactor after being tapped from the cupola. The gas stirring improves mixing between the reactant and the metal, increasing the reaction surface area and liquid phase mass transfer, resulting in higher rates of reaction. The system is continuous but otherwise somewhat similar to desulfurization in a steel or hot metal ladle for which there has been a considerable amount of work done to understand the effect of gas stirring.

In a recent publication, Kim and Fruehan[1] reviewed much of the previous work. Briefly, there has been a considerable amount of work on mixing times in the major phase, representing the metal, as a function of the amount and method of gas stirring. As shown by Matway[2], mixing times reflect one phase mixing which may not be related to two phase mixing in all cases. There has also been some work on two phase mixing. Typically in these model studies, water and paraffin oil are used to simulate the metal and slag respectively and $\beta$-napthol is the tracer. These studies showed that the mass transfer parameter ($K$), the mass transfer coefficient ($k$) times the interfacial area ($A$), is proportional to the flow rate ($Q$) to the power ($n$)

$$K \propto Q^n \qquad (1)$$

The exponent $n$ changes abruptly at some critical flow rate usually associated with two phase mixing. The problem with the $\beta$-napthol system is that the distribution coefficient between oil and water is less than one indicating that the rate is controlled by mass transfer in the oil or the simulated slag phase. In metallurgical systems the distribution coefficient is usually much greater than one and therefore the rate is controlled mass transfer in the metal phase.

In a recent publication, the authors reported the results of an investigation in which the effects of tuyere pattern, gas flow rate and the presence of a second phase on the mixing times were investigated.[1] In addition, two phase mass transfer was determined as a function of flow, tuyere pattern, oil properties, oil-to-water ratio, tuyere size and type of injector using a system in which the distribution coefficient is similar to that found in metallurgical systems. The results of that study are briefly discussed in this paper.

## EXPERIMENTAL TECHNIQUES AND RATE EQUATIONS

A scale model of a steelmaking ladle with approximately a 45 cm bottom diameter and a 2.5° taper was used. Compressed air was blown through a 0.2 or 0.48 cm diameter tuyeres or a lance. The mixing times were determined using a conventional conductance method with the mixing time defined as the time to reach ± 5% of the final uniform concentration.

Two phase mass transfer was determined using a water modeling technique developed by Matway[2] in which thymol is the tracer and a mixture of paraffin and cotton seed oil is the second phase. The distribution ratio for this system is about 300 as in metallurgical systems, and therefore the rate is controlled by transfer in the simulated metal phase. The integrated form of the rate equation is given by

$$\ln\frac{\left\{\left(1+\frac{hV_w}{V_o}\right)\frac{C_w}{C_w^o} - h\frac{V_w}{V_o}\right\}}{1+h\frac{V_w}{V_o}} = \frac{-kA}{V_w}t \qquad (2)$$

where $h$ is the distribution coefficient, $C_w$ and $C_w^o$ are the concentrations of thymol in the water at time ($t$) and the initial concentration, respectively, and $V_w$ and $V_o$ are the volumes of the water and oil respectively. For convenience the left hand side of Equation 2 is called L.H.S.

# TWO PHASE MASS TRANSFER

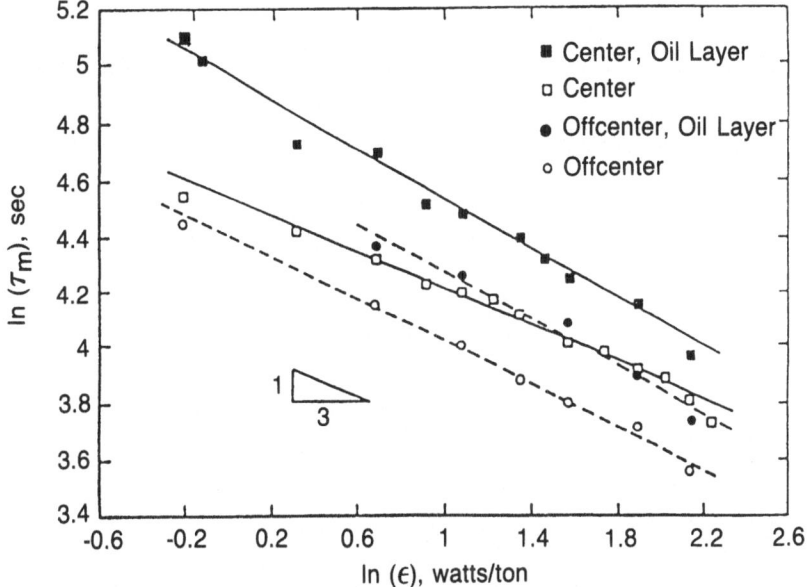

Figure 1. Mixing time versus stirring energy density for two different tuyere patterns with and without oil layer.

This technique only measures $kA$, defined as the mass transfer parameter, because it is not possible to separate $k$ and $A$. In an attempt to measure $k$ the rate of dissolution of solid benzoic acid rods were measured in which the interfacial area was known. From this value of $k$ the value of $A$ in the liquid-liquid experiments can be estimated.

## RESULTS AND DISCUSSION

The effect of a second phase and tuyere position on mixing time is shown in Figure 1. The mixing time is proportional in the mixing energy to the $-0.32$ power, which is in reasonable agreement with previous work.[3-5] The interesting effect is that the presence of the second phase increases mixing times and the off-center tuyere improves mixing. However, as will be shown later, off-center tuyeres do not improve two phase mass transfer but actually decrease the rate.

The L.H.S. of Equation 2 for mass transfer is plotted versus time in Figure 2 for several gas flow rates; the relationship is linear, indicating the validity of the equation. The mass transfer parameter is shown as a function of flow rate in Figure 3, indicating three flow rate regimes. These regimes can be explained in terms of the breaking up of the oil layer, thus increasing the surface area; a similar behavior occurs in metallurgical systems.

*References p. 301*

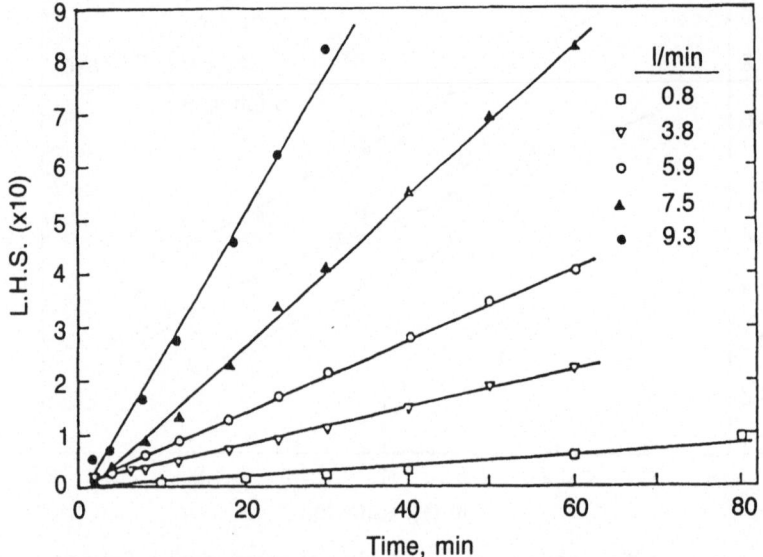

Figure 2. Relationship between L.H.S. of mass transfer equation and time for a center tuyere injection.

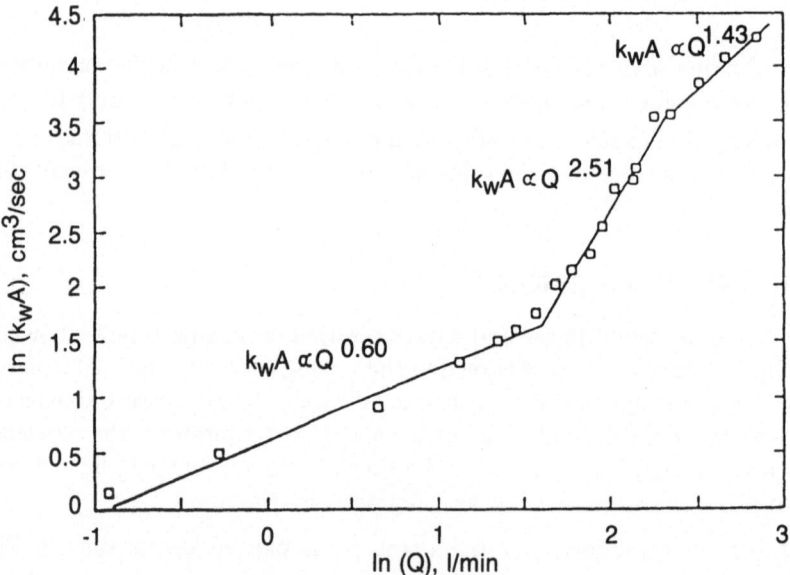

Figure 3. Relationship between $k_w A$ and Q in terms of the order dependence, n, for a center tuyere injection.

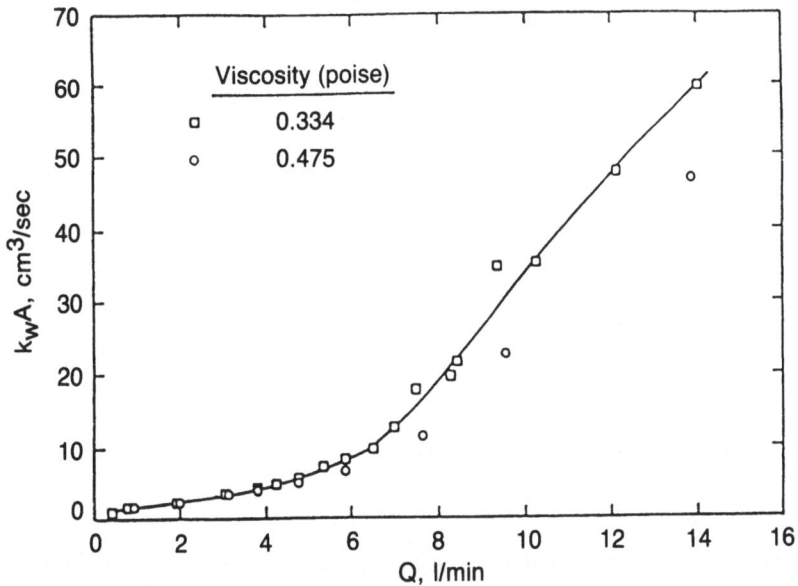

Figure 4. Effect of oil viscosity on the mass transfer parameter for a center tuyere injection.

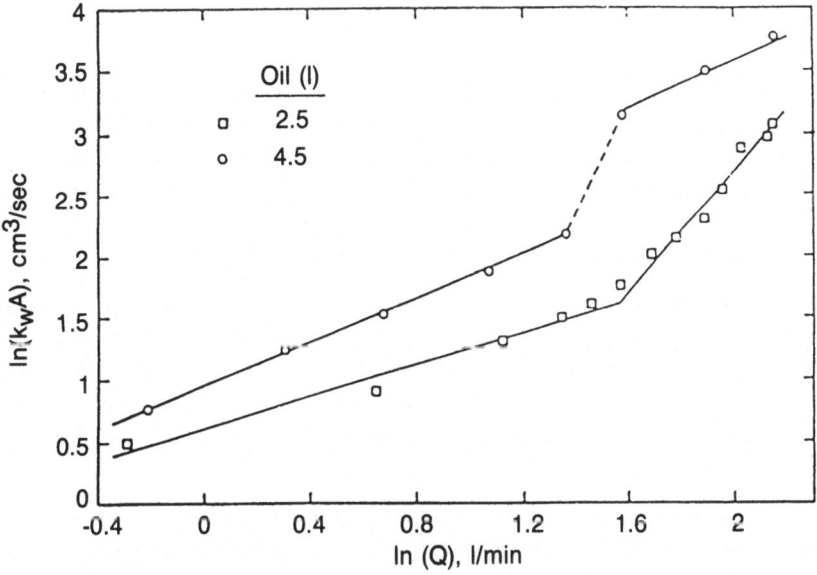

Figure 5. Effect of oil volume on the mass transfer parameter for a center tuyere injection.

*References p. 301*

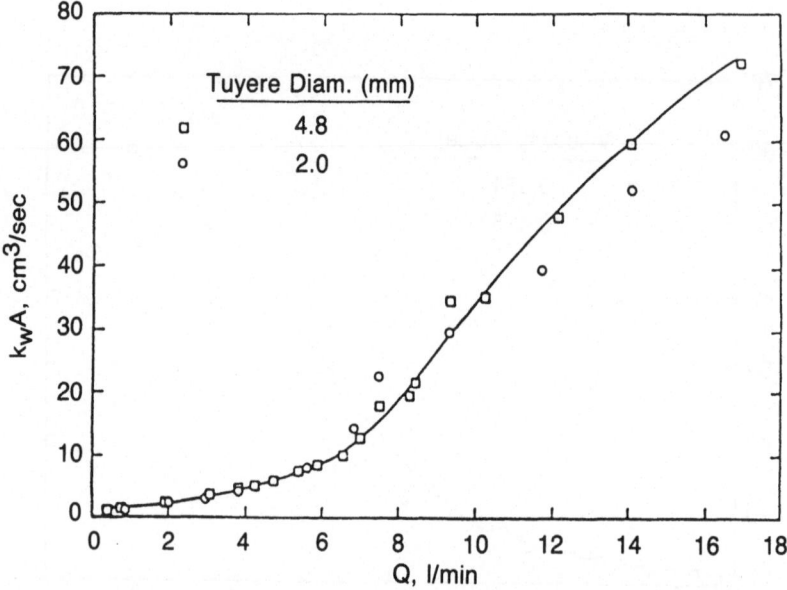

**Figure 6.** Effect of tuyere diameter on the mass transfer parameter for a center tuyere injection.

As shown in Figure 4, oil vicsocity has a small effect on mass transfer at high flow rates. Increasing the amount of oil increases the mass transfer parameter as shown in Figure 5 due to an increase in the interfacial area. As shown in Figure 6, tuyere diameter did not significantly affect the mass transfer parameter. Furthermore, as shown in Figure 7, the mass transfer parameter with lance injection is the same as with a bottom tuyere for equal stirring energy.

The most interesting result was the effect of tuyere number and position. As shown in Figure 8, a single center tuyere gave better mass transfer than an off-center tuyere or multiple tuyeres for equal total gas rates. This was contrary to the case for mixing times. This behavior can be explained by the mixing patterns shown in Figure 9. The active oil area which breaks into droplets is larger in the case of a single center tuyere.

The increase in the mass transfer parameter at a critical flow rate is primarily due to the increase in the area factor in Equation 2. This was confirmed by the measurements of $k$ using the solid rods. From these values of $k$, which were converted to the liquid-liquid system, and the mass transfer parameter, the interfacial area was estimated and is shown in Figure 10. The area increases rapidly at flow rates above the critical flow rate.

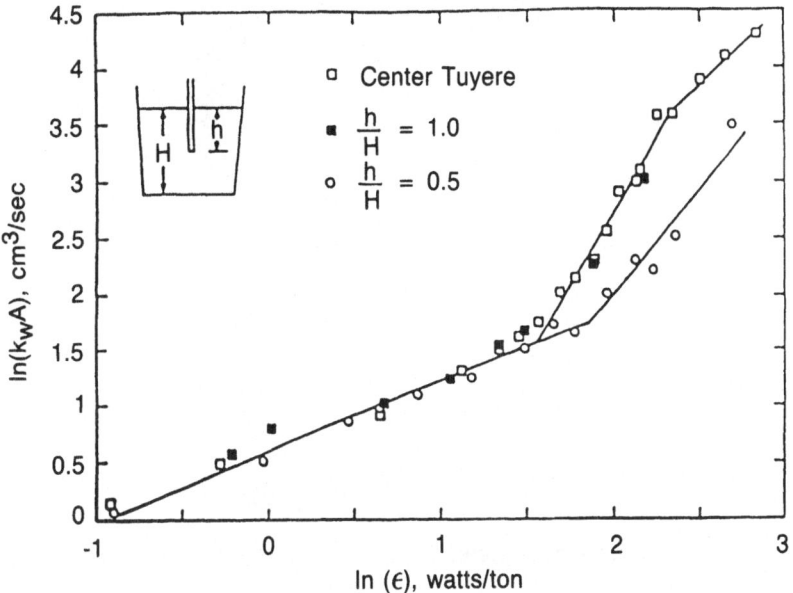

Figure 7. A plot of $k_wA$ as a function of stirring energy density for two different depths of lance injection.

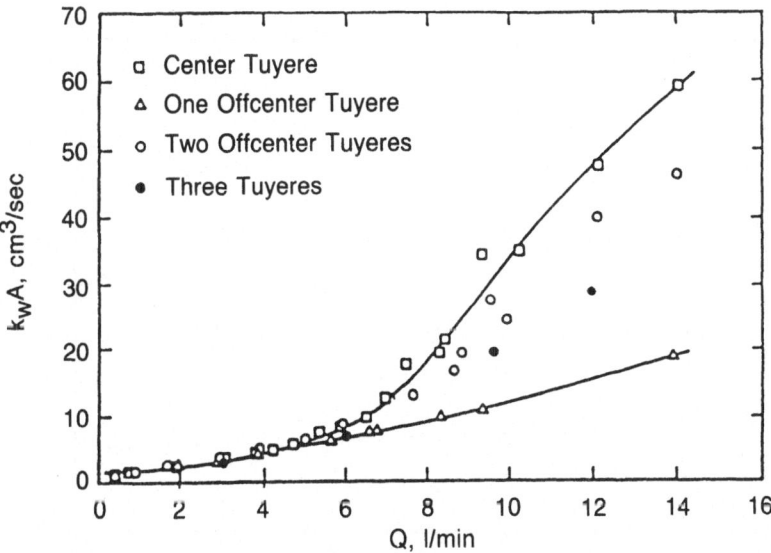

Figure 8. Effect of tuyere position on the mass transfer parameter.

*References p. 301*

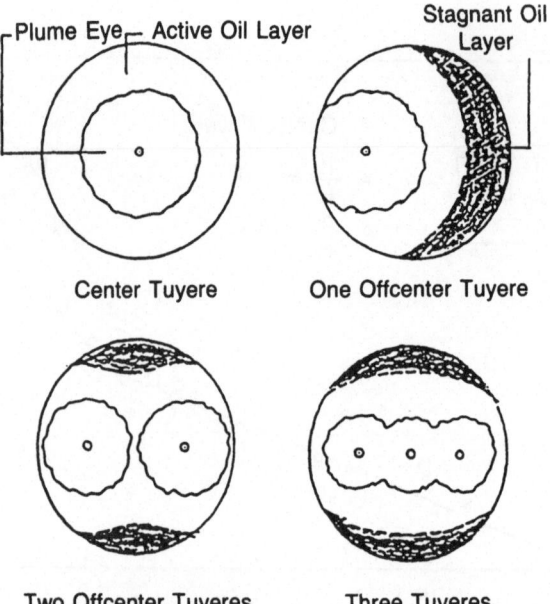

Figure 9. A schematic diagram of top views of the oil layer during gas injection of high flows according to the different arrangements of tuyere(s).

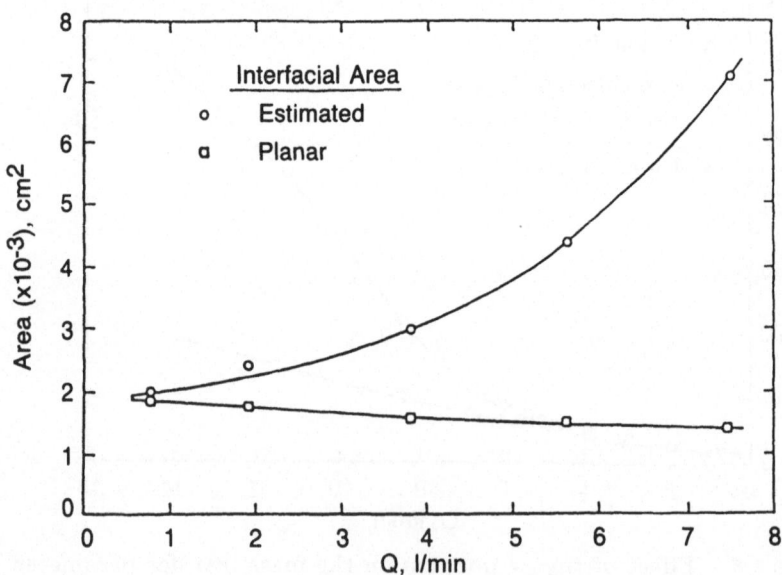

Figure 10. A comparison between the estimated and planar interfacial area for a center tuyere injection.

## SUMMARY

Mixing times and two phase mass transfer rates were determined as a function of operating variables. Whereas, mixing times are favored by off-center tuyeres, the same is not necessarily true for mass transfer. It was previously generally assumed that there was a correlation between mixing times and mass transfer. Tuyere size and oil properties have only a small effect on mass transfer. The change in the mass transfer parameter can generally be explained in terms of the changes in the interfacial area. Currently work is being carried out to extrapolate these results to real systems using dimensional analysis.

## ACKNOWLEDGEMENTS

Support of this research by the Center for Iron and Steelmaking Research which is supported by member companies and NSF Grant 8421112 is gratefully acknowledged. Also, the authors wish to thank R. Matway (CMU) and R. I. L. Guthrie (McGill University) for their helpful discussions.

## REFERENCES

1. S-H Kim and R. J. Fruehan, Submitted to *Met. Trans. B*.
2. R. Matway, PhD Thesis, Carnegie Mellon University, 1986.
3. S. Asai, T. Okamoto, J. He and I. Muchi, *Trans ISIJ*, *23*, 1983, 43.
4. T. Lehner, *Ladle Treatment of Carbon Steel*, McMaster University Symp., Hamilton, Ontario, Canada, 1979.
5. Mazumdan, PhD Thesis, McGill University. 1985.

## DISCUSSIONS

**K. Lange** *(Technical Univ. of Aachen)*

Did the mixing times Professor Fruehan mentioned refer to the metal phase?

**A.** Yes. For mixing times the oil phase is essentially inert except for its influence on fluid flow at the surface.

**Craig Landefeld**

The previous question prompts a question for either of the speakers. Suppose you put a tracer in both the slag and metal phases, stirred them, and measured mixing times for each. Would you expect the ratio of the mixing times to correlate with the mass transfer coefficient between the two phases?

**A. R. J. Fruehan** No. **S. Asai** No. If we know the mixing time we can calculate circulation time because the product of circulation time and mixing time is a nearly constant. So once we measure the mixing time and the capacity coefficient we can evaluate the value of the Stanton number, which tells whether the slower mass transfer step occurs within the bulk metal or at the interface.

**R. J. Fruehan**

My impression is that one-phase mass transfer is rapid in metallurgical processes, such as desulfurization, dephosphorization in the combined blowing vessel and desulfurization in an iron reactor. One phase stirring by gas is fairly violent and gives very good mixing and one-phase mass transfer. The critical thing is two-phase mass transfer, and mixing times don't tell much concerning the rates of refining reactions.

**N. Lillybeck** *(Deere & Co.)*

What would be the result if you conducted these experiments in a square vessel rather than a round vessel?

**A. R. J. Fruehan** I don't think that in terms of two phase mass transfer there would be any measurable difference between a square vessel and a round vessel. Rod Guthrie and Yogesh Sahai have looked at energy dissipation quite extensively. At least for two phase mass transfer I feel confident that it would not be a function of the intricate geometry of the vessel.

# THE PRINCIPLES OF GAS AND POWDER INJECTION FOR IRON REFINING

### G. A. IRONS
McMaster University
Hamilton, Canada

### ABSTRACT

Powder injection techniques have been widely adopted in the steel industry for economically carrying out ladle metallurgy operations such as desulfurization, dephosphorization and alloying. However, in general, the foundry industry has not fully utilized this technology. One of the obstacles to this has been the lack of understanding of the physical and chemical phenomena which occur during powder injection. In this paper, a number of studies incorporating physical and mathematical modelling of injection as well as large scale studies of calcium carbide injection into iron are summarized. These studies may provide a basis for the design of powder injection processes to reduce operating costs and improve quality and consistency in the foundry industry.

## INTRODUCTION

In recent years the pneumatic injection of powders into iron and steel has become a very successful way to accomplish desulfurization, desiliconization, and dephosphorization. The major advantages of powder injection are:

1. the interfacial area of the injected particles in contact with the melt is much greater than with a top slag and therefore reactions by injection are faster;

2. the carrier gas generally keeps the ladles well-mixed;

3. the rate and total quantity of reagent can be precisely controlled;

4. a wide variety of powders can be injected so that (i) sequential operations of refining or alloying can be accomplished easily during a heat. or (ii) reagents can be substituted to take advantage of new reagents or changes in relative price of reagents;

5. the atmosphere can be controlled, for example, to eliminate oxygen, which allows powerful desulphurizers, such as magnesium, calcium, and calcium carbide, to be used; and finally

6. Injection stations and the associated equipment are not excessively expensive.

The main disadvantage of powder injection is that the residence times of the particles are quite short (a few seconds); thus the reagent efficiencies are often poor. Consequently, the trajectories of the particles are of fundamental importance to these processes.

In the present paper, several physical and mathematical modelling studies of the physical interactions between the gas, liquid and solid phases are summarized. On the basis of this information it is possible to design the injection conditions for a wide variety of reactions such as desulphurization, submerged combustion and alloying.

Experimental work and mathematical modelling of iron desulphurization by calcium carbide injection was also carried out. In these studies the importance of oxygen activity, the top slag reactions and the injection conditions were highlighted. These studies provide a general framework for quantitatively understanding and designing new or improved injection processes. It is hoped that this information will stimulate the development of more injection processes for the foundry industry.

## PHYSICAL INTERACTIONS BETWEEN PARTICLES, GAS AND LIQUID

There have been many studies in which powder injection into iron or steel has been simulated by injection into water,[1-6] lead[7] and lead alloys.[8] It has been found that fine particles at low solid-to-gas loadings promote jets of particles piercing bubbles (Figure 1a) while at higher loading homogeneous gas-particle jets form (Figure 1b). With coarse particles the bubbling regime is more predominant (Figure 1c). At high loadings of coarse particles it is difficult to make an homogeneous jet, and some bubbles are still produced (Figure 1d).[9]

The transition from bubbling to jetting for vertically downward injection can be explained by considering the state of flow in the conveying line. (Angled injection is discussed later). With high loadings of fine particles, the gas and particles are traveling at near the same velocity as shown in Figure 2c. The momentum boundary layer around each particle is relatively large, so that most of the gas travels with the particles in a coupled state. On contact with the liquid, the gas continues to flow with the particles which have considerable momentum. Thus, a gas-particle-liquid jet forms.

When the particles are coarse, such as in Figure 2a, the particles travel considerably more slowly than the gas, and the boundary layers are relatively thin. Thus most of the gas is not in the boundary layers (uncoupled flow), and upon contact with the liquid, the gas does what it would do in the absence of particles, that is, form bubbles. The transition is taken as the critical loading (expressed as

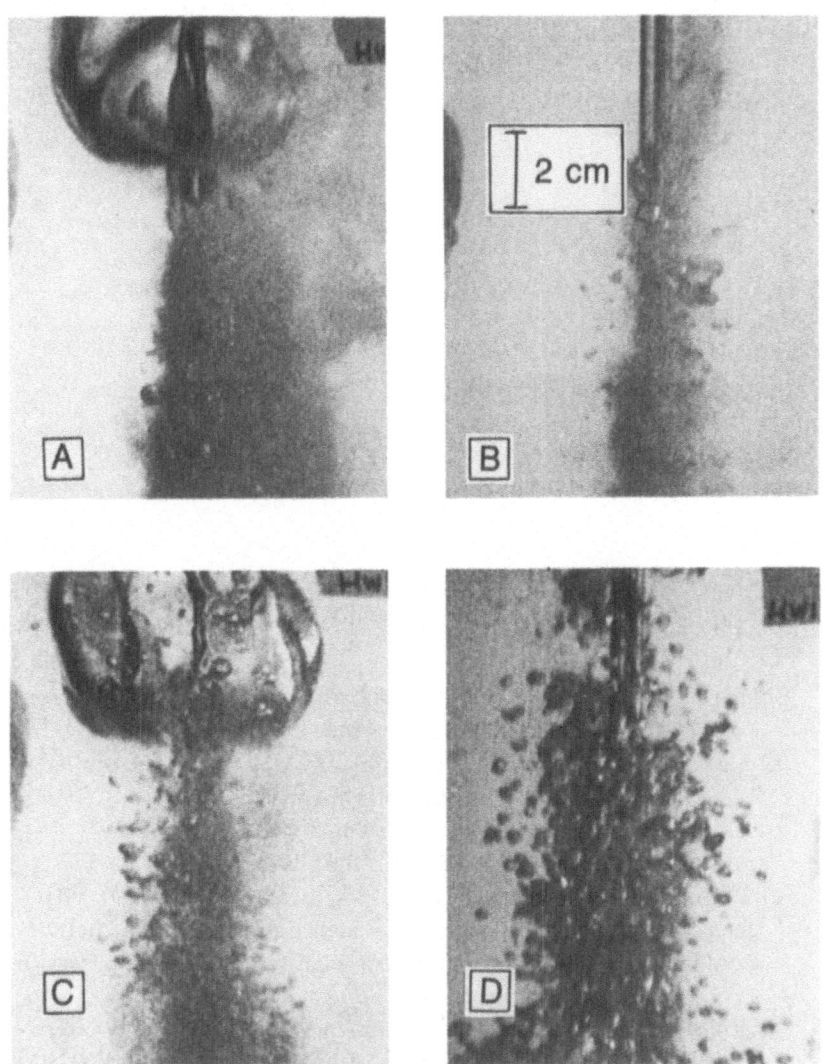

Figure 1. Behaviour in the He-sand-water system.
a) Uncoupled flow,
$d_p = 115\ \mu m$, $d_0 = 7$ mm, $Q = 124$ slpm, $m_p = 0.636$ kg/min, $\theta_p = 0.006$.
b) Coupled flow,
$d_p = 115\ \mu m$, $d_0 = 7$ mm, $Q = 42$ slpm, $m_p = 2.4$ kg/min, $\theta_p = 0.08$.
c) Uncoupled flow,
$d_p = 450\ \mu m$, $d_0 = 7$ mm, $Q = 150$ slpm, $m_p = 0.702$ kg/min, $\theta_p = 0.0078$.
d) Partially coupled flow,
$d_p = 450\ \mu m$, $d_0 = 7$ mm, $Q = 217$ slpm, $m_p = 4.91$ kg/min, $\theta_p = 0.046$.

*References pp. 327–328*

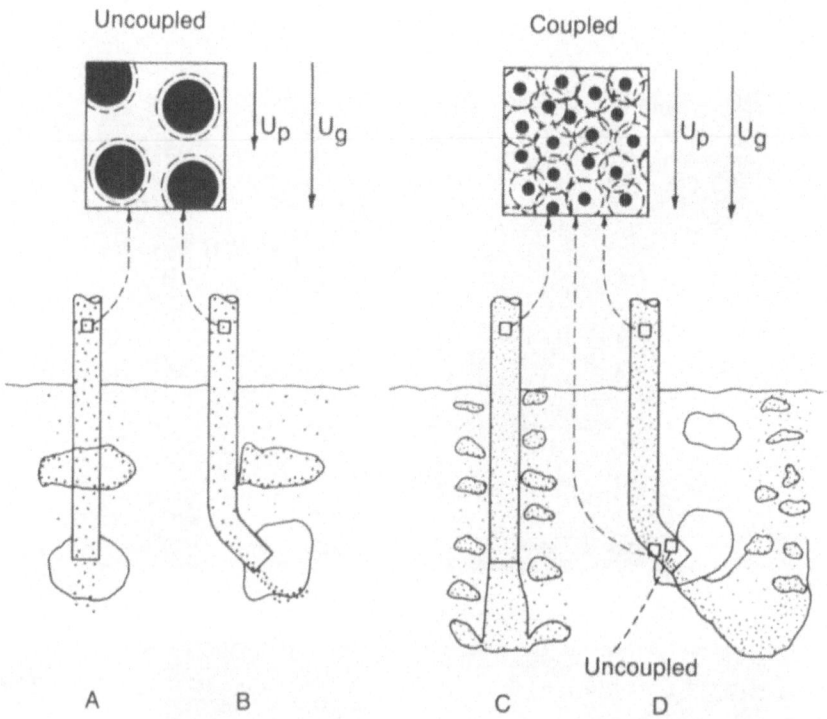

**Figure 2.** Schematic diagram illustrating the coupling between gas and solid phases for various configurations.
a) Coarse particles injected straight downward are uncoupled so that bubbles form.
b) Coarse particles injected at an angle segregate to the bottom of lance and bubbling is produced.
c) Fine particles injected straight downward are coupled with the gas and therefore form jets which penetrate until their momentum is dissipated. Bubbles with particles inside rise from this point.
d) Fine particles injected at an angle segregate to the bottom of the lance and therefore a jet of particles and some gas penetrates into the bath. At the lance tip most of the gas is in the top part of the lance and therefore the gas is uncoupled from the powder and therefore bubbles form at the lance tip as well.

a volume fraction of powder in the mixture) at which adjacent boundary layers overlap. The boundary layer is, in turn, a function of the relative velocity between the gas and particles and the particle diameter which can be grouped as a Reynolds Number for the particle, $Re_p$. The critical loading is[7,9]:

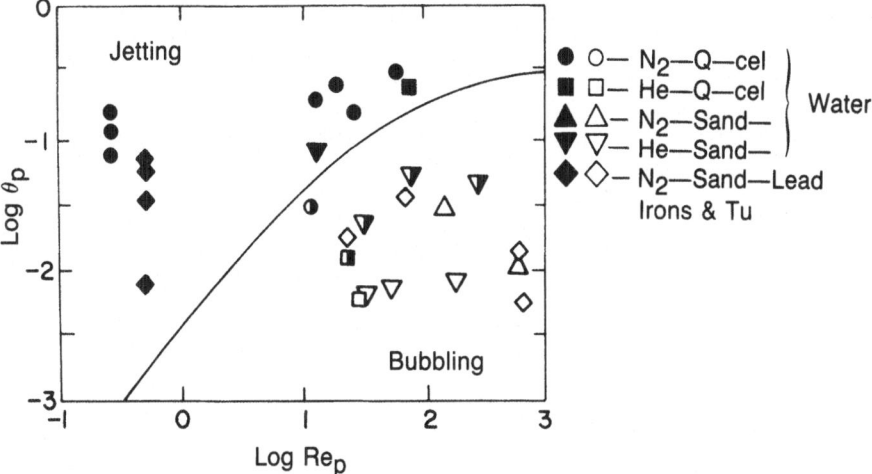

Figure 3. Particle relative Reynolds number versus critical particle volume fraction for coupled flow. Jetting results are indicated with solid shapes, bubbling with hollow shapes, and transitions with half-shaded shapes.

$$\Theta_c = \frac{\Pi}{6\left(\frac{6}{\sqrt{2Re_p}} + 1\right)^3} \qquad (1)$$

When relative gas and particle velocities are measured $Re_p$ can be directly calculated, however in the absence of such instrumentation the Reynolds Number at the terminal settling velocity for a single particle in stagnant carrier gas may be used. This relation explains the transition from bubbling to jetting seen in Figure 3 for experiments conducted in water and liquid lead.[9] The same transition is expected in iron and steel, and similar flow regime maps have been made for magnesium and calcium based reagents, Figures 4 and 5, respectively.[10] For these diagrams, the more familiar solid-to-gas loading, $L$, and particle diameter are used. These diagrams show that reagents such as fine lime, lime-magnesium mixtures and calcium carbide fall in the jetting regime, while salt-coated magnesium and coarse calcium silicide are expected to be in the bubbling regime. It has been shown that bubbling promotes clogging because liquid can flow up the lance between bubbles, whereas fine particles at high loadings produce steady-state jets which penetrate deeply into the melt without the opportunity for liquid to enter the nozzle. This phenomenon may explain the lance-clogging problems associated with magnesium and calcium silicide granules.[10]

The penetration distance of gas-particle jets can be found by solving differential equations balancing the buoyancy of the jet and its momentum.[1,2,9] In Figure 6, the dimensionless jet penetration in numbers of nozzle diameters is plotted as

*References pp. 327–328*

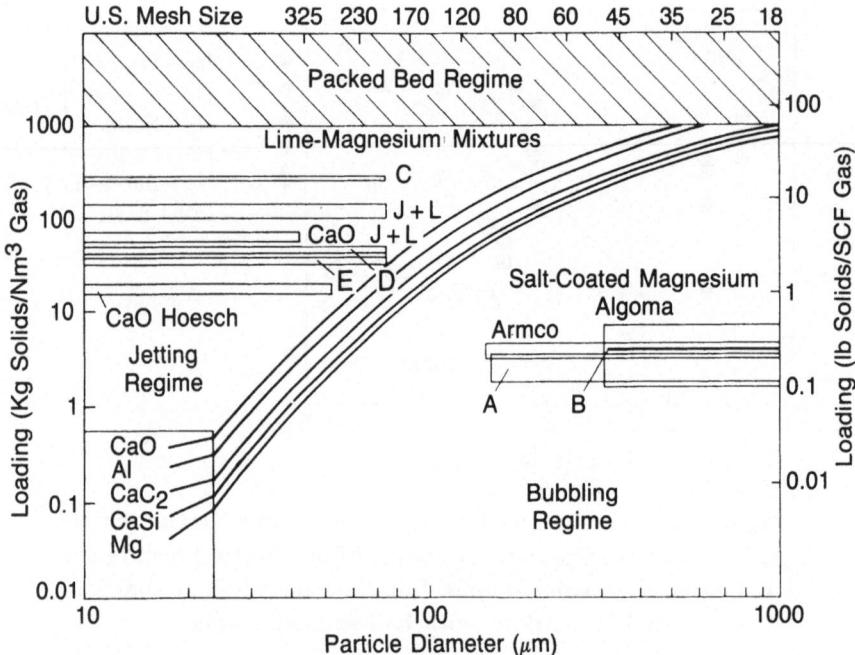

Figure 4. Solids loading as a function of particle size for several industrial processes based on magnesium reagents. The calculated transition from bubbling to jetting is shown as well. References for each process are:
Salt-coated magnesium: Armco,[13] Algoma[16]
Lime-magnesium mixtures: J&L (Jones and Laughlin)[12]
Lime: J&L,[12] Hoesch[17]

a function of the Froude Number, $N_{Fr}$, for various loadings, $\gamma$.[9] These predictions agree with experimental results in water and lead over a wide range of conditions. For most commercial injection conditions the Froude Number can be calculated approximately from the gas flow rate:

$$N_{Fr} = \frac{U_m^{o^2}}{d_i g} \doteq \left(\frac{4W_g}{\pi \rho_g^o}\right)^2 \frac{1}{d_i^5 g} \qquad (2)$$

and the loading approximated from the solid to gas ratio:

$$\gamma = \frac{\rho_\ell}{\rho_m^o} \doteq \frac{\rho_\ell}{L} \qquad (3)$$

For example, to remove 0.05% sulphur from 10 tonnes of iron with 80% pure calcium carbide at 20% efficiency in 3 minutes requires a solids flow rate of 20 kg/min. If this were conveyed with 0.285 Nm³/min of nitrogen through a

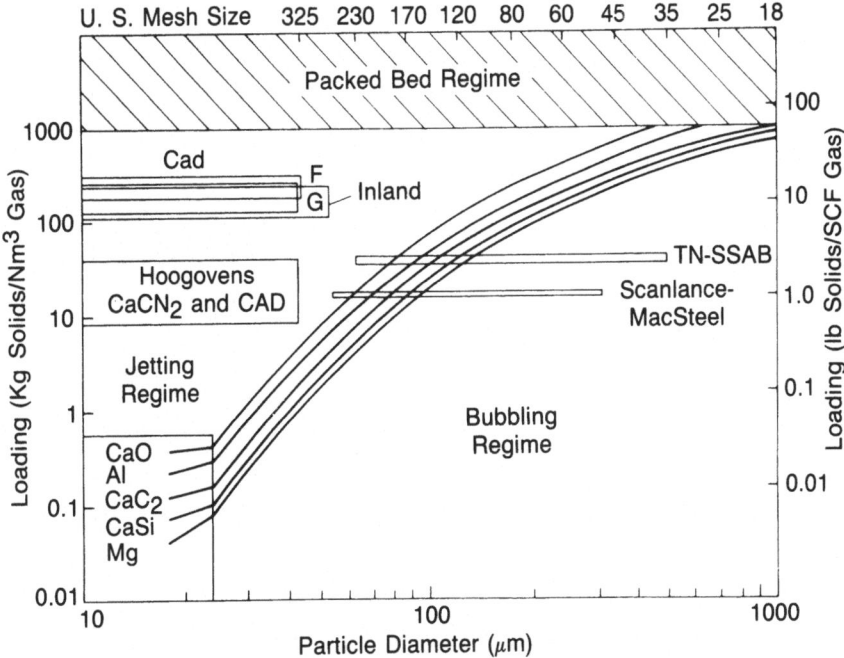

Figure 5. Solids loading as a function of particle size for several industrial processes based on calcium reagents. The calculated transition from bubbling to jetting is shown as well. References for each process are:
CAD: mixtures of $CaC_2$ and $CaCO_3$, Inland,[11] Hoogovens[18]
Calcium silicide: Scanlance,[14] TN.[15]

10 mm diameter lance the loading, $L$, would be 70 kg/Nm$^3$, $\gamma$ would be 100 and the Froude Number would be 37400. The dimensionless penetration from Figure 6 is 50 lance diameters which is 0.5 m. Such calculations can be used to position the lance tip so that maximum particle penetration is achieved without excessive erosion of the refractories on the bottom of the vessel.

The various regimes of flow for vertically downward injection are illustrated in Figure 7. In a simple way it shows the various regimes that are expected, as a function of gas velocity and particle size and type. Other factors such as particle velocity, loading and densities and viscosities of the various phases are all included in the relations used to determine transitions in regime indicated by the letters A to I as discussed in the original paper.[9] The transitions are largely based on the fundamentals of multi-phase fluid dynamics, and if properly applied can yield the scaling criteria for full-scale processes. For the lowest gas velocities (Ma< 0.05) the particles may not penetrate the bubble interface, however these velocities are so low that the particles cannot be pneumatically conveyed. Therefore this regime is of academic interest only. Under most ladle injection conditions (0.03<Ma<0.5),

*References pp. 327–328*

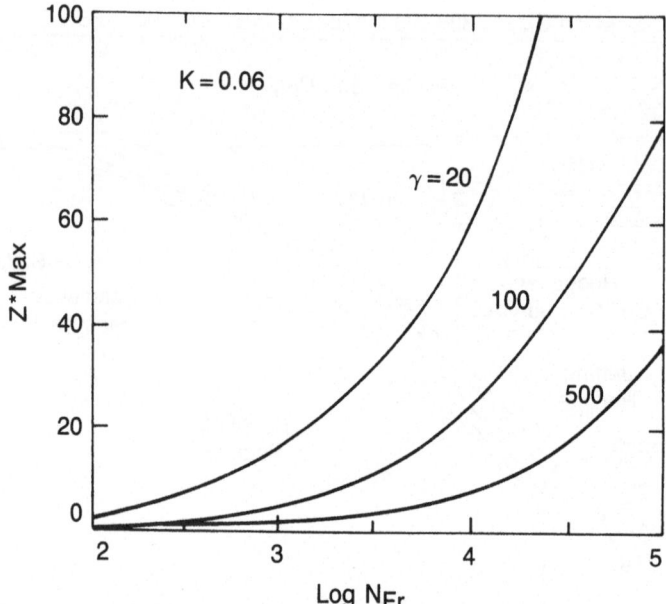

Figure 6. Predicted penetration against Froude Number for various liquid to mixture density ratios, $\gamma$.

bubbles or jets may be formed (Figures 4 and 5). At higher velocities ($0.3 < Ma < 1.0$), characteristic of magnesium granule injection into ladles or lime injection into bottom-blown oxygen furnaces, even coarse particles produce a jet near the nozzle which subsequently forms bubbles.[19] To generalize even further, one could say that jetting is favored by high gas momentum. For gas-only this is achieved by high velocity or high gas density. Another way to increase the "gas" momentum is to couple it effectively to a denser phase such as fine particles. In this case only low velocities are required for jetting.

For inclined and "hockey-stick" lances, which are widely used industrially, the situation is slightly different as depicted in Figure 2. It has been found that even with fine particles at high loading, bubbles are formed at the lance tip.[8,20] Figure 2d shows that fine particles tend to segregate to the bottom of the lance, leaving the gas and particles uncoupled at the top of the lance. Consequently, gas bubbles can form at the top while a coupled jet of gas and particles exits from the bottom. It has also been shown that angled injection improves particle liquid contact.[20]

## CHEMICAL INTERACTIONS BETWEEN PARTICLES, GAS AND SOLIDS

Calcium carbide is an interesting reagent to study for several reasons. In the Foundry Industry this reagent is most often placed on top of "tea-pot" ladles for

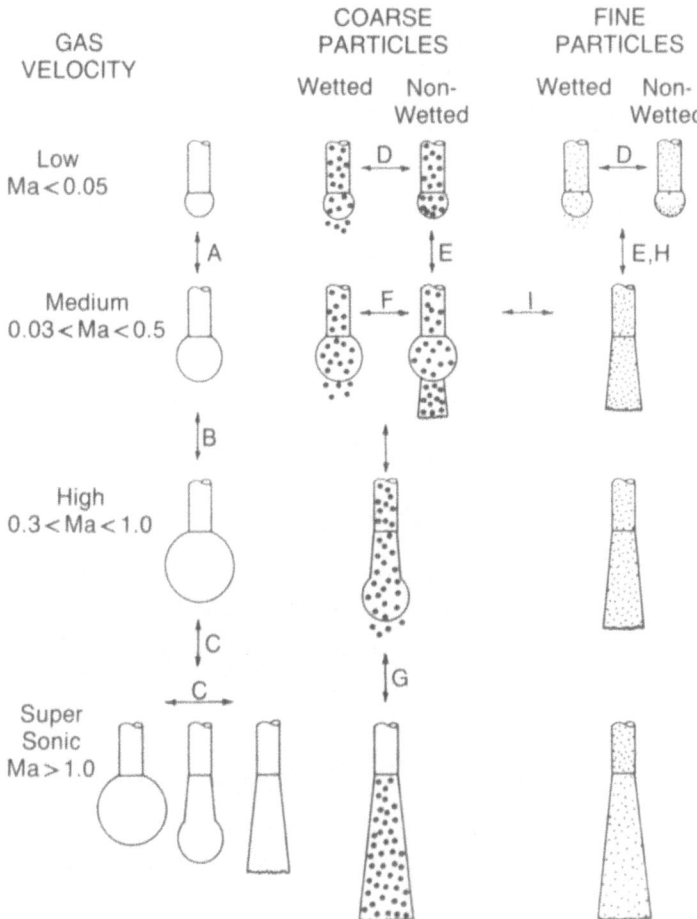

Figure 7. Summary of the different possible regimes in injection of gas and gas-powder mixtures into liquids.

continuous desulphurization, whereas in the steel industry it is injected deeply into the melt. The reaction rates are fast, generally permitting greater than 3 tonnes of iron per minute to be treated. However, the reagent utilization is low, approximately 20% (based on stoichiometric utilization of all $CaC_2$, $CaO$ and $CaCO_3$ in the mixture), but is higher with injection. Therefore there is great potential for process improvement. In the steel industry there is great interest in achieving final sulphur contents of 0.005% S, which is quite difficult with current practice. With these problems in mind, desulphurization of hot metal using DSR-100, an 80% calcium carbide reagent, was studied using the 3-tonne induction melting facilities at the Stelco Research Centre located in Burlington, Ontario. Calcium carbide efficiency and rate were examined as functions of the top slag quality (basicity and fluidity), the hot metal oxygen activity, the reagent flow rate, and the gas-to-solids ratio.[21]

*References pp. 327–328*

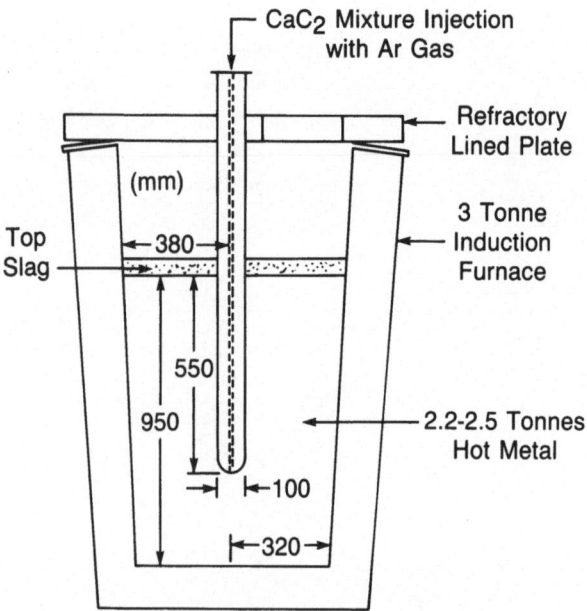

Figure 8. Cross-sectional view of the pilot plant induction furnace at the Stelco Research Centre.

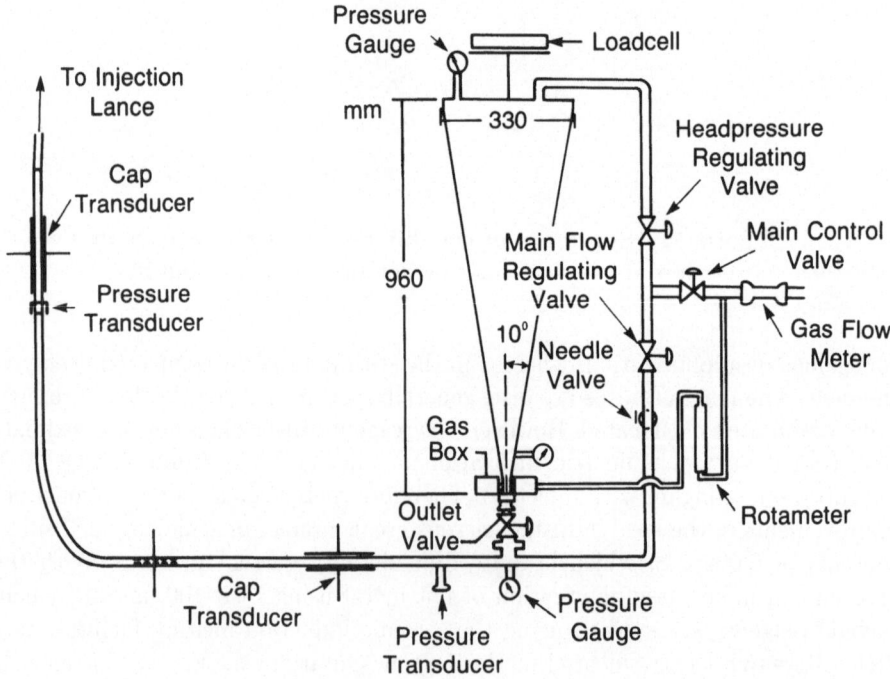

Figure 9. Injection system used for the pilot plant tests.

# PRINCIPLES OF GAS AND POWDER INJECTION

## EQUIPMENT AND PROCEDURE

All trial injection tests were carried out using a 3-tonne induction melting furnace. A cross section of the furnace showing the size of the melt and orientation of the injection lance is provided in Figure 8.

A scaled-down injection system was used to add the calcium carbide. The injection equipment was adjacent to the induction furnace with the dispenser being connected to the lance via a flexible hose. Details of the powder dispenser design and instrumentation are shown in Figure 9.

Pig iron was melted in the induction furnace to provide the molten iron required for each desulphurization test. The typical composition of the iron was 3.5% C, 0.3% Si, 0.5% Mn, and 0.04% P. The melt temperature was controlled to 1350 ±5°C. The iron was resulphurized prior to each test by plunging pure sulfur into the melt. The range of initial sulphur levels evaluated during the trials was 0.018 to 0.118%.

DSR-100, a non-limestone-bearing reagent, was selected for the pilot plant study. This reagent simplifies the chemical reactions in the system by eliminating the limestone and its products, $CO_2$ and lime. The mixing required to achieve the desulphurization reactions was supplied solely by the injection gases. Under these simplified conditions, the chemical efficiency of calcium carbide desulphurization was examined as a function of the hot metal oxygen activity, the slag-metal reactions, and the powder and gas flow rates.

To examine the predominant chemical reactions taking place during desulphurization and their impact on calcium carbide efficiency, four top slag conditions were selected (Table 1).

## RESULTS AND DISCUSSION

**Basic Top Slag Additions**—Iron without a top slag present, or with an unreactive, solid, lime-silica slag, desulphurized slowly at the start of injection (Figure 10a). Frequently, the hot metal sulphur level did not decrease significantly for the initial 20 to 30% of the injection period. This inefficiency was termed an incubation period, a period during which the reagent had to create the appropriate top slag before it could begin to remove sulphur.

The application of the liquid mold powder slags or the lime-silica slag which was "liquified" by dissolved alumina (aluminum-modified slag condition) eliminated the incubation period, resulting in more rapid desulphurization (compare Figure 10a and Figure 10b). The rate of sulfur removal shown in Figure 10a for the no top slag condition is somewhat lower than that for the dry slag conditions because the initial sulphur level is lower (0.047 compared to 0.070% S). Calcium carbide efficiency generally improves as the initial sulphur increases because of the high concentration of dissolved sulphur.

Comparison of the results for each pilot-scale slag condition indicates that the

*References pp. 327–328*

## TABLE 1
## Top Slag Conditions used during Pilot Plant Desulphurization Trials

| Slag Condition | Details |
| --- | --- |
| No Slag | No additions to the iron surface before injection starts |
| Dry Slag | Addition of 5.9 kg/tHM of a mechanical mixture of lime and silica powder: $CaO/SiO_2 = 1.2$ |
| Liquid Slag | Addition of 5.9 kg/tHM of STG continuous casting powder: $CaO/SiO_2 = 1.3$, melting point $= 1135°C$ |
| Aluminum-modified slag | The plunging of 1.7 kg/tHM of aluminum into the molten iron followed by the addition of 5.9 kg/tHM of dry slag. |

reagent efficiency for the liquid slag conditions is higher than that for cases of no slag or dry slag (conditions which lead to incubation). Results covering an initial sulphur range from 0.02 to 0.09% S are summarized in Figures 11 and 12. The sulphur range was restricted so that a proper comparison could be made among the slag conditions. Furthermore, this sulphur range is typical of sulphur levels experienced in the industrial desulphurization of hot metal.

The incubation period which appeared at the initial stages of injection for both no slag and dry slag initial conditions (Figure 10a) can be explained by the inability of the slag to physically or chemically absorb the partially reacted calcium carbide particles. With no initial top slag, the melt surface is susceptible to oxidation, and a liquid layer of iron oxide probably forms. This slag can then oxidize the calcium sulphide produced during the desulphurization process, reverting to sulfur into the iron melt. The net effect is that no desulphurization is observed, i.e., an incubation period. As the partially reacted $CaC_2/CaS$ particles continue to build up on the surface during injection, conditions will become more reducing thereby favoring desulphurization. The slag that forms is an effective barrier to oxygen entering the system from the atmosphere.

The situation is similar for the dry slag condition because the mechanical mixture of CaO and $SiO_2$ is initially unreactive. Thus an iron oxide-rich slag can form which must be deoxidized and become more basic before effective desulphurization can proceed.

# PRINCIPLES OF GAS AND POWDER INJECTION

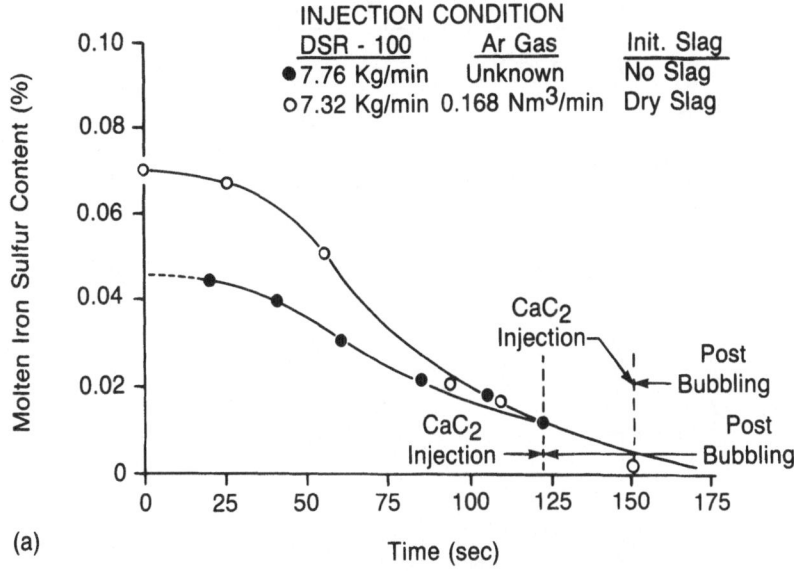

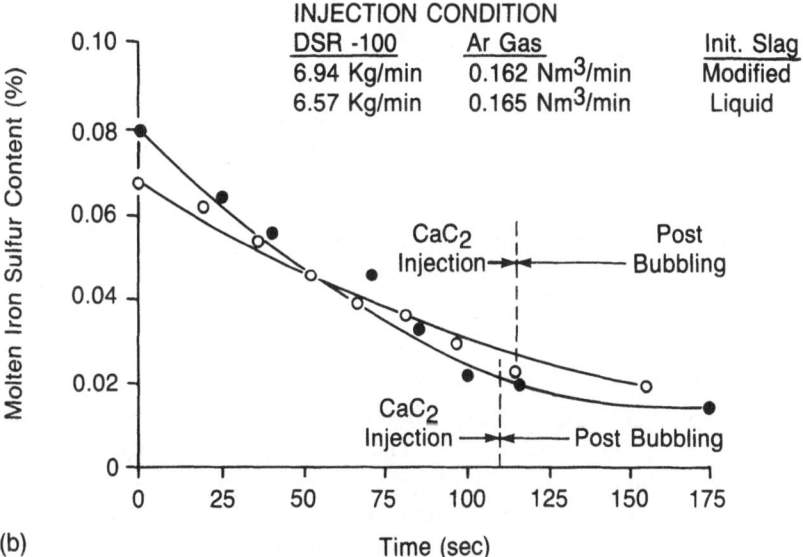

Figure 10. The impact of slag conditions on the rate of sulfur removal from hot metal; (a) No top slag and dry top slag condition; (b) Liquid and aluminum-modified top slag conditions.

*References pp. 327–328*

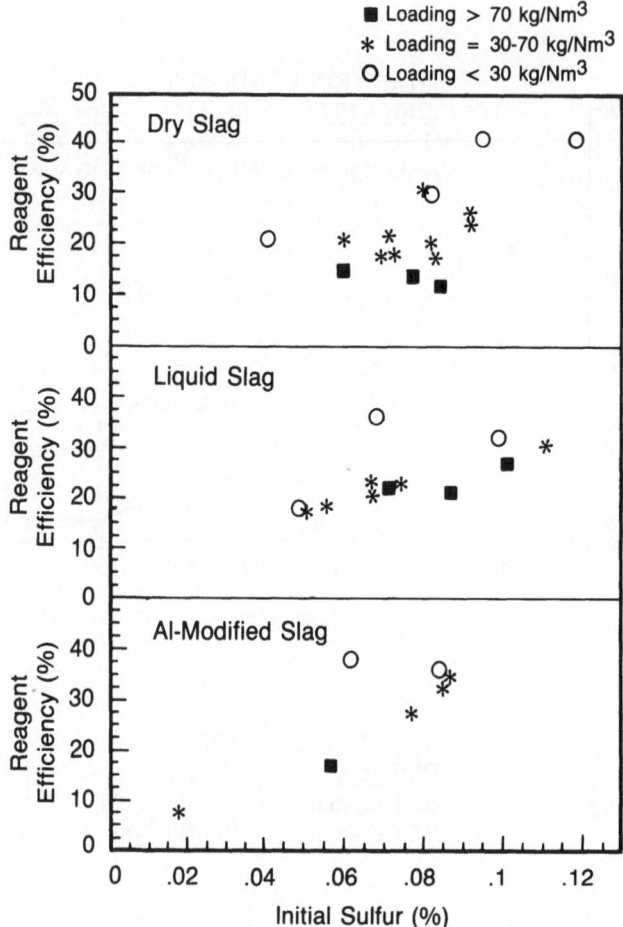

Figure 11. Improvement in reagent efficiency as the solids loading level is reduced.

In contrast, the liquid slag and the modified dry slag have the ability to immediately dissolve the CaS formed during desulphurization due to the sufficiently high slag volume, basicity and fluidity. The dry slag modified by the aluminum addition improved the reaction rate, reagent efficiency, and final sulphur level, due to both the deoxidation of the hot metal and the formation of molten calcium aluminum silicates with high sulphide-carrying capacity.

**Control of Hot Metal Oxygen Activity**—Measurements of hot metal oxygen activity in the pilot plant using the Celox II/CLL oxygen probe in untreated molten iron indicated that the oxygen activity averaged 2.5 ppm, decreasing to

# PRINCIPLES OF GAS AND POWDER INJECTION

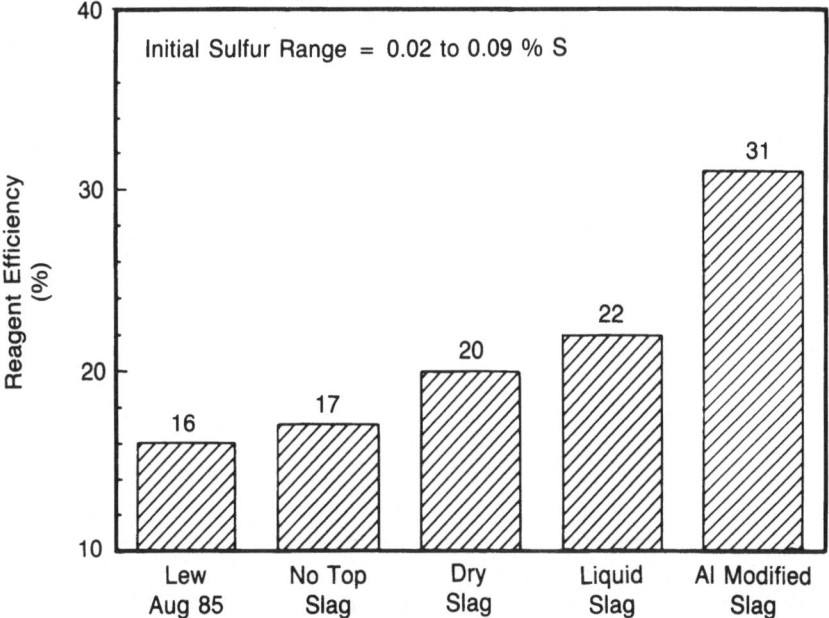

Figure 12. Improvement in calcium carbide reagent efficiency with control over top slag quality and hot metal oxygen activity.

2.3 ppm after treatment. Analysis of the oxygen activity before and after calcium carbide injection indicates that it is controlled by the dissolved silicon/oxygen equilibrium ($a_{SiO_2} = 1$) (Figure 13), and that this oxygen activity is larger than that generally assumed from the dissolved carbon and oxygen equilibrium ($P_{CO} = 1$ atm).

Aluminum was added to the hot metal in the pilot plant studies to reduce oxygen activity to lower levels, since it dissolves easily in iron and has an extremely high affinity for oxygen. Aluminum was added by plunging aluminum ingots into heats covered with the dry slag, producing the aluminum-modified slag (Table 1). These additions reduced the oxygen activity to well below 1 ppm (Figure 14).

A comparison of non-aluminum-treated and aluminum-treated hot metal under similar injection conditions shows the efficiency improvement with reduced oxygen activity (Figure 11). The aluminum-modified slag reduces calcium carbide consumption through two mechanisms:

(1) low melting point (1350–1400°C) lime-aluminum-silica slags with high sulphide capacities are produced at the surface (molten slags were observed during the test work), and

(2) residual aluminum dissolved in the hot metal scavenges any incoming oxygen which would otherwise react with the calcium carbide.

*References pp. 327–328*

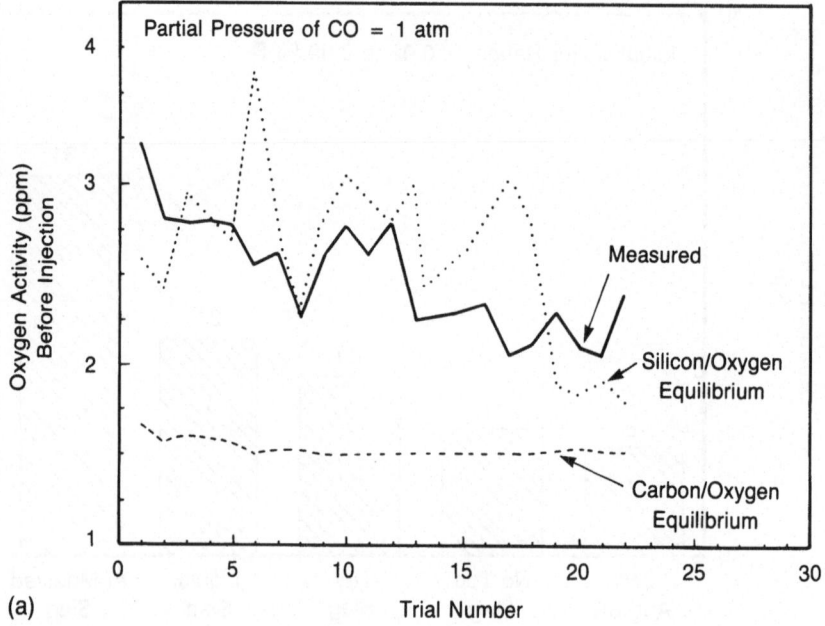

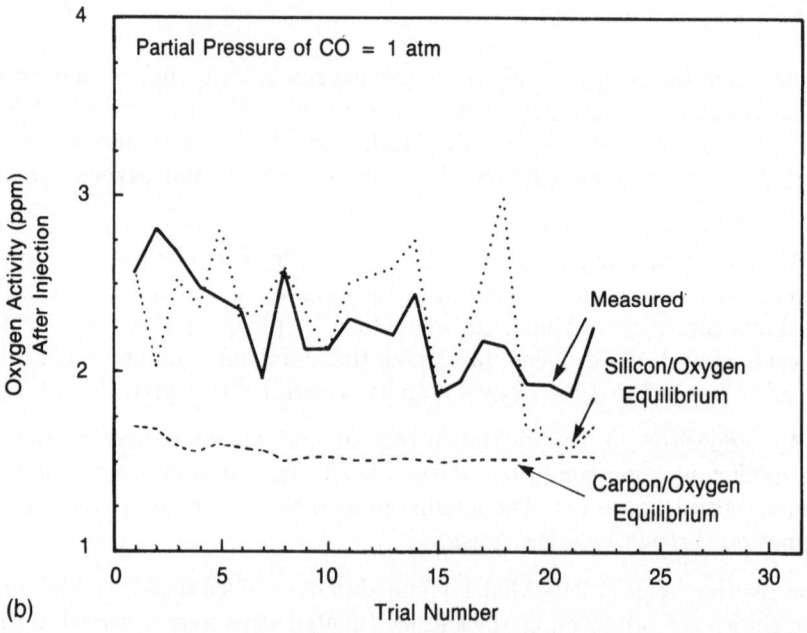

Figure 13. Measured molten iron oxygen levels compared to values predicted by silicon-oxygen and carbon-oxygen equilibria; (a) Before injection; (b) After injection.

# PRINCIPLES OF GAS AND POWDER INJECTION

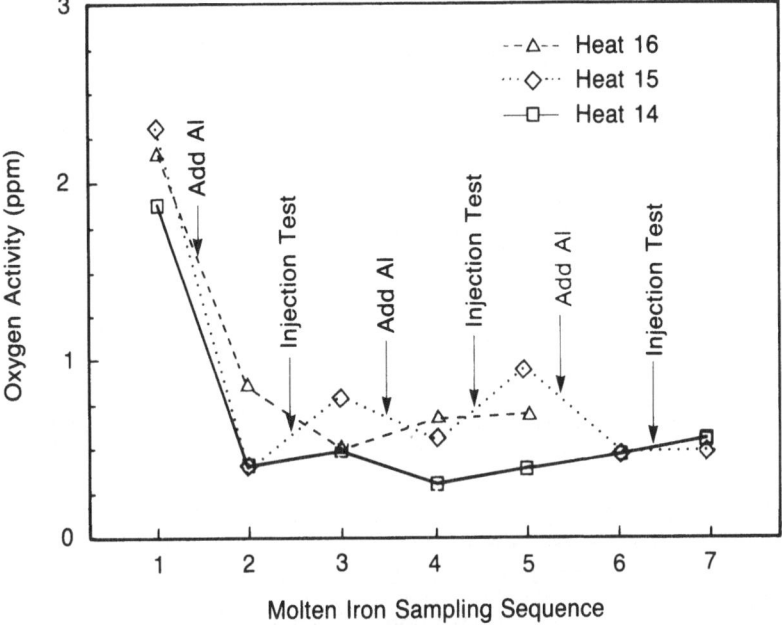

**Figure 14.** Reduction in molten iron oxygen activity with aluminum additions.

The oxygen activity in equilibrium with the measured dissolved aluminum content is about 0.001 ppm, two orders of magnitude below the measured activity. This discrepancy can probably be attributed to inaccuracies in the measured oxygen activity below 1 ppm, and influxes of oxygen from the refractories or other sources.

The lowest achievable hot metal sulphur content can be predicted from the competition between oxygen and sulphur for calcium carbide, and is governed by

$$CaS + O \rightarrow CaO + S \qquad K = \frac{h_s}{h_o} \qquad (4)$$

The theoretical minimum sulphur-to-oxygen activity ratio (one weight percent Henrian scale) for desulphurization to proceed is plotted as a function of temperature in Figure 15. The pilot plant experiments were conducted at 1350°C, at which this ratio is 90. For the average measured oxygen activity of 2.5 ppm (without aluminum additions) and a sulphur activity coefficient of 2.7, the predicted minimum sulphur content is 0.008% which is close to that obtained experimentally.

To reach ultra-low sulfur contents, consistently, with low calcium carbide consumption, the oxygen activity in the hot metal and top slag must be reduced. Work is continuing in this area. The limestone injected with the calcium carbide on a commercial basis has been identified as a possible source of oxygen. However,

*References pp. 327–328*

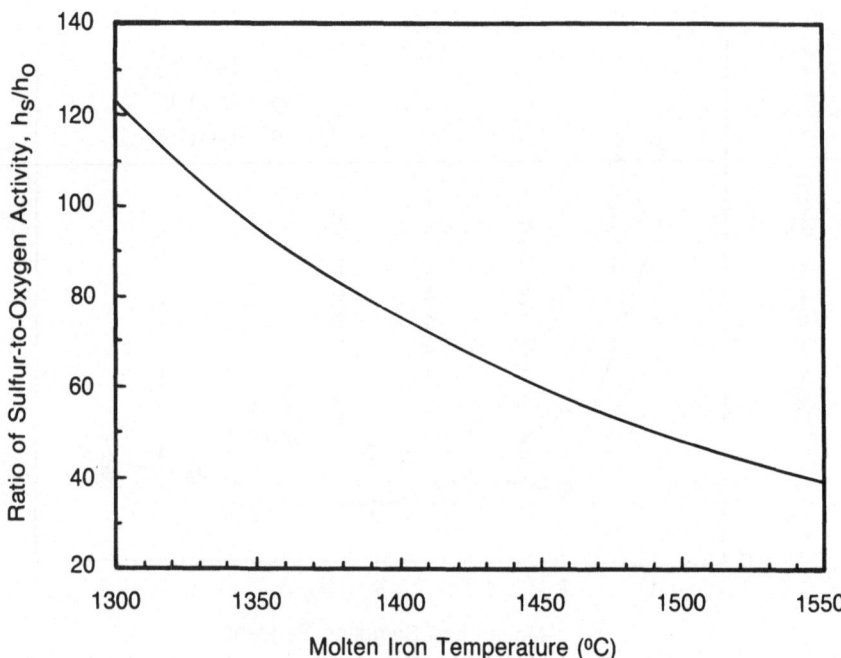

**Figure 15.** Sulfur/oxygen activity ratio in equilibrium with lime and calcium sulfide.

as discussed in the next section, the stirring is essential and, therefore, practical nonoxidizing gas release agents are under investigation.

**Solids to Gas Ratio**—Reducing the solid-to-gas ratio [i.e. lower loading (kg/Nm$^3$)] increases the amount of gas available to mix the molten iron and calcium carbide. This improved agitation increases the calcium carbide's probability of reacting with the dissolved sulphur. The effect of solids loading on calcium carbide efficiency during the pilot plant trials is presented in Figure 11. For all slag conditions, reduced solids loading resulted in higher calcium carbide efficiency. Loading levels greater than 70 kg/Nm$^3$ consistently yielded the lowest efficiency for all initial sulphur levels and slag conditions.

The positive effect of decreased solids loading on desulfurization efficiency is caused by:

(1) intense gas stirring which provides favorable kinetics for slag/metal desulphurization, and

(2) increased mixing in the plume and greater bubble interface which improves the plume reaction kinetics.

Desulphurization of the hot metal occurs in the ascending plume and at the top slag-metal interface (Figure 16). A jet of calcium carbide and argon issues from the lance tip and penetrates until its momentum is dissipated. The penetration depth is calculated from the injection conditions as described earlier. At the bottom of

# PRINCIPLES OF GAS AND POWDER INJECTION

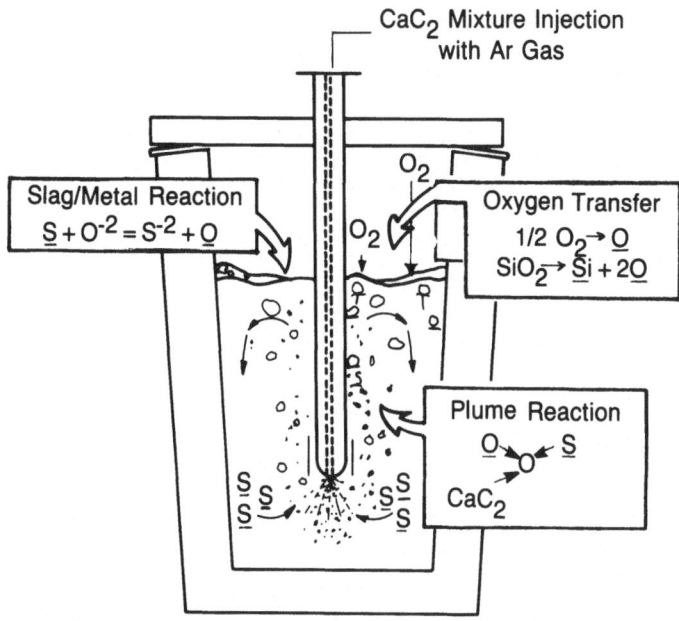

Figure 16. Physical and chemical aspects of the model for calcium carbide desulfurization.

the jet, a plume forms comprising gas, particles, and liquid, which rises to the bath surface.

The position of the particles (i.e. in the bubbles, on the bubble surface, or in the liquid) strongly influences the calcium carbide efficiency and the desulphurization rate. Once the particles reach the bath surface they are incorporated in the top slag and are then included in the top slag rate equations. The effects of the top slag and plume reactions were separated by analysis of the desulphurization results immediately after calcium carbide injection stopped, i.e. top slag desulphurization only. The plume reactions were found to depend heavily on the injection conditions, specifically the gas and solids flow rates. These results are discussed with reference to a mathematical model to describe the plume reactions.

The present model extends the model of Farias and Irons[22] for momentum and heat transfer in the plume to include mass transfer and chemical reaction. The model calculates the volume fractions, velocities, temperatures, and compositions of the gas, liquid, and solid phases as they rise in the plume. All of these variables are averaged across the plume. The model differs from previous plume models[23,24] in that these variables are allowed to change with vertical position and the plume is treated as a three-phase region, not a single phase region of intermediate density. The model is general in that various momentum, heat, and mass transfer coefficients can be employed; this aspect will be used to evaluate different hypotheses concerning

*References pp. 327–328*

particle position with respect to the bubbles.

The fluid flow and heat transfer equations are solved simultaneously with the mass balance equation for sulphur in the plume:

$$\frac{d\left(C_s^{P\ell} A_{p\ell} U_\ell\right)}{dZ} = \frac{C_S^B}{\rho_\ell} \frac{d\left(P_\ell A_{p\ell} U_\ell \theta_\ell\right)}{dZ} + (1-f)\theta_p k_{ov} A_{p\ell} \left(C_S^{p\ell} - C_{S,eq}^{CaC_2}/\eta\right) \quad (5)$$

where

$$k_{ov} = \left(\frac{V_p}{k_p A_p} + \frac{V_p(r_o - r_i)}{4\pi r_i r_o D_{\text{eff},S}}\right)^{-1} \quad (6)$$

Equation 5 considers that the rate will be controlled by a combination of three consecutive steps as follows:

(1) Pumping control—The first term on the right-hand side is the convective flux of sulphur into the plume which is principally controlled by the carrier gas flow rate. For example, at low gas flow rates there may be insufficient flow of sulphur to the plume, resulting in low rates of desulphurization and low calcium carbide efficiency. The second term contains the overall mass transfer coefficient, $k_{ov}$, which accounts for the two remaining steps:

(2) Contact Control—The sulphur pumped into the plume must then diffuse to the particles of calcium carbide. Thus this second step depends on the number of particles in contact with the melt and the mass transfer coefficient to the particles. Five particle contact patterns will be compared with the experimental results.

(3) Product Layer Control—As the reaction proceeds, a layer of calcium sulphide grows on the surface of the particles. Talaballa et al.[25] studied iron desulfurization in calcium carbide crucibles. An analysis of their results indicates that the rate is controlled by diffusion of calcium vapor through the product layer. This flux of calcium was converted mathematically to an equal flux of sulfur vapor, so that consecutive steps of sulphur transport could be used for convenience. The equivalent diffusivity of sulphur vapor through the calcium sulphide layer, $D_{\text{eff},S}$, was deduced from their results.

Results of the model for particular conditions are shown in Figure 17 in which the fraction of control for each step is plotted as a function of the distance from the bottom of the plume. At the bottom, control lies mainly with the pumping and contact steps. As the reaction proceeds, moving up in the plume, the product layer thickens and control shifts to product layer control. For most conditions studied, product layer control dominated the reaction rate. Therefore, calcium carbide desulphurization could be improved if the rate of this step were enhanced, either by thinning the product layer or by increasing the diffusivity of the calcium vapor.

The contact control step was investigated using five different cases for particle-liquid contact (Table 2). Calcium carbide is not wetted by liquid iron, so case 1

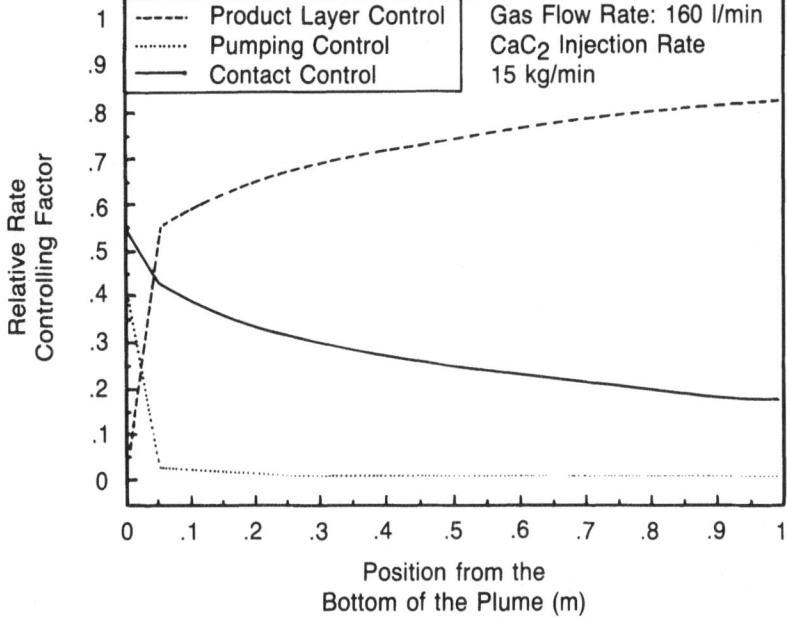

Figure 17. Fraction of control for each mass transfer step as a function of distance from the bottom of the plume.

corresponds to all particles trapped on the interface by surface tension; the particles are assumed to be free to circulate on the interface, and thus all particles react. This is similar to a model proposed by Engh[26] for liquid fluxes. Case 2 is based on the findings of Irons and Farias[20] that the bath cooling rate for powder injection was only 30% of that which would be expected if all the particles were in the melt. Turbulent mass transfer coefficients were used.[27] Case 3 was chosen to show the effects of very slow particle recirculation on the bubble interface. Cases 4 and 5 represent estimates for the fastest and slowest rates, respectively.

The various contact control cases are compared with the experimental first order rate constants for the plume reactions for two carrier gas flow rates (50 and 160 L/min, Figures 18 and 19). All the data falls between the limiting cases of all particles in the liquid (Case 4) and stagnant particles on the bubble interface (Case 5). It is difficult to choose among cases 1, 2 and 3 due to the scatter in the data. Nevertheless, a significant fraction of the particles are obviously on the bubble interface and, therefore, techniques to improve particle-liquid contact should improve desulphurization rate and calcium carbide efficiency. For example, Irons and Farias have found that the use of inclined or "hockey-stick" lances and the use of gas release agents (i.e. $CaCO_3$ in the case of $CaC_2$ injection) improve particle-liquid contact as measured by bath cooling rates.[20]

*References pp. 327–328*

## TABLE 2
### Cases Used to Investigate Particle-Liquid Contact Control in Plume Desulfurization

| Case | Description | Contact Control Factor |
|---|---|---|
| 1 | 100% of particles on bubble interface; particles circulate | diffusion of sulfur to bubble interface |
| 2 | same as 1, except 30% of particles are dispersed in liquid | diffusion of sulfur to bubble interface and to particles in liquid |
| 3 | same as 2, except particles on interface are stagnant | same as 2, except that only a monolayer of particles on bubble reacts |
| 4 | 100% particles in liquid | diffusion to particles |
| 5 | 100% of particles as stagnant layer on bubble interface | diffusion to bubbles with monolayer reaction |

Increasing the carrier gas flow rate from 50 to 160 L/min enhanced the plume reactions as indicated by Figure 18 and 19, the result of better hot metal pumping and more bubble surface area at increased flow rates.

## SUMMARY

In this paper, the physical and mathematical models for the fluid dynamics associated with submerged powder injection into iron have been reviewed. It is now possible to predict with reasonable certainty the flow regimes for various injection conditions, and on this basis design the best conditions for a given process.

A study of calcium carbide injection into iron was also described. The most important variables were the condition of the top slag, the oxygen activity in the iron and the solid-to-gas injection ratio. By choosing the optimum conditions, the reaction speed and reagent efficiency could be improved by approximately 50%. The results were also interpreted with a mathematical model to describe desulphurization in the plume in which it was found that the most important resistance to mass transfer was diffusion through the calcium sulphide layer on each particle. It was also apparent that significant portions of the particles rise with the gas bubbles and do not react with the iron.

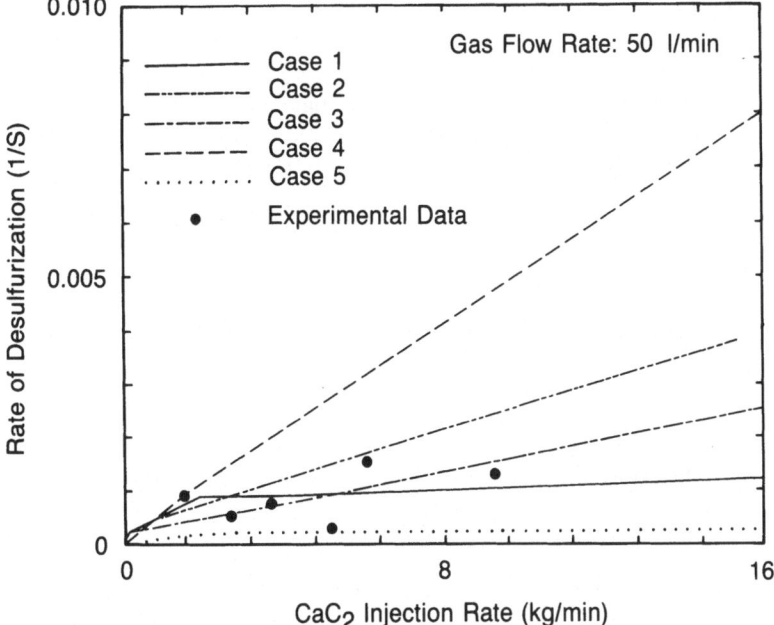

Figure 18. First-order rate constants for desulfurization in the plume from experimental data and the present model for various solid injection rates and powder dispersion conditions (50 L/min).

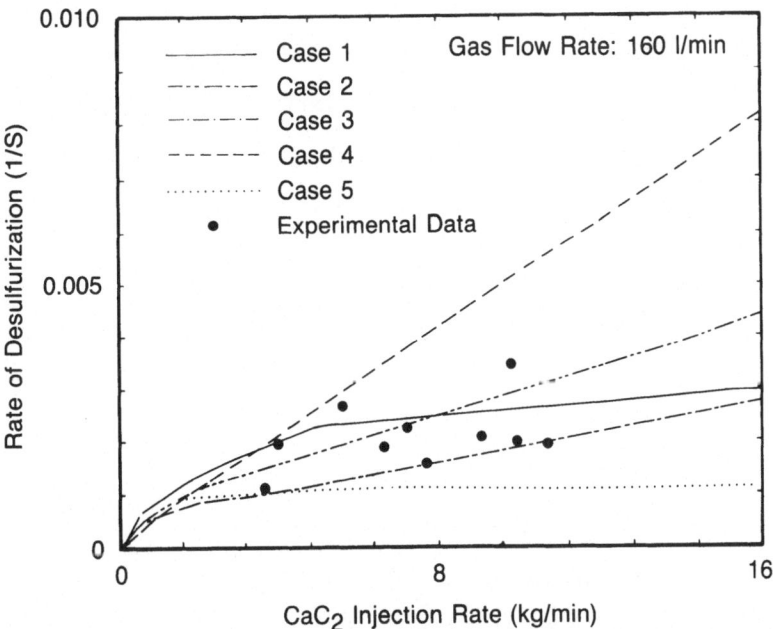

Figure 19. First-order rate constants for desulfurization in the plume from experimental data and the present model for various solid injection rates and powder dispersion conditions (160 L/min).

## NOMENCLATURE

$A_{p\ell}$ = cross-sectional area of plume, m²

$A_p$ = surface area of calcium carbide particle, m²

$C_S^{p\ell}$ = sulfur concentration in the plume, mole/m³

$C_{S,eq}^{CaC_2}$ = equilibrium sulfur concentration with $CaC_2$, mole/m³

$C_S^B$ = sulfur concentration in the bulk, mole/m³

$d_i$ = inner lance diameter, m

$D_{\text{eff},S}$ = equivalent effective diffusivity of gaseous sulfur through the reacted layer, m²/s

$f$ = fraction of powder inside gas bubbles

$g$ = gravitational acceleration, m/s²

$k_{ov}$ = overall desulfurization rate constant, (1/s)

$k_p$ = mass-transfer coefficient, m/s

$L$ = solid-to-gas loading, kg/Nm³

$Ma$ = nominal Mach Number

$Re_p$ = Reynolds Number of particles with respect to gas

$r_i$ = radius of unreacted core of calcium carbide, m

$r_o$ = outer radius of calcium carbide, m

$U_\ell$ = liquid velocity in the plume, m/s

$U_m$ = mixture velocity, m/s

$V_p$ = volume of single calcium carbide particle, m³

$W_g$ = gas flow rate, kg/s

$Z$ = vertical distance from bottom of plume, m

$\rho_\ell$ = density of hot metal, kg/m³

$\Theta_c$ = critical volume fraction of powder

$\theta_p$ = phase volume of calcium carbide in plume

$\theta_\ell$ = phase volume of hot metal in plume

$\eta$ = sulfur equilibrium partition ratio, $C_{Sg}/C_S$

# REFERENCES

1. T. A. Engh and H. Bertheussen: *Scan. J. Met.*, 1975, *4*, 241–249.
2. T. A. Engh, K. Larsen, and K. Venas: *Ironmaking and Steelmaking*, 1979, *6*, 268–73.
3. D. G. C. Robertson, D. S. Conochie, and A. H. Castillejos: *Proceedings of Scaninject II Conference*, Luleå, Sweden, June 12–13, 1980, Mefos and Jernkontoret, 1980, 4.1–4.36.
4. M. J. McNallan: *Proceedings of Scaninject II Conference*, Luleå, Sweden, June 12–13, 1980, Mefos and Jernkontoret, 1980, 8.1–8.10.
5. D. N. Ghosh and K. W. Lange: *Ironmaking and Steelmaking*, 1982, *19*, 3, 136–141.
6. L. R. Farias and D. G. C. Robertson: *Proceedings of 3rd Process Technology Conference*, Pittsburgh, PA, Mar. 28–31, 1982, ISS of AIME, 1982, 206–220.
7. G. A. Irons and B. H. Tu: *Proceedings of Scaninject III Conference*, Luleå, Sweden, June 15–17, 1983, Mefos and Jernkontoret, 1983, 11.1–11.29.
8. M. J. McNallan, J. O. Park and Y. W. Chang: *Proceedings of Scaninject III Conference*, Luleå, Sweden, June 15–17, 1983, Mefos and Jernkontoret, 1983, 9.1–9.17.
9. L. R. Farias and G. A. Irons: *Metal. Trans. B*, 1985, *16B*, 211–225.
10. G. A. Irons: *Transactions of the Iron and Steel Society*, 1984, *5*, 33–45.
11. M. A. Palmer and J. S. Becker, *Proceeding of the McMaster Symposium on External Desulfurization of Hot Metal*, 12.1 to 12.15, May 22–23, 1975, McMaster University, Hamilton, ed. W.-K. Lu.
12. H. T. Kossler et al., *Proceedings of McMaster Symposium on Developments in Hot Metal Preparation for Oxygen Steelmaking*, May 25–26, 1983, McMaster University, Hamilton, ed. W.-K. Lu.
13. A. M. Smillie, Chapter 19, *Blast Furnace Ironmaking*, 1980, McMaster University, Hamilton, ed. W.-K. Lu.
14. G. W. Sova et al., *Steelmaking Conference Proceedings*, *64*, 1981, 76–80, March 29–April 1, Toronto, ISS of AIME.
15. B. Tivelius and T. Sohlgren, *Proceedings of McMaster Symposium on Ladle Treatment of Carbon Steel*, 1979, McMaster University, Hamilton, ed. J. S. Kirkaldy.
16. A. McGowan et al., *Proceedings of McMaster Symposium on Developments in Hot Metal Preparation for Oxygen Steelmaking*, May 25–26, 1983, McMaster University, Hamilton, ed. W.-K. Lu.
17. U. Puckoff and H. Kister, *Proceedings of McMaster Symposium on External Desulfurization of Hot Metal*, 1975, pp. 8.1 to 8.20, May 22–23, McMaster University, Hamilton, ed. W.-K. Lu.
18. R. Boom and R. F. Beisser, *Proceedings of McMaster Symposium on Developments in Hot Metal Preparation for Oxygen Steelmaking*, May 25–26, 1983, McMaster University, Hamilton, ed. W.-K. Lu.
19. Y. Ozawa and K. Mori, *Injection Phenomena in Extraction and Refining*, Conference Proceedings, Newcastle, April 21–22, 1982, A. E. Wraith, ed., 1982, 11–118.
20. G. A. Irons and L. R. Farias, unpublished research, McMaster University, 1984.
21. L. K. Chiang, G. A. Irons, W.-K. Lu and I.A. Cameron: *Proceedings of 5th International Iron and Steel Congress*, Washington, D.C., April 1986.
22. L. R. Farias and G. A. Irons: *Metall. Trans.*, *17B*, 1986, 77–85.

23. N. El-Kaddah and J. Szekely: *Ironmaking and Steelmaking*, 6, 1981, 269–278.

24. S. Ohguchi and D. G. C. Robertson: *Ironmaking and Steelmaking*, 11, 5, 1984, pp. 262–273.

25. M. Talaballa et al: *AFS Transactions*, 1976, 775–786.

26. T. A. Engh et al: *Scand J. Metall*, 1972, *1*, 103.

27. Y. Sano et al: *J. Chem. Eng.*, 1974, 7, 255.

# DISCUSSION

### Question

In the water experiments, did water wet the particles that you used?

**A.** No, the particles were not wetted.

### Gino Sovran *(General Motors)*

You have a three-phase system—gas, solid, and liquid. Do you expect scaling problems in trying to go from a laboratory scale unit up to a larger size?

**A.** The water, lead and iron experimental work were on a smaller scale, however I would like to point out that the solids flow rates were generally between 1 and 10 kg/min, whereas in full-scale Ladle Metallurgy stations in the Steel Industry the flow rates are usually 10 to 100 kg/min. Therefore in terms of the phenomena occurring at the lance tip we are within an order of magnitude in flow rate. Our liquid volumes were several orders of magnitude smaller than full-scale, so that phenomena in the single-phase region far from the lance are not modelled. In most Foundries, the scale is much smaller than the Steel Industry, and, as the calculations in the text demonstrate, the flow rates would be similar to those in our models. The equations which we use to model these processes are based on multi-phase models to understand the physics behind injection. The models, which are largely based on fundamentals, have been verified in water and liquid lead over a wide range of parameters, so that we have some confidence in their predictive capability.

### Gino Sovran

I had recent experience with a group of people in this country who are experts in multi-phase flow, which I am not, and they said their biggest problem in multi-phase flow is scaling. They have to scale up from a laboratory size unit up to commercial equipment for the chemical and nuclear industries. They have identified scaling as the single largest problem that they have.

**A.** I think that's true, and in the metallurgical field it's very difficult to measure fluid mechanic parameters, because of the temperatures and large scale. Professor Szekely yesterday showed some data that had been obtained in the full scale, but it's a very limited amount; so, as he said in his talk yesterday, we have to proceed with a combination of mathematical models and as much experimental verification as possible.

**Gary Ruff** *(General Motors)*

Can you comment on the efficiencies of wire injection versus powder injection and the fact that certain companies are producing materials with magnesium in combination with calcium carbide? Are there optimum ratios for desulfurization?

A. Regarding the comparative efficiencies of wire-based and powder-based reagents, the most experience has been obtained with calcium reagents used for modification of inclusions in steel. It is generally observed that the efficiencies are higher with wire injection, however the cost of the wire is usually higher than this efficiency improvement. Wire feeding equipment is less expensive than powder injection equipment. Therefore when only a few heats need to be treated, wire feeding is cost effective. On the other hand, for heavy injections where the reagent cost is a significant fraction of the operating cost, powder feeding is generally cheaper.

Regarding the optimum ratio of calcium carbide to magnesium, there has been a considerable amount of work on hot metal desulfurization with these reagents, and I think they have looked at this ratio as a variable; I cannot remember the exact ratio. (See Bieniosek, T., Hot Metal Desulfurization by Co-Injection of Calcium Carbide and Magnesium, *Steelmaking Proceedings*, Volume 69, 1986, pp. 349–356, ISS of AIME, for a discussion of this practice). Not very much magnesium is required because magnesium acts mainly as a deoxidizer, just as aluminum did in the work I described in my presentation. This allows the calcium carbide, which is the major fraction of the mixture, to desulfurize. Now, magnesium is also a powerful desulfurizer, but in this particular case magnesium cannot account for very much desulphurization because it is only a minor component.

**Sy Katz** *(General Motors)*

You were talking about efficiencies and that the hockey stick design was for better separation of the gas and solid. Have you actually measured in the case of calcium carbide what the improvement in efficiency is between the straight lance and the hockey stick?

A. All the tests with the calcium carbide were done with the straight lances. The work that I was referring to that showed that the hockey stick lances were better were some studies that we did in liquid lead, in which we measured the bath cooling rates as a function of the solids injection rate. With straight lances we found the bath cooling rate was only about 30% of that expected if the particles were heated up to bath temperature. So, this means that the particles tend to stay with the gas phase and are not heated. When we used the hockey stick lance, the bath cooling rate was increased by about a factor of 2, so more particles were in contact with the metal. (Irons, G. A. and Farias, L. R., The Influence of Lance Orientation and Gas Evolution on Particle-Liquid Contact during Submerged

Powder Injection, in press in the *Canadian Metallurgical Quarterly* and Irons, G. A., Fundamental and Practical Aspects of Lance Design for Powder Injection Processes, *Proceeding of SCANINJECT IV*, June 11–13, 1986, Lulea, Sweden, Mefos and Jernkontoret, pp 3:1–3:25.

**Sy Katz**

So then you haven't tried it with actual desulfurization?

A. No, we haven't. Hockey-stick lances are used in some plants, but straight, inclined lances are most often used.

**D. G. C. Robertson** *(Chairman)*

A T-shaped lance with a horizontal exit is less liable to clog as well and is very commonly used.

**Peter Wieser** *(Case-Western Reserve)*

You made a brief comment about nozzle clogging with salt-coated magnesium. Could you elaborate on that?

A. We studied lance clogging phenomena in liquid lead, and found that two conditions were necessary for lance clogging. First, one must be operating in the bubbling regime in which case there is the opportunity of liquid to enter the lance between bubbles. The second condition is that the inner surface of the lance must be below the melting point of the liquid, so that the liquid that surges into the lance will freeze on the inside. The temperature of the inside wall of the lance depends upon the heat extracted from the lance by the gas and particles inside the lance, and by the heat entering from the liquid. We've developed a model for the heat transfer in this situation which works very well; the onset of clogging occurs when the measured and predicted wall temperatures fall below the melting point of the liquid lead (Irons, G. A., Heat Transfer and Lance Clogging during Submerged Powder Injection, in press *Metallurgical Transactions B*, March 1987). For the case of salt-coated magnesium, we would expect it to operate in the bubbling regime as shown in Figure 4 because of its coarse particle size. It has been found that considerable lance clogging occurs unless the gas velocity is high enough to undergo transition G in Figure 7 to jetting. However these high flow rates cause excessive iron splashing.

**Ara Hacetoglu** *(Cynamid)*

I have two comments. One is an answer to a question that came from General Motors Central Foundry on carbide and magnesium reagents. Yes, there is an optimum carbide/magnesium ratio, and it is determined by the size of the ladle and its shape, i.e. transfer ladle versus torpedo ladle. The other comment concerns Dr. Irons' last comments on how to apply this steel injection technology to the foundry application. There are two major differences between the foundry and steel applications: one is the size and the other is the process. Steel injection is a batch operation, while the foundry is porous plug in a continuous operation. Injection into a continuously running melt will be difficult. One, you need a lance material that can withstand temperatures for extended periods and two, there's a continuous deslagging of the ladle. If these two problems can be resolved, you can get definitely better efficiencies from calcium carbide, which would give two benefits. One is to reduce the desulfurizing costs and two is reducing retained or residual carbide in the slag, which is causing disposal problems.

**Gino Sovran**

Most of the activity is in the vicinity of the lance tip, and the hockey stick seemed better. Why can't you put more than one hockey stick on this vertical lance, that is, use two, or three or four around the vertical.

**A.** All sorts of lance designs have been proposed. As Dr. Robertson mentioned, there are straight lances which separate into either 2 or 4 streams at a "tee" at the lance bottom. There are also straight, inclined lances and hockey-stick lances. The reasons for many of these lance designs are not particularly clear, however some of them are justified on the basis that mixing times in water models are reduced. Our work shows that improved mixing in the bulk of the ladle is only a small part of improved desulphurization performance and that factors such as the gas-particle separation at the lance tip, particle position in the bubbles, diffusion through reaction product layers and the control of oxygen activity are much more important issues.

# PHYSICO-CHEMICAL ASPECTS OF THE LADLE DESULPHURIZATION OF IRON AND STEEL

## H. GAYE, C. GATELLIER and P.V. RIBOUD

*IRSID, Maizieres-les-Metz - France*

### ABSTRACT

Some physico-chemical aspects of the reactions occuring in iron (hot-metal) and steel external desulphurization processes are discussed. In steelmaking, these processes consist of batch treatments performed in torpedo cars (iron) or open ladles (steel and in some cases iron).

For hot-metal desulphurization, very potent reagents are available (magnesium-based mixtures, calcium carbide and lime-based mixtures, sodium carbonate). As the efficiency of the treatment usually hinges on kinetic limitations, the development of injection techniques has been an important asset.

For killed-steels, desulphurization is the result of a reaction between metal and top slag. Requirements in terms of slag control and stirring needs are analyzed.

## INTRODUCTION

Steelmakers have long recognized that optimal operation of blast furnace and converter do not allow full metal desulphurization to be obtained in these two reactors. As a consequence, they have developed specific external desulphurization techniques: on hot-metal for mass production of standard steels, and in addition, on liquid steel for the production of low to ultra low-sulphur steel grades. It is now possible, with the most efficient of these treatments, to obtain, whenever desired, hot-metal with sulphur contents as low as 20 to 50 ppm, and to desulphurize steel well below 10 ppm.

A very comprehensive review of the numerous processes which have been developed, and of the current status of worldwide iron and steel desulphurization, has been recently published by Pehlke and Fuwa[1]. In this paper, we will merely present typical industrial results for these treatments, and discuss the thermodynamic and kinetic aspects of the reactions involved.

## AVAILABLE REACTIONS

The reagents most widely used in steelmaking for iron and steel desulphurization are alkali or more frequently alkaline-earth metals and their compounds, that is Na used as sodium carbonate, Mg based mixtures, and Ca used as calcium alloys, calcium carbide or lime based mixtures.

In all cases, there is competition between sulphur and oxygen to combine with this reactive metal. This is true in contact with the reactive metal itself, pure or dissolved in iron, as well as for oxides and sulphides already formed in the melt, and oxides from the refractory and top slag. This can be represented by two reaction schemes, depending on the initial state of the products:

$$\text{a) } [X] + \underline{S} = (XS)$$

with the side reaction

$$[X] + \underline{O} = (XO)$$

$$\text{b) } [XO] + \underline{S} = (XS) + \underline{O}$$

where X stands for 2Na, Mg or Ca. In these relations, [ ] indicates that the reactive metal can be pure or fixed in a compound, ( ) that the oxide and sulphide can be pure or dissolved in a slag, and underlining indicates that the element is dissolved in liquid metal.

In all cases, sulfur removal will be easier when:

- the activity coefficient of sulphur in liquid metal is high (it is close to 1 in unalloyed steels and reaches a value around 5 in C-saturated iron);
- the oxygen potential at the reaction site can be maintained at a very low value. This is more readily obtained in hot-metal (even though the oxygen potential explaining the observed sulphur distribution might be up to one order of magnitude higher than expected from the carbon equilibrium, either in the cupola[2] or in the blast-furnace[3]) than in deoxidized steel.

These two facts explain that the bulk of desulphurization is performed on hot-metal (and no other specific treatments are then required for most steel grades), whereas steel desulphurization is generally devoted to the production of low to ultra low-sulphur steel grades.

**General thermodynamic considerations**—The complete thermodynamic description of desulphurization reactions requires, in addition to the equilibrium constants of reactions a) and b), the evaluation of component activities in the liquid metal and of sulphides and oxides in the compounds and slags in which they are combined. The formalism of interaction coefficients to describe the properties

## TABLE 1

**Equilibrium constants for various reactions relevant to iron and steel desulphurization.**

| Reaction | log (K) | Reference |
|---|---|---|
| $2Na\ (g) + S = Na_2S\ (l)$ | $24620/T - 13.0$ | [6] |
| $2Na\ (g) + O = Na_2O\ (l)$ | $21195/T - 12.48$ | [6] |
| $Mg\ (g) + S = MgS\ (s)$ | $20680/T - 8.62$ | [6] |
| $Mg\ (g) + O = MgO\ (s)$ | $32020/T - 10.85$ | [6] |
| $Mg\ (g) = \underline{Mg}\ (pure\ Fe)$ | $\log(\%\underline{Mg}/P_{Mg}) = 1980/T - 2.28$ | [9–10] |
| $Mg\ (g) = \underline{Mg}\ (C\text{-saturated Fe})$ | $\log(\%\underline{Mg}/P_{Mg}) = 11840/T - 7.8$ | [11] |
| $\underline{Mg} + \underline{S} = MgS\ (s)$ | $13615/T - 6.04$ | [8] |
| $\underline{Mg} + \underline{O} = MgO\ (s)$ | $26110/T - 8.24$ | [8] |
| $Ca\ (l) = Ca\ (g)$ | $-8026/T + 4.55$ | [6] |
| $Ca\ (l) + S = CaS\ (s)$ | $21342/T - 4.05$ | [6] |
| $Ca\ (l) + O = CaO\ (s)$ | $27284/T - 5.79$ | [6] |
| $Ca\ (l) = \underline{Ca}\ (pure\ Fe)$ | $\%\underline{Ca}\ (sat) = .032$ at 1600 C | [12–13] |
| $\underline{Ca} + \underline{S} = CaS\ (s)$ | $19980/T - 5.9$ | [8] |
| $\underline{Ca} + \underline{O} = CaO\ (s)$ | $25655/T - 7.65$ | [8] |
| $Ca\ (l) + 2C\ (gr.) = CaC_2\ (s)$ | $19200/T - 7.7$ | [6] |
| $2\underline{Al} + 3\underline{O} = Al_2O_3\ (s)$ | $62680/T - 20.54$ | [14] |
| $\underline{Mn} + \underline{O} = MnO\ (s)$ | $15050/T - 6.7$ | [14] |
| $\underline{Si} + 2\underline{O} = SiO_2\ (s)$ | $31040/t - 12.0$ | [14] |

of metal phases is now widely used and exhaustive compilations of these coefficients are available[4-5]. For oxides and sulphides phases, the various thermodynamic tools which may be used will be outlined when each specific case is presented.

The equilibrium constants, for various reactions which may take place between Na, Mg and Ca and sulphur and oxygen dissolved in pure or C-saturated Fe are given in Table I. The values taken from Reference [6] are based on calculations made from data on the free energies of formation of the compounds from the elements[7]. For calcium and magnesium, we have also reported the values derived from direct measurements of the solubility products in steel[8]. The solubility products from direct measurements are much higher (in particular for calcium) than those derived from the accepted stabilities of the compounds. It is difficult to guarantee that the direct solubility products measurements are not slightly biased by remaining inclusions, but it seems that, at least for CaO and CaS, the standard free energy of formation should be reconsidered.

*References pp. 353–355*

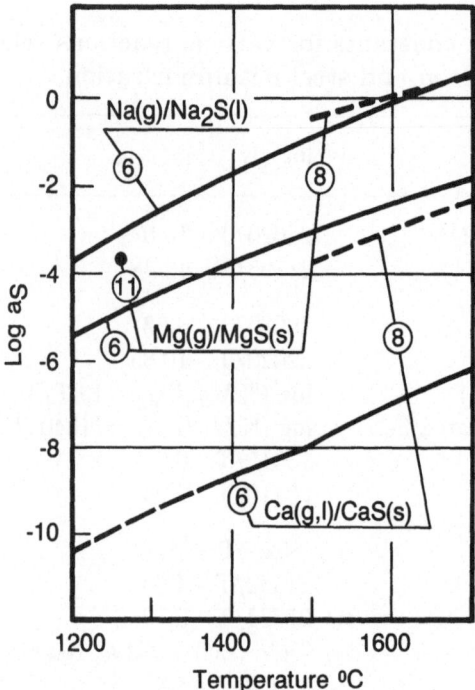

**Figure 1.** Activity of sulphur dissolved in liquid iron at equilibrium with pure Na, Mg and Ca, and their pure sulphides.

Based on these values, we have represented:

• in Figure 1, the activity of sulphur that would be reached in liquid Fe at equilibrium with the pure reactive metal under its stable form and the pure sulphide,

• in Figure 2, the ratios of activities of sulphur and oxygen dissolved in liquid Fe at equilibrium with the pure sulphide and oxide.

In spite of the considerable uncertainty of the thermodynamic data, unequivocal conclusions can be drawn. Sodium can be an efficient desulphurizer in hot-metal, but not in steel if used alone. In the case of magnesium, and to a lesser extent of calcium, the oxygen content of the metal will have to be maintained at a very low value for desulphurization to progess. In fact, pure MgS can be formed only in hot-metal, and even though pure CaS (stable in the case of hot-metal desulphurization) might be formed at high sulphur content in liquid steel, sulphur removal in steel can be efficient only if the sulphide formed is dissolved in a CaO-rich slag.

When a slag phase is present (that is, in practice, for hot-metal desulphurization with sodium carbonate, or for steel desulphurization), the thermodynamics of sulphur distribution can be expressed using the "slag sulphide capacity, $C'_s$" concept[15]:

$$(\%S)/\%\underline{S} = C'_s \cdot f_s/a_O \tag{1}$$

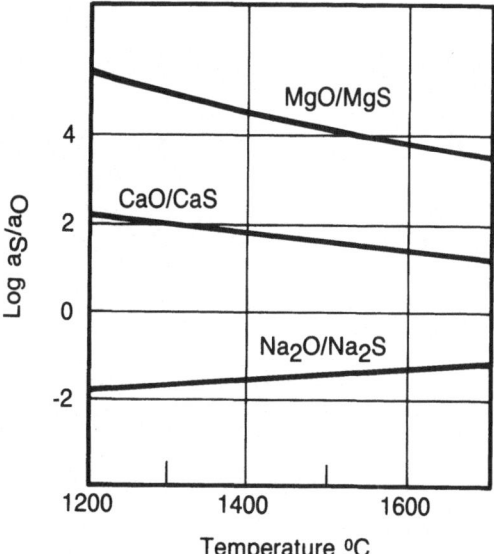

**Figure 2.** Ratios of the activities of sulphur and oxygen dissolved in liquid iron at equilibrium with the pure sulphides and oxides of Na, Mg and Ca.

Slags rich in $Na_2O$ and $CaO$ have high sulphide capacities. Correlations in terms of "optical basicity index" have been shown to represent adequately the sulphide capacities of a wide range of slag compositions[16-17].

In addition to these elements, barium, rare-earth elements and zirconium are also powerful deoxidizers. For economical reasons, they have had no notable industrial applications.

**General kinetic considerations**—As a general rule, the chemical reaction itself between dissolved sulphur and the reagent is potentially very fast. The kinetic limitations may arise from:

- introduction of the reagent in the metal,
- transport of reactants to reaction sites,
- elimination of reaction products.

Introduction of the reagent: Maintaining a large contact area between liquid metal and reagent is a necessary condition for optimal operation. The most important technological developments in iron desulphurization concern this domain (powder injection and reentrainment of the products with mechanical paddles). Several studies have been made in the last few years on the behaviour of pneumat-

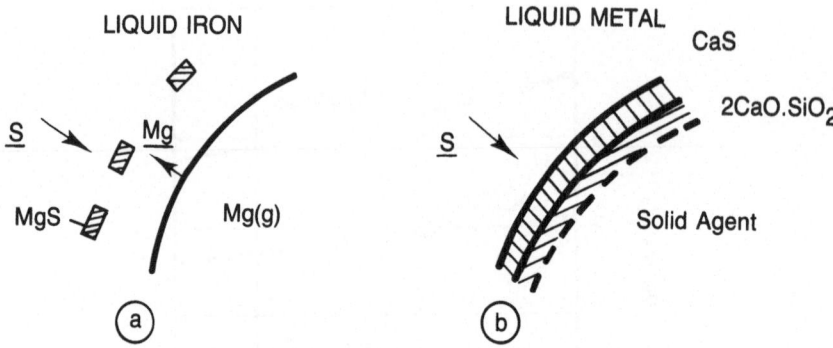

Figure 3. Sketch of sulphur transfer near the interface between liquid metal and a bubble of Mg(a), or a particle of CaO or $CaC_2$(b).

ically injected particles. A critical review of these and an investigation by physical modeling and by multi-phase fluid dynamics have recently been published by Farias and Irons[18]. Their general conclusion is that, in the conditions of most industrial operations, fine particles, such as lime or calcium carbide powders at high loadings, form jets and may have difficulties penetrating through the gas-liquid interfaces, whereas coarse particles, such as magnesium powders, lead to a bubbling regime and separate easily from the carrier gas.

Transport of reactants to reaction sites: Once in contact with the metal, the concentrated reagent (then solid, liquid or gaseous) has to react with dissolved sulphur. Sulphur transport in the metal is always one of the components of the global kinetics, but additional resistance can occur (Figure 3) when the reactant is a solid compound or a slag. Furthermore, several reaction sites are often available (plume, interface between top slag and metal), and liquid phase mixing may also be a limiting step. Various models based on fluid dynamics calculations have been developed[19-20] to describe, in particular, reactions in the plume and liquid phase mixing. Ohguchi and Robertson[21] proposed a framework to analyse the relative importance of transitory reaction at the injected particles surface, permanent reaction at slag-metal interface and bulk phase mixing. They suggest that, in usual industrial conditions, bulk phase mixing can be a limiting step only at very low sulphur contents or in large capacity torpedo ladles, and that, for lime injections, the desulphurization rate in the early stage can be improved by forming a large enough quantity of liquid top slag at the beginning.

Elimination of reaction products: In the case of iron desulphurization, flotation of the formed sulphides is strongly influenced by the type of reagent used:

• for reactions with metals (Mg desulphurization of iron), and especially at low sulphur content, dissolved sulphur and magnesium react at some distance from Mg bubbles interfaces. Small MgS particles are then formed[22] and their flotation is slow;

- for reactions with injected compounds, the flotation is much easier: the finer particles have had no chance of penetrating the metal and are entrained by the carrier gas, whereas the flotation of coarser particles which have penetrated the metal is naturally fast.

In the case of steel desulphurization, the slag droplets which may have been torn away from the liquid slag are rather large due to the high slag-metal interfacial tension, and their entrapment in the steel is not to be feared[23].

Complete emersion of the sulphides is an important step of sulphur elimination from iron. When no slag layer is present at the surface, the stable position of sulphides is that of partial emersion. They will pile-up at the surface, trapping metal droplets between them. These are lost at the final slagging-off. Complete emersion of the sulphides is possible only in the presence of a liquid surface slag.

## IRON DESULPHURIZATION

A wide variety of processes are industrially used, either in open ladles or torpedo cars. The most important factor of evolution in the last few years has been the generalization and mastering of injection techniques.

**Desulfurization with sodium carbonate**—The use of sodium carbonate as a desulphurizing agent is one of the earliest processes known for treating liquid iron. The simple addition during ladle filling, used traditionally, leads to rather erratic results. Desulphurization ratios around 60% are obtained on the average, for initial sulphur contents around 0.04% and soda ash consumption of about 8 to 10 kg/ton. A marked improvement is brought by deep injection of the product[24]: fumes emission is largely reduced, and final sulphur contents below 0.01% can be obtained in the same conditions as above. It is worth noting that considerable effort has recently been devoted in Japan and Europe on the use of soda ash to dephosphorize low silicon hot-metal. Sulphur contents as low as 20 ppm are obtained in the process[25].

For all these treatments, the overall desulphurization reaction can be globally written (considering that the reducing agent is essentially the silicon dissolved in iron) as:

$$Na_2CO_3 + Si + S = (Na_2S + SiO_2) + CO$$

Some of the sodium carbonate is decomposed to liberate $Na_2O$ which can in part create a slag with the silica formed and in part release sodium vapor. The release of sodium vapor will be fast at the beginning of the contact of sodium carbonate with hot metal, and then decrease as $Na_2O$ is neutralized by silica. It is possible to estimate, from the known thermodynamic data on $Na_2O$-$SiO_2$ that at 1300 degrees Celsius, $P_{Na}=0.1$ atm would be in equilibrium with a slag containing 48% $Na_2O$: at that point, the release of sodium vapor would be reasonably slow. Indeed, slags resulting from the treatment have approximately equal $Na_2O$ and $SiO_2$ contents, whatever the contamination by residual blast-furnace slag (for dephosphorization of low-Si hot metal, this same condition would correspond approximately to the neutralization of $Na_2O$ by $SiO_2+3P_2O_5$, which is about the stoichiometry observed at the end of industrial treatments).

Two routes are possible to reach the global desulphurization written above:

• direct reaction with sodium vapour to form pure $Na_2S$ that will in turn dissolve in the slag. It will be predominant at the beginning of the contact between soda ash and liquid metal.

• reaction with the formed slag. It will be predominant later on and fix the conditions of the final equilibrium that might be reached.

In practice, the final sulphur contents are somewhat erratic and far above the expected equilibrium. One of the reasons, as mentioned before, is the contamination with left over blast-furnace slag. Another reason is undoubtedly unsatisfactory kinetic conditions, in particular with the pouring technique. Note, however, that the very low sulphur contents reached during dephosphorization treatments indicate a much better approach of equilibrium. The kinetic conditions of top slag-metal reaction will be discussed later, on the example of steel desulphurization.

The final slags are usually liquid and free of entrapped metal; they can be skimmed-off conveniently and with little metal losses after addition of a thickening agent. The heavy fumes formation and refractory wear, however, as well as the difficulty in obtaining the requisite low sulphur contents usually looked for in low-P hot-metal, are severe handicaps of these methods.

**Desulphurization with Mg-based products**—The use of magnesium-impregnated coke (Mag-coke) dunked under a bell in molten iron was introduced about 1970. Since then, however, injection processes have taken a primary role due to lower operating costs and higher efficiency. The injected products can be salt-coated granules, or pure magnesium powder mixed with lime (lime-mag: 80 % CaO - 20 %Mg) or basic granulated slag (USIRMAG 2: 20% slag - 80% Mg). This conditioning of the magnesium powder was found necessary in order to prevent lance clogging.

In all these treatments, desulphurization occurs mainly through reaction of dissolved sulphur with Mg vapour (or Mg dissolved from the vapour) to form pure MgS (with a contribution of the large quantity of lime used in the case of lime-mag). The two types of processes differ not so much by the low levels of sulphur contents that can be reached (both can guarantee sulphur contents between 20 and 50 ppm) than by the magnesium yield which is substantially higher in injection processes[26].

In Figure 4 are presented experimental values of final %$\underline{Mg}$ and %$\underline{S}$ obtained in treatment with Mag-coke[27] and powder injection[28], respectively. In the treatment with Mag-coke at temperatures around 1260-1340 C, points have been taken just after the end of the normal dunking treatment and after flushing with inert gas. There is a large decrease in sulphur content between the two samples (more than could be expected from the slight temperature decrease), evidence of the flotation of MgS particles. For both processes, the final values after bubbling correspond to %$\underline{Mg}$.%$\underline{S}$ products very close to the value determined experimentally by Speer and

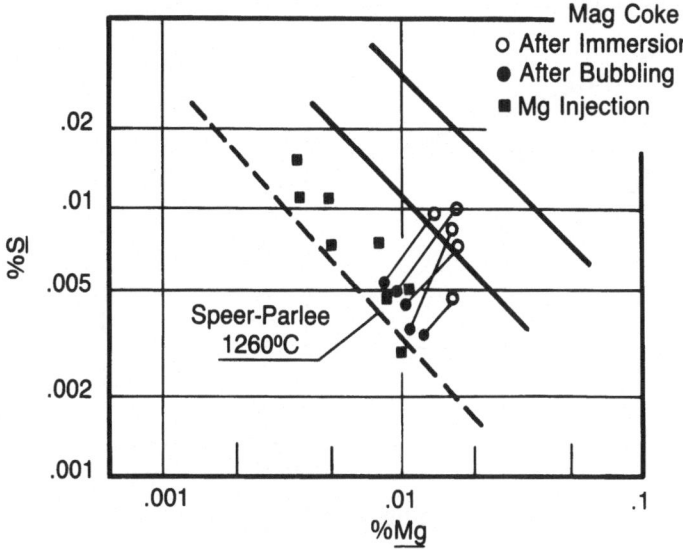

Figure 4.  Sulphur and magnesium contents in hot-metal after a Mag-Coke[27] or Mg-injection[28] treatments.

Parlee[11] (taking into account that the powder injections were made at temperatures around 1350 C). It appears that flotation of MgS particles is accelerated in injection processes, for which the carrier gas is maintained after the injection.

The marked difference in efficiency between injection and dunking can be understood on the basis of Mg bubbles formation and dissolution kinetics. Firstly, a large amount of the Mg, typically 15%[22], is brutally vaporized as soon as the bell is immersed, whereas injection guarantees smooth generation of Mg. The second point concerns the initial bubble size: it is expected to be much larger when escaping from the bell than during the injection. As seen in Figure 5, which represents the results of a discrete bubbles-liquid transfer model applied to the dissolution of magnesium, although magnesium dissolution is fairly fast, the maximum size admissible to obtain complete dissolution within the bath decreases sharply as temperature increases or injection depth decreases. In most conditions, bubbles resulting from injection would have sufficient time to dissolve completely, but this is not necessarily the case for Mag-coke. A summary of computed repartition of Mg from Mag-coke[22] is presented in Figure 6: the losses become quite large at high temperature.

**Desulphurization with Ca-containing compounds**—Calcium is used as calcium carbide or lime-based mixtures, either injected or reacted with the use of mechanical stirrers. As for treatments with magnesium, sulphur contents below 50 ppm can be achieved, provided some precautions are taken, especially in the case of lime. Typical results (initial sulphur content of the metal around 0.04%),

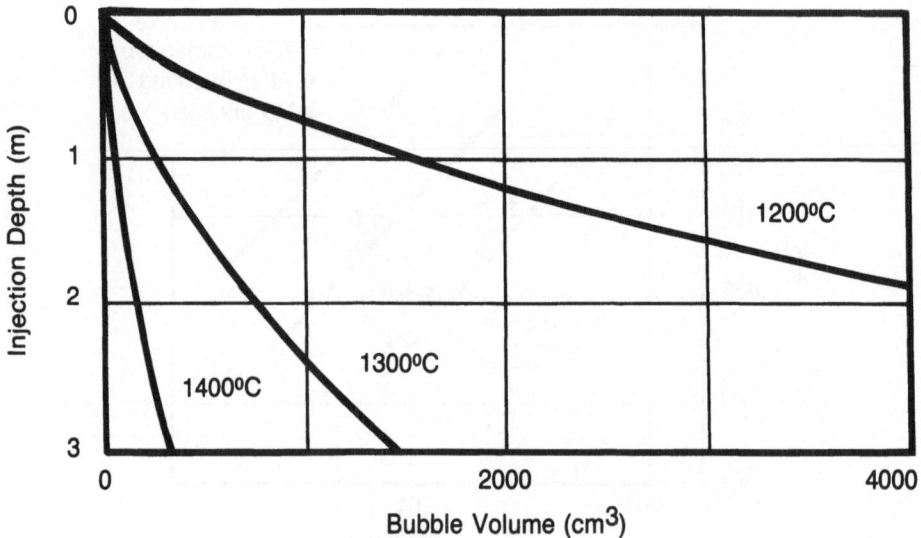

Figure 5. Maximum initial size of Mg bubbles dissolving completely within the metal bath, as a function of injection depth and temperature.

expressed as the degree of desulphurization as a function of used lime are shown in Figure 7. Before discussing these results, let us consider the reaction mechanisms.

For both calcium carbide and lime, the degree of utilisation of the product can be rather low, partly due to the entrainment of the very fine particles used with the carrier gas, and partly to slow kinetics at the metal-reagent interface. Let us look at the conditions at the interface (Figure 3):

- $CaC_2$ is stable in C-saturated iron at usual treatment temperatures and presents no solid solubility with CaS. It is possible to show, using the values of Table 1, that the sulphur activity at equilibrium with pure CaS and $CaC_2$ is below $10^{-4}$. The reaction will therefore proceed with formation of a CaS shell around the carbide particle.

- In contrast, desulphurization with lime requires that the oxygen potential remains very low (Figure 2). This can be achieved if the oxygen released by the reaction combines with a strong deoxidant. In practice, silicon contained in the hot-metal plays that role. However, at usual silicon contents, very low equilibrium sulphur contents can be achieved only if the silica formed is neutralized by lime. The reaction can then be written:

$$2\underline{S} + \underline{Si} + 4CaO = 2CaS + 2CaO - SiO_2$$

leading to the configuration shown in Figure 3 (after reference 29).

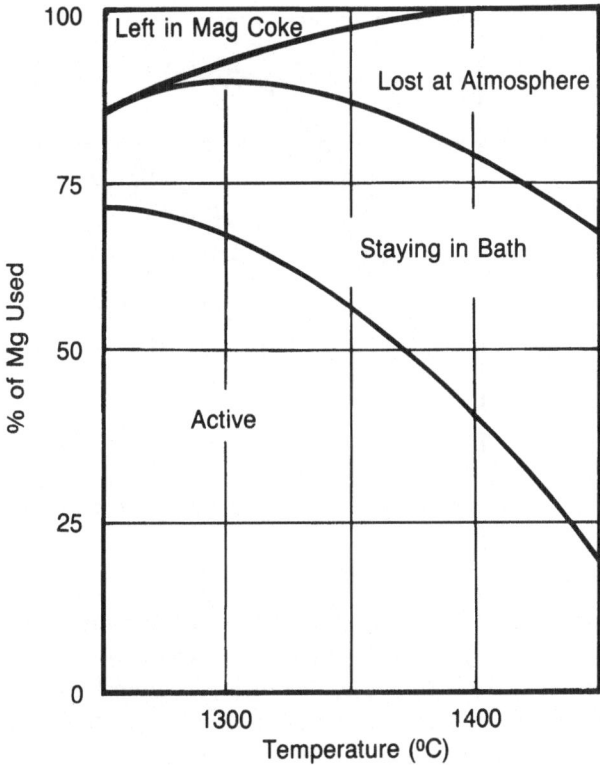

Figure 6. Example of calculated repartition of Mg during a Mag-coke treatment (after [22]).

Clearly, the resistance to sulphur transfer due to the shell will be more constraining in the case of lime, so that we will concentrate the discussion on this case.

A typical diagram of combination of the resistance to liquid phase transfer and to transfer through the shell is shown in Figure 8. For the case considered on this figure (residence time of each particle around 1 min), liquid-phase transfer would control the kinetics below about 0.03%S, and solid phase transfer would be predominant above. Landefeld and Katz[30] have recently made experiments with longer residence times (and thus thicker shells) and found that solid-phase transfer was limiting at much lower sulphur contents (below 0.01 %S). The same limit was proposed by Taoka et al[31], using industrial results obtained with $CaCO_3$-CaO-C-$CaF_2$ fluxes. In this case, however, they observed a large increase in surface area, due to $CaCO_3$ decomposition.

On the basis of the lower values of the shift in transfer control, an explanation of our results[32] of Figure 7 can be attempted:

- the presence of Al in hot metal will lead to the formation of liquid aluminates

*References pp. 353–355*

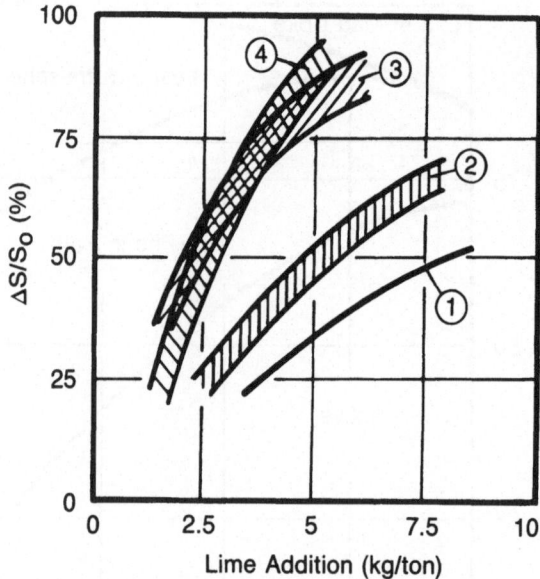

Figure 7. Typical results of hot metal desulphurization with injection of lime-based mixtures:
(1) Pure lime powder in 6 ton open ladle
(2) Lime-alumina as a mixture or premelted, in 6 ton open ladle
(3) Pure lime powder in Al-containing hot metal
(6 ton open ladle and 450 ton torpedo car)
(4) $CaCO_3$-$CaO$-$C$-$CaF_2$ powdered flux in 350 ton torpedo car (recalculated as CaO amount after [31]).

or alumino-silicates instead of dicalcium silicate and therefore improve the kinetics down to low sulphur contents. Indeed, a lesser efficiency of the treatment was observed below 1360 C, which might be close to the solidus temperature of the lime alumino-silicates;

- when only alumina is used, even premelted with the lime, the initial lime-aluminate may well become saturated with the silica formed with the oxygen released by the sulphur removal.

A similar explanation has been proposed for the fluxing effect of $CaF_2$[33].

## LADLE DESULFURIZATION OF STEEL

A large number of ladle treatment results in French plants and literature data were analyzed in terms of oxygen and sulphur exchange reaction between metal and top slag[34].

$$\underline{S} + (O^{2-}) = \underline{O} + (S^{2-})$$

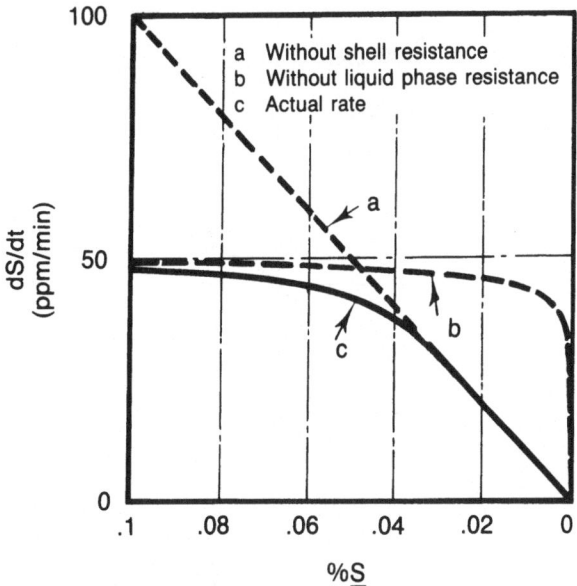

Figure 8. Example of calculated rate of desulphurization of iron with lime (after [29])
(a) Virtual rate in the absence of solid-phase transfer resistance
(b) Virtual rate in the absence of liquid-phase transfer resistance
(c) Actual rate (1300°C - lime grains of radius 0.07 mm reacted with mechanical stirrer under methane).

Indeed, this reaction is, in current practice, the only efficient one for sulphur removal. Calcium metal, when injected, has essentially a role in intensifying slag-metal stirring and in some instances, it can be helpful in the fast reduction of the top slag, but it does not to a large extent contribute directly to sulphur extraction[35] (although calcium is of utmost importance for inclusion morphology control, and so are rare-earth elements used after an efficient treatment of steel desulphurization[36]).

**Thermodynamic analysis**—The equilibrium sulphur partition coefficient between slag and metal can be expressed as:

$$L_s = (\%S)/\%\underline{S} = C'_s \cdot f_s/a_O \tag{2}$$

where $C'_s$ represents the sulphide capacity of the slag, $f_s$ is the activity coefficient of sulphur in the metal, and $a_O$ is the oxygen activity at the slag metal interface.

Provided desulphurization kinetics is not limited by oxygen transfer (see later the discussion on this point), the oxygen activity at the slag-metal interface can be computed assuming equilibrium between deoxidants dissolved in the bulk metal and the corresponding oxides dissolved in the bulk slag. For this calculation, we used, in addition to the values of solubility products listed in the accompanying

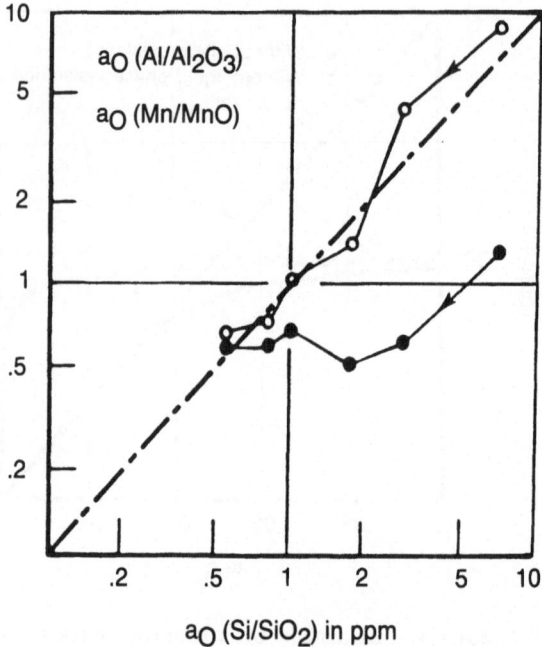

**Figure 9.** Evolution of oxygen activity computed according to, respectively, the Si, Al and Mn equilibria, during a treatment of slag-metal stirring under vacuum.

Table, and to the now well established formalism of interaction coefficients to compute metal component activities, a statistical thermodynamics model of the slag[37], which provides both oxide component activities and phase diagram and can be applied for complex iron and steelmaking slags. An example of the good accuracy of this estimation is illustrated in Figure 9 which represents the redistribution of deoxidants between metal and slag during a treatment of slag-metal pneumatic stirring under vacuum[38]. The repartition of each one of the deoxidants, Al, Mn and Si between metal and slag is characterized by the oxygen activity corresponding to the individual equilibria. The initial samples were taken right after an aluminum addition, time at which manganese and silicon were mutually close to the computed equilibrium. At the end of the treatment, the computed oxygen activities for all three equilibria are quite coherent, which gives credibility to the calculation, in particular that of oxide activities in the slag.

To evaluate the slag sulphide capacity from slag composition, a correlation established by Duffy et al.[16], in terms of "optical basicity index," was used. Incorporating temperature effects deduced from data by Venkatradi and Bell[39], Abraham and Richardson[40], and Ozturk and Turkdogan[41], the following equation, for the slags under consideration, was obtained:

$$\log(C'_s) = B/A - 13300/T + 2.82 \tag{3}$$

where

$$B = 5.62\,\%CaO + 4.15\,\%MgO - 1.15\,\%SiO_2$$
$$+ 1.46\,\%Al_2O_3$$
$$A = \%CaO + 1.39\,\%MgO + 1.87\,\%SiO_2 + 1.65\,\%Al_2O_3$$

A more general expression of the temperature effect has recently been proposed[17]. For the limited range of slag composition interesting this study, however, the above correlation gives a slightly better fit of extant experimental data.

Obviously, the slag-metal equilibrium sulphur partition coefficient, for a given slag composition, will depend on the oxygen potential, hence on metal composition (in particular aluminum and silicon contents). This result is illustrated in Figure 10, where the computed equilibrium sulphur partition ratio is represented for liquid steels in which $a_{Al} = 0.03$ and $a_{Si} = 0.4$, respectively.

It is clear, from these diagrams, that the domains of liquid slag compositions leading to high partition ratios are rather limited, and the efficiency of the sulphur removal treatment will rely on the ability to reach these domains. The aimed compositions should be close to lime saturation (CaO + MgO around 60% with preferably MgO < 8%), with $SiO_2 < 15\%$ for Al-killed steels and $SiO_2 < 10\%$ for steels in which silicon is the only deoxidant.

The kinetics study described below shows that sulphur removal is strongly hindered when the slag is heterogeneous. Although an increase in treatment temperature or $CaF_2$ additions do not appreciably improve the desulphurization power of the slag (thus, Ca fixed to F should be subtracted in the slag analysis when computing the sulphide capacity), these actions can be useful in helping the formation of the basic liquid slag.

**Kinetic analysis**—The kinetics of desulphurization, in the case of a well deoxidized liquid slag of high equilibrium sulphur partition coefficient in contact with well stirred steel containing deoxidants (a necessary condition to avoid limitations due to oxygen transfer, and, hence, poor desulphurization results), is controlled by sulphur mass transfer in the metal. The following relationship was used to describe a large number of pilot plant and industrial results, obtained in ladles ranging from 5 to 300 tons capacity:

$$d\%\underline{S}/dt = -k_s \cdot (A/V) \cdot [\%\underline{S} - (\%S)/L_s] \tag{4}$$

*References pp. 353–355*

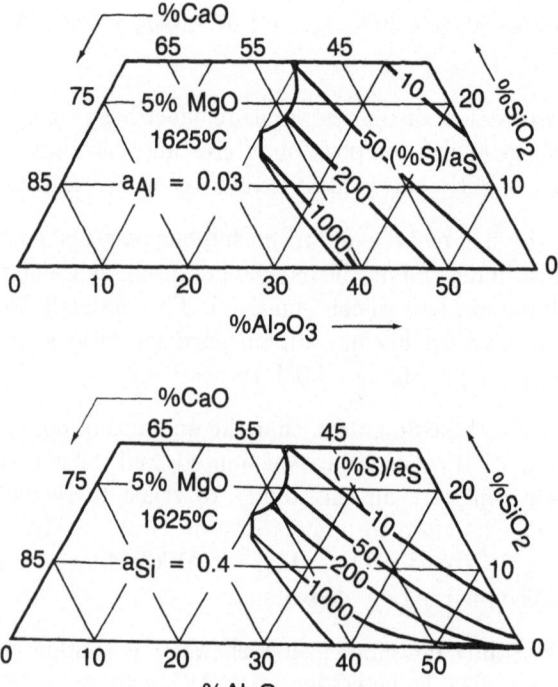

Figure 10. Sulphur partition coefficient at equilibrium between slags of the CaO-Al$_2$O$_3$-SiO$_2$-MgO system and steel with a$_{Al}$ = 0.03 or a$_{Si}$ = 0.4, at 1625°C.

where

%$\underline{S}$ and (%S) represent the metal and slag sulphur contents at time t,

$k_s$ is the slag-metal mass transfer coefficient $(m \cdot s^{-1})$,

A is the ladle cross-section area $(m^2)$,

V is the volume of metal $(m^3)$,

$L_s$ is the equilibrium sulphur partition coefficient.

When using Equation 4 integrated over the treatment time to evaluate $k_s$, a very accurate estimation of $L_s$ (and of its variation during the treatment) is necessary. In particular, neglecting the term $(\%S)/L_s$ can lead to meaningless values of the mass transfer coefficient.

For pneumatic stirring, it was possible to express the mass transfer coefficient as:

$$k_s = \beta \cdot [D_s \cdot Q/A]^{1/2} \qquad (5)$$

where

$D_s$ is the diffusion coefficient of sulphur in the metal $(m^2 \cdot s^{-1})$,

Q/A is the specific volumetric gas flow-rate across the interface $(m \cdot s^{-1})$,

$\beta$ is an empirical parameter $(m^{-1/2})$.

On the basis of our desulphurization data, as well as data reported by Usui et al[42], it was possible to show that a value of 500 $m^{-1/2}$ for the empirical parameter $\beta$ allows a fair description of the results. The same value was derived to describe dephosphorization during inert-gas stirring in the converter[43]. This value is about two orders of magnitude higher than experimental values reported on a laboratory scale[33,44]. It should be emphasized that this analysis is valid only for deep gas injections evenly spread over the interface between metal and liquid slag. In particular, the desulphurization results are drastically deteriorated when the slag contains substantial amounts of solid phases or when the slag cover is so thin that the plume uncovers a substantial part of the ladle cross-section area.

For induction stirring, it is possible to estimate the value of k from the calculation of flow patterns in the metal[45]. The mass transfer coefficient is strongly a function of the velocity field in the vicinity of the slag-metal interface (it is proportional to the 3/2 power of the friction velocity at the interface), which depends on position and design of the induction coil and on current frequency. For a properly designed coil, kinetic coefficients as high as those usually reached with pneumatic stirring can be obtained when using electromagnetic stirring alone. This

*References pp. 353–355*

is, in particular, the case for treatments made in IRSID's 6 ton induction-heated ladle[45]. In most existing large capacity industrial induction-heated ladles, however, the mass transfer coefficients obtained using induction stirring alone are too small, and additional pneumatic stirring has to be used for appropriate desulphurization results.

When Equation 4 is combined with a sulphur balance between slag and metal, one obtains the following expression of the desulphurization ratio (neglecting initial sulphur content of the slag and sulphur departure to the atmosphere):

$$R = [\%\underline{S}_O - \%\underline{S}]/\%\underline{S}_O = [1 - \exp(-B - B/\lambda)]/[1 + 1/\lambda] \qquad (5)$$

where

$\lambda = L_S \cdot M_L$ is a thermodynamic parameter, product of the equilibrium sulphur partition coefficient by the slag specific consumption in tons/ton of steel,

$B = k_S \cdot A/V \cdot t$ is a kinetic parameter, characteristic of the conditions and duration of slag-metal contact.

This relationship allows a critical analysis of the requirements to be met in order to ensure a desired desulphurization performance. This is illustrated on Figure 11 which represents the desulphurization ratio as a function of treatment characteristics. On this chart, the treatment time scale would correspond to a 140 ton ladle stirred by argon bubbling at a flow-rate of 0.03 $m^3 \cdot s^{-1}$, and the two slag weight scales to average equilibrium sulphur partition ratios of 200 and 1000, respectively. For a point in shaded area I, the treatment can be improved only if B is increased (increase in treatment time and/or gas flow-rate), whereas for a point in shaded area II, an increase of $L_s$ and/or $M_L$ is necessary.

Calcium alloys are sometimes injected in association with the stirring treatment. It has been shown[34] that the observed effect, in industrial practice, is what might be expected from the increase in stirring gas flow-rate due to vaporization of most of the calcium.

Injection of low-melting synthetic desulphurizing slag has also been attempted[46], resulting in an increase in reaction rate. An important acceleration in kinetics can however be obtained only if the effect of transitory reaction at the injected slag particle surface and permanent-contact reaction at the top slag-metal interface are cumulated[21]. In practice, the rate of feed of slag powder cannot be high enough, so that part of the slag has to be added at the top. In most performing plants[42], surface stirring alone is judged sufficient.

**Limitations due to side reactions**—Oxygen transfer: When the slag is initially rich in FeO (due to slag carry-over from the refining vessel), the situation

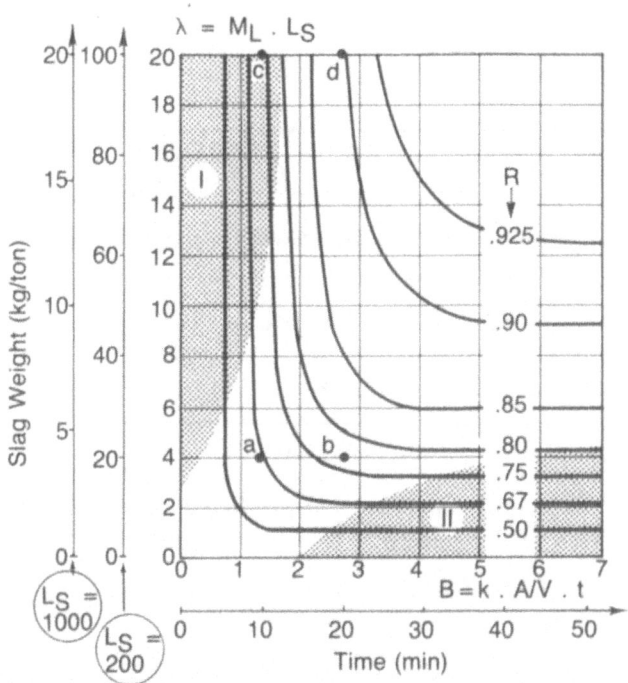

Figure 11. Chart for estimating steel desulphurization ratio R as a function of thermodynamic parameter $\lambda$ and kinetic parameter B. *Treatment time scale*: It corresponds to pneumatic stirring in a 140 ton open ladle, with an argon flow-rate of 0.03 m$^3$·s$^{-1}$. *Slag weight scales*: They correspond to average equilibrium sulphur partition coefficients of 200 and 1000, respectively.

shown in Figure 12 may arise[47]. The precipitation of deoxidation products may occur below the slag-metal interface as a result of counter-current transfer of deoxidant from the metal and oxygen from the slag. The oxygen activity at the slag-metal interface then remains high (and in consequense the equilibrium sulphur partition coefficient low) as long as the iron oxides of the slag are not fully reduced. The kinetics of slag reduction has been analyzed with the assumption that it is limited by the transport of aluminum and oxygen dissolved in the metal[34], and equation 5 for the mass transfer coefficients was shown to represent adequately the experimental data of Pluschkell et al.[47]

Slag condition: Slag phase resistance may become predominant when the slag is not a homogeneous liquid. Even with slags of liquidus temperature well below metal temperature, too weak a slag stirring may provide insufficient renewal of the slag surface: radiation losses are such that a solid crust tends to develop. This is probably part of the explanation for erratic results and apparently lower kinetic

*References pp. 353-355*

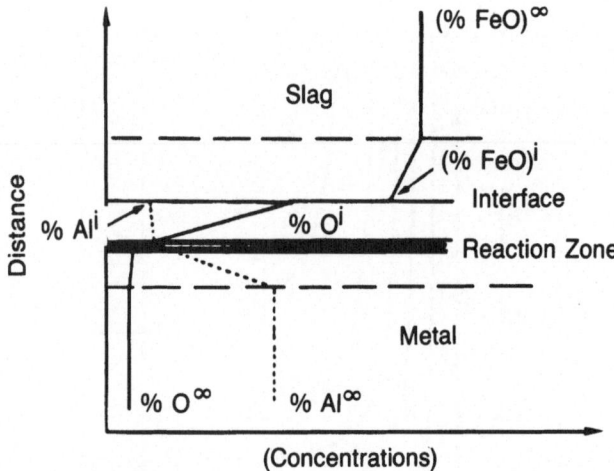

**Figure 12.** Sketch of concentration profiles at slag-metal interface during reduction of the slag by a strong deoxidant dissolved in steel.

constants observed in the conditions of low stirring energies. Fluorspar additions may help in these cases.

Refractories: Although the deterioration of slag composition arising from high alumina refractory wear can be compensated by increased lime addition (indeed, satisfactory results have been obtained in some plants with acid lining in the lower part and corrosion resistant bricks at slag level), slag composition has to be so strictly defined when ultra-low sulphur contents are aimed at (<20ppm) that basic refractories provide much more reliable operation.

Atmosphere: The beneficial effect of sulphur losses to an oxidizing atmosphere can be quite significant when aiming at medium low sulphur contents[23]: minimum slag weight has then to be formed. However, at low sulphur levels (<20ppm) in an oxidizing atmosphere, reoxidation in the plume vicinity leads to local sulphur reversion and can slow down the overall sulphur removal. Protective atmosphere is therefore advisable for peak performance.

## CONCLUSIONS

Iron desulphurization has now become a routine practice worldwide. Very powerful reagents are available, and the most important improvements, these last few years, have been in increasing the efficiency of introduction of the reagents and improving the reliability of attainment of very low sulphur contents. Magnesium and calcium carbide have now become agents very easy to handle. More economical lime and calcium carbonate can almost achieve the same low sulphur levels, provided they are used in optimal conditions.

For the ladle desulphurization of steel via reaction with the top-slag, the important points for the obtention of reproducible results are:

- a very careful adjustment of slag composition to provide adequate equilibrium slag-metal sulphur partition ratio. This requires a good knowledge of the quantities of added fluxes, scorified deoxidants and refractory wear. In addition, the top slag should be carefully deoxidized: carry-over of converter or electric-furnace slags rich in FeO and MnO should be minimised, and these oxides should be reduced rapidly through intense stirring of the slag and deoxidizer-rich metal;

- efficient stirring conditions not only to provide adequate renewal of sulphur at the slag-metal interface, but also to maintain the whole slag at high temperature and thus prevent the formation of a solid crust at the surface.

Once these conditions are fulfilled, it is possible to optimize the amount of added fluxes and treatment time to achieve the desirable desulphurization ratio.

## REFERENCES

1. R. D. Pehlke and T. Fuwa: *Int. Metals Review*, *30*, (1985), 125–140.
2. S. Katz and H. C. Rezeau: *AFS Transactions*, *87*, (1979), 367–376.
3. E. T. Turkdogan: *Metall. Trans. 9B*, (1978), 163–179.
4. A. Rist, M. F. Ancey-Moret, C. Gatellier and P. V. Riboud: *Techniques de l'Ingenieur*, M1730a, (1974).
5. G. K. Sigworth and J. F. Elliott: *Metal Science*, *8*, (1974), 298–310.
6. P. V. Riboud and M. Olette: Congres "Physico-Chimie et Siderurgie." Versailles, France, (1978). *Revue de Metallurgie CIT.* (1979), 559–596.
7. *JANAF Thermochemical Tables* 2nd Ed., (1971), US Dept. of Commerce, National Bureau of Standards.
8. M. Nadif and C. Gatellier: *Revue de Metallurgie CIT*, (1986), 377–394.
9. B. P. Burylev: *Russian Castings Production*, (1966), 466–467.
10. Yu. A. Ageev and S. A. Archugov: *Izv. Akad. Nauk. Metally*, (1984), 78–80.
11. M. C. Speer and N. A. D. Parlee: *AFS Cast Metal Research Journal*, *8*, (1972), 122–128.
12. D. C. Sponseller and R. A. Flinn: *Trans. AIME*, *230*, (1964), 876–888.
13. H. J. Engell, M. Kohler, H. J. Fleisher, R. Thielmann and E. Schurmann: *Stahl und Eisen*, *104*, (1984), 443–449.
14. C. Gatellier and M. Olette: *Revue de Metallurgie CIT*, (1979), 377–386.
15. F. D. Richardson and C. J. B. Fincham: *JISI*, *178*, (1954), 4–15.
16. J. A. Duffy, M. D. Ingram and I. D. Sommerville *J. Chem. Soc. Faraday Trans.*, *174*, (1978), 1410–19.
17. D. J. Sosinsky and I. D. Sommerville: *Metall. Trans.*, *17B*, (1986), 331–337.
18. L. R. Farias and G. A. Irons: *Metall. Trans*, *16B*, (1985), 211–225.
19. N. El Kaddah and J. Szekely: *Ironmaking and Steelmaking*, *8*, (1981), 269–278.

20. L. R. Farias and G. A. Irons: *Metall. Trans, 17B*, (1986), 77–86.

21. S. Ohguchi and D. G. Robertson: *Ironmaking and Steelmaking, 11*, 262–273 and 274–282.

22. M. H. Rand: Foseco Congres. Hamburg, (FRG). (1975).

23. P. V. Riboud and M. Olette: 7th ICVM, Iron and Steel Institute of Japan, (1982).

24. Y. Bienvenu, G. Denier, B. Dubois, D. Coisplet and M. Poire: *Rev. Metall., 77*, (1980), 201–211.

25. J. C. Grosjean et al: Congres Acierie, Juin 1984, Srasbourg.

26. P. J. Koros, R. J. Petrushka and R. G. Kerlin: *Iron and Steelmaker, 6*, 1977, 34.

27. M. S. Yamagushi: Foseco Congres. Hamburg, (FRG). (1975).

28. Y. Bienvenu and G. Denier: 36th Annual World Magnesium Conference, Oslo, 1979.

29. F. Oeters, P. Strohmenger and W. Pluschkell: *Archiv Eisenhuttenw., 44*, (1973), 727–733.

30. C. F. Landefeld and S. Katz: "A shrinking core model of desulphurization by CaO" *5th Int. Iron and Steel Congress*, Washington, D.C., 1986.

31. K. Taoka, H. Morishita, S. Yamada and F. Sudo: 67th Steelmaking Conference Chicago, 1984.

32. F. Leclercq, J. P. Reboul, C. Gatellier, A. Chevaillier, P. Gugliermina, A. Dufour: *Scaninject, 1983*, Lulea.

33. C. F. Landefeld and S. Katz: "Kinetics of desulfurization by $CaO-CaF_2$" *5th Int. Iron and Steel Congres*. Washington D.C., April 1986.

34. P. V. Riboud and R. Vasse: *Revue de Metallurgie CIT*, (1985), 801–810.

35. M. Olette, C. Gatellier and R. Vasse: Intern. Symposium on the Physical Chemistry of Iron and Steelmaking, Toronto, (1982).

36. C. Gatellier and M. Nadif: *Revue de Metallurgie CIT*, (1985), 103–110.

37. H. Gaye and J. Welfringer: *Proc. of the 2nd Intern. Symp. on Metallurgical Slags and Fluxes*. Lake Tahoe, Nev., 1984. AIME Publ., edited by H. A. Fine and D. R. Gaskell, 357–375.

38. M. Faral and H. Gaye: *Proceedings of the 2nd Intern. Symp. on Metallurgical Slags and Fluxes*. Lake Tahoe, Nev., 1984. AIME Publ., edited by H. A. Fine and D. R. Gaskell, 159–179.

39. A. S. Venkatradi and H. B. Bell: *J. Iron and Steel Inst., 207*, (1969), 1110–1113.

40. K. P. Abraham and F. D. Richardson, *J. Iron and Steel Inst., 196*, (1960), 313–319.

41. B. Ozturk and E. T. Turkdogan: *Metal Science, 18*, (1984), 299–305 and 306–310.
42. T. Usui, K. Yamada, Y. Miyashita, H. Tanabe, M. Hanmyo and K. Tagushi: *Scaninject*, Lulea, (1980).
43. H. Gaye and J-C. Grosjean: Steelmaking Conf. AIME, Pittsburgh, (1982).
44. F. D. Richardson, D. G. C. Robertson and B. B. Staples: *Darken's Conference on Physical Chemistry in Metallurgy.* US Steel 1976, 25–48.
45. R. Vasse, H. Gaye, Y. Fautrelle and J-P. Motte: 7th ICVM, Iron and Steel Inst. of Japan, (1982).
46. Y. Ohnishi, T. Komai, H. Siino, Y. Mizukami, I. Kobayashi and S. Fujino: *Ironmaking and Steelmaking*, (1985), 29–34.
47. W. Pluschkell, B. Redenz and E. Schurmann: *Scaninject*, Lulea, (1980).

## DISCUSSIONS

**David Sponseller** *(Amax)*

I have a comment and two questions. It is an interesting coincidence that the two main centers in the world which have measured solubilities of Ca and Mg under pressure are represented at this conference. We measured these solubilities at the University of Michigan 25 years ago, and more recently Dieter Janke and the Max-Planck group in Düsseldorf confirmed these results. Regarding the discrepancy between the measured solubility product and the thermodynamic prediction, were your measurements made under pressure?

**A.** No, they were made in Fe-Ni and Fe-Cr alloys.

**D. Sponseller**

Do you believe the lack of pressure could help explain the discrepancies? For example, 15 bars Mg pressure is needed to maintain liquid Mg.

**A.** I should describe the experimental procedures and interpretation of results. Gatellier made many measurements of the CaO and CaS solubility products in iron alloys.

For the CaO measurements he measured $a_O$ with a thoria probe, as well as the total $\underline{O}$ and $\underline{Ca}$ contents in successive samples. In each experiment the $\underline{Ca}$ content continually decreases by evaporation, with a consistent increase in $\underline{O}$ content. Typically, $a_O$ is from <1 to a few ppm, total oxygen is not much greater, and total Ca is larger (up to several hundred ppm in Fe-Ni alloys). He then subtracted the Ca in inclusions from the total Ca to get dissolved Ca; that was computed by the difference between total and dissolved oxygen. He then calculated $a_{Ca}$, using in particular the data of Sponseller and Flinn. He thus obtained a solubility product $a_{Ca}a_O$ which was reasonably constant during an experiment and from one experiment to another. These solubility products are about $10^4$ those expected from thermochemical data on $\Delta G^\circ_{f,CaO}$.

For the CaS measurements he measured total S and total Ca, whose levels varied greatly among experiments. Care was taken to prevent formation of a top slag, so $a_{CaS} \simeq 1$. In this case, too, the solubility product $a_{Ca}a_S$ was about $10^4$ the values expected from $\Delta G^\circ_{f,CaS}$.

Possible explanations which could reconcile the experimental and calculated solubility products include errors in $a_O$, a strong effect of $\underline{O}$ on the $\underline{Ca}$ activity coefficient, and a large amount of CaS inclusions in the sampled metal. Such errors are not likely great enough to explain the discrepancies, however. Based partly on the experimental solubility products, we have modelled the equilibrium precipitation of calcium aluminates and CaS in steels, and the predictions agree well with the compositions of inclusions removed from steel samples by electrolytically dissolving the Fe matrix.

We are led, therefore, to suspect the accuracy of $\Delta G^\circ_{f,CaO}$ values. The original measurements date to 1890. $\Delta G^\circ_{f,CaS}$ was measured relative to CaO. The relative stability of CaO and CaS from thermochemical data is almost identical to that implied by the solubility product measurements, which further suggests that $\Delta G^\circ_{f,CaO}$ should be redetermined.

## D. Sponseller

My second question—you state strongly that Mg cannot desulfurize steel. Have you tried this? I know some American shops are using Mg in a limited production basis to desulfurize steel.

**A.** They are mainly stirring and deoxidizing with Mg.

## Dieter Janke *(Max-Planck-Institut für Eisenforschung)*

It is interesting to consider mixtures of Ca and Fe. I know there is a commercial product of this type, a mixture of Ca and Fe powders which are pressed into cylindrical pellets, which can be plunged into the liquid metal. Have you any experience or information about the effectiveness of such a product used in this way for desulfurization?

**A.** I have no information about that. We use Ca for steel, and it is important for inclusion modification. In iron I have no experience. I have experience with people using Mg wire to desulfurize, but it is very expensive compared to injection.

## Klaus Lange *(Technical University of Aachen)*

I refer to your Equation 5. It is quite clear, because of the size of the mass transfer coefficient, that there is some emulsification.

# THERMODYNAMIC ASPECTS OF REMOVING IMPURITY ELEMENTS FROM CARBON-SATURATED IRON

### NOBUO SANO

*The University of Tokyo*
*Bunkyo, Tokyo, Japan*

## ABSTRACT

The author and his associates have been interested in the thermodynamics of removal of impurity elements from iron and steel by using highly basic slags. We have worked with sodium compound slags as well as CaO-bearing slags. The CaO-CaF$_2$ or CaO-CaF$_2$-SiO$_2$ systems have been studied extensively to examine their refining capacities for a variety of impurities. The Na$_2$S-FeS system has been investigated to evaluate the thermodynamic possibility of removing copper from carbon-saturated iron.

## INTRODUCTION

If impurity elements in iron are to be removed by oxidation using oxide fluxes, the chemical reactions can be expressed by Equation 1 or 2, depending upon whether the element is stable as a cation or an anion in fluxes

$$M + xO_2 = M^{(4x)+} + 2xO^{2-} \tag{1}$$

$$M + xO_2 + yO^{2-} = MO_{2x+y}^{2y-} \tag{2}$$

where $M$ has the charge of $+4x$ as an ion. Which ionic species is predominant is determined by the activity of oxygen ions, $a_{O^{2-}}$ as seen in Equation 3, derived by the combination of Equations 1 and 2.

$$M^{4x+} + (2x+y)O^{2-} = MO_{2x+y}^{2y-} \tag{3}$$

The $a_{O^{2-}}$ is the most exact measure of slag basicity but unfortunately can not be measured in principle by experiments. (The basicity, $B$, can be defined as $B = \log a_{O^{2-}}$, analogous to pH for aqueous solutions).

According to Equation 3, the predominant species in basic slags is $MO_{2x+y}^{2y-}$ and $M^{4x+}$ is more stable in acidic slags. Certainly both ions can coexist in neutral slags. The tendency for impurity elements to behave in this way depends upon how amphoteric their oxides are, and the oxides of transition metals such as Ti, V, Cr, Fe etc. having multi-valencies are more or less typical. On the other hand, the elements whose oxides are acidic, for example, Si, P, S, are stable as anions in fluxes.

Some impurity elements in IVa, Va and VIa groups in the periodical table can have either positive or negative charges. For example, the valency of sulfur changes widely from +6 to −2, so that $SO_4^{2-}$ and $S^{2-}$ can be the predominant species in slags depending upon the prevailing oxygen partial pressure in surroundings, as demonstrated by Equation 4.

$$S^{2-} + 2O_2 = SO_4^{2-} \qquad (4)$$

Impurity elements which are electrochemically more noble than iron can not be oxidized preferentially to iron but there remains a chance to remove them by reduction. In this sense, sulfur has been a good example to iron and steel engineers and this principle may be applicable to remove tramp elements from iron, as stated below.

Apart from oxide slags, there may be the possibility to make use of other compounds, such as sulfide fluxes. In this case the greater affinity of an impurity to sulfur than that of iron is employed.

With these physicochemical backgrounds, the author and his associates have been interested in the removal of impurity elements from iron through the thermodynamic approach. The following summarizes the results, excluding those for dephosphorization which foundry engineers are not very much interested in.

## THERMODYNAMIC PROPERTIES OF SOME ELEMENTS

**The $Na_2O$-$SiO_2$ System** [1,2]—This system has been brought to light due to its capability to remove sulfur and phosphorus simultaneously from hot metals, using its strongly basic property. However, this will also help us to remove other impurities by the reason stated earlier if they are stable as anions in the flux after oxidation, as seen in Equation 2.

Figure 1 shows the distribution of Nb, V, and Mn between $Na_2O$-$SiO_2$ melts and carbon-saturated iron at 1200°C as a function of the flux composition. The measurement was made by employing the chemical equilibration technique for the flux and molten iron in a graphite crucible under a CO atmosphere. When the mole fraction of $Na_2O$ in the flux exceeded 0.5, a Pb-Na reservoir was used to compensate

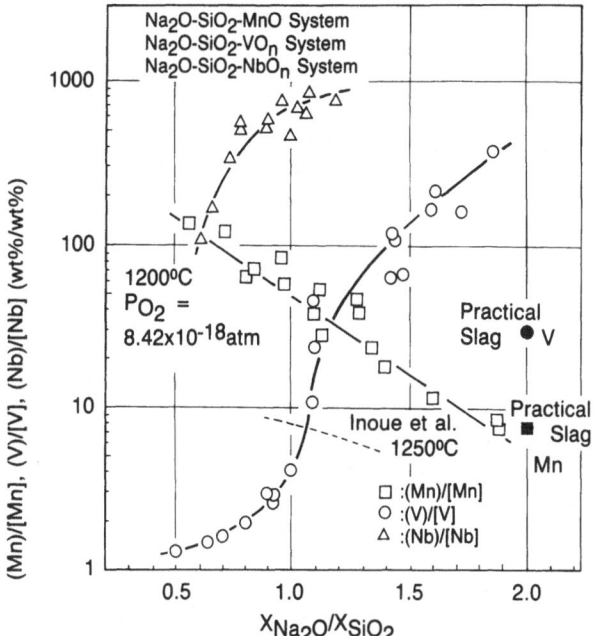

Figure 1. Relation between distribution ratios and $X_{Na_2O}/X_{SiO_2}$ for manganese, vanadium and niobium at 1200°C for the $Na_2O - SiO_2$ system.

sodium vapor loss. By this technique the activity of $Na_2O$ was also determined by measuring $Na$ content in lead. As seen in Figure 1, the V and Nb distributions rise with increasing $X_{Na_2O}/X_{SiO_2}$, whereas the Mn distribution decreases. In the light of Equations 1 and 2, this indicates that V and Nb are stabilized as anions in this particular flux and Mn as a cation. In other words we can remove and recover V and Nb from carbon-saturated iron preferentially to Mn by using the $Na_2O$-$SiO_2$ system as a flux.

The temperature dependences of these distributions are shown in Figure 2. Figure 3 demonstrates the effect of flux composition on the charge of ions of Nb, V, and Sb. As seen in the figure most Nb is present as an anion with $Nb^{5+}$ such as $NbO_x^{-(2x-5)}$ even under very low oxygen partial pressure of $8.42 \times 10^{-18}$ atm at 1200°C. It is interesting that the valency of V increases from $V^{2+}$ to $V^{5+}$ with an increase in $X_{Na_2O}/X_{SiO_2}$. Taking into consideration that the stable vanadium

*References p. 373*

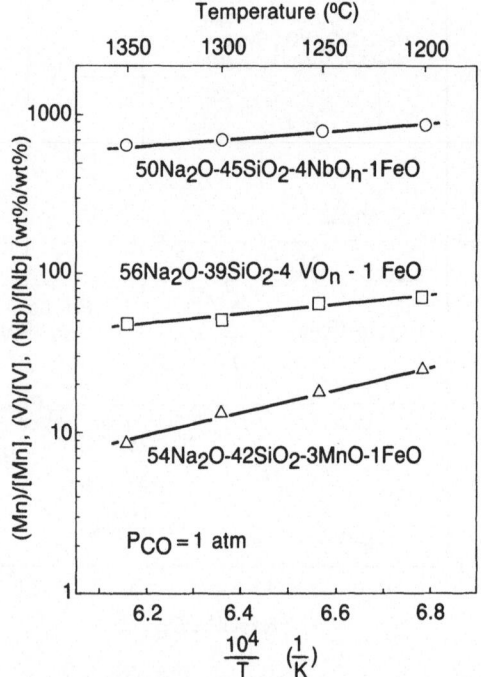

Figure 2. Temperature dependences of manganese, vanadium and niobium distribution ratios between 1200°C and 1350°C for the $Na_2O - SiO_2$ system.

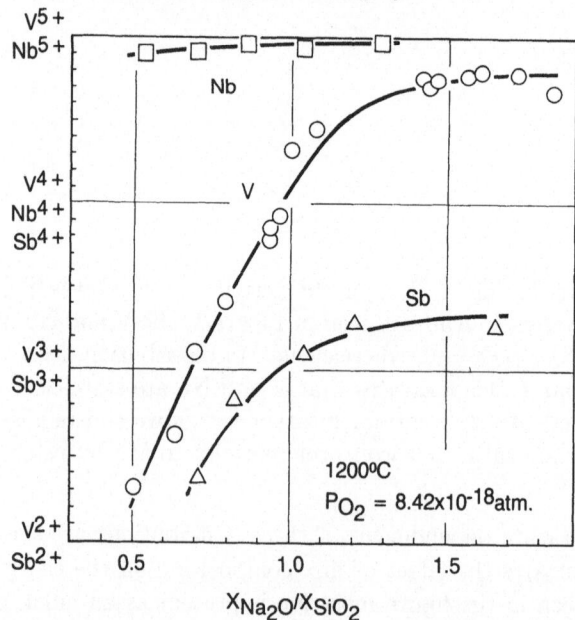

Figure 3. Change in valency of vanadium, niobium and antimony with the compositions of the $Na_2O - SiO_2$ system at 1200°C.

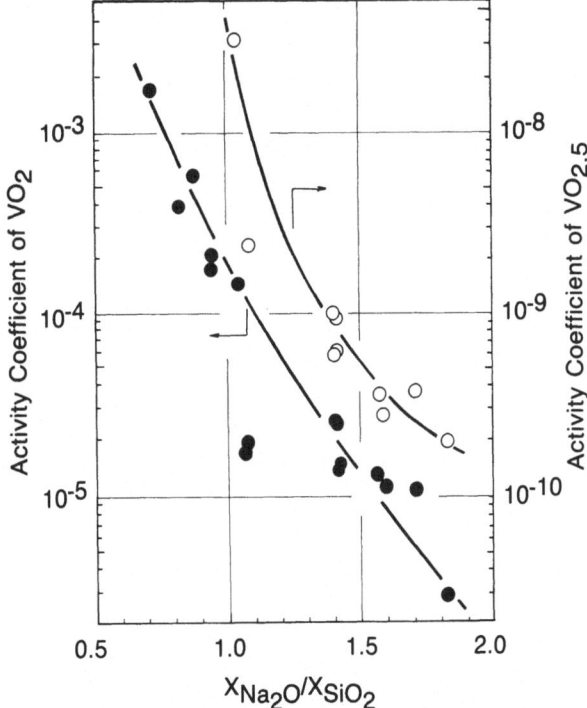

Figure 4. Activity coefficients of $VO_2$ and $VO_{2.5}$ for the $Na_2O - SiO_2 - VO_n$ system at 1200°C.

oxide under this condition is VO, V is more oxidized in the flux than outside the flux and is stabilized as an anion. The activity coefficients of $VO_2$ and $VO_{2.5}$ were calculated and are shown in Figure 4 in which the acidic properties of both oxides are clearly observed. According to Figure 4, Sb in the flux is present as $Sb^{2+}$ and $Sb^{3+}$.

Figure 5 shows the distribution of Sb between $Na_2O$-$SiO_2$ melts and carbon saturated iron. Interestingly the distribution decreases with an increase in $Na_2O$ content of the flux up to about 57% and then increases beyond this critical content. This indicates that $Sb^{3+}$ or $Sb_2O_3$ behaves amphoterically. Unfortunately the absolute values of distribution are not favorable to remove this impurity even using such a very basic flux so that an alternative way must be considered, as described later.

*References p. 373*

**The Na$_2$S-FeS System**[3]—It is well known[4,5] that copper in carbon-saturated iron may be removed by sulfidation to some extent, using Na$_2$S or Na$_2$SO$_4$. In order to establish the thermodynamic background of this phenomenon, the Na$_2$S-FeS-Cu$_2$S melts were equilibrated with carbon-saturated iron under a CO atmosphere. For the same reason as described for the Na$_2$O-SiO$_2$ system, a Pb-Na reservoir was used, thus enabling us to determine the Na$_2$O activity and the distribution of copper between lead and carbon-saturated iron. The information of the latter is useful because the transfer of copper from iron to lead has been considered to be an alternative method to remove copper.

Figure 6 shows the iso-distribution contours of copper in the ternary diagram at 1200°C. The Cu distribution $(=(\%Cu)/[\%Cu])$ of 17 was obtained with Na$_2$S 90wt%, FeS 10wt%. This is not very favorable for the industrial application but is far larger than 2.3 for the distribution of Cu between Pb and Fe. Unexpectedly the sulfur content was as low as 0.01 wt%. In Figure 7 is shown the temperature dependence of $(\%CuS_{0.5})/[\%Cu][\%S]^{0.5}$, yielding $\Delta H° = -105$ kJ/mole as the heat of reaction (5),

$$[Cu] + 1/2[S] = (CuS_{0.5}) \tag{5}$$

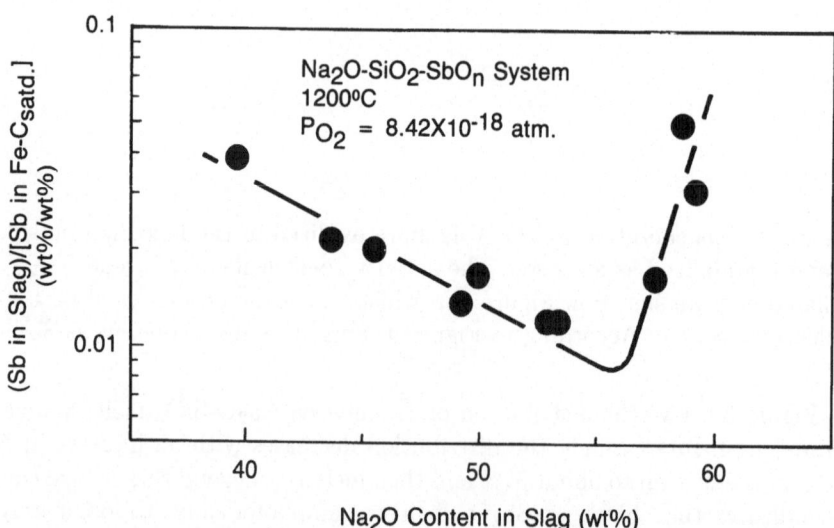

Figure 5.  Effect of Na$_2$O content on distribution of antimony between Na$_2$O – SiO$_2$ – SbO$_n$ and carbon-saturated iron at 1200°C.

*References p. 373*

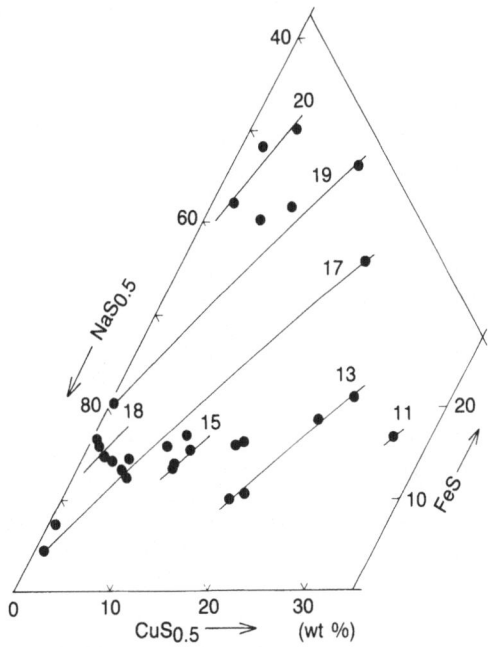

Figure 6. Iso – Cu distribution contours for the $Na_2S - FeS - Cu_2S$ system at 1200°C.

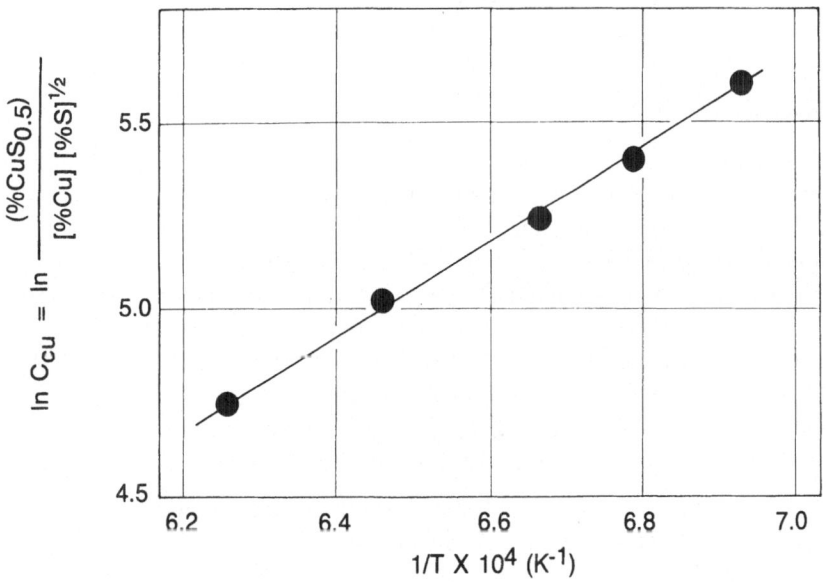

Figure 7. Temperature dependence of $C_{Cu}$ for the $NaS_{0.5}$-FeS system. (%$NaS_{0.5}$=79.9~88.5, %FeS=10.7~19.4, %$CuS_{0.5}$=0.55~0.76)

*References p. 373*

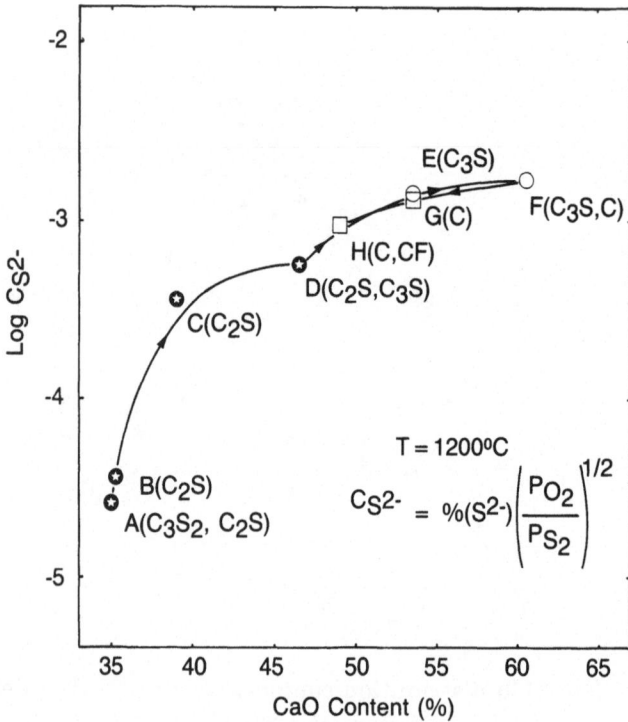

**Figure 8.** Sulfide capacities of $CaO - CaF_2 - SiO_2$ melts as a function of wt%CaO at 1200°C.

## THE $CaO$-$CaF_2$(-$SiO_2$) SYSTEM

This system has been extensively studied by our group because it is likely to be the most basic among CaO-based fluxes, judging from the $CO_2$ solubility as a measure of slag basicity. Although our major interest has been in dephosphorization by using this system, the sulfide capacity was also measured by equilibrating the flux with molten silver under controlled oxygen partial pressures along the liquidus at 1200°C as shown in Figure 8.[6] As the silica content increases, the coexisting solid phase changes successively from CaO to $3CaO \cdot SiO_2$, $2CaO \cdot SiO_2$, $3CaO \cdot 2SiO_2$ and $CaO \cdot SiO_2$ in this order. According to Figure 8, the double saturation of flux melts with CaO and $3CaO \cdot SiO_2$ gives the maximum sulfide capacity in this system. This is exactly in accord with the trend of phosphate capacity.[7]

Similarly, the same system was studied by measuring the sulfur content in carbon-saturated iron in equilibrium with the flux saturated with CaS, under a CO atmosphere at 1300°C.[8]

$$[S] + C + CaO = CaS + CO \qquad (6)$$

As seen in Equation 6, the activity of CaO can be calculated from the measured

[%S] and it was found to be in good agreement with the calculation. For example, since the sulfur content in metals was 11.7ppm at the CaO saturation, the CaO activity of melts saturated with $3CaO \cdot SiO_2$ and $2CaO \cdot SiO_2$ is estimated as 11.7ppm/13.9ppm = 0.84 from 13.9ppm of [S] for this flux composition. This agrees approximately with the thermodynamic estimation, 0.97. Similarly the CaO activity at the double saturation of $2CaO \cdot SiO_2$ and $3CaO \cdot 2SiO_2$ is estimated as 0.13 against 0.16 of the thermodynamic estimation. The CaS solubility was 3 to 4% regardless of the flux composition.

Judging from the fact that a little addition of $Na_2O$ to the flux has been observed to enhance the phosphate capacity significantly, $CaO$-$CaF_2$-$SiO_2$ melts with 2 to 3% $Na_2O$, doubly saturated with CaO and $3CaO \cdot SiO_2$ would be the most favorable flux among CaO-based fluxes as far as we know. In fact, the $Na_2O$ activity estimated from Na content in the coexisting lead was found to be equal to that of 50wt%$Na_2O$-$SiO_2$.[7]

## THERMODYNAMIC BEHAVIOR OF TRAMP ELEMENTS UNDER STRONGLY REDUCING CONDITIONS

As stated earlier, the possibility to remove tramp elements would be to reduce and absorb them into fluxes, using their negative charges under strongly reducing conditions. Accordingly, the impurities in the VIa, and Va groups have such a possibility, while Cu, Ni, Co etc. would be hopeless in applying this principle.

Although our initial interest was in removing phosphorus from Cr containing steels as $P^{3-}$,[9] we extended the established technique to understand the thermodynamic behavior of Sb, As, Sn and Pb in $CaO$-$CaF_2$ melts. Namely the flux and metals containing one of these impurities in a graphite crucible were equilibrated under a flow of CO-Ar mixture at 1500°C, the oxygen partial pressure being regulated. In the Sb case, the Cu-5%Sb was used as a metal.[10] The distribution of Sb between the CaO-saturated $CaF_2$ and the metal is plotted against $P_{O_2}$ in Figure 9, yielding a straight line of a slope of $-0.52$ against $-0.75$ of the following theoretical expectation. The left end of the plot is for the $CaC_2$ saturation, which gives the lowest $P_{O_2}$ for the experimental conditions.

If Sb is present in the flux as $Sb^{3-}$ or $Ca_3Sb_2$, the presence of this compound being well known, the reaction under reducing atmosphere may be expressed by Equation 7 or 7'.

$$[Sb] + 3/2O^{2-} = Sb^{3-} + 3/4O_2 \qquad (7)$$

$$[Sb] + 3/2CaO = Ca_{1.5}Sb + 3/4O_2 \qquad (7)'$$

Consequently the slope of a straight line is expected as 0.75 when $\log L_{Sb} (\equiv \log (Sb)/[Sb])$ in fluxes is plotted against $\log P_{O_2}$, if the Henrian behavior of Sb in fluxes is assumed. As summarized later in Figure 16, we can extrapolate these findings to the lower $P_{O_2}$ range in order to estimate the realistic conditions to remove Sb from carbon-saturated iron.

*References p. 373*

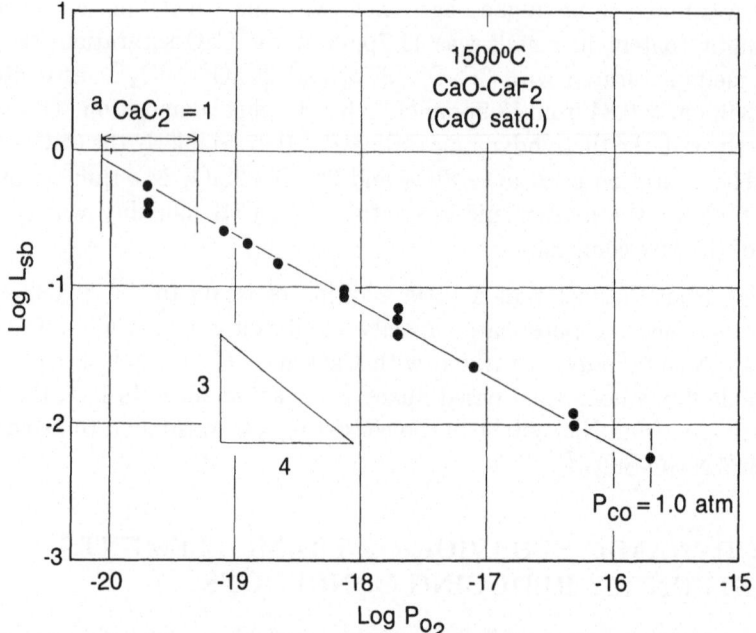

Figure 9. Dependence of the Sb distribution between Cu − 5%Sb and CaO − CaF$_2$ melts on the oxygen partial pressure at 1500°C.

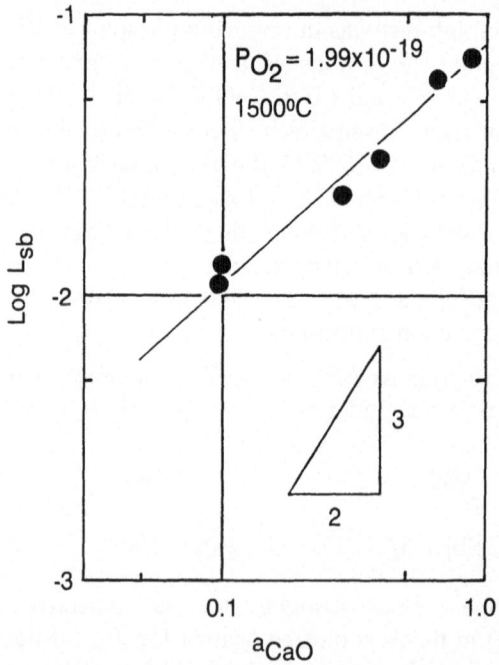

Figure 10. Dependence of the Sb distribution between Cu and CaO − CaF$_2$ melts on the CaO activity at 1500°C.

The dependence of Sb distribution on the CaO activity is shown in Figure 10. As the slope of a straight line, 1.2 was yielded against 1.5 expected from Equation 7 or 7'. From the temperature dependence of Sb distribution, the heat of reaction for Equation 7' was calculated as +561 kJ/mole.

Cu and **CaO – CaF$_2$** melts on the CaO activity at 1500°C.

Arsenic, which is a relative of Sb, was investigated similarly at 1500°C using Ag containing 0.1 to 0.6% As.[11] The results are demonstrated in Figure 11 and Figure 12 for the CaO activity and $P_{O_2}$ dependence, respectively. The As distribution, $L_{As}(= (\%As)/[\%As])$ is expressed by Equations 8 and 9.

$$\log L_{As} = 1.61 \log a_{CaO} + 0.47 \text{ at } P_{O_2} = 1.99 \times 10^{-18} \text{atm} \quad (8)$$

$$\log L_{As} = -0.68 \log P_{O_2} - 11.58 \text{ at CaO saturation} \quad (9)$$

These are consistent with the expression of Equation 10

$$As + 3/2 \text{ CaO} = Ca_{1.5}As + 3/4 \text{ O}_2 \quad (10)$$

These data can be used to assess $L_{As}$ for carbon-saturated iron, using the thermodynamic properties of the Ag-As and Fe-C systems. The heat of reaction of Equation 10 was found to be 565 kJ/mole, which happens to equal exactly with that for Sb.

Sn and Pb belong to the IVa group and their most stable compounds with calcium are $Ca_2Sn$ and $Ca_2Pb$ whereas the stable compounds for members of the Va group are $Ca_3M_2$ (M: N, P, As, Sb, Bi) as demonstrated above. The behavior of Sn in CaO-CaF$_2$ melts was studied by using the same technique as mentioned above, except that molten tin was used as a metal because the stability of its compounds with Ca had been predicted to be weaker than $Ca_3M_2$.[10]

Figure 13 shows the (Sn) content in CaO-CaF$_2$ melts saturated with CaO as a function of the partial pressure of oxygen at 1500°C. The slope of the straight line is 0.99, indicating that the reaction can be expressed by Equation 11.

$$Sn + 2CaO = Ca_2Sn + O_2 \quad (11)$$

From the temperature dependence of the Sn content, the heat of reaction for Equation 11 was calculated as +791 kJ/mole.

Pb is now under investigation[11] and it can be said from the tentative data that $Ca_2Pb$ is likely to be less stable than $Ca_2Sn$. This is reasonable, considering the difference in heats of formation of both compounds.

When Cu-5%Sb was used for studying the Sb distribution, we found that Cu also dissolved into CaO-CaF$_2$ melts. Figure 14 shows the Cu content vs. the partial pressure of oxygen, indicating Equation 12, by the same analytical method as before.

$$Cu + CaO = CaCu + 1/2O_2 \quad (12)$$

*References p. 373*

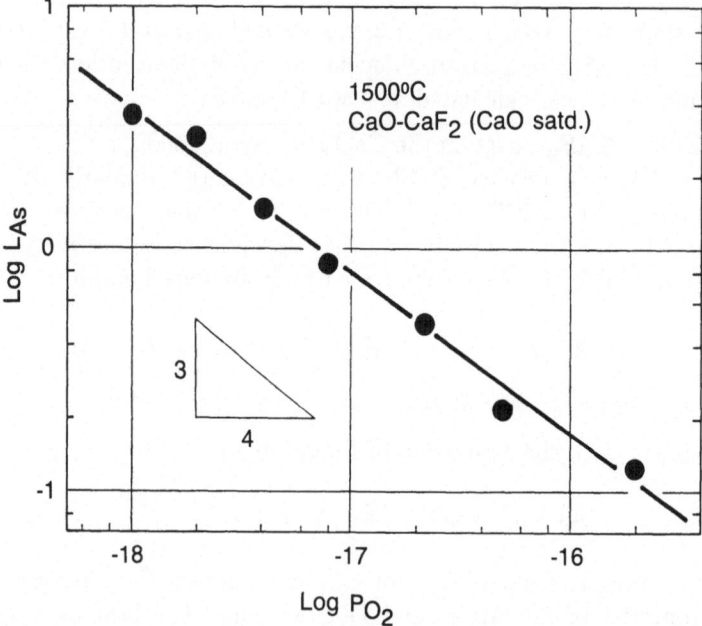

Figure 11. Dependence of the As distribution between Ag and CaO – CaF$_2$ melts on the oxygen partial pressure at 1500°C.

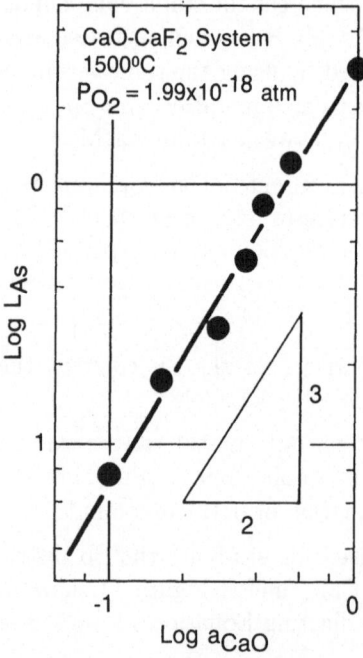

Figure 12. Dependence of the As distribution between Ag and CaO – CaF$_2$ melts on the CaO activity at 1500°C.

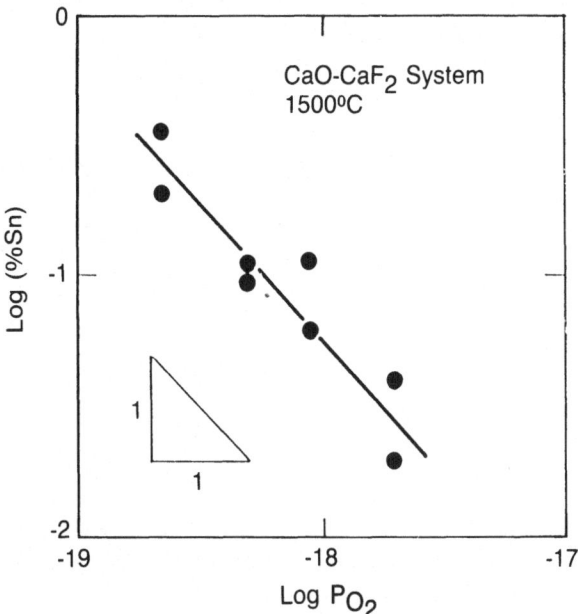

Figure 13. Dependence of Sn content in $CaO_{sat.} - CaF_2$ melts in equilibrium with Sn on the partial pressure of oxygen at 1500°C.

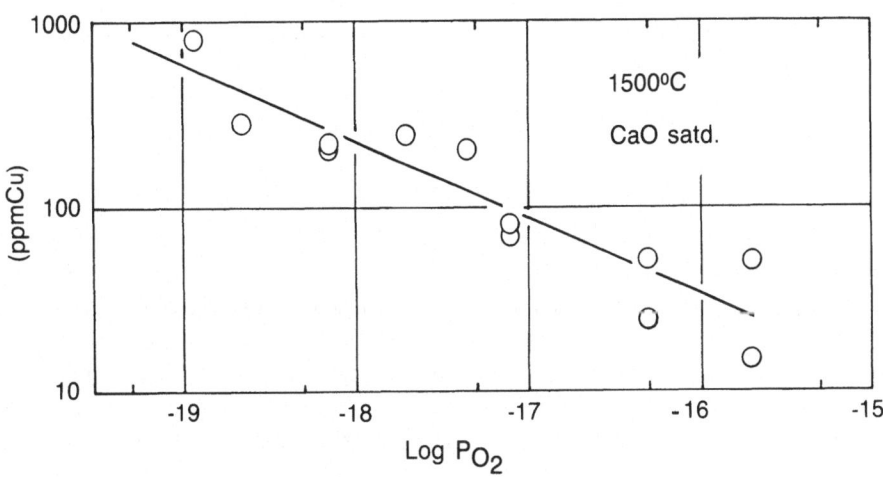

Figure 14. Dependence of Cu content in $CaO_{sat.} - CaF_2$ melts in equilibrium with 5%Sn − Cu on the partial pressure of oxygen at 1500°C.

However, since CaCu is not a stable compound according to the phase diagram and Cu is not likely to be present in fluxes as a negative ion, the mechanism of dissolution is totally unknown. Similar observations were made by Köhler[12] without much discussion, and the results of the present work approximately agree with his findings.

## THERMODYNAMIC PROPERTIES OF CALCIUM COMPOUNDS

The thermodynamic properties of calcium compounds which appear in the foregoing reactions are not well known so that we have started measuring them in series. The following is a part of our results on $Ca_2Sn$.[13]

Since $Ca_2Sn$ is not a stable compound, Equation 13 was studied at temperatures ranging from 1000°C to 1120°C.

$$2CaC_2(s) + [Sn]_{in\ Ag} = Ca_2Sn(s) + 4C(s) \tag{13}$$

equilibrating Ag, $CaC_2$, $Ca_2Sn$ in a graphite crucible, the total system being sealed in a silica capsule. Since the activity of Sn is fixed at a fixed temperature, it can be calculated by using the Sn content in silver as a solvent plus its activity coefficient which was predetermined separately by using Equation 14.

$$CaC_2(s) = [Ca]_{in\ Ag} + 2C(s) \tag{14}$$

Namely, since the activity of Ca is fixed on equilibrating $CaC_2$, Ag and C, the change in calcium content with Sn addition yielded $-5.7$ as the interaction parameter, $\varepsilon_{Ca}^{Sn}$ (atom fraction concentrations). This enables us to calculate the activity of Sn in Equation 14. Thus the Gibbs free energy of formation for $Ca_2Sn$, according to Equation 15 can be expressed as

$$2Ca(l) + Sn(l) = Ca_2Sn(s) \tag{15}$$

$$\Delta G° = -352,300(\pm 1760) + 83.05(\pm 1.26)T\ \text{J/mole}(1000-1200°C) \tag{16}$$

Equation 16 confirms the estimation[14] made from low temperature measurements for $\Delta H_f°$, as shown in Figure 15. Since the solubility of Ca in the Fe-C-B system is very limited, Ag was replaced by this system at a later stage of investigation, eliminating a chance for errors due to the effect of Ca on the Sn activity to be introduced. The square symbols in Figure 15 were obtained by using this system and combining with the activity coefficient of Sn in the system, 9.93 which was separately measured. The results are on the extrapolated line for data measured with Ag.

Since the foregoing experiments on the thermodynamic properties of tramp elements in $CaO$-$CaF_2$ melts were carried out using different metals suitable for each case, the results were extrapolated to lower oxygen partial pressures to be applicable to assessing their removals from carbon-saturated iron, although the validity of such extensive extrapolation is open to question. In Figure 16, the distribution of

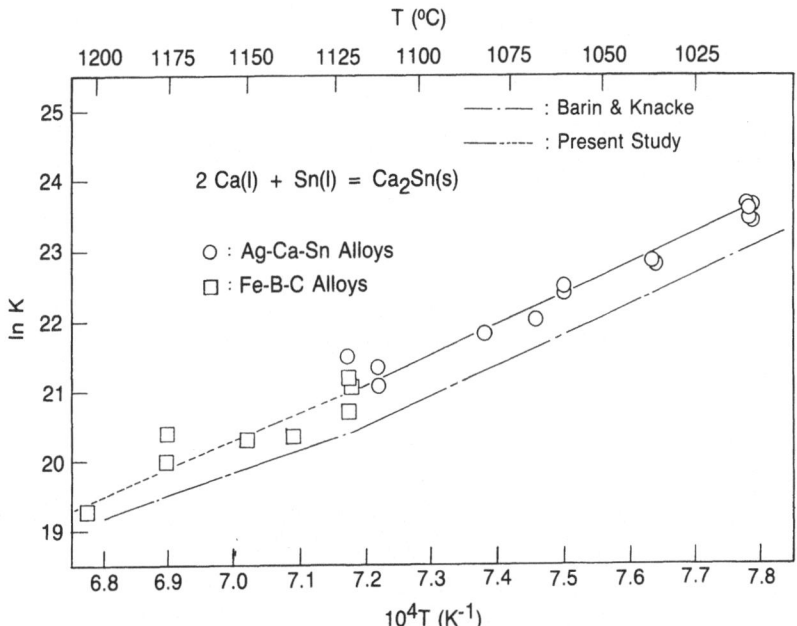

Figure 15. Equilibrium constant for $2\text{Ca}(l) + \text{Sn}(l) = \text{Ca}_2\text{Sn}(s)$ as a function of temperature.

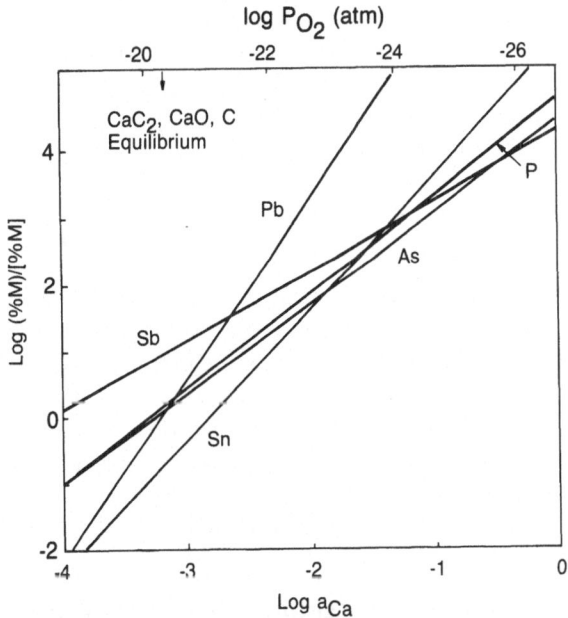

Figure 16. Estimated distributions of five impurity elements between carbon-saturated iron and CaO-saturated $\text{CaF}_2$ melts at 1500°C as a function of the calcium activity, $a_{\text{Ca}}$, or the oxygen partial pressure.

*References p. 373*

an element M between the metal and the CaO saturated $CaF_2$ at 1500°C, $L_M$, is plotted against the activity of Ca, $a_{Ca}$, or the oxygen partial pressure. The difference in $L_M$ values is not large from metal to metal and depends on $P_{O_2}$, although the general order follows that of $\Delta H_f^\circ$. It is indicated from Figure 16 that if $a_{Ca}$ in fluxes exceeds 0.03 or $P_{O2}$ is lower than $2 \times 10^{-24}$ atm, most impurities can be eliminated from the metal in a thermodynamic sense.

According to the literature[15,16] the values of $L_M$ in general are one or two orders of magnitude larger than the directly measured ones for carbon-free iron, although the difference in temperature must be taken into account. This is not clearly explained but the relative magnitudes of $L_M$ among various impurities are consistent with the results in practical operations found by many investigators. If the standard free energy of formation of CaO recently proposed by Kay and Subramanian (D. A. R. Kay and S. V. Subramanian: *Proceedings of International Symposium on the Control and Effects of Inclusions and Residuals in Steels*, sponsored by the Iron and Steel and the Engineering Section of the Metallurgical Society of CIM, pp. I-25–I-43, August 17-20, 1986, Toronto) is used to calculate $a_{Ca}$, the situation will be entirely changed. The axis of abscissa will be shifted to the left approximately by two orders of magnitude. This is rather too much for the discrepancy to disappear. More details will be described in reference 11.

## CONCLUSIONS

Thermodynamics of impurity removal has been discussed on the basis of experimental findings of the author's laboratory. The $Na_2O$-based fluxes have a potential to remove most less noble impurities from iron by oxidation, but many problems remain to be overcome such as choice of refractory materials, recovery of $Na_2CO_3$, its cost etc. On the contrary, CaO-based fluxes have a long history in steelmaking and are less expensive. Based on our findings, it is proposed to add a small amount of $Na_2CO_3$ to CaO-saturated fluxes in order to attain significant enhancement of their refining capacities.

The removal of tramp elements which are more noble than iron was discussed. The only chance seems to be in reducing and absorbing them into fluxes, but this technique also has many problems to be solved before it can be applied in industry. Namely the refining efficiency is limited, costly $CaC_2$ or Ca based alloys must be used, and the reaction products which are hazardous to human health must be treated before disposal, etc. Nevertheless, the fundamentals and applications of this technique should be further investigated, because specifications of steel products become increasingly severe and the chance of contamination from scrap increases as its volume accumulates.

The author expresses his sincere gratitude to his former and present associates for their valuable contributions.

# REFERENCES

1. F. Tsukihashi, A. Werme, F. Matsumoto, K. Kasahara, M. Yukinobu, T. Hyodo, S. Shiomi and N. Sano: *Metallurgical Slags and Fluxes*, Edited by H. A. Fine and D. R. Gaskell, The Metallurgical Society of AIME, 1984, 89–106.
2. F. Tsukihashi, A. Werme, A. Kasahara, M. Okada and N. Sano: *Tetsu-to-Hagane*, 1985, 71, 831–838.
3. T. Imai and N. Sano: *Tetsu-to-Hagane*, 1986, 72, S.962. Presented at the meeting of Iron and Steel Institute of Japan, October, 1986.
4. Y. A. Topkaya and W-K. Lu: *Metal-Slag-Gas Reaction and Processes*, The Electrochemical Society, 1975, 111–139.
5. F. C. Langenberg, R. W. Lindsay and D. P. Robertson: *Blast Furnace and Steel Plant*, 1955, 43, 1142–1147.
6. K. Susaki, M. Maeda and N. Sano: to be published.
7. M. Muraki, H. Fukushima and N. Sano: *Trans. Iron and Steel Institute of Japan*, 1985, 25, 1025–1030.
8. S. Tanaka: Thesis for B. Eng., March, 1986.
9. S. Tabuchi and N. Sano: *Metall. Trans.*, 1984, 15B, 351–356.
10. T. Isawa, K. Noguchi and N. Sano: *Tetsu-to-Hagane*, 1985, 71, 932.
11. T. Isawa, T. Wakasugi, K. Noguchi and N. Sano: To be published.
12. M. Köhler, H-J. Engell and J. Janke: *Proceedings of the 8th International Vacuum Metallurgy*, sponsored by Eisenhütte Österreich, Papers of the Melting Session, Vol. 1, 851–868, September 30—October 4, 1985.
13. D. J. Min and N. Sano: To be presented at the Fall Meeting of The Japan Institute of Metals, October, 1986.
14. I. Barin, O. Knacke: *Thermochemical properties of Inorganic Substances*, Springer-Verlag, Berlin, New York, 1973.
15. M. Köhler and H-J. Engell: The same as reference 1, 483–496.
16. Y. Zhu, Y. Dong, Y. Peng and S. Wei: *Proceedings of The 3rd China-Japan Symposium on Science and Technology of Iron and Steel*, April, 1985, 187–197.

# DISCUSSION

### D. G. C. Robertson

I would like to use my Chairman's prerogative and ask Dr. Sano to comment on the interesting results on sodium sulfide to remove copper. How practical is sodium sulfide as a reagent? It's obviously corrosive, and is probably an expensive compound to manufacture.

**A.** I don't know how good the industrial application would be.

### Dieter Janke *(Max-Planck-Institut)*

Dr. Sano, I am very pleased to see your results about the reactions of calcium with antimony, etc. You confirmed what we found when studying refining reactions with calcium under elevated pressure in a high pressure furnace. When we plotted our tin over calcium or antimony over calcium concentrations (residual contents in the metal), we found these slopes that you found, for instance -2.1 for calcium stannite. There was evidence that this compound exists as a reaction product. I'm very happy to see that you did this very exactly now under different experimental conditions.

# THE EFFECT OF BISMUTH IN GRAY CAST IRON AND THE CHEMISTRY OF ITS NEUTRALIZATION WITH RARE EARTH METALS

### CARL R. LOPER, JR.

*Professor*
*Metallurgical and Mineral Engineering*
*The University of Wisconsin-Madison*
*Madison, Wisconsin*

## INTRODUCTION

The cast metals industry depends upon recycled material for most of the metal consumed and, as a result, changes in the nature of the charge materials available can have a profound effect on the metallurgical aspects of the products produced. Recently there has been an effort to increase the amount of free machining steel used in the manufacturing of various components. Economic considerations have been responsible for these changes, and have been accompanied by environmental concerns which have encouraged a shift from lead to bismuth treatment to manufacture these specialized steels. Furthermore, a change in foundry melting procedures toward greater use of induction melting has caused the presence of these steels in the charge to warrant more attention.

Bismuth additions are commonly made in the production of malleable irons, and are frequently encountered in ductile iron production in order to counteract pin hole formation, increase nodule count and improve nodularity by offsetting the effect of rare earths. However, gray cast irons comprise the bulk of the ferrous casting alloys, and bismuth is an undesirable additive to gray cast irons[1-11]. Increased chill, and the formation of undercooled graphite, accompany bismuth in these alloys.

The ability for rare earths to counteract the effect of bismuth has been studied, particularly in ductile irons which contain extremely low sulfur contents[6,8,12-16]. However, the effect of this method of neutralization in the high (about 0.10%) sulfur content of conventional gray cast irons has not been documented.

In general, the current understanding of this effect can be summarized as follows:

1. Bismuth is known to increase the amount of undercooling occurring during the solidification of cast irons, thereby causing the graphite shape to change from Type A to Type B to Type D to carbide. As little as 0.003% bismuth addition can exert this deleterious effect.

2. The effect of bismuth may be counteracted by the addition of rare earths to the melt, the amount being dependent upon the amount of bismuth and sulfur present. In gray cast irons the greatest chill reduction was encountered at a base iron sulfur level of 0.021 to 0.031%, while at 0.15% S the treatment was not effective.

3. The level of bismuth present in the melt can be reduced significantly through vaporization. While significant amounts of bismuth can be removed from the melt in commercial foundries, this procedure is not effective in certain melting operations and does not usually reduce the bismuth content to acceptable levels for quality gray iron production.

Accordingly, the effect of bismuth in gray cast irons and its possible neutralization with rare earths were evaluated.

## EXPERIMENTAL PROCEDURE

A series of gray cast irons was produced representing variations in the bismuth and sulfur contents of these irons, the superheat temperature and the level of rare earths added as mischmetal in an attempt to counteract the effect of the bismuth, Table 1. For the most part, bismuth was introduced through the use of bismuth-containing free machining steel in the metallic charge. These steels contained 0.14% Bi and 0.33% S, or 0.22% Bi and 0.35% S. In a few instances, in order to control the sulfur level of the iron, bismuth was also introduced as commercially pure shot. Details concerning the production of these melts have been presented elsewhere[16], but the chemical analyses of the various irons studied are presented in Table 2.

The first set of heats, A, was produced to confirm the effect of variations in the bismuth content on the microstructure and mechanical properties of gray iron castings. These heats were superheated to 1482 C (2700 F), with bismuth additions of 0.002 to 0.040%. Set B was essentially a repeat of Set A, but utilized a higher superheat temperature, 1554 C (2830 F). The recovery of bismuth was related to the superheat temperature and the bismuth addition as shown in Figure 1, and can be expressed as follows:

1482C (2700F)   $X = (4.11 \times 10^{-4}) + 0.534\ Y$   Corr. Coef. = 0.973
1554C (2830F)   $X = (6.21 \times 10^{-4}) + 0.183\ Y$   Corr. Coef. = 0.950

Where: $X$ = Bismuth content of the iron produced,
       $Y$ = Amount of bismuth added.

On the basis of these results, neutralization studies were conducted at a bismuth addition level of 0.03%. Set C was produced at a sulfur level of 0.1%,

# TABLE 1

## Chemistry and Superheat Conditions for Heats Studied

| Heat Set | Bismuth Added | Bismuth Content | Sulfur Content | Rare Earth Addition | Rare Earth Content | Superheat Temperature |
|---|---|---|---|---|---|---|
| A | 0.002% | 0.0018% | 0.091% | | | 1482C (2700F) |
| | 0.004 | 0.0032 | 0.093 | | | ,, |
| | 0.006 | 0.0034 | 0.097 | | | ,, |
| | 0.008 | 0.0063 | 0.100 | | | ,, |
| | 0.010 | 0.0046 | 0.103 | | | ,, |
| | 0.020 | 0.0091 | 0.120 | | | ,, |
| | 0.030 | 0.0159 | 0.136 | | | ,, |
| | 0.040 | 0.0231 | 0.152 | | | ,, |
| B | 0.002 | 0.0006 | 0.083 | | | 1554C (2830F) |
| | 0.004 | 0.0019 | 0.088 | | | ,, |
| | 0.006 | 0.0017 | 0.090 | | | ,, |
| | 0.008 | 0.0033 | 0.095 | | | ,, |
| | 0.010 | 0.0032 | 0.096 | | | ,, |
| | 0.020 | 0.0032 | 0.093 | | | ,, |
| | 0.030 | 0.0046 | 0.106 | | | ,, |
| | 0.040 | 0.0075 | 0.104 | | | ,, |
| | 0.070 | 0.0144 | 0.104 | | | ,, |
| C | 0.03 | 0.0019 | 0.110 | 0.01% | 0.025% | ,, |
| | 0.03 | 0.0044 | 0.109 | 0.03 | 0.033 | ,, |
| | 0.03 | 0.0036 | 0.106 | 0.05 | 0.035 | ,, |
| | 0.03 | 0.0044 | 0.108 | 0.07 | 0.039 | ,, |
| | 0.03 | 0.0024 | 0.098 | 0.09 | 0.041 | ,, |
| | 0.03 | 0.0033 | 0.108 | 0.10 | 0.039 | ,, |
| D | 0.03 | <0.01 | 0.11 | 0.01 | 0.006 | ,, |
| | 0.03 | <0.01 | 0.11 | 0.03 | 0.032 | ,, |
| | 0.03 | <0.01 | 0.11 | 0.05 | 0.042 | ,, |
| | 0.03 | <0.01 | 0.097 | 0.07 | 0.032 | ,, |
| | 0.03 | <0.01 | 0.10 | 0.09 | 0.056 | ,, |
| | 0.03 | <0.01 | 0.10 | 0.10 | 0.058 | ,, |
| E | 0.03 | <0.01 | 0.009 | 0.01 | 0.008 | ,, |
| | 0.03 | <0.01 | 0.009 | 0.03 | 0.018 | ,, |
| | 0.03 | <0.01 | 0.01 | 0.05 | 0.046 | ,, |
| | 0.03 | <0.01 | 0.014 | 0.07 | 0.048 | ,, |
| | 0.03 | <0.01 | 0.01 | 0.09 | 0.060 | ,, |
| | 0.03 | <0.01 | 0.009 | 0.10 | 0.072 | ,, |
| F | 0.02 | 0.0031 | 0.053 | 0.02 | 0.0114 | ,, |
| | 0.02 | 0.0021 | 0.051 | 0.04 | 0.0168 | ,, |
| | 0.02 | 0.0036 | 0.054 | 0.06 | 0.0290 | ,, |
| | 0.02 | 0.0046 | 0.047 | 0.08 | 0.0284 | ,, |

*References pp. 390–391*

## TABLE 2
### Chemical Analyses of the Heats

| Heat No. | %C | %Si | %S | %P | %Mn | %Bi | %RE |
|---|---|---|---|---|---|---|---|
| Base | 3.45 | 2.21 | 0.071 | 0.08 | 0.47 | — | — |
| A1 | 3.35 | 1.93 | 0.091 | 0.03 | 0.56 | 0.0018 | — |
| A2 | 3.37 | 1.88 | 0.093 | 0.03 | 0.55 | 0.0032 | — |
| A3 | 3.32 | 1.88 | 0.097 | 0.03 | 0.55 | 0.0034 | — |
| A4 | 3.35 | 1.92 | 0.100 | 0.03 | 0.57 | 0.0063 | — |
| A5 | 3.29 | 1.90 | 0.103 | 0.03 | 0.57 | 0.0046 | — |
| A6 | 3.33 | 1.87 | 0.120 | 0.03 | 0.59 | 0.0091 | — |
| A7 | 3.26 | 1.81 | 0.136 | 0.03 | 0.60 | 0.0159 | — |
| A8 | 3.29 | 1.85 | 0.152 | 0.04 | 0.61 | 0.0231 | — |
| B1 | 3.25 | 1.95 | 0.083 | 0.03 | 0.53 | 0.0006 | — |
| B2 | 3.35 | 1.88 | 0.088 | 0.03 | 0.57 | 0.0019 | — |
| B3 | 3.33 | 1.83 | 0.090 | 0.03 | 0.55 | 0.0017 | — |
| B4 | 3.36 | 1.85 | 0.095 | 0.03 | 0.55 | 0.0033 | — |
| B5 | 3.33 | 1.88 | 0.096 | 0.03 | 0.55 | 0.0032 | — |
| B6 | 3.32 | 1.79 | 0.093 | 0.03 | 0.52 | 0.0032 | — |
| B7 | 3.38 | 1.80 | 0.106 | 0.03 | 0.54 | 0.0046 | — |
| B8 | 3.39 | 1.82 | 0.104 | 0.03 | 0.48 | 0.0075 | — |
| B9 | 3.30 | 1.67 | 0.104 | 0.03 | 0.41 | 0.0144 | — |
| C1 | 3.55 | 1.98 | 0.110 | 0.04 | 0.53 | 0.0019 | 0.025 |
| C2 | 3.55 | 1.84 | 0.109 | 0.04 | 0.53 | 0.0044 | 0.033 |
| C3 | 3.60 | 1.84 | 0.106 | 0.04 | 0.52 | 0.0036 | 0.035 |
| C4 | 3.58 | 1.81 | 0.108 | 0.04 | 0.54 | 0.0044 | 0.039 |
| C5 | 3.60 | 1.99 | 0.098 | 0.04 | 0.54 | 0.0024 | 0.041 |
| C6 | 3.61 | 1.86 | 0.108 | 0.04 | 0.57 | 0.0033 | 0.039 |
| D1 | 3.60 | 1.68 | 0.11 | 0.034 | 0.50 | <0.01 | 0.006 |
| D2 | 3.64 | 1.74 | 0.11 | 0.035 | 0.51 | <0.01 | 0.032 |
| D3 | 3.69 | 1.69 | 0.11 | 0.036 | 0.51 | <0.01 | 0.042 |
| D4 | 3.57 | 1.75 | 0.097 | 0.035 | 0.51 | <0.01 | 0.032 |
| D5 | 3.64 | 1.83 | 0.10 | 0.035 | 0.56 | <0.01 | 0.056 |
| D6 | 3.57 | 1.78 | 0.10 | 0.037 | 0.56 | <0.01 | 0.058 |
| E1 | 3.54 | 1.75 | 0.009 | 0.027 | 0.56 | <0.01 | 0.008 |
| E2 | 3.56 | 1.78 | 0.009 | 0.027 | 0.57 | <0.01 | 0.018 |
| E3 | 3.63 | 1.70 | 0.01 | 0.026 | 0.59 | <0.01 | 0.046 |
| E4 | 3.54 | 1.47 | 0.014 | 0.025 | 0.54 | <0.01 | 0.048 |
| E5 | 3.66 | 1.70 | 0.01 | 0.026 | 0.58 | <0.01 | 0.060 |
| E6 | 3.53 | 1.87 | 0.009 | 0.027 | 0.57 | <0.01 | 0.072 |
| F1 | 3.55 | 1.92 | 0.053 | 0.024 | 0.38 | 0.0031 | 0.0114 |
| F2 | 3.56 | 1.87 | 0.051 | 0.024 | 0.36 | 0.0021 | 0.0168 |
| F3 | 3.58 | 1.86 | 0.054 | 0.028 | 0.44 | 0.0036 | 0.0290 |
| F4 | 3.50 | 1.99 | 0.047 | 0.024 | 0.38 | 0.0046 | 0.0284 |

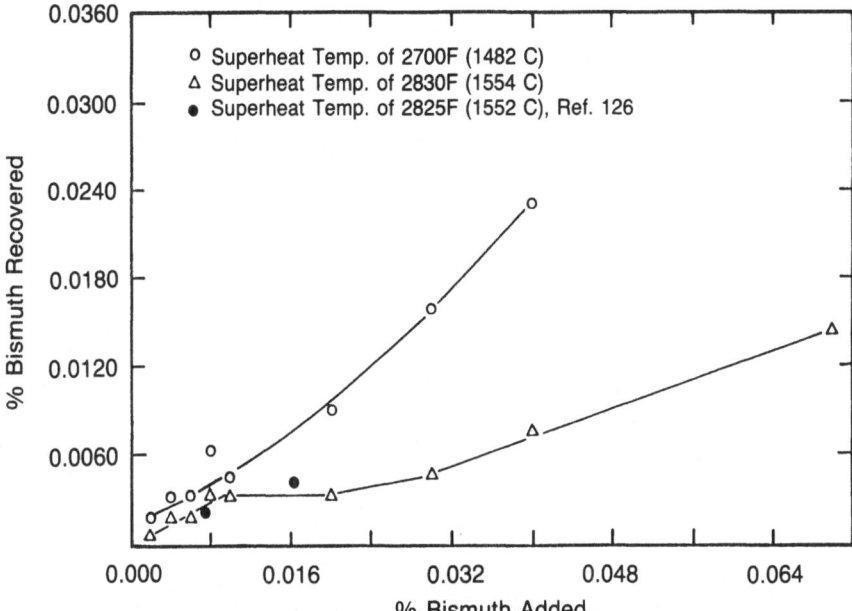

Figure 1. Effect of superheat temperature on bismuth recovery.

with the bismuth additions set at 0.03% and the level of rare earths added varied from 0.01 to 0.10%. Set D was essentially a repetition of Set C. Set E was again a repetition of Set C, but was prepared at a sulfur level of 0.01%. The final series, Set F, was prepared at a 0.06% S level and a 0.02% Bi addition with rare earth addition levels of 0.02, 0.04, 0.06 and 0.08%.

Cylindrical bars and plates were cast from these heats in green sand molds, while air set molds were used to produce a set of ASTM No. 2 and 4 wedges, and a series of six pins from 3.2 to 12.7 mm (0.125 to 0.5 in.) diam. The cylindrical bars varied from 12.7 to 63.5 mm (0.5 to 2.5 in.) diam and the plates from 3.2 to 38.1 mm (0.125 to 1.5 in.) in thickness. Tensile test coupons (ASTM-A48) were machined from the 38.1 mm (1.5 in.) diam bar castings.

## EXPERIMENTAL RESULTS

**Chilling Tendency**—Perhaps the first result of the presence of bismuth in gray cast irons is observed in the chilling tendency of the iron, measured in this work in terms of the depth of chill obtained in the ASTM wedge castings and the fracture characteristics of the pin castings. This chilling tendency is apparent in the data of Figures 2 and 3 for Sets A and B, respectively. An increase of chill with increasing bismuth content is expected, however, the reduction in chill encountered at very high bismuth levels was not anticipated.

The addition of rare earth metals (REM) was found to offset the chilling effect of bismuth, and a reduction in the depth of chill was observed, Figures 4a,

*References pp. 390–391*

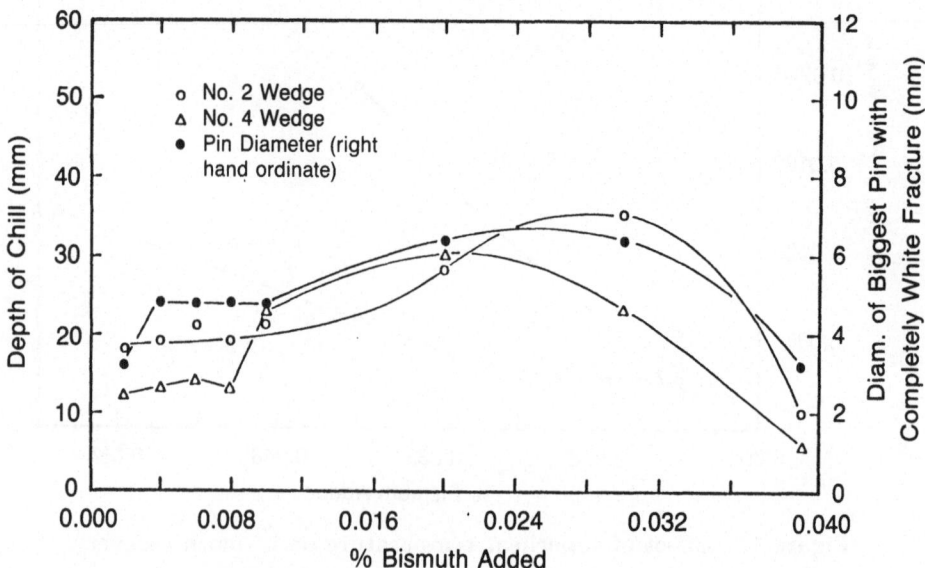

Figure 2. Effect of bismuth addition on chilling tendency (Set A).

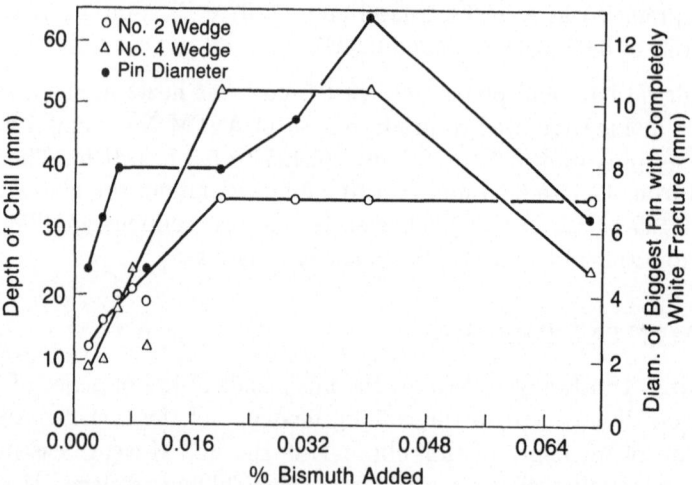

Figure 3. Effect of bismuth addition on chilling tendency (Set B).

b and c. The chill level obtained at 0.03% Bi in Set B was used as the basis for comparison of the wedge results depicted in Figures 4a and b. The depth of chill decreases with rare earth content and at about 0.07% mischmetal the effect of 0.03% bismuth seems to have been neutralized. At low sulfur levels, however, high rare earth contents increase the chilling tendency, Figure 4c. This is also evident when the fracture surfaces are examined, Figure 5. The gray fracture of the base heat becomes completely white with the addition of 0.03% Bi. At 0.1% S and 0.3% Bi, the wedge fracture is completely gray with a 0.07% mischmetal addition. At the low sulfur content, 0.01%, the fracture surface reverts to completely white at rare earth additions over 0.09%. When not balanced by sulfur, excess rare earth induce chill. This effect was also observed in the casting microstructures and casting properties.

**Casting Hardness**—The effect of increasing bismuth content on the Brinell hardness of a range of casting diameters is presented in Figure 6 (Set B). Note that the effect of bismuth on increasing chill is observed most significantly in the thin casting sections. The effect of rare earth additions to bismuth containing heats (C, D, E) is presented in Figures 7-9. An appreciable increase in the hardness is apparent with excess rare earths at a lower sulfur level (Set E, Figure 9).

**Tensile Properties**—Tensile test coupons were machined from 38.1 mm (1.5 in.) diam cylindrical bars to yield the data presented as a function of bismuth content in Figures 10 and 11. Relative hardness (HR), a value used to compare measured hardness to that calculated from the measured tensile strength, is depicted in these figures and is defined as:

$$HR = HB/[100 + 3.03 \times 10^{-3} \times UTS]$$

It may be noted that low levels of bismuth have caused a slight increase in the tensile strength of the gray irons, but that this is reduced markedly as the bismuth content exceeds about 0.004%. The tensile properties of the untreated gray irons were not fully recovered when rare earths were added, Figure 12.

**Microstructural Effects**—The variety of graphite morphologies which can exist in cast irons is well documented, and it is generally appreciated that these graphite shapes can be changed by progressively altering certain process parameters, e.g., solidification cooling rate, composition, melting conditions, inoculation, the presence of certain elements, etc. Accordingly, a continuous change can be expected in the "ladder" of graphite morphology: flake, undercooled flake, compacted, deteriorated forms of spheroidal graphite, and spheroidal graphite.

The base iron of this study contained typical Type A, randomly distributed flake graphite. The addition of bismuth introduced a mesh type graphite where the graphite was more "noodle like" than the characteristic "corn flake" shape of Type A graphite. In addition, Type B and Type D graphite appeared as a result of undercooling caused by bismuth. The increased surface area of the graphite formed was associated with an increase in the ferrite content of the matrix, and a resultant decrease in hardness and tensile strength.

Excessive rare earths effected a change in graphite from flake to undercooled types to compacted graphite and to the formation of some graphite spheroids.

*References pp. 390-391*

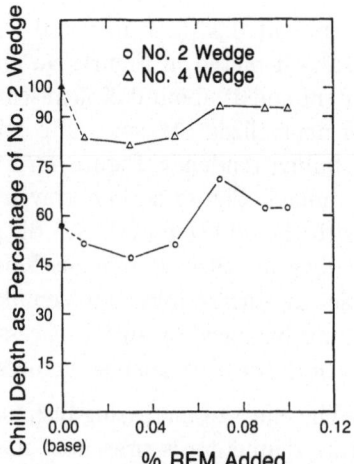

**Figure 4a.** Effect of rare earth addition on chill reduction (Set C).

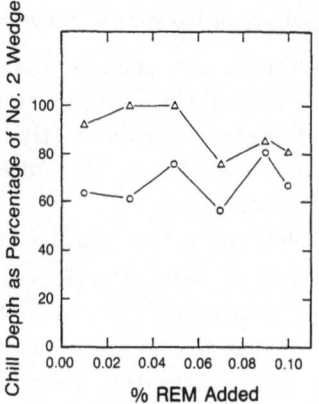

**Figure 4b.** Effect of rare earth addition on chill reduction (Set D).

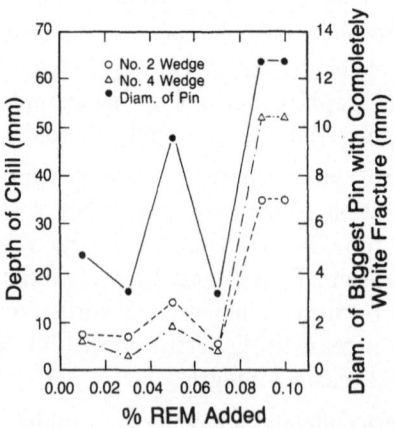

**Figure 4c.** Effect of rare earth addition on chill reduction (Set E).

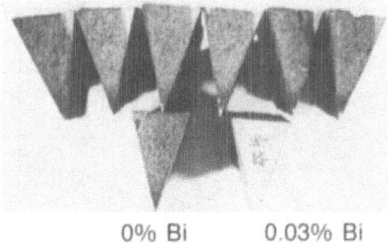

0% Bi    0.03% Bi

Figure 5a.  Top row: The neutralization of gray cast irons with 0.03% Bi and 0.10% S (Set C, Table 1) by the addition of rare earths, as illustrated by the ASTM No. 4 wedge fracture surface (from left: 0.01, 0.03, 0.05, 0.07, 0.09 and 0.10% rare earth addition). Bottom row: The effect of bismuth on the ASTM No. 4 wedge fracture surface. (Set B, Table 1)

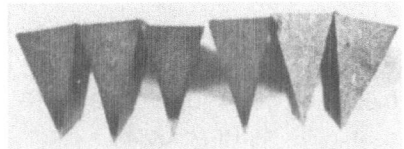

Figure 5b.  The neutralization of gray cast irons with 0.03% Bi and 0.01% S (Set E, Table 1) by the addition of rare earths, as illustrated by the ASTM No. 4 wedge fracture surface (from left: 0.01, 0.03, 0.05, 0.07, 0.09 and 0.10% rare earth addition).

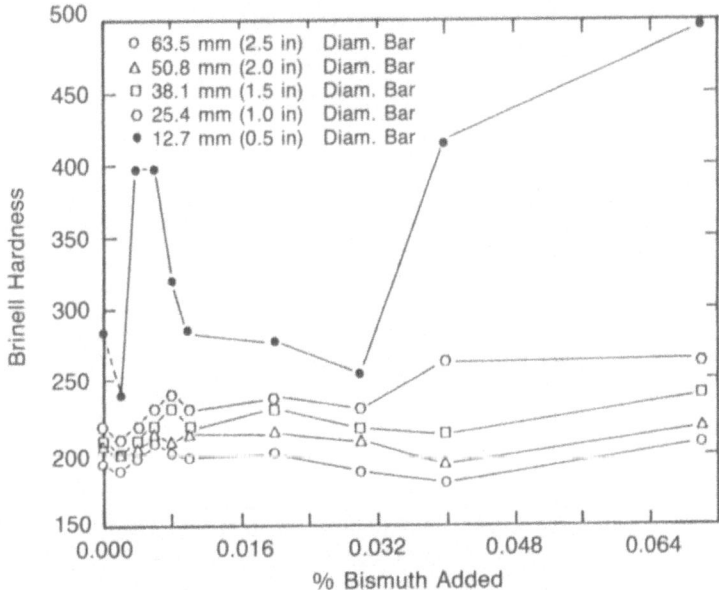

Figure 6.  Effect of bismuth addition on the hardness of cylindrical bars (Set B).

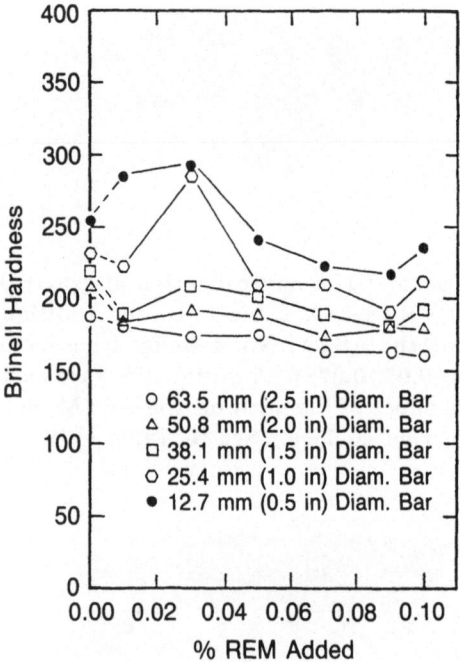

Figure 7. Effect of bismuth addition on the hardness of cylindrical bars (Set C).

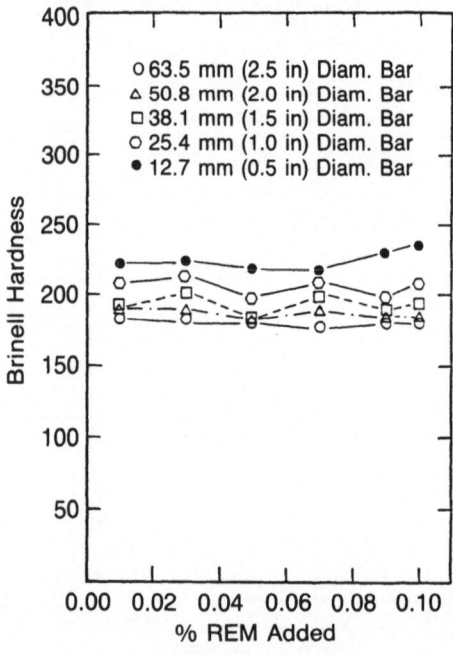

Figure 8. Effect of rare earth addition on the hardness of cylindrical bars (Set D).

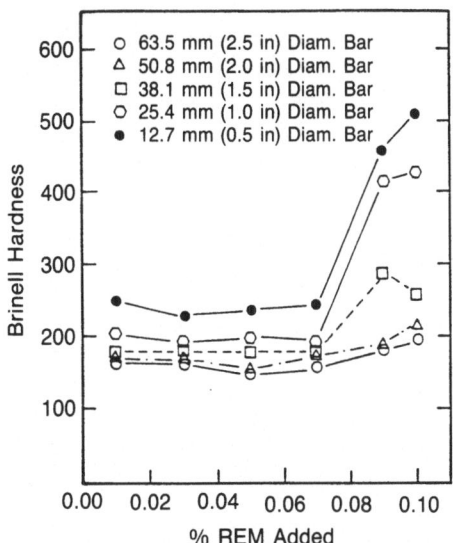

Figure 9. Effect of rare earth addition on the hardness of cylindrical bars (Set E).

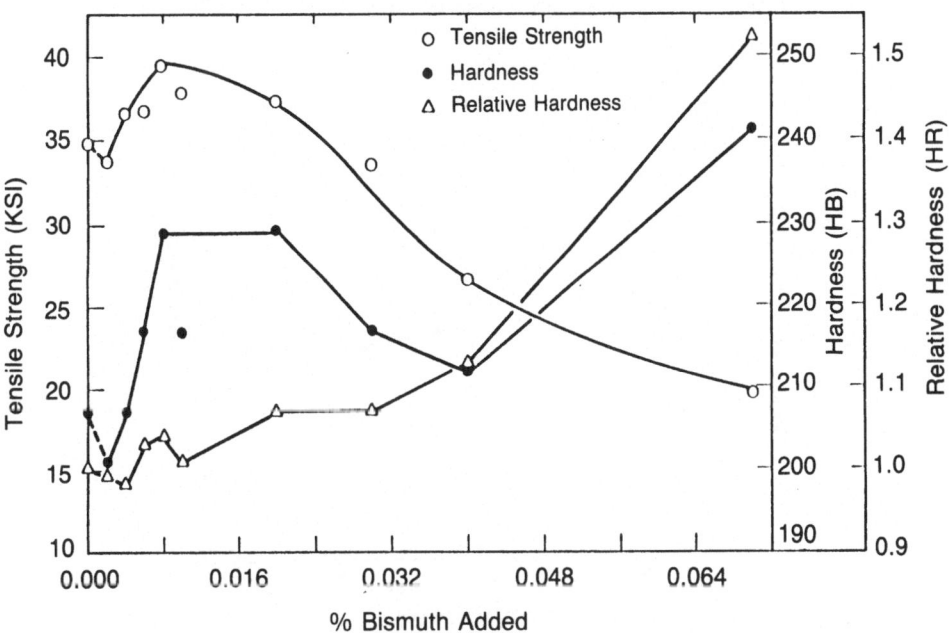

Figure 10. Effect of bismuth addition on the tensile strength, hardness and relative hardness of gray cast irons (Set B).

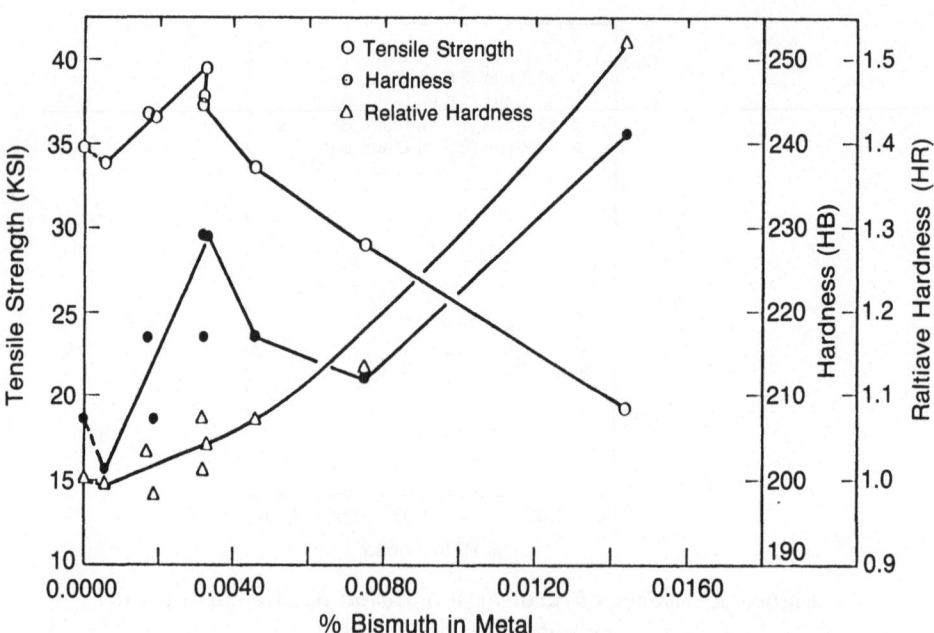

Figure 11. Effect of bismuth content in the iron on the tensile strength, hardness and relative hardness of gray cast irons (Set B).

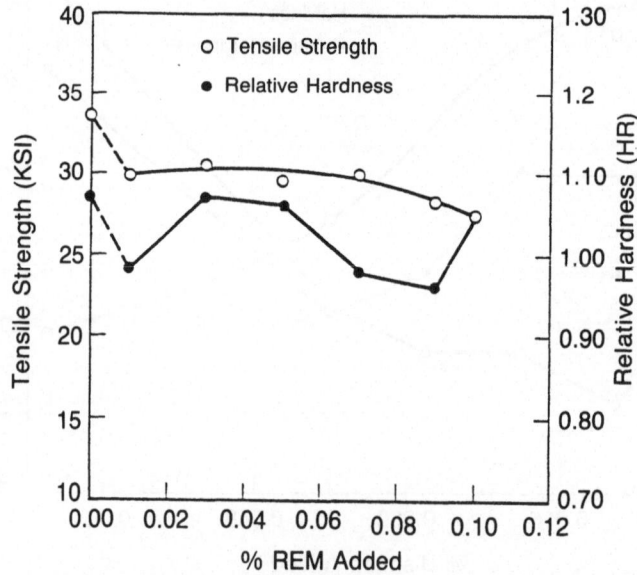

Figure 12. Effect of rare earth addition on the tensile strength and relative hardness of gray cast irons.

## DISCUSSION

On the basis of the observation of microstructures, hardness and tensile strength values of gray cast irons it was determined that the amount of rare earths added as mischmetal required to neutralize the effect of bismuth could be given by the following relationship:

$$\%RE = 0.013 + 4.18 \times \%Bi + 0.358 \times \%S$$

A nomograph expressing this relationship is presented in Figure 13. This is substantially less rare earths than would be required to form the most stable rare earth compounds of sulfur and bismuth, e.g., $Ce_2S_3$ and BiCe, in typical gray cast irons. The stoichiometric requirements for neutralization would be expressed by:

$$\%RE = 1.12 \times \%Bi + 2.91 \times \%S$$

The experimental data has shown that bismuth (in the range of about 0.03%) combines with a larger share of the rare earths, and sulfur (in the range of 0.10%) combines with a smaller share of rare earths than predicted stoichiometrically. This is further complicated by the realization that the sulfides would appear as complex oxysulfides, that mischmetal contains a substantial amount of lanthanum (about 30%) in addition to cerium (about 50%) and other rare earths (about 20%), and that the inclusions observed in the microstructure of these castings were complex rare earth-bismuth-sulfur compounds.

Both cerium and lanthanum (the principle constituents of mischmetal) are known to form a variety of compounds with bismuth. These are apparent in their respective phase diagrams, Figures 14 and 15. The melting points (congruent and incongruent) of the most stable of these phases are summarized below:

| | | | | | |
|---|---|---|---|---|---|
| BiCe | 1525 C | (2777 F) | $La_2Bi$ | 1252 C | (2286 F) |
| $Bi_3Ce_4$ | 1685 C | (3065 F) | $La_5Bi_3$ | 1350 C | (2462 F) |
| $BiCe_3$ | 1400 C | (2552 F) | $La_4Bi_3$ | 1670 C | (3038 F) |
| | | | LaBi | 1615 C | (2939 F) |

The heat of formation of rare earth-bismuth compounds has been estimated at $-251 \pm 84$ KJ/mol[17].

It must be recognized, however, that the most stable rare earth compounds are the oxides, and these are followed by oxysulfides, sulfides, rare earth-bismuth compounds, rare earth-nitrogen compounds, and rare earth carbides. The rare earths are effective desulfurizing agents, a compilation of the standard free energy of formation of some rare earth sulfides is presented in Table 3[18]. But, one would expect rare earth desulfurization of bismuth containing cast irons to result in the formation of oxysulfides, or complex oxysulfides prior to the neutralization of bismuth.

The inclusions present in the untreated gray irons were primarily manganese sulfides (or oxysulfides). On the other hand, those irons which were treated with

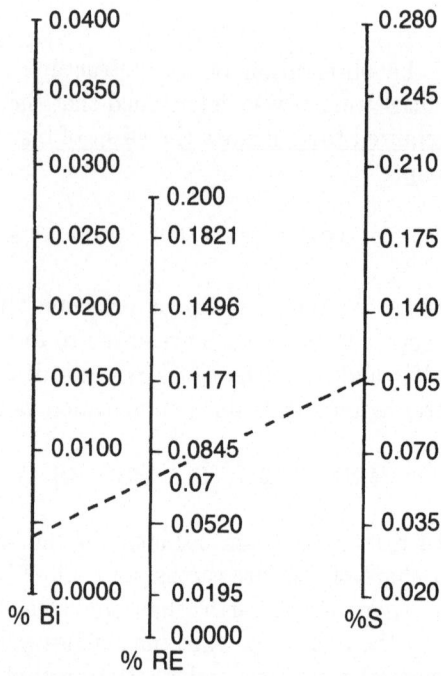

Figure 13. Nomograph to determine rare earth requirements to neutralize the effect of bismuth at various sulfur contents.

rare earths contained complex, duplex inclusions in the matrix consisting of a core which was rich in rare earths and an outer layer which was manganese rich. Sulfide inclusions were also observed associated with the graphite flakes. Typical compositions were as follows:

|  |  | % Mn | % Ce | % La | % S |
|---|---|---|---|---|---|
| Matrix: | Core | 16.89 | 27.43 | 6.87 | 19.25 |
|  | Outer | 52.71 | 5.74 | 1.77 | 26.27 |
| Graphite Flake |  | 7.61 | 14.37 | 3.02 | 12.15 |

While bismuth was observed by energy dispersive x-ray analysis in the inclusions within the flake graphite, this semi-quantitative analysis did not record that element in this compound. Electron microprobe traverses of the cast specimens revealed that both bismuth and rare earths were primarily concentrated in the graphite flakes, and that except for the sulfide inclusions found in the matrix detectable concentrations of these elements were not observed there. Inclusions could be identified by energy dispersive x-ray analysis with the SEM or using polarized light optical microscopy[17].

It is possible that the neutralization of bismuth by rare earths involves the chemical combination of rare earth with bismuth to form high melting point

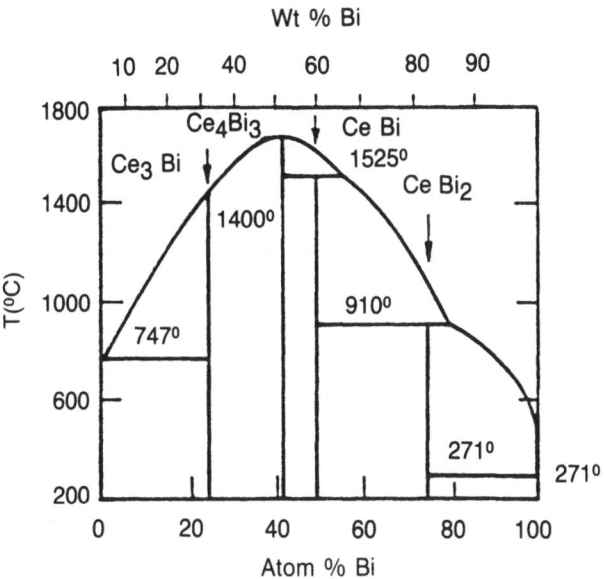

Figure 14. The bismuth-cerium phase diagram.

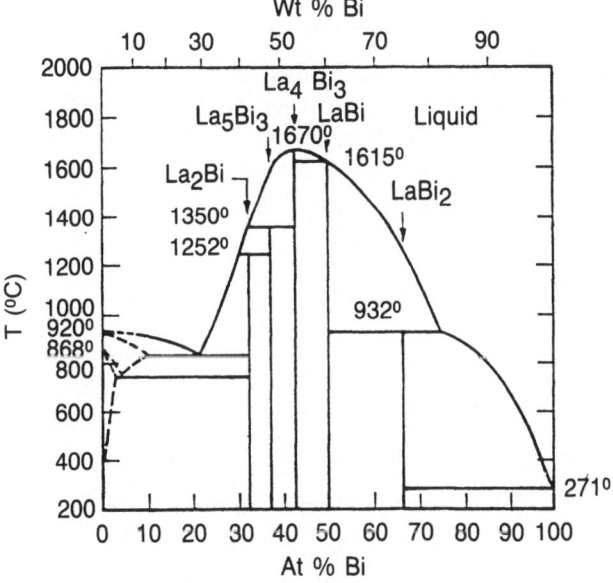

Figure 15. The bismuth-lanthanum phase diagram.

## TABLE 3

### Standard Free Energy of Formation of Some Rare Earth Sulfides[18]

Standard Free Energy of Formation

(kilojoules per mole of rare earth)

| Sulfide | 1100-2000K (1455-3075F) | 1827K (2829F) | Melting Point |
|---|---|---|---|
| LaS | -515.678 + 0.106T | -322 | 1971C (3580F) |
| CeS | -554.296 + 0.104T | -364 | 2099C (3810F) |
| $Ce_3S_4$ | -673.875 + 0.138T | -422 | |
| $Ce_2S_3$ | -734.710 + 0.159T | -444 | 2149C (3900F) |
| $Ce_2O_2S$ | | | 1950C (3540F) |
| MnS | | -145 | 1620C (2950F) |

intermetallic compounds which are dispersed throughout the structure. These compounds may then form sulfides which can serve as effective substrates for graphite nucleation and were observed within graphite flakes. Thus, the rare earths not only remove the bismuth from the iron, but provide increased graphite nucleation resulting in greater amounts of Type A graphite.

In addition to the complex thermodynamics of this type of neutralization, the kinetics of these reactions warrant further study. Rare earths have been shown to be effective in reducing chill and undercooled graphite in gray cast irons when added to the melt late in its processing, just prior to casting, but the effectiveness of this treatment after reasonable or delayed holding of the treated iron has not yet been evaluated.

## REFERENCES

1. E. K. Smith and H. C. Aufderhaar, "Effect of Adding Bismuth to Cast Iron," *The Iron Age*, July 9, 1931, 96–100.

2. A. L. Boegehold U. S. Patent Gazette No. 2, 1945.

3. T. H. Chu and L. C. Chiung, "The Influence of Bismuth on the Graphitization of Cast Iron During its Solidification," *Scientia Sinica*, VII, 10, 1958, 1061–1077.

4. J. Pelleg, "Influence of Bismuth on Mechanical Properties of Cast Irons," *AFS Trans.*, 72, 1964, 25–35.

5. R. A. Kukina and G. S. Timefeeva, "Bismuth in Gray Cast Iron," *Russ. Casting Prod.*, May 1974, 224.

6. C. E. Bates and J. F. Wallace, "Effects and Neutralization of Trace Elements in Gray and Ductile Irons," *AFS Trans.*, 75, 1967, 815–838.
7. C. R. Loper, Jr., "Tramp Elements in Cast Iron," *Fdry. M. & T.*, Oct. 1977, 57.
8. J. Sawyer and J. F. Wallace, "Effects and Neutralization of Trace Elements in Gray and Ductile Iron-II," *AFS Trans.*, 76, 1968, 386–404.
9. R. Wray, "The Occurrence of Trace Elements in Pig Iron and Their Effects on Iron Castings" *Castings*, 8, Nov. 1962, 19.
10. P. Basutkar, S. A. Yew and C. R. Loper, Jr., "Effect of Certain Additions to the Melt on the As Cast Dendritic Microstructure of Gray Cast Iron," *AFS Trans.*, 77, 1969, 321–328.
11. C. R. Loper, Jr., "The Solidification and Graphitization of Certain Cast Irons," Ph.D. Thesis, The University of Wisconsin-Madison, 1961.
12. C. E. Bates and J. F. Wallace, "Trace Elements in Gray Iron," *AFS Trans.*, 74, 1966, 513–524.
13. H. Morrough, "The Harmful Influence of Some Residual Elements in Magnesium Treated Nodular Cast Irons and Their Neutralization by Cerium," *BCIRA Jnl.*, April 1952, 292.
14. M. J. Lalich, "Effects of Rare Earths on Structure and Properties of Cast Iron," *Fdry. M & T*, March 1978, 118–129.
15. Z. Bofan and E. W. Langer, "The Mechanism of Interaction of Pb, Bi and Ce in Ductile Iron," *Scan. Jnl. Met.*, 13, 1984, 15–22.
16. D. M. Stefanescu and C. R. Loper, Jr., "Effect of Lanthanum and Cerium on the Structure of Eutectic Cast Iron," *AFS Trans*, 89, 1981, 425–436.
17. U. H. Udomon, "The Effects of Bismuth on the Structure and Mechanical Properties of Gray Cast Iron and its Neutralization with Rare Earth Metals," Ph.D. Thesis, The University of Wisconsin-Madison, 1986.
18. W. G. Wilson, D. A. R. Kay and A. Vahed, "The Use of Thermodynamics and Phase Equilibria to Predict the Behavior of the Rare Earth Elements in Steel," *Jnl. Metals*, May 1974, 14.

## DISCUSSION

**Ned Rehder** *(University of Toronto)*

Carl, was the melting done in an induction furnace or a cupola?

**A.** An induction furnace. In the cupola Bi really isn't a problem, except for the bismuth that's carried off in the effluent; it is not retained in the iron that's produced unless the bismuth levels are really high.

**G. J. Kipouros** *(General Motors)*

What was the composition of the rare earth metals that were used?

A. Conventional mischmetal was used to make the rare earth additions. The
analysis of this material was:

|  |  |  |
|---|---|---|
| 50.0 % Ce | 2.72% Fe | 0.0018% Ca |
| 46.99% Other rare earths | 0.14% Si | 0.1234% Mg |
|  |  | 0.018 % C |

**Charles Bates** *(Southern Research Institute)*

Professor Loper, hasn't there been a bismuth-containing inoculant patented and perhaps marketed? And why is bismuth added in the inoculant and then countered in this work?

A. Yes there is a specialized inoculant which contains bismuth, and that product is marketed to obtain increased nodule counts in ductile irons which have been treated with a rare earth containing alloy. This incoulant contains both bismuth and titanium, the amount of bismuth being very small and the addition of elements which can produce a counteracting effect when added to the melt, and where that counteracting effect induces nucleation. The mere presence of these same elements in the melt analysis is observed to be ineffective in this inoculation process.

# CHEMICAL IMPURITIES IN ALUMINUM

## J. H. L. VAN LINDEN, R. BACHOWSKI and R. E. MILLER

*Aluminum Company of America*
*Alcoa Technical Center*
*Alcoa Center, Pennsylvania 15069*

## ABSTRACT

The rapid growth of aluminum recycling has raised the awareness for chemical contamination in both the foundry and wrought product sectors. The spectacular success of UBC recycling was only possible because novel technology was developed to solve difficult contamination issues.

Potential and novel technologies, including premelt separation, separation during melting and postmelt impurity removal are discussed to set the stage for a new contaminant—namely lithium. It is quite likely that the growing use of Al-Li alloys for aerospace applications will increase the likelihood of Li contamination in other alloy systems. Existing evidence shows that serious adverse effects on molten metal quality will be encountered at low Li levels. Additional work is required to completely understand the effect of only a few ppm on the properties of the finished casting.

Potential methods of postmelting lithium removal are presented along with a potential in situ real time lithium monitoring technique.

## INTRODUCTION

The revolutionary growth of aluminum recycling during the last twenty years has exacerbated the chemical impurity problem. In other words more metal is being remelted more frequently, and consequently reexposed to contamination sources. If not addressed this results in an ever increasing level of contamination. For this reason, the issue of chemical contamination must be recognized and resolved in the major aluminum remelting operations involved with both foundry and wrought production.

Traditionally the aluminum foundryman has had to be aware of a number of these contaminating elements and their potential effects. These issues are generally well known in the industry as evidenced by the following examples: the adverse effect of excessive Fe on tensile properties of premium castings; sludging associated with high levels of Cr, Mn and Fe; the poisoning effect of several ppm of P on Na or Sr modification in hypoeutectic Al-Si alloys; and dross generation and embrittlement associated with low levels of Na in 5XX alloys.

The insidious nature of this problem is not only the concern of secondary aluminum remelters. The technical challenges posed by the ever increasing demand to recycle UBCs (used beverage containers) have dramatically demonstrated the extent of the problem to the primary producers, i.e., Alcoa, Reynolds, etc.

## UBC RECYCLING

From a metal quality control point of view, the UBC is a most undesirable raw material. It is comprised of two inseparable parts of different alloys, has a very high surface-to-volume ratio, is coated, and often carries with it a variety of organic and inorganic contaminants, i.e., plastic, paper, sand, iron, lead, brass, zinc, etc. The two alloys are 3004 (a low Mg, high Mn alloy used for the bodies) and 5182 (a high Mg, low Mn alloy used for ends). Obviously, if these two alloys are simply remelted as returned, the resulting composition will not be right for directly recycling back into either of the originals. This incompatibility is made even more difficult by the fact that the maximum acceptable Fe and Si levels are considerably lower for the end alloy than for the body stock.

If not done carefully, scrap processing steps themselves can significantly increase the problem. To appreciate this, it is necessary to trace the recycling process from the point of UBC receipt at the remelt location. First, incoming 400 kg bales are shredded, creating many fines which contribute to excessive skim generation. Next, a thermal delacquering process inevitably oxidizes some metal to further increase skim generation. The final step is the actual remelting, which due to the extremely low density requires a special charge method. Conventional melting techniques would yield unacceptably high skim generation combined with unacceptably low production rates.

Skim or dross is the term commonly used in the industry to describe the spongy floating mass on the surface of an aluminum melt. This seemingly "dry" substance often initially contains up to 95% metal entrained in a network of oxides, pigments ($TiO_2$), sand ($SiO_2$), etc. However, frequently this skim will ignite, generating extremely high localized temperatures promoting the reduction of $TiO_2$ and $SiO_2$ to elemental Ti and Si, and generating more oxides of aluminum. These elements (in addition to Fe) are commonly referred to as "tramps" as they are unaffected by normal recycling technology and, therefore, build to successively higher levels during each recycling circuit.

With these varied and nontrivial contamination challenges it is easy to appreciate the original skepticism about the viability of grand scale UBC recycling

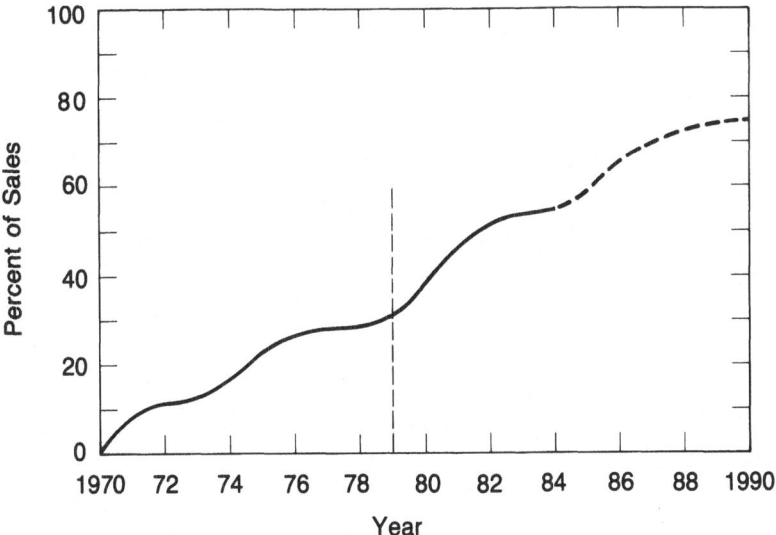

Figure 1. Cans recycled vs cans sold.

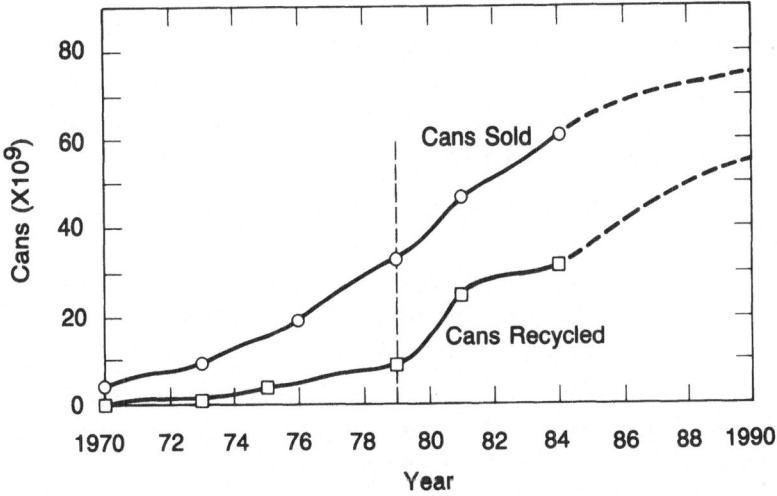

Figure 2. Growth of can sales and recycling.

into new can stock. As a result of Alcoa's 1979 recycling policy initiative, the rate of UBC return accelerated even faster than anticipated[1] (Figures 1 and 2). To deal with this influx, a model was created in 1980 to predict the limits of recycling UBCs back into the two original alloys, subject to the constraints of the above mentioned factors[2].

*References p. 406*

This computer simulation predicted several important items:

- The recycling rate for a unialloy can (body and ends from same alloy) would be limited by the accumulating silicon level.
- Without a technique for physically separating the alloys (two alloy can) before melting, the maximum attainable recycling rate is limited to 80%.
- With an 80% efficient alloy separation technique, there are no limits to the recycling rate.
- Skim fires should be promptly separated from the melt because titanium could very easily become the rate controlling element.

The UBC recycling success story stands as a reminder to the industry that novel technology can and has been developed to solve what seemed to be impossible recycling contamination problems.

## AVOID AND/OR REMOVE

In general, impurities are dealt with either prior to, during or after the melting event. Techniques employed throughout the industry on a variety of impurity types will be reviewed.

**Premelt Separation**—Scrap preparation techniques aimed at upgrading scrap quality range from very mundane to fairly sophisticated.

- *Simple screening* can effectively remove sand, clay and other fine dirt, provided the scrap does not contain too many metallic fines.
- *Magnetic separators* are routinely applied to prevent contamination from ferrous metals.
- *Air knives* can drastically reduce problems caused by small, dense, foreign nonferrous scrap, inadvertently mixed with aluminum scrap, e.g., lead tire weights, zinc, copper or stainless steel automotive parts.
- *Eddy current separators* are effective at removing nonmetallic material from a para- and/or diamagnetic metal stream. They are presently used in municipal refuse reclamation to recover small amounts of nonferrous metals from large refuse streams.
- *Heavy media separating processes* making use of characteristic density differences between the components of mixed scrap can create a number of highly concentrated scrap forms.
- *A thermal-mechanical separation process* has been developed to detach aluminum can ends from their bodies by closely controlling the temperature of a mechanically agitated UBC stream. Under controlled conditions, grain boundary melting occurs in one of the alloys, resulting in fragmentation into a statistically significant small size fraction which is separated by screening. This process can also be applied to upgrade certain wrought/cast alloy scrap mixtures.

It should be stressed that whenever possible an impurity-avoiding premelt treatment is the preferred approach. Inevitably, postmelting "cures" are not only more costly and complex, but often contribute detrimental side effects. Unfortunately, it is not always possible to prevent contaminants from finding their way into melts. Several methods can be employed to avoid or minimize contamination during the melting stage.

**Separation During Melting**—Remelting of today's lighter and dirtier scraps is increasingly effected in continuous or side bay melters. Their advantage is that charging is confined to a relatively small chamber which allows for concentrated treatment such as:

- *Molten salt flux additions.* A typical example is the use of halide fluxing in continuous UBC melting. In spite of all possible premelt cleanup steps, the scrap generally contains about 1 wt % $TiO_2$ (the paint pigment) and varying amounts of $SiO_2$ (trapped sand or glass bottle fragments). It has been demonstrated that additions of as little as 1/2% of a eutectic NaCl-KCl salt mixture with a small amount of fluoride can practically eliminate Si and Ti pickup from these sources.

- *Flotation.* Entrained oxide particles and other inclusions can be floated out with the assistance of inert gas fluxing either in the charge compartment, or in an adjacent chamber in the continuously recirculating melter. The nonmetallic particles attach to the surface of the rising gas bubbles, collect on the surface and are skimmed off.

- *Sweat melting.* This is the classic crude solid/liquid separation process for alloys which are inseparable in the solid state but have substantially different melting points. With the availability of more sophisticated temperature and process control technology this method could be refined to effect separation of aluminum alloys with different melting ranges.

These methods are obviously not effective at removing impurities introduced or formed in the melt independent of the conditions of the scrap, e.g., Si pickup due to reaction with container materials, or byproducts from alloying additions. In most cases, once the contaminant has entered the molten alloy in elemental form (either from dissolution or following a redox reaction), the removal problem becomes much more difficult.

### Postmelt Impurity Removal—

- *Reaction gas fluxing.* This well-known technique of bubbling chlorine or chlorine/argon mixtures through molten metal removes alkali metals. Applications of this principle include conventional furnace fluxing, side bay fluxing with or without downstream bed filters and the well-known Alcoa 622 process reactor (Figure 3). Since liquid salts may be formed (especially when magnesium is present) and since oxides are readily wetted by these molten salts, a new species of inclusion can be formed. Unfortunately this oxide/salt agglomerate is not readily

*References p. 406*

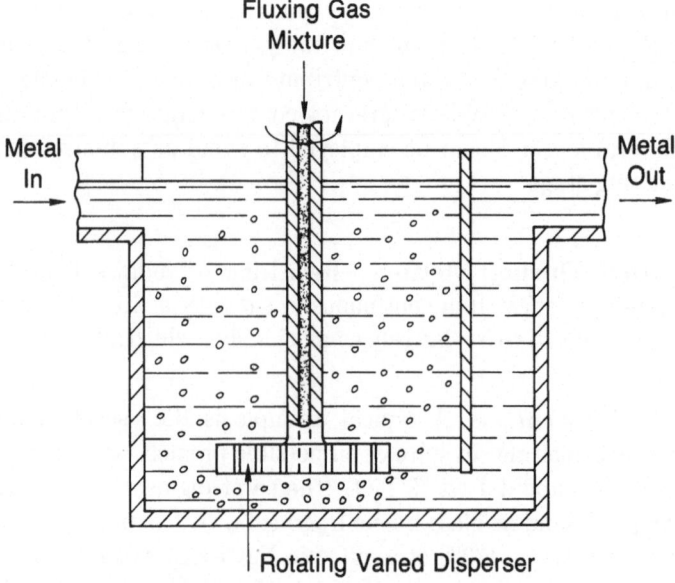

Figure 3. Alcoa 622 Process.

removed from the molten alloy stream by standard filtration systems. This is a classic example where solving one problem may create another.

- *Intermetallic compound formation.* If one or more of the undesirable elements is capable of forming a primary constituent, it is possible to precipitate, cultivate (grow) and collect it as a sludge, thus purifying the melt. Although aimed primarily at removing Fe (and Si), the principle can also be used to remove vanadium, titanium and zirconium by adding boron. The resulting borides are insoluble and can be removed. The drawback is that excess B is required for good results and it is possible to trade one contaminant for another by reaching an unacceptable boron level.

- *Solid salt reactions/complex formation.* This is a very attractive way to remove elements because the solid salt particles can be used either as a bed to provide large contact areas, or stirred in for an extended reaction time. Subsequently, the solid reaction products can be easily filtered for total removal. Generally, the environmental consequences are minimal because there is no off-gas to be scrubbed, and little spent reagent. A typical example of this method is the process of removing alkali elements from potroom metal by stirring $AlF_3$ particles into a melt containing low levels of Na, Li and/or Mg. In the case of sodium the reaction product would be $Na_3AlF_6$. Typically, the sodium content of a 3 ton crucible at 800°C can be lowered from about 20 ppm to <3 ppm in less than 10 minutes.

- *Electrolytic transport.* This technique relies on ionic transport of the most electropositive cation in the alloy to the cathode, and is generally accomplished using a molten salt layer as the transport medium. Major drawbacks to this process are the severe materials of construction/containment demands, and inherently high energy costs.

## LITHIUM—A NEW CONCERN

Today a relatively new contaminant is appearing on the horizon—lithium. During the last several years, lithium has become a very important alloying element in commercial aerospace alloys. A predictable consequence is the growing concern about potential lithium contamination in the scrap loops of other commercial alloy systems. The following examples illustrate that, although data are incomplete and experience is limited, there is good reason to be concerned about lithium.

1. Initial laboratory melts of 333 alloy were prepared with Al-Li scrap to provide 0.1-0.2% Li. In each case appreciably more dross formed on the Li melts than on the base alloy standard, and fluidity was reduced by 10% as determined by the standard spiral arm test. Permanent mold test bars cast from these melts showed a substantial increase in gas porosity which significantly reduced elongation. The only good news was that these Li levels provided a moderate degree of modification of the eutectic silicon particles.

2. In another set of trials, production size melts of 2024 alloy were prepared with sufficient additions of Li bearing scrap to reach a 0.03% Li level. Subsequent chlorine fluxing reduced the Li to 0.001-0.003%. Even at this level excessive oxide formed in the furnace and troughing, and vacuum density samples indicated a very high inclusion content. Rectangular rolling ingot cast by the DC process from these melts had a somewhat thicker oxide layer on the surface, but rolled normally after scalping. The rolled product showed no effect of the residual Li content on either mechanical properties or corrosion performance.

3. In a somewhat similar situation production melts of 5083 alloy were prepared with additions of lithium bearing scrap to reach a 0.014% Li level. Conventional metal treatment methods were then used to reduce Li content to 0.0003% before casting into DC ingots. During the cast thick layers of wet skim accumulated in the transfer system. In addition, a ceramic foam filter plugged very rapidly, indicating that the inclusion content of the metal was high. Although these ingots were not fabricated it is likely that this level of Li in a high magnesium alloy like 5083 is sufficient to generate excessive edge cracking during hot rolling.

These experiences dictate that caution must be exercised when dealing with potential Li contamination, since even very low concentrations were sufficient to cause excessive skim generation and high inclusion loadings. Increased hydrogen content and lower fluidity are also possible. The effect of Li contamination on mechanical properties is not quite so clear, since effects ranged from significant to nil depending on alloy and Li content. In general, these results agree with a recent Russian publication[3]. As indicated previously, considerably more work is required to fully appreciate the consequences of lithium contamination.

*References p. 406*

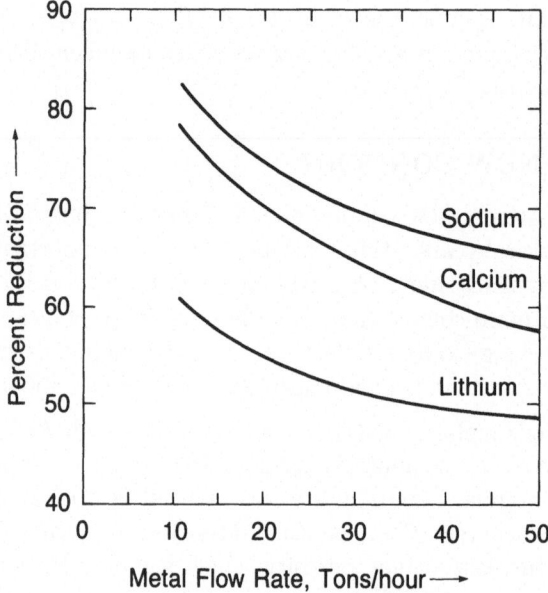

Figure 4. Single stage 622 — relative trace element reduction.

## LITHIUM REMOVAL FROM ALUMINUM ALLOYS

Lithium is an alkali element and as such can be expected to respond similarly to sodium with respect to metal purification technology. However, its higher heat of dissolution and stronger negative deviation from ideality in very dilute solutions suggests that removal at the ppm level will be even more difficult. This was confirmed in removal tests using the Alcoa 622 process (Figure 4). Its relatively high vapor pressure suggests (and calculations have confirmed) that vacuum removal can be a practical method. Lithium is a much smaller atom than sodium and has a strong tendency to intercalate in a variety of solid structures such as graphite. This property might be used to develop a selective lithium removal technique. Furthermore, lithium can be electrolytically removed directly from the alloy in a transport cell, or indirectly via halogenation and reduction in a salt bath. There may be other potential methods which can facilitate the recycling of aluminum-lithium alloys into traditional alloy systems, but in this paper only the more readily available techniques will be considered.

**Chlorination**—At the Alcoa Technical Center it has been shown that lithium can be selectively reduced from 1.6% to 0.001% (10 ppm) in Al-Li-Mg alloys, with good chlorination rates and high efficiencies[4]. Below that point appreciable amounts of $MgCl_2$ and $AlCl_3$ are formed. On the negative side are long reaction times, undesirable containment material reactions, and relatively low chlorination efficiencies near the end of the run. At this stage it appears that standard chlorination would only be viable as the first leg of a two step process.

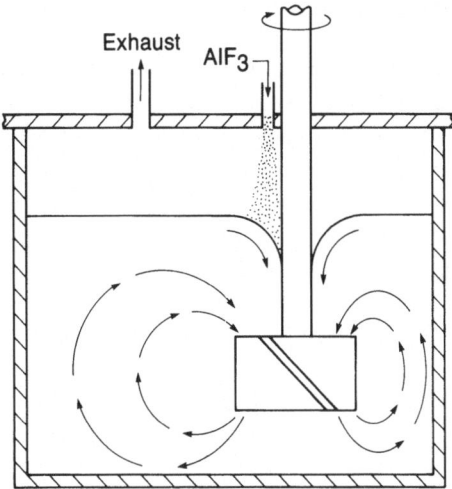

Figure 5. Schematic diagram of TAC Process.

**Carbonation**—It is known that $CO_2$ is an effective cover gas for protecting molten Al-Mg alloys against oxidation. Therefore, it seemed a logical candidate for protecting Al-Mg-Li alloys, but to the contrary, it proved to be a selective lithium removal agent. The reaction has been shown to be fast, especially at lower temperatures, i.e., < 750°C. For example, the lithium content of a 750 g Al-Mg-Li alloy was reduced from 1.8% to <0.001% in 15 minutes using 0.03 m³ of $CO_2$[5]. In that work no attempts were made to determine the feasibility of reducing the lithium any further.

**Reaction with Solid Halides**—An attractive removal method is one based on Alcan's TAC process in which $AlF_3$ powder is stirred into crucibles containing potroom metal with up to 20 ppm lithium[6,7,8] (Figure 5). The governing reaction is:

$$2AlF_{3(s)} + 3[Li] \rightarrow Li_3AlF_{6(s)}$$

The $Li_3AlF_6$ reaction product is skimmed off and recycled in the potroom for increased current efficiency. Although this process has some minor disadvantages (crucible cleaning is more difficult, the use of a mechanical stirrer, and process efficiency is inversely proportional with temperature) it appears to be cost effective and operationally and environmentally sound. The same process can be used for remelted scrap provided the lithium level is not too high (cost of $AlF_3$) and the magnesium level as low as possible (magnesium activity adversely influences lithium activity)[9].

*References p. 406*

## LOW LITHIUM LEVEL MEASUREMENT

Regardless of which removal method is selected, a matter of growing importance is the accuracy and speed of determining the residual lithium concentrations. At present, quantometer results at the ppm level are considered reliable within 50%, which is acceptable for most operating conditions. However, because it is a very time consuming process there is a great need for an in situ, real time lithium monitoring technique. One method showing promise uses a solid electrolyte to measure the potential of a simple electrochemical cell of the type

$$M(\text{ref})|M^+\text{conductor}|M(\text{metal})$$

where $M(\text{ref})$ is the reference material of known activity, $M^+$ conductor is an ionic conductor of $M^+$ ions and $M(\text{metal})$ is the activity of $M$ in solution in the metal. The potential across the cell is:

$$E = -\frac{RT}{ZF}\ln\frac{a_{M(\text{metal})}}{a_{M(\text{ref})}}$$

where $E$ is the potential to be measured, $Z$ is the charge, $F$ is Faraday's constant, $R$ is the gas constant, $T$ is the temperature and $a_{M(\text{metal})}$ and $a_{M(\text{ref})}$ are the activities of $M$ respectively in the metal and the reference. The relationship between the concentration and the activity of $M$ in the metal is given by

$$a_M = \gamma_M X_M$$

where $\gamma_M$ is the activity coefficient and $X_M$ the atom fraction. The proper choice of electrolyte and reference is crucial for successful measurements. Fray et al. developed a lithium probe based on the solid electrolyte $Li_{3.6}Si_{0.6}P_{0.4}O_4$ (actually a mixture of lithium orthosilicate $Li_4SiO_4$ and lithium orthophosphate $Li_3PO_4$[10,11]. The selected reference material was $Li_2Ti_3O_7$ (actually a two-phase mixture of $Li_4Ti_5O_7$ and $TiO_2$, which has a fixed Li activity when exposed to a fixed oxygen potential such as air).

Dubreuil and Pelton have proposed a probe consisting of lithium-beta-alumina ($Li_2O \cdot xAl_2O_3$) as solid electrode and a mixture of $Li_3AlF_6$ and LiF and alumina as reference material. Since lithium-beta-alumina is very hygroscopic, they decided to use sodium-beta-alumina for their experimental work. They demonstrated good agreement with the calculated Nernst slope for the relatively lithium rich range of 0.5-0.7%[12].

In order to confirm low level lithium removal a probe provided by Dr. Fray is being used in an experimental simulation of a multipass plug flow reactor (recirculating side bay system) using $CO_2$ as the fluxing gas (Figure 6).

The crucible is divided by baffle (4) into a "flux bay" and a "return zone." The apparent density in the flux bay is lower than in the return zone because of intimate gas mixing. Therefore, the level in the flux bay is higher than in the return zone

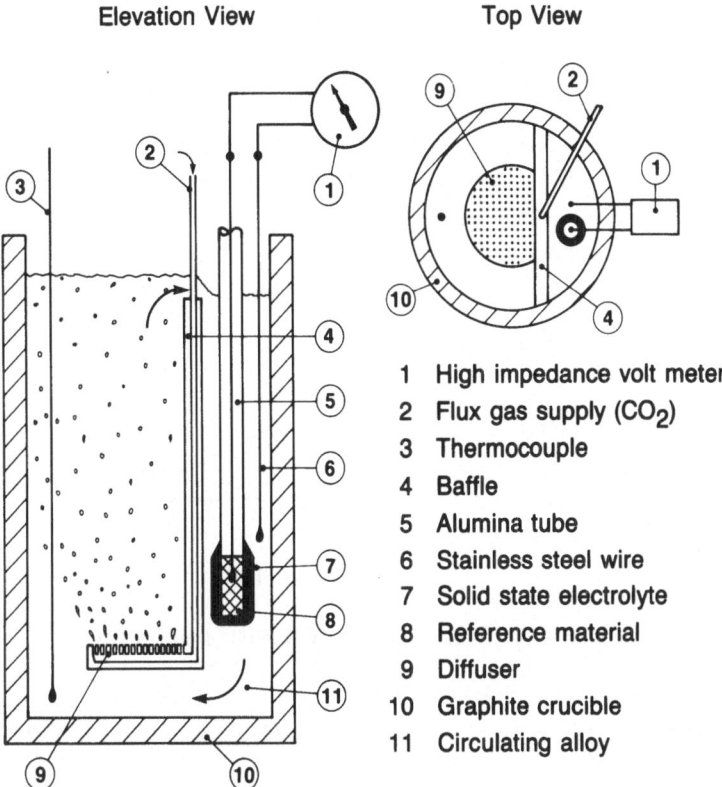

Figure 6. Test apparatus for lithium removal and activity measurement.

creating a continuous flow over the baffle as indicated by the arrows. The probe is placed in the return zone to avoid $CO_2$ interference at the electrode-metal interface.

The change in measured cell EMF represents total lithium removal rate. Since lithium has a significant vapor pressure at the test temperature and also tends to intercalate into the container material, there is some uncertainty as to whether the measured removal is achieved entirely by the reaction gas. To separate the effect of these "contributors" a control test is run simultaneously with metal of identical starting composition, but fluxed with dry argon at the same flow rate. Any lithium loss measured in the control test is then deducted from the results of the $CO_2$ test to determine the true effectiveness of the fluxing procedure.

## DISCUSSION OF TEST RESULTS

**Probe Performance**—Of the three probes provided by Dr. Fray, one cracked upon insertion in molten alloy and another probably failed soon after insertion judging from the drifting of the measured EMF. The third probe stabilized at 560 mV in pure aluminum, which is in agreement with Fray's measurements.

*References p. 406*

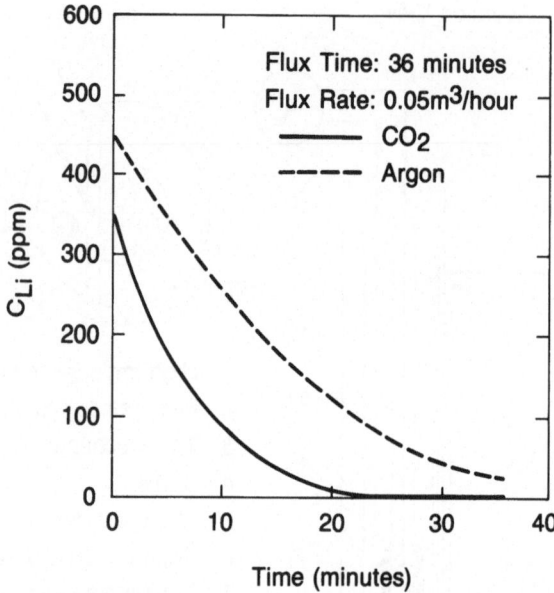

Figure 7. Lithium removal with gas fluxing.

The addition of lithium to create a 17 ppm level dropped the EMF unexpectedly to 430 mV.

Fluxing with $CO_2$ had the predicted effect of steadily increasing the EMF. Analysis of the limited number of samples taken during the run was in accord with the changing EMF values, which stabilized at 556 mV. Quantometer samples taken at that moment indicated that all lithium had been removed.

In spite of the fact that the probes have a mechanical weakness and that the combination of test conditions and limited measurements did not allow confirmation of Fray's data, or Nernst slope at the operating temperature of 760°C, the preliminary results are encouraging and warrant efforts to improve probe durability.

**$CO_2$ Fluxing**—Figure 7 compares the results of $CO_2$ and argon fluxing at the same flow rates, and identical sampling procedures. It shows clearly that lithium "loss" through inert gas fluxing and/or intercalation is significant, but that $CO_2$ fluxing is more effective and efficient.

In Figure 8 the comparison of two different flow rates of $CO_2$ is depicted. The test with a 40% lower flow rate needed approximately 40% more actual flux time until the quantometer samples indicated 0.0000% Li.

A point of interest was the behavior of magnesium during the argon/$CO_2$ comparison test. Starting at 200 ppm (about half of the lithium level) the magnesium level dropped steadily but slowly to about 100 ppm. In the $CO_2$ test, the lithium content is then 1 ppm and in the following sample neither element was detected. In the argon test, the magnesium level was 70 ppm at test end.

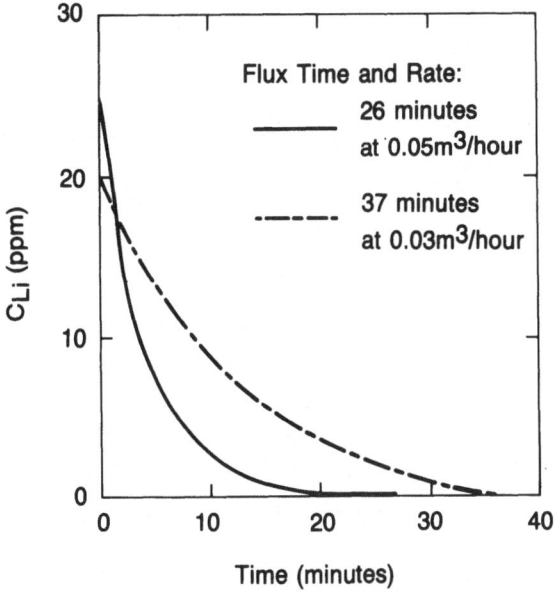

Figure 8. Low level lithium removal with $CO_2$.

This suggests that magnesium can be removed by $CO_2$ as well and that the presence of lithium retards the process. More work is needed to quantify the interaction of these elements.

In summary, $CO_2$ fluxing appears to be an attractive lithium removal method, considering that the gas is inexpensive compared to chlorine (A622) or $AlF_3$ (TAC), has no environmentally unacceptable byproducts, and the rate of removal is in the same range as the other processes. The only negative point is the relatively high skim generation. Preliminary tests indicate, however, that high metal recovery from the skim can be expected.

## SUMMARY

From an extensive literature review and careful analysis of several experiments, it is safe to conclude that lithium contamination can be a significant concern. Even at the several parts per million level, its presence has been shown to degrade properties and complicate conventional processing techniques.

Numerous separation or removal processes have been considered for possible application to the lithium contamination problem. Unfortunately most of the standard techniques must be ruled out because of the unique characteristics of lithium. However, several processes based on promoting halide reactions (gaseous or solid) have demonstrated a degree of effectiveness which argues strongly for further optimization specific to this problem. Unfortunately, each of the several variants has its own set of operational disadvantages and limitations.

*References p. 406*

A new removal technique utilizing $CO_2$ as a dynamic fluxing gas has shown encouraging preliminary results. Much more work must and will be done to further investigate, and perhaps develop a viable process utilizing this concept.

In conclusion, it bears repeating that in light of the extreme difficulty of effecting lithium removal after molten alloy contamination, every effort must be made to properly identify and isolate lithium bearing alloys *before* remelting for eventual recycling back into themselves.

## REFERENCES

1. J. H. L. van Linden, *Recycle and Secondary Recovery of Metals*, Met. Soc. of AIME, 1985 December, Warrendale, PA, 35–45.
2. J. H. L. van Linden and R. E. Hannula, *Light Metals 1981*, G. M. Bell, ed., Met. Soc. of AIME, Warrendale, PA 813–825.
3. A. S. Fedosov and A. V. Kurdyumov, *Tsvetyne Metally*, *12*, 1985 December, 62–63.
4. R. A. Christini, "Al-Li Chlorination Experiments", ATC Letter Report, 1986 June 18.
5. C. N. Cochran, "Oxidation and Removal of Lithium From Molten 2020 Alloy in Air, Oxygen and Carbon Dioxide", ATC Division Report 8-63-9.
6. P. Achim and G. Dube, *Light Metals 1982*, J. E. Anderson, ed., Met. Soc. of AIME, Warrendale, PA, 903–918.
7. G. Dube and V. J. Newberry, *Light Metals 1983*, E. M. Adkins, ed., Met. Soc. of AIME, Warrendale, PA, 991–1003.
8. B. Gariepy, G. Dube, C. Simoneau and G. LeBlanc, *Light Metals 1984*, J. P. McGeer, ed., Met. Soc. of AIME, Warrendale, PA, 1267–1279.
9. J. Kruger and F. Patak, *Erzmetall*, 36 (1983), Nr. 4, 186–191.
10. R. W. Hardeman and D. J. Fray, "Intercalation of Li from Al-Li Alloys Into $Li_2O \cdot TiO_2$", to be published.
11. P. C. Yao and D. J. Fray, *Met. Trans.*, *16B*, 1985 March, 41–46.
12. A. A. Dubreuil and A. D. Pelton, *Light Metals 1985*, H. O. Bohner, ed., Met. Soc. of AIME, Warrendale, PA, 1197–1205.

## DISCUSSION

**E. F. Ryntz** *(General Motors)*

Since Li is beneficial for forging alloys, consider leaving the Li in and casting the alloy in that condition. What are the properties of aluminum castings containing Li?

**A.** Alcoa is considering research on that subject, but we have no information now.

**J. Puckett** *(Nelson Metal Products)*

Have you or can you die cast Al-Li alloys?

**A.** Alcoa has not, and I don't know of anyone who has. I think it would have to be kept well degassed because it really picks up hydrogen, and consequently porosity was a problem in casting ingots. I don't know whether sticking would be a problem. Oxidation would be a major concern.

**Question**

I asked because you mentioned the rather rapid oxidation, and I wondered whether that would cause casting defects. You know Mg alloys can be die cast.

**A.** Al and Li metals are another world.

**M. D. Hanna** *(General Motors)*

Are you considering adding some elements to stabilize Al-Li alloys? For example, with Si a very stable Al-Li-Si compound forms which is corrosion resistant.

**A.** Not that I am aware of, although I am not in the alloy development area. How much Si is used?

**Hanna**

It depends on other things. You can use 5 or 6% or even up to the 12.6% eutectic, but Li must be about 1.5%.

**R. J. Fruehan** *(Carnegie Mellon Univ.)*

I'm not familiar with the term "skim." What is it? Did you analyze your skim?

**A.** Skim is the material left at the top of the melt after melting a solid charge. It is usually oxides.

**N. Jarrett** *(Alcoa)*

When skim is first taken off, it is about 90% Al metal. Some metal is easily extracted, taking it down to about 50% Al metal, balance oxide.

**Fruehan**

Did you analyze the skim for a lithium compound?

**A.** (Bachowski) Yes, after $CO_2$ fluxing, and we found no lithium compounds.

**Fruehan**

You found no Li in either the skim or melt?

**A.** That's right.

**C. R. Loper** *(Univ. of Wisconsin)*

When you ran the carbonization test, did you bubble the $CO_2$ through the melt, and for how long?

**A.** Yes. We bubbled a 25 kg melt for about 20 minutes and took periodic samples.

**I. L. Edwards** *(General Motors)*

Continuing the previous question, was the bubbling done with the 622 process or with a lance?

**A.** With a lance.

**Z. Zurecki** *(Air Products & Chemicals)*

Some explanation might be found in a patented Elkem process in which Al-Li alloy is heat treated under the right $CO_2$ atmosphere. There is another finding that $Li_2CO_3$ is extremely stable at the temperature of melt treatment. I think that injection of $CO_2$ into pure Al will simply produce alumina and graphite. Assuming that alumina forms on the bubble surface, it could further react with the dissolved Li, and that product would transfer Li to the surface where it could evaporate.

**G. K. Sigworth** *(Cabot Corp)*

In a paper that Thorvald Engh and I published in *Scandinavian Journal of Metallurgy* Vol. 11 (1982) p. 143–149 a thermodynamic explanation of your findings is offered. It predicts that a few percent of Li enters the alumina, forming something like beta alumina. Chemical analysis would reveal the Li, but x-ray diffraction analysis would show only alumina. By this mechanism Li can be removed to quite low levels. When we were writing the paper someone from Norsk Hydro named Franz Patak, I think, did some experiments similar to yours. So I think you were actually oxidizing the Li.

**D. R. Gaskell** *(Purdue Univ)*

To what extent would say 20 ppm of Li increase hydrogen solubility under one atmosphere of $H_2$ pressure?

**A.** We don't know. We would like to have that kind of data measured at universities. Our experiences of unusually great porosity suggest that Li increases hydrogen solubility.

**N. P. Lillybeck** *(Deere & Co.)*

Do you relate the edge cracking to hydrogen or something else?

**A.** That has been a question for many years. We find a good correlation between edge cracking and the alkali elements, but there is always some doubt about the interaction of alkali elements and hydrogen concentration. Excessive concentrations of the alkaline elements - Ca, Na, Li - will always cause edge cracking.

**M. D. Hanna** *(General Motors)*

Is Alcoa doing anything to remove other tramp elements, like Sb?

**A.** No.

# THE THERMODYNAMICS AND KINETICS OF GAS DISSOLUTION AND EVOLUTION FROM IRON ALLOYS

### R. J. FRUEHAN

*Metallurgical Engineering and Materials Science Department*
*Carnegie-Mellon University*
*Pittsburgh, Pennsylvania 15213*

## ABSTRACT

Evolution of hydrogen and nitrogen during the solidification of cast iron may be a cause of pinholes or porosity in the casting. The fundamental thermodynamics and kinetics of the absorption and desorption of the gases are reviewed. Recent experimental results on the effect of S, P, Bi, Sn, Pb and Te on the rate of the nitrogen reaction with carbon saturated iron are presented. Several of these elements are apparently surface active on the melts and retard the rate of the chemical reaction significantly. The nitrogen in the liquid metal may come out of solution during solidification causing porosity. Even though the solubility of nitrogen is higher in austenite than in the Fe-C melts containing more than 2% C, nitrogen may evolve during solidification because the carbon enriched liquid has a lower solubility for nitrogen. Hydrogen can contribute to the total gas pressure and the thermodynamic pressure for the combined effect of hydrogen and nitrogen is given. The possibility of removing nitrogen by argon bubbling was explored using a mixed control model and the rate of removal is estimated to be slow. However, it may be possible to remove hydrogen by this method.

## INTRODUCTION

Gas evolution during the solidification of cast iron may cause porosity in the casting. There are several possible causes of pinholes or blowholes in the casting. For example, water in the sand contained in the mold can react with dissolved carbon to form CO and $H_2$. This can result in what are usually referred to as blowhole type defects. Similarly, dissolved silicon or aluminum can react with water releasing hydrogen also causing porosity in the casting. Another possible cause of pinhole type defects is the evolution of dissolved gases, hydrogen and nitrogen, during solidification. A recent investigation by Katz and Landefeld[1] indicates that castings from iron produced in a cupola are nearly saturated with nitrogen and therefore nitrogen is of particular importance.

It is the purpose of this paper to review the fundamentals of the reactions of hydrogen and nitrogen with carbon-saturated iron. The solubility of hydrogen and nitrogen and the kinetics of gas metal reactions are reviewed. Recent experiments on the rate of the dissociation of nitrogen on carbon saturated iron melts are presented. In this work, the effect of sulfur, phosphorus, lead, tin, bismuth and tellurium on the rate were determined. The reactions that occur during solidification which may cause gas evolution are examined. Finally, the possibility of removing nitrogen or hydrogen from the liquid metal by argon bubbling is briefly explored.

## SOLUBILITY OF HYDROGEN AND NITROGEN

Hydrogen and nitrogen dissolve in liquid iron alloys as atomic species, according to the reactions,

$$\frac{1}{2} H_2 = \underline{H} \tag{1}$$

$$\frac{1}{2} N_2 = \underline{N} \tag{2}$$

The solubilities of both elements in liquid and solid iron alloys have been measured extensively and the results have been adequately summarized[2]. The solubility of nitrogen and hydrogen in liquid iron is decreased by carbon. The nitrogen and hydrogen contents in iron carbon alloys in equilibrium with one atmosphere pressure of the respective gases are shown in Figure 1 as a function of carbon content. The solubility of nitrogen at 1550°C decreases from about 450 ppm for pure iron to about 80 ppm for an alloy containing 4.5%C. The solubilities of hydrogen for the same metals are 24 and 10 ppm, respectively. The hydrogen solubility decreases with temperature and at eutectic temperature is 6.5 ppm; temperature has only a small effect on the solubility of nitrogen. Silicon decreases the solubility of nitrogen even further. For a 4%C-1%Si melt the solubility of nitrogen is about 75 ppm.

When the cast iron solidifies it forms austenite and graphite or cementite. The solubility of nitrogen in austenite[3] is also shown in Figure 1. For austenite containing about 1.8% C the solubility is about 150 ppm. Therefore when the liquid iron freezes the solubility in the solid is higher than the liquid; this interesting phenomena will be discussed later in detail. There is some solubility of nitrogen in cementite.[3] The exact amount is not known but is about the same as in austenite. The solubility of hydrogen in austenite[2] is about 7 ppm; therefore the solubility of hydrogen in the liquid and solid are about equal and there is a little segregation of hydrogen during solidification.

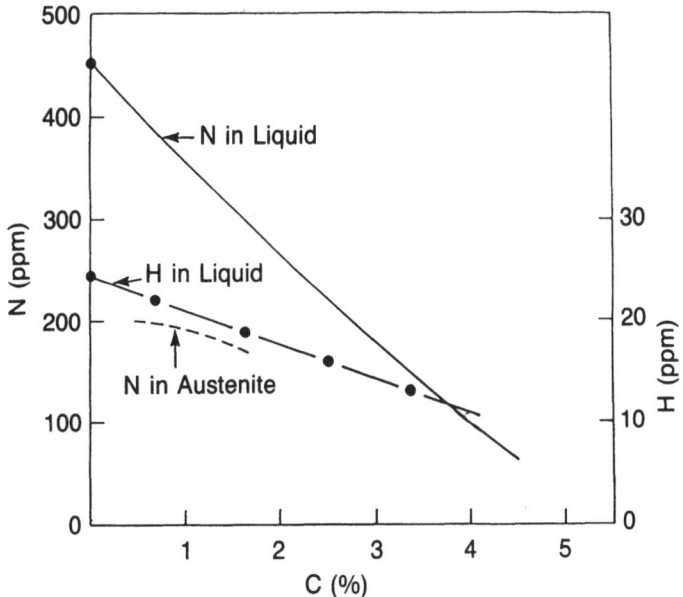

Figure 1. Solubility of hydrogen and nitrogen in Fe-C alloys at 1450°C in the liquid and 1225°C for austenite.

## KINETICS OF GAS-LIQUID REACTIONS

The absorption or desorption of hydrogen and nitrogen is controlled by one of the following steps:

1. Diffusion of the gas to the surface.
2. Chemical reaction on the surface.
3. Diffusion of the element in the liquid away from the surface.

For the desorption of a gas the same three steps are important but occur in reverse order. The process need not be controlled by only one of the above steps; it can be controlled by two or more steps in series. For liquid cast iron nitrogen is the more important of the gases and it will be considered in detail.

Gas diffusion is usually not rate controlling for nitrogen absorption from the atmosphere because of the pressure of $N_2$ and consequently the driving force for diffusion is high. Consequently one of the other steps is slower and rate controlling. However, for nitrogen or hydrogen removal by an inert gas, diffusion of $N_2$ or $H_2$ must be considered. The flux of nitrogen ($J_{N_2}$) away from the surface is given by,

$$J_{N_2} = \frac{m}{RT}\left(p^S_{N_2} - p^B_{N_2}\right) \quad (3)$$

where $m$ is the mass transfer coefficient, $p^B_{N_2}$ is the pressure of nitrogen in the bulk gas, and $p^S_{N_2}$ is the surface pressure of nitrogen in equilibrium with the melt where:

$$p^S_{N2} = K_N^2 f_N^2 (\%N)^2 \quad (4)$$

$K_N$ is the equilibrium constant for the nitrogen removal reaction and $f_N$ is the activity coefficient of nitrogen with respect to the 1wt% standard state.

However, as is the case for nitrogen absorption, gas diffusion for this reaction is generally faster than the chemical reaction or liquid phase diffusion and can be neglected in most cases.

The rate of the chemical reaction is controlled by the dissociation of the nitrogen molecule on the surface and the rate is given by,

$$\frac{dn_{N_2}}{dt} = k_B A (1 - \theta)(p_{N_2} - p^e_{N_2}) \tag{5}$$

where $k_B$ is the rate constant for pure iron, $A$ is the surface area, $p^e_{N_2}$ is the equilibrium nitrogen pressure given by an expression similar to Equation 4, and $(1 - \theta)$ is the number of vacant sites not occupied by surface active elements. Certain elements are surface active on liquid iron in that they lower the surface tension of iron by covering most of the surface. For example, oxygen and sulfur are surface active on iron and for bulk concentrations of 0.03% of oxygen or sulfur, over 90% of the surface sites will be covered by oxygen or sulfur. These elements therefore retard the rate of the chemical reaction. At high coverage by the surface active element $(1 - \theta)$ is inversely proportional to the activity $(a)$ of the element. Therefore the rate is given by,

$$\frac{dn_{N_2}}{dt} = \frac{k_B k_A A}{a}(p_{N_2} - p^e_{N_2}) \tag{6}$$

where $k_A$ is related to the adsorbtion coefficient of the element on the surface, the quantity $\frac{k_B k_A}{a}$ is the overall rate constant $k$. For dilute solution the activity of the surface active element is proportional to its concentration. Therefore, the overall rate constant is inversely proportional to the weight percent of the surface active element in iron.

If the rate is controlled by liquid phase mass transfer the flux of nitrogen atoms is given by,

$$J_N = m_N (C^B_N - C^S_N) \tag{7}$$

where $m_N$ is the liquid phase mass transfer coefficient of nitrogen, and $C^B_N$ and $C^S_N$ are the bulk and surface concentrations of N. The integrated form of Equation 7 is,

$$\ln \frac{\%N - \%N_e}{\%N_o - \%N_e} = -\frac{A \rho m_N t}{W} \tag{8}$$

where $\rho$ is the density of iron, $W$ is the weight of the metal and $\%N_e$ and $\%N_o$ are the equilibrium and initial nitrogen contents respectively.

Very often the rate is controlled by two processes in series; most often the chemical reaction and liquid phase mass transfer. The rate in this case can be obtained by equating the fluxes obtained from Equations 6 and 7 and solving for the surface concentration as shown by Fruehan et al.[4]

The rate of the nitrogen reaction with iron alloys containing oxygen, sulfur, chromium and other elements has been measured by many investigators and there is good agreement in most cases.[5-10] However, the rate for carbon saturated iron containing other elements was not investigated until recently by Tsukihashi and Fruehan;[11] this work is briefly reviewed below.

The rate was measured using an isotope exchange technique described previously in detail.[8] A 30 gram iron-carbon alloy containing varying amounts of alloying elements of interest was heated in an alumina crucible to 1450°C in an induction furnace. A gas mixture containing 1% of the nitrogen isotope $N_2^{30}$ in normal $N_2^{28}$ was jetted onto the surface of the melt. The concentration of $N_2^{30}$ in the gas was measured before and after the reaction with an inline mass spectrometer. The overall reactions occurring on the melt surface are given by,

$$N_2^{30} = 2N^{15} \tag{9}$$

$$N^{15} + N^{14} = N_2^{29} \tag{10}$$

Therefore, from a measure of the change in the $N_2^{30}$ concentration the rate of dissociation of the $N_2$ molecule, which is the rate controlling step for nitrogen absorption into iron, can be determined.

Iron alloys containing S, P, Pb, Sn, Bi, or Te were investigated. In order to avoid a decrease in the concentration of Bi and Pb, due to vaporization, 10 grams of a Ag-Bi or Ag-Pb alloy was also in the crucible to maintain a constant Bi or Pb activity. In the case of the alloys containing sulfur, its activity on the surface was maintained constant by having a small amount of the equilibrium $H_2$-$H_2S$ gas mixture with the reaction gas.

The rate constant for Fe-$C_{SAT}$-S alloys are presented as the reciprocal of the sulfur activity at 1450 and 1600°C in Figure 2. The activity coefficient of sulfur in carbon saturated iron is 6.3 relative to the 1 wt% standard state, therefore the concentration of sulfur is considerably lower. Sulfur decreases the rate as expected. The effect of carbon at constant sulfur activity shown in Figure 3 indicates that carbon does not affect the rate except in increasing the activity of sulfur. The effect of lead, phosphorus, bismuth and tellurium are shown in Figures 4 through 7 respectively. Small amounts of Pb, Bi and Te greatly reduce the rate. As little as 50 ppm Te reduces the rate by 90% and 50 ppm Bi by 80%. Tin did not affect the rate significantly.

The results indicate a very small amount of sulfur, lead, bismuth or tellurium can reduce the rate of the nitrogen reaction significantly. This information can be helpful in controlling the nitrogen reaction. For example, if it is desired to remove nitrogen by argon gas flushing, the concentrations of these elements should be as low as possible. On the other hand, it may be possible to retard nitrogen evolution during solidification by the deliberate addition of these elements.

*References p. 422*

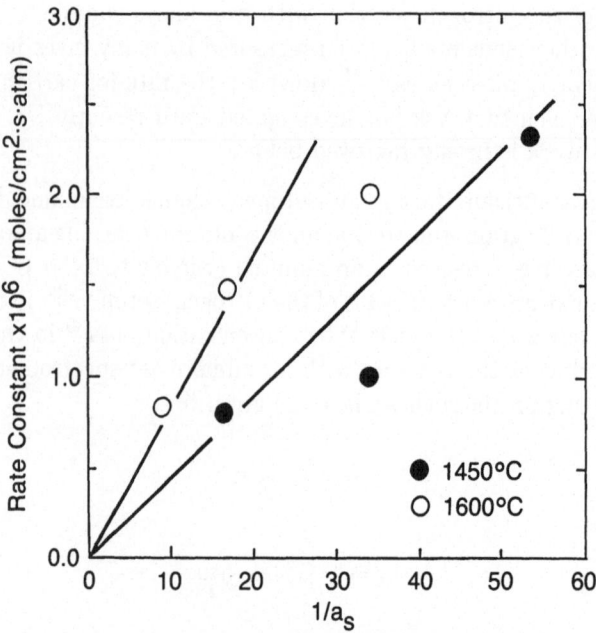

**Figure 2.** The effect of sulfur on the rate of the nitrogen reaction in Fe-$C_{SAT}$-S alloys.[11]

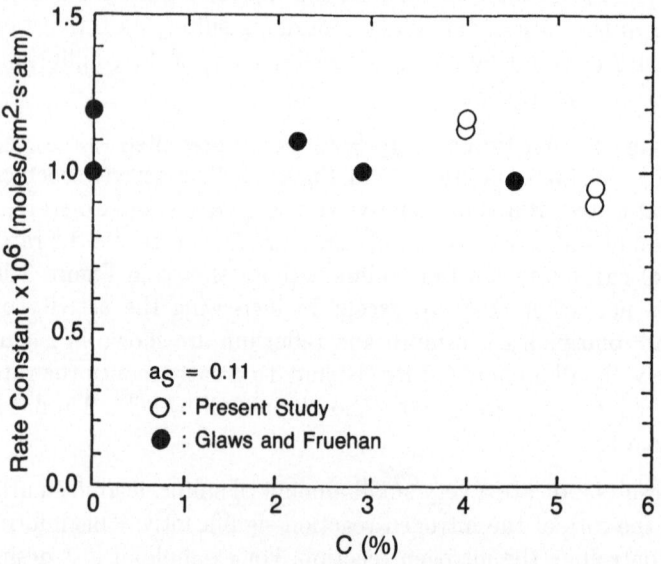

**Figure 3.** The effect of carbon on the nitrogen reaction in Fe-C-S alloys at constant sulfur activity.[11]

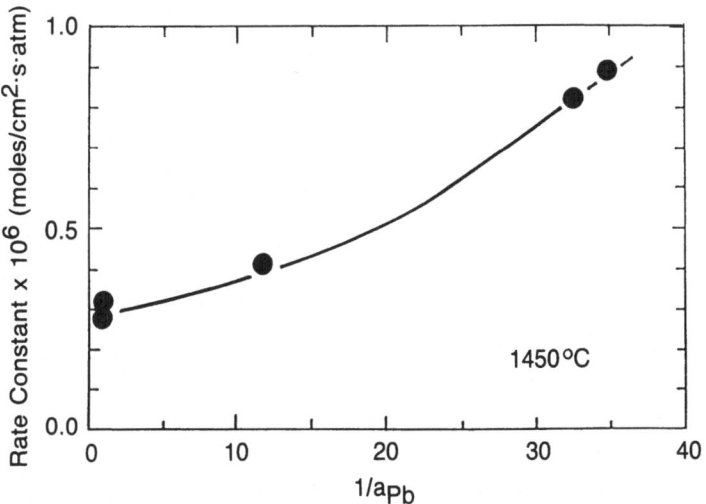

Figure 4. The effect of lead on the nitrogen reaction with Fe-$C_{SAT}$-Pb alloys at 1450°C.[11]

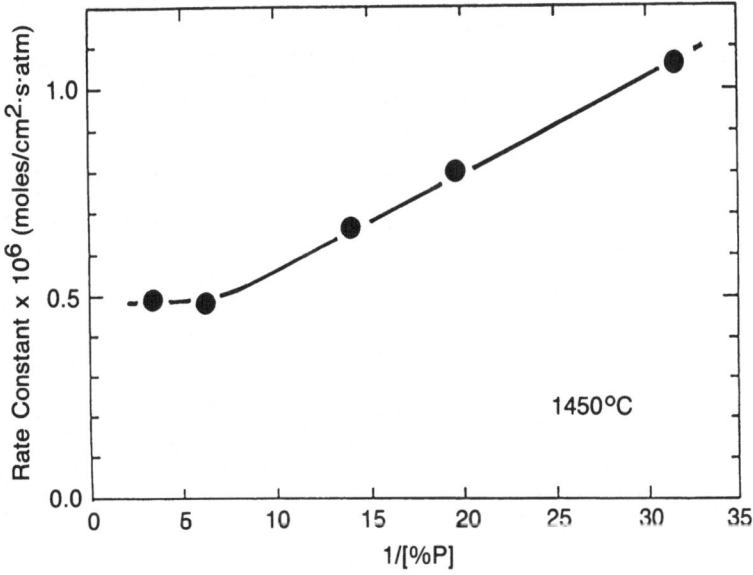

Figure 5. The effect of phosphorous on the nitrogen reaction with Fe-$C_{SAT}$-P alloys at 1450°C.[11]

## REACTIONS DURING SOLIDIFICATION

During the solidification of most simple iron alloys there is enrichment of the alloying element because there is greater solubility in the liquid than in the solid. The difference in the solubilities is expressed by the partition ratio ($k^{S/L}$) which is

*References p. 422*

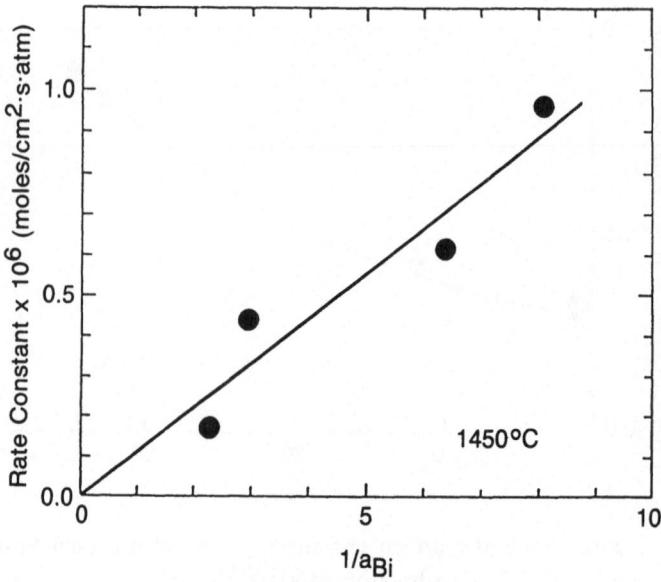

**Figure 6.** The effect of bismuth on the nitrogen reaction with Fe-$C_{SAT}$-Bi alloys at 1450°C.[11]

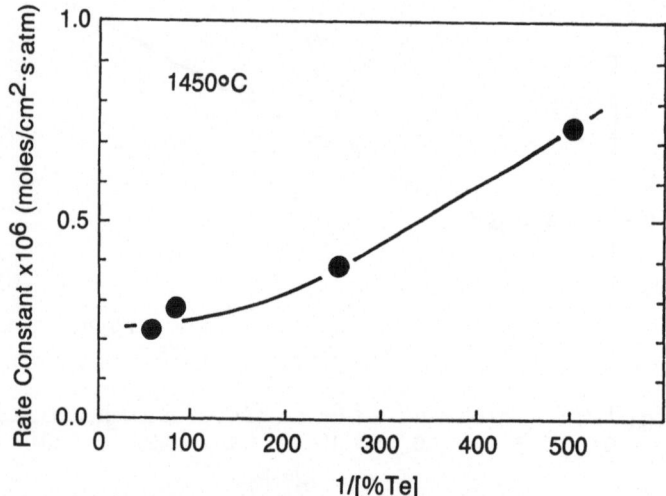

**Figure 7.** The effect of tellurium on the nitrogen reaction with Fe-$C_{SAT}$-Te alloys at 1450°C.[11]

determined from the slopes of the solidus to liquidus lines on the phase diagram. For most simple alloys the partition ratio is less than one, resulting in liquid enrichment during solidification. For example for an Fe-0.02%N alloy the last portion of liquid to solidify will be enriched and will have a concentration of 0.045%N which is the solubility of nitrogen in liquid iron at one atmosphere.[12] Therefore, even though the liquid alloy initially is far from being saturated with nitrogen, during solidification the concentration will increase to the point where the solubility is exceeded and nitrogen gas will be evolved.

Kagawa and Okamoto[3] have measured the partition ratio for nitrogen in iron alloys of high carbon content. They found the partition to be 1.9 and 2.2 for stable and metastable eutectic solidification indicating there is greater solubility in the solid than the liquid. Therefore in this case, as the alloy solidifies, nitrogen is actually enriched in the solid and the nitrogen content decreases in the liquid. It therefore may appear that since there is no enrichment during solidification, the thermodynamic pressure of nitrogen cannot increase. The thermodynamic pressure is defined by an expression similar to Equation 4. However, since carbon greatly increases the activity coefficient of nitrogen in the liquid and decreases it's solubility, the thermodynamic pressure may increase since the carbon content of the liquid is increasing during solidification.

For example, for an alloy containing 3.8% C that is saturated with 1500°C with nitrogen (110 ppm) just prior to eutectic solidification, the concentration in the liquid will decrease to 97.5 ppm due to enrichment in the austenite. However, the solubility of nitrogen in the liquid has decreased to 90 ppm primarily due to the increase in the carbon content. Consequently, the thermodynamic pressure of nitrogen exceeds one atmosphere and the evolution of nitrogen may cause a pinhole.

When considering the possibility of pinhole formation, one must consider the total thermodynamic pressure of all the gases and if the total pressure ($p_T$) exceeds one atmosphere, a gas pinhole may result. For example, for hydrogen and nitrogen the total pressure is given by,

$$p_T = K_H f_H^2 (\%H) + K_N^2 f_N^2 (\%N)^2 \tag{11}$$

where the $K_i$ is the equilibrium constant for the gas reaction and the $f_i$ is the activity coefficient. The concentrations will be that at the end of solidification. Even a small amount of hydrogen may be important. For example, consider a Fe-3.8%C alloy containing only 4 ppm hydrogen. The hydrogen content in the liquid will increase slightly, due to the enrichment, to about 4.5 ppm, just prior to the eutectic reaction. However, even for this small amount of hydrogen, it's thermodynamic pressure is 0.4 atm. Therefore, if the thermodynamic pressure of nitrogen exceeds 0.6 atmosphere a gas pinhole may result. Calculations indicate that as low as 80 ppm nitrogen in the iron with 4 ppm hydrogen may have a gas pressure exceeding 1 atm.

Figure 8 is a plot of the total pressure of hydrogen and nitrogen just prior to final solidification equal to one atmosphere as a function of bulk nitrogen and hydrogen contents in an Fe-3.8%C alloy. If the hydrogen and nitrogen contents are such that they are below the line the pressure will not exceed one atmosphere and a pinhole due to gas evolution will not occur; if above the line and the pressure exceeds one atmosphere, a pinhole from this source is possible. For pinholes to develop it is also necessary to nucleate a bubble.

*References p. 422*

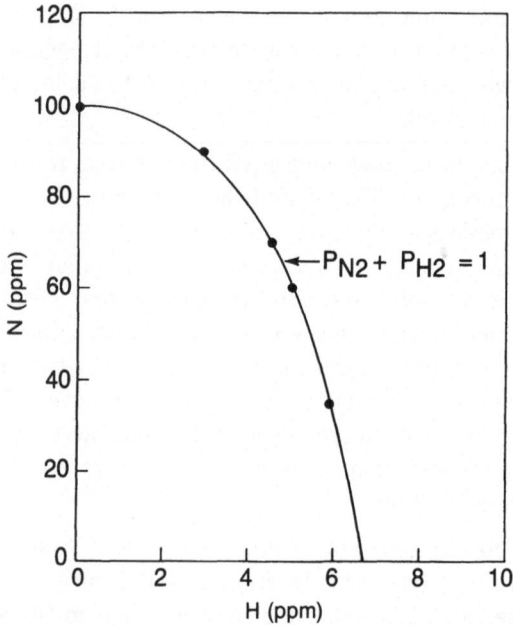

Figure 8. The individual thermodynamic pressures of nitrogen and hydrogen for an Fe-3.8%C alloy at the end of solidification where the combined pressures are equal to one atmosphere.

## NITROGEN AND HYDROGEN REMOVAL BY INERT GAS FLUSHING

The preceding discussion indicates that nitrogen and hydrogen evolution during solidification may be a cause of pinholes in castings. Therefore, it would be desirable to remove part of these gases if possible. One possible method is by argon bubbling in the desulfurization reactor or ladle. Desulfurization in the foundry is carried out in both continuous and batch modes. However, for the present calculation a batch reactor will be assumed. Since this is only an order of magnitude calculation this assumption is not too unreasonable for either mode.

When argon is bubbled through the melt, nitrogen atoms combine on the bubble surface to form $N_2$. The rate is controlled by the chemical reaction on the surface and liquid phase mass transfer of nitrogen to the surface in series. For the purpose of the present calculations a 10 tonne reactor, one meter deep, an argon flow rate of 0.28 $Nm^3$/min (10 scfm) through the melt, an initial nitrogen content of 100 ppm and 6 cm diameter bubbles were assumed. The rate constant for iron containing a relatively small amount of sulfur is about $1.5 \times 10^{-6}$ moles/$cm^2 \cdot s \cdot$ atm and the mass transfer coefficient ($m_N$) is about 0.1cm/s. The gas velocity, retention time and surface area were estimated as done previously.[4] Equations 6 and 7 were used to obtain the results given in Figure 9. The calcu-

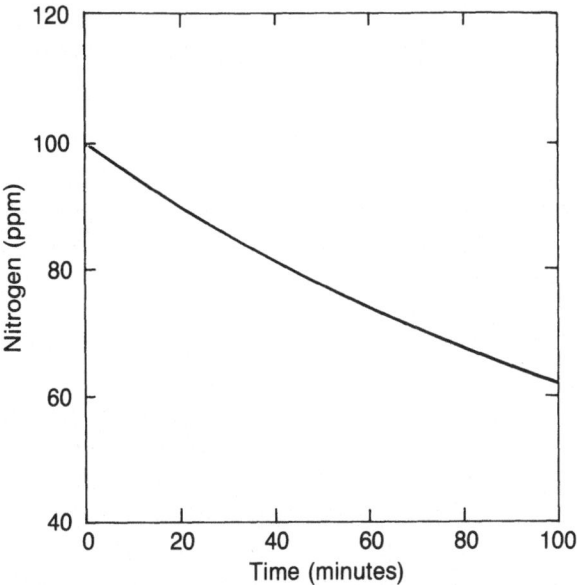

**Figure 9.** The rate of removal of nitrogen from Fe-3.8%C alloy with argon bubbling at 0.28 Nm³/min (10 scfm) in a 10 tonne reactor.

lations indicate that the rate is truly mixed control as indicated by the surface concentration being between zero and the bulk concentration. The results indicate it would take about 40 minutes to remove 20 ppm; if the argon flow was doubled it would still take over 20 minutes to remove 20 ppm. Whereas these calculations are rather crude, they indicate it would be difficult to remove nitrogen by argon bubbling.

For hydrogen the chemical reaction and mass transfer is considerably faster than for nitrogen. If equilibrium between the metal and the gas bubbles leaving the system is assumed it is possible to calculate the amount of hydrogen that can be removed by gas bubbling from thermodynamic considerations alone. The development of the equations are similar to those previously developed for steel.[12] The equation takes the form,

$$\frac{1}{[H]} - \frac{1}{[H_o]} = k_H \, V$$

where [H] and [H$_o$] are the respective hydrogen content after bubbling and the initial hydrogen content and $V$ is the total volume of gas used per tonne of metal. The constant $k_H$ is related to the solubility of hydrogen and therefore depends on the alloy composition. For example, 10 tonnes of an Fe-4.5%C melt at 1500°C containing 4 ppm hydrogen bubbling argon for twenty minutes at 0.28 Nm³/min (10 scfm) will reduce the hydrogen content to 1.6 ppm. The above calculation assumes equilibrium and therefore the fastest rate possible and may be an over estimate of

*References p. 422*

the amount of hydrogen removed. However, the calculation does indicate that it is possible to remove significant amounts of hydrogen by gas bubbling.

## SUMMARY AND CONCLUSION

Evolution of nitrogen and hydrogen during solidification can be a cause of pinholes in castings. Carbon decreases the solubility of hydrogen and nitrogen in iron alloys. The rate of absorption or desorption of nitrogen is controlled by the dissociation or formation of the nitrogen molecule on the surface. The rate of the reaction is reduced by sulfur, bismuth, lead tellurium and phosphorus while tin has only a small effect. Even though the solubility of nitrogen is higher in austenite, than in the liquid iron containing more than 2% carbon, gas evolution can occur during solidification. As the carbon content in the liquid increases during solidification the solubility of nitrogen decreases and nitrogen may be evolved. Hydrogen may also have a signficant effect on the possibility of pinhole formation. Calculations indicate that the rate of removal of nitrogen by argon bubbling is controlled by liquid phase mass transfer and the chemical reaction in series and that the removal of nitrogen by this method will be slow. However, gas bubbling may remove a significant amount of hydrogen.

## ACKNOWLEDGEMENT

The author wishes to thank the Center for Iron and Steelmaking Research at CMU, its member companies and the National Science Foundation (Grant 8421112) for support of the experimental work.

## REFERENCES

1. S. Katz and C. Landefeld, Private Communication, General Motors Research, Warren, Michigan.
2. *Making, Shaping and Treating of Steel*, United States Steel Corporation, Pittsburgh, Pennsylvania 1979.
3. A. Kagawa and T. Okamoto, *Trans. of Japan Institute of Metals*, 22, 2, 1981, 137.
4. R. J. Fruehan, B. Lally and P. C. Glaws, *Fifth International Iron and Steel Congress Process Technology Proceedings*, Washington, D.C., 1986, ISS of AIME, 339.
5. R. D. Pelke and J. Elliott, *Trans. TMS-AIME* 227, 1963, 894.
6. M. Inouye and T. Choh, *Trans. ISIJ*, 8, 1968, 134.
7. R. J. Fruehan and L. J. Martonik, *Metall. Trans. B*, 11B, 1980, 615.
8. P. C. Glaws and R. J. Fruehan, *Metall. Trans. B*, 16B, 1985, 551.
9. P. C. Glaws and R. J. Fruehan, *Metall. Trans. B*, 17B, 1986, 317.
10. M. Byrne and G. R. Belton, *Metall. Trans. B*, 14B, 1983, 441.
11. F. Tsukihashi and R. J. Fruehan, submitted to *Trans. JISI*.
12. R. J. Fruehan, *Ladle Metallurgy, Principles and Practices*, ISS of AIME, 1985.

## DISCUSSION

**R. D. Pehlke** *University of Michigan*

I enjoyed your presentation very much, and I intend to rely on it as an introduction to my presentation this afternoon. I would challenge you on one point. That is, your last suggestion that inert-flush degassing looks promising for hydrogen. This stems from your assertion that the effectiveness of inert-flush degassing can be gaged by guidelines based on equilibrium calculations. Several years ago we studied inert-flush degassing of aluminum using argon. We found the rate was extremely slow.

Considering iron melts, the rate of degassing will be governed by the rate of hydrogen diffusion to the gas bubble. This is a very slow process despite the large diffusion coefficient for hydrogen, because the concentration gradient in the liquid metal, adjacent to the bubble is very small, perhaps a few parts per million. On this basis, I think that inert-flush degassing for removal of hydrogen, particularly as you get down into the lower concentration levels, really doesn't have much promise.

**A.** You think degassing is going to be limited by diffusion in the liquid boundary layer. That's possible. I should qualify our curve, it represents the very fastest possible rate and as such, it looks promising. In flushing hydrogen out of steel, the equations developed hold reasonably well. I think it would be of interest for someone to examine the situation with respect to cast iron. However, I agree with you, liquid phase mass transfer could be a problem. It should also be noted that argon flushing of aluminum for hydrogen is a common practice and therefore the rates must be reasonable.

**K. W. Lange** *(Technical University of Aachen)*

I would like to say that scavenging dissolved gases is a discouraging business. I don't think it can be accomplished as easily as you suggest. In the case of hydrogen scavenging the bubbles are not saturated. There is a process that might be helpful. If you decrease the pressure over the melt the flushing action is increased, nearly in proportion to the decrease in the outer pressure. In addition, you can more easily nucleate bubbles of the dissolved gases. If CO is generated, as it is for steel, there will be additional scavenging due to the additional gas flow.

**A.** I agree with your comments, but if we're going to be concerned by liquid phase mass transfer as a possible rate controlling mechanism for the hydrogen removal as Professor Pehlke indicated, then changing the pressure will not help the process.

**K. W. Lange**

At least the amount of gas you need to get a certain flushing action is decreased. If the same amount of gas is used, the bubble volume and the interfacial surface for catching the dissolved gas is going to get larger.

**A.** Correct, but if it's controlled by diffusion of hydrogen in the metal, it doesn't matter how large a reservoir for hydrogen is available; the gas will not absorb the hydrogen due to slow diffusion in the metal.

**E. Kato** *(Waseda University)*

I have a comment. I have had experience with the examination of nitrogen and hydrogen in cast iron melted in cupolas and induction furnaces and there appears to be no problem with formation of blowholes. However, when a pitch coke was used for raising the carbon concentration of induction furnace iron, the nitrogen concentration increased drastically. Another case, where nitrogen blowholes were observed was when the mold binder contained nitrogen compounds. At the temperature of liquid iron, the nitrogen activity in these compounds is very high and it can lead to a great increase in nitrogen absorption. In my experience you don't need to degas if you don't generate excessive levels of nitrogen in the iron.

**A.** The plot I presented is the thermodynamic pressure that has to be exceeded for a pinhole to form; it's a necessary but not sufficient condition for pinhole formation.

**T. DebRoy** *(Pennsylvania State University)*

In the isotopic exchange technique that you used, the concentration of $N_2^{29}$ is very small. What is the accuracy of the isotope concentration measurement?

**A.** That's an excellent question. We use an in-line mass spectrometer to avoid problems with sampling. It also gives us the ability to take a large number of samples, so we can obtain a good statistical average. The reproducibility of the measurements using our standard gas is within three or four percent. During our experiments we take about 50 to 60 measurements so we can do a good statistical analysis. A bigger problem is monitoring the flow rate accurately because the rate constant depends on the amount of gas that you're putting through. Considering all sources of error I would say that 10 to 15% would be a reasonable estimate of accuracy.

**G. K. Sigworth** *(Reading Foundry Products)*

Concerning the rate of degassing under mass transfer controlled conditions, Enge and I developed a dimensionless correlation a few years ago which might be useful for such calculations. It might also be possible to modify our model to make it useful under conditions where degassing is controlled by surface reaction.

I also have a question. I know there is experimental evidence from the early work of Dr. Turkdogan in the 1960s to show that the rate at which nitrogen is absorbed into steel is determined by the presence of surface active elements, but is there experimental evidence to show that the opposite is true? That in degassing the surface active materials play a role?

**A.** Yes, there is experimental evidence that surface active elements retard nitrogen desorption as well.

# FORMATION OF POROSITY DURING SOLIDIFICATION OF CAST METALS

## ROBERT D. PEHLKE

*Department of Materials Science and Engineering*
*The University of Michigan*
*Ann Arbor, Michigan 48109*

## ABSTRACT

Porosity forms in castings by shrinkage resulting from density differences between liquid and solid, and by evolution of dissolved gases. These effects can manifest themselves independently, or they can develop porosity synergistically.

With emphasis on basic principles underlying mechanisms involved in generation of porosity during solidification, the roles of nucleation and growth processes are outlined. Examples of individual mechanisms are presented based on practical systems.

Modeling of porosity formation during solidification utilizing analyses of basic mechanisms is outlined. The use of modeling to predict distribution, amount and pore size is discussed. Examples of plate castings of Al-4.5% Cu and steel are compared with predictions from a numerical model.

Empirical rules utilized to assure sound castings have been developed. These rules are supported by fundamental considerations. Minimization of gas content and maximization of mold chilling capability to assure strong directional solidification are required to avoid porosity. Casting design which incorporates these considerations and meets productivity requirements is vital to a successful foundry operation.

## INTRODUCTION

Two major fundamental effects contribute to the formation of porosity in solidifying metals: shrinkage resulting from the volume decrease in going from liquid to solid, and gas evolution resulting from the decrease in solubility in solid metal compared to the liquid. These phenomena can occur simultaneously and act synergistically to develop porosity in solidifying metals.

Castings can have porosity defects as a result of shrinkage and gas evolution. Shrinkage porosity on a macroscopic basis has been predicted from heat flow considerations[1-4]. Other studies have predicted shrinkage porosity resulting from limited liquid metal flow during dendritic solidification[5-8]. Recently, Kubo and Pehlke have presented a mathematical model which calculates porosity formation in solidifying alloys considering both shrinkage and gas evolution[9].

An initial approach at predicting porosity formation is detailed. The model considers the effects of shrinkage and gas evolution, and also incorporates the influence of liquid metal feeding, segregation, and in the case of steel, deoxidation. An overview of an algorithm to accomplish this analysis is presented, and a detailed step-by-step review of the integrated calculation is detailed.

## POROSITY FORMATION

Shrinkage on solidification is a primary source of porosity formation in solidifying castings. Volumetric shrinkage in metals on transforming from the liquid to solid state may range from 3 to 10% with 5 to 8% being typical of most cast alloys. Shrinkage can manifest itself as piping in a riser, and risering systems are designed to confine the shrinkage porosity to the upper portion of risers. In the event that risers are unable to provide the required metal feeding, local porosity will form in the casting in the last regions to solidify.

Shrinkage porosity also occurs on a much smaller scale in the form of dispersed, so-called "microshrinkage" or "microporosity," porosity which forms in the interstices of dendritic solidification regions. This type of shrinkage porosity is found in alloys with large differences between liquidus and solidus temperatures. Liquid metal feeding in the dendritic solidification zone plays a key role in formation of this type of porosity.

Gas evolution also can contribute to formation of porosity. Gases generally are more soluble in liquid metals than in the solid. This decrease in solubility can result in gas evolution on solidification, not only because of the change in solubility on phase transformation, but also because the liquid phase is continuously enriched in the gaseous component as the solid forms and rejects solute species to the liquid. These effects are exhibited by hydrogen in aluminum, copper and iron base alloys. Nitrogen can cause porosity in iron alloys, but because the diffusion coefficient is much lower than that of hydrogen (by about an order of magnitude), nitrogen is less active as a participant in porosity formation. Compound gases, such as CO, in iron alloys (steels in particular) and $H_2O$ in copper alloys also can generate porosity.

Nucleation of a gas bubble is required before growth of gas generated porosity can occur. However, when shrinkage porosity forms, the large energy requirements for nucleation of a gas bubble are overcome.

## ROLE OF INTERDENDRITIC FEEDING

Possible feeding mechanisms proposed by Campbell are shown schematically in Figure 1. Several types of feeding are indicated including liquid feeding which

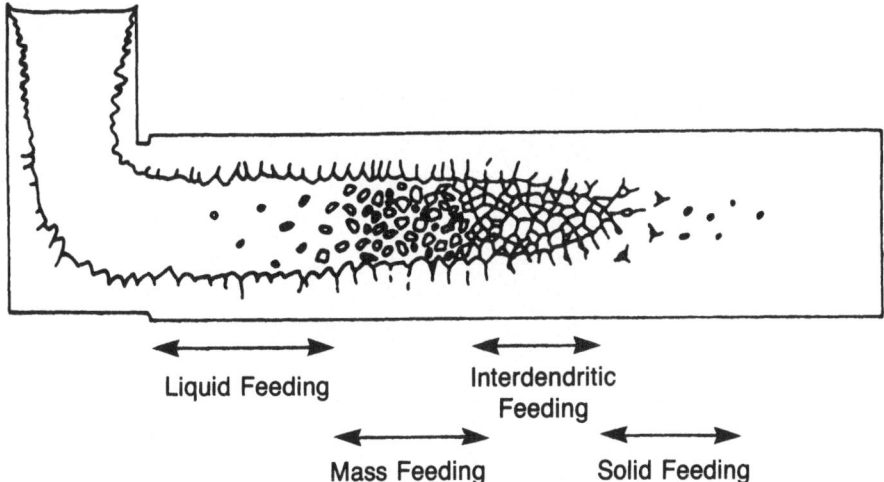

**Figure 1.** Schematic Representation of Possible Feeding Mechanisms by Campbell.[6]

occurs at the initial stage of solidification. The second is mass feeding which may occur, but is not well understood. The third is interdendritic feeding which occurs at a later stage of solidification and may have direct effects on formation of porosity defects. The fourth is solid feeding which occurs at the last stage of solidification and is related to distortion of castings.

Porosity defects are caused by the limitations of these feeding mechanisms. Since the resistances to liquid feeding and mass feeding are small because they occur at low solid fraction, interdendritic feeding is considered to be the most important stage for creation of porosity defects.

Kubo and Pehlke[9] have presented typical examples of porosity in castings of Al-4.5%Cu and Cu-8%Sn which are shown in Figure 2. In Al-4.5%Cu, round porosity defects are observed between secondary dendrite arms and at grain boundaries. In Cu-8%Sn, dispersed small pores are found at the roots of the secondary arms, and continuous porosity is found at grain boundaries. These types of porosity are considered to form not only by lack of interdendritic feeding, but also by gas rejection into the liquid during solidification.

The growth process in porosity formation is shown in Figure 3 where gas porosity nucleates at the base of dendrite arms Figure 3a.[9] The synergism between shrinkage porosity and gas porosity formation overcomes the large negative free energy required to form a gas-metal surface, and porosity nucleates easily at the location shown in Figure 3a.

With the progress of solidification, the gas dissolved in the liquid increases resulting in a higher potential for gas evolution, and the porosity grows. Since the radius of the porosity becomes large enough to decrease the contribution of interfacial energies, the porosity can detach from the dendrite as shown in Figure 3b. Buoyant and convective forces promote this detachment. Furthermore, at a

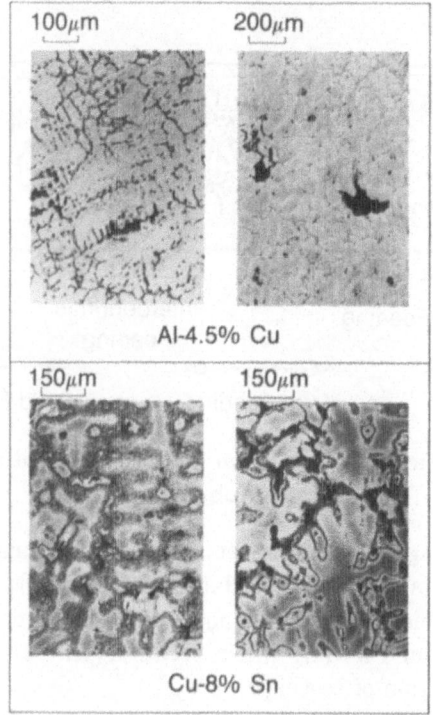

Figure 2. Porosity in Solidified Alloys.[9]

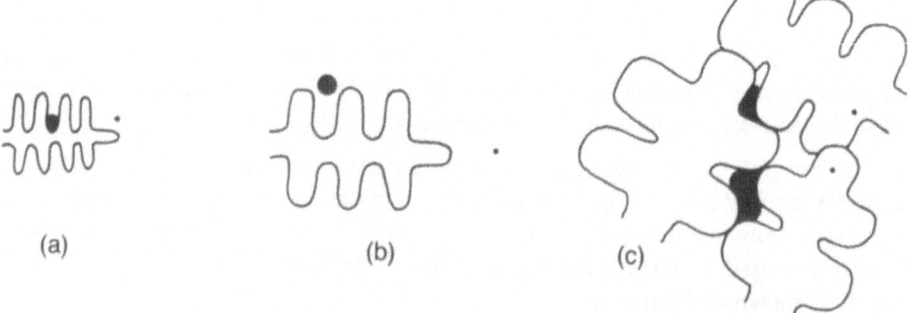

Figure 3. Growth Process of Porosity Formation.[9]

# FORMATION OF POROSITY

later stage of solidification, the neighboring dendrites collide, so that interdendritic feeding becomes difficult. At that point, the porosity is assumed to grow to compensate for solidification shrinkage.

## QUANTITATIVE ANALYSIS

**Heat Balance**—In the analysis of plate castings, two dimensional heat flow was considered. The heat balance equation can be expressed as:

$$C\rho \frac{\partial T}{\partial t} + f_L C \rho \left( u \frac{\partial T}{\partial x} + v \frac{\partial T}{\partial y} \right) = \lambda \left( \frac{\partial^2 T}{\partial x^2} + \frac{\partial^2 T}{\partial y^2} \right) - \rho L \frac{\partial f_L}{\partial t} \qquad (1)$$

where $\rho$ is an average density of liquid and solid being expressed as $f_L \rho_L + f_S \rho_S$. Here $\rho_L$ and $\rho_S$ are used as $\rho$ for the fully liquid and solid regions, respectively. Since the specific heat and the thermal conductivity of solid and liquid at the melting point for most casting alloys are not known well, they are assumed to be independent of temperature and equal for the two phases. The first term on the left side of Equation 1 is the heat accumulation, and the second term is the heat flux by interdendritic flow. The first term on the right side of Equation 1 is the heat flux by conduction, and the second term is the evolved latent heat. The temperature gradients are small in the mushy zone and the liquid fraction is relatively small, and therefore, the temperature terms for interdendritic flow can be neglected. Also, the effects of slight changes in solute concentration on the energy balance are negligible. Equation 1 reduces to

$$\frac{\partial T}{\partial t} = \frac{\lambda}{C\rho} \left( \frac{\partial^2 T}{\partial x^2} + \frac{\partial^2 T}{\partial y^2} \right) - \frac{L}{C} \frac{\partial f_L}{\partial t} \qquad (2)$$

Solution of Equation 2 requires a relationship between liquid fraction, $f_L$, and temperature. Assuming equilibrium, the lever rule can be used

$$f_L = \frac{C_0 m_L - k_0 (T - T_m)}{T - T_m - k_0 (T - T_m)} \qquad (3)$$

Alternatively, the Scheil equation which ignores solid diffusion could be used

$$f_L = \left( \frac{T - T_m}{m_L C_0} \right)^{1/(k_0 - 1)} \qquad (4)$$

In the examples presented, Equation 3 was used for steel, based on the high diffusivity of carbon in solid iron, and Equation 4 was used for the Al-4.5%Cu alloy.

**Continuity Equation**—From a mass balance for a volume element, the continuity equation requires

$$\left( \frac{\rho_S}{\rho_L} - 1 \right) \frac{\partial f_L}{\partial t} - \frac{\partial f_L}{\partial x} u - \frac{\partial f_L}{\partial y} v + \frac{\partial f_V}{\partial t} = 0 \qquad (5)$$

*References p. 440*

The first term on the left side of Equation 5 is the amount of shrinkage, and the second and third terms are the amount of liquid input by interdendritic flow. The last term is the amount of porosity growth. Equation 5 indicates that the shrinkage during solidification is compensated by interdendritic flow and the growth of porosity.

**Motion Equation**—The motion equation describing interdentritic flow can be expressed from Darcy's law for horizontal flow as[10]

$$u = -\frac{k}{\mu}\frac{\partial P}{\partial x} \tag{6}$$

and for vertical flow with a gravitational term as

$$v = -\frac{k}{\mu}\frac{\partial P}{\partial y} - \frac{k\rho g}{\mu} \tag{7}$$

where $k$ is permeability, being calculated from the liquid fraction, $f_L$ and dendrite cell size, $d$ as[11,12]

$$k = \frac{f_L^3 d^2}{180(1 - f_L)^2} \tag{8}$$

When $f_L$ is larger than 0.7, the value obtained by substituting 0.7 into Equation 8 has been used in calculations. When $f_L$ is less than 0.01, that obtained by substituting 0.01 has been used. The region where feeding is controlled by interdendritic flow has not been distinguished from other regions. The following relationship between the dendrite cell size and the local solidification time has been used to calculate the permeability by Equation 8.

$$d = a\Delta\theta_f^n \tag{9}$$

Before solidification is initiated, a time increment for the numerical calculations is substituted in Equation 9.

Gas pressure in the porosity should balance with the metal pressure and the liquid-gas surface energy as[10]

$$P_g = P + \frac{2\sigma_{LG}}{r} \tag{10}$$

In this development, the diameter of porosity forming first is assumed to be the same as the dendrite cell size. When the solidification rate is extremely large, Equation 10 is not satisfied because of very small values of $r$. Generally in dendritic solidification, the cell size is more than ten micrometers, so it is rare that gaseous solute is trapped completely in the solid, except at very low gas contents.

**Equations for Conservation of Gas Content: Hydrogen in an aluminum alloy**—Hydrogen alone can dissolve in an aluminum alloy. The diffusivity of hydrogen in solid aluminum at 933 K is relatively large. In this case, assuming complete equilibrium, the conservation equation for gas content is expressed as

$$[H_0] = (1 - f_L)[H_S] + f_L[H_L] + \alpha_H \frac{P_g f_V}{T} \qquad (11)$$

The left side of Equation 11 is the initial hydrogen content. The first, second, and third terms on the right are the amounts of hydrogen in the solid, liquid, and porosity, respectively. The hydrogen content in the solid and liquid are expressed as

$$[H_S] = K_{SH} P_g^{\frac{1}{2}} \qquad (12)$$

$$[H_L] = K_{LH} P_g^{\frac{1}{2}} \qquad (13)$$

**Equations for Conservation of Gas Content: CO in steel**—CO gas is the main origin of gas porosity in steel. Although nitrogen may also be a source, it normally exhibits minor effects. In the present case, CO gas is assumed to be the only source of gas porosity. However, CO is a chemical compound, and the treatment of CO solubility is somewhat complex.

At first, CO gas is formed by the reaction of dissolved carbon and oxygen in steel as

$$[C_L] + [O_L] = CO \qquad (14)$$

Then, the gas pressure of CO is expressed as

$$P_g = [C_L][O_L]/K_{CO} \qquad (15)$$

Assuming complete equilibrium, a mass balance for carbon and oxygen can be expressed as

$$[C_0] = [C_L]f_L + [C_S]f_S + \alpha_C \frac{P_g f_V}{T} \qquad (16)$$

$$[O_0] = [O_L]f_L + [O_S]f_S + \alpha_O \frac{P_g f_V}{T} + \beta \Delta SiO_2 \qquad (17)$$

The carbon and oxygen contents of the solid are expressed as

$$[C_S] = k_{Fe-C}[C_L] \qquad (18)$$

$$[O_S] = k_{Fe-O}[O_L] \qquad (19)$$

As Turkdogan[13] reported, with the progress of solidification the oxygen content in the liquid increases, so deoxidation reactions occur, as for example in the case of formation of silica.

$$[Si] + 2[O_L] = SiO_2 \qquad (20)$$

*References p. 440*

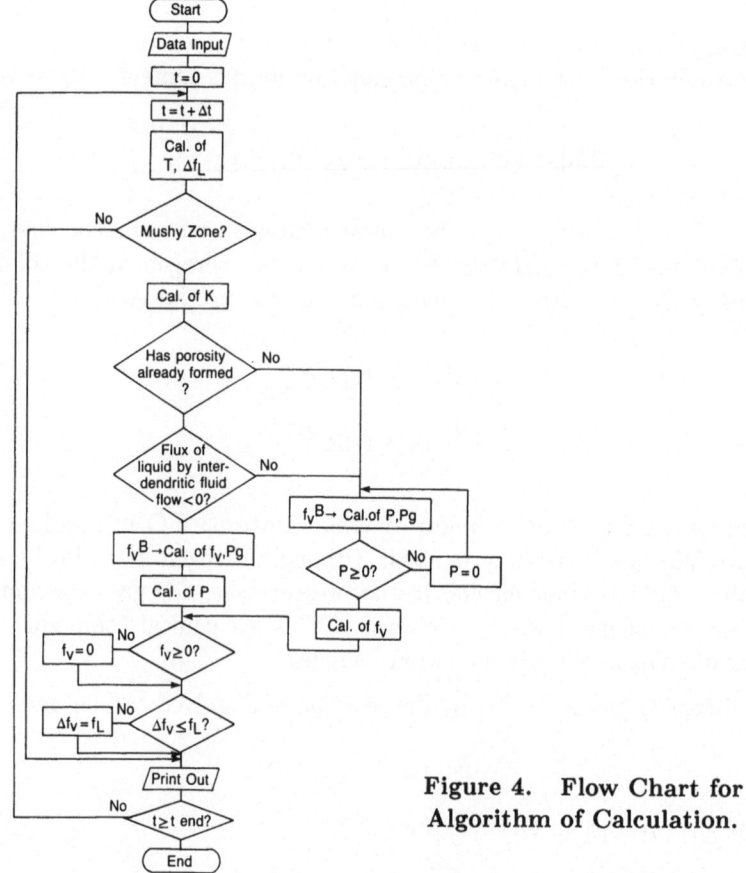

**Figure 4.** Flow Chart for Algorithm of Calculation.

In this work the effect of manganese has been ignored to simplify this illustrative example. In Equation 20, the equilibrium partition ratio is assumed to be unity. The oxygen and silicon contents in the liquid can be expressed as

$$[O_L]^2 = K_{SiO_2}/[Si] \tag{21}$$

$$[Si] = [Si_0] - \beta \Delta SiO_2 \tag{22}$$

As described later, Equations 15 through 17, 21 and 22 are solved simultaneously to obtain $P_g$ or $f_V$.

## ALGORITHM FOR CALCULATION

Kubo and Pehlke have presented a mathematical model for describing the distribution, amount, and radius of porosity in solidifying alloys accounting for gas evolution and interdendritic flow.[9] The algorithm for the calculation has been detailed, including incorporation of a heat balance, volumetric continuity, interdendritic flow of liquid and conservation equations for gas content, and is outlined in Figure 4.

At first the temperature of a volume element and the change of solid fraction are calculated from Equation 2 by an implicit finite difference method. If the volume element is in the mushy zone, the variables are calculated as follows:

If no porosity has formed, the metal pressure, $P$, is calculated from the equations of continuity and motion. The gas pressure, $P_g$, is calculated from Equation 10, assuming the diameter of porosity forming first is the dendrite cell size. Subsequently, a new amount of porosity is calculated from the equations for conservation of gas content.

If porosity exists, the flux of liquid by interdendritic fluid flow is calculated from the equations of continuity and motion. When the flux of liquid is positive, $P$ and $P_g$ are calculated from these equations and Equation 10, using the amount of porosity before the current time increment. A new value for the volume fraction of porosity is calculated from the equations for conservation of gas content. When the flux of liquid is negative, the metal pressure before the current time increment is used in the equations of continuity and motion to calculate the volume fraction of porosity. $P_g$ is then calculated from the equations for conservation of gas content. Then, a new value for metal pressure, $P$, is determined from Equation 10.

If the volume fraction of porosity and the metal pressure are calculated to be negative, they are set equal to zero. The calculated volume fraction of porosity is not allowed to exceed the current fraction of liquid.

This procedure is repeated over each volume element and each time step until the total time of simulation is reached. Convergence of the calculation is shown when lower values of the time and space increments yield the same solution.

The constants and properties used in the calculations have been summarized.[9] The castings analyzed in two dimensions were plates with an end riser. A mold-metal heat transfer coefficient was used to describe the heat flow at the mold-metal interface; values of 42 and 420 W/m²·K, corresponding to sand and metal molds, respectively, were used.

## RESULTS

Kubo and Pehlke[9] considered the cases of hydrogen enhanced porosity in an aluminum-copper alloy, and carbon monoxide formation in a silicon (manganese effects were not considered for simplicity) deoxidized steel.

Comparisons are made for Al-4.5%Cu plate castings in Figure 5 for amount of dissolved gas (0.2 and 0.3 cc/100 gm), cooling rate (heat transfer coefficients of 42 and 420 W/m²·K), and section size (plate thickness of 3 and 1.5 cm) effects on the calculated amount and distribution of porosity. With the larger hydrogen content and a lower heat transfer coefficient, corresponding to a sand mold (Figure 5a), the calculated porosity is more than 3% throughout the casting. This is used as a base case. A decrease in dissolved gas content shows a substantial decrease in porosity (Figure 5b). With a higher heat transfer coefficient, corresponding to a metallic mold, the porosity decreases substantially (Figure 5c). A decrease in the

*References p. 440*

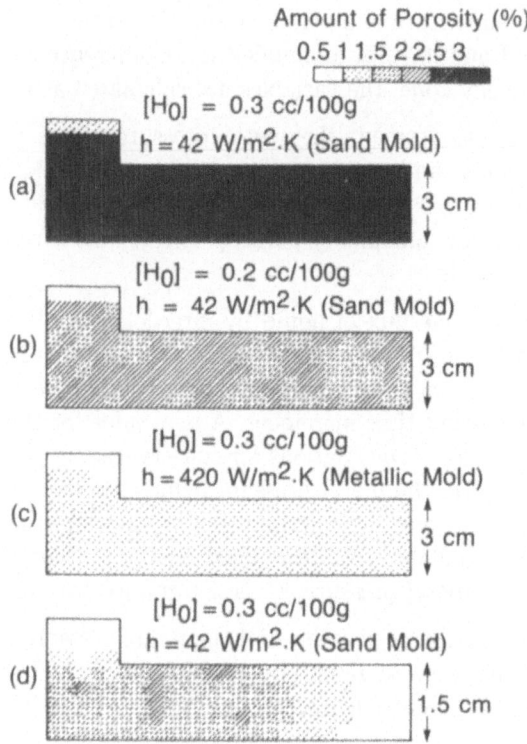

Figure 5. Calculated Amount of Porosity in Al-4.5% Cu Plate Castings.[9]

plate thickness (Figure 5d) also decreases the porosity, particularly near the end of the plate. Good qualitative agreement is noted with the work of Deoras and Kondic[14], Nishi and Kurobuchi[15], and Nishi, et al.[16]

Calculated radii of porosity in these Al-4.5% Cu plate castings are shown in Figure 6. The nominal radius ranges from 10 to 40 $\mu$m. The calculated radius of porosity in a thick plate is large regardless of hydrogen content (Figure 6a base case, and Figure 6b). The size of the porosity is predicted to decrease for a higher heat transfer coefficient (Figure 6c) and lower plate thickness (Figure 6d). These results are in good qualitative agreement with the experimental results of Kubo, et al.[17]

A calculation was also made for steel plate castings (0.14% C, 0.3% Si) poured into sand molds. The calculated results are presented in Figure 7.[9] The selected thickness of the plate, $D$, is 3 cm, and the ratios of $D$ to the length from the end to the riser, $L$, are 3, 4, and 6. When $L/D$ is 3 or 4, severe porosity defects are not predicted. However, when $L/D$ is 6, porosity up to 3 pct is predicted near the centerline. The calculation indicates that porosity defects do not occur within $2.5D$ from the end of the plate regardless of the value of $L/D$. Once porosity defects

# FORMATION OF POROSITY

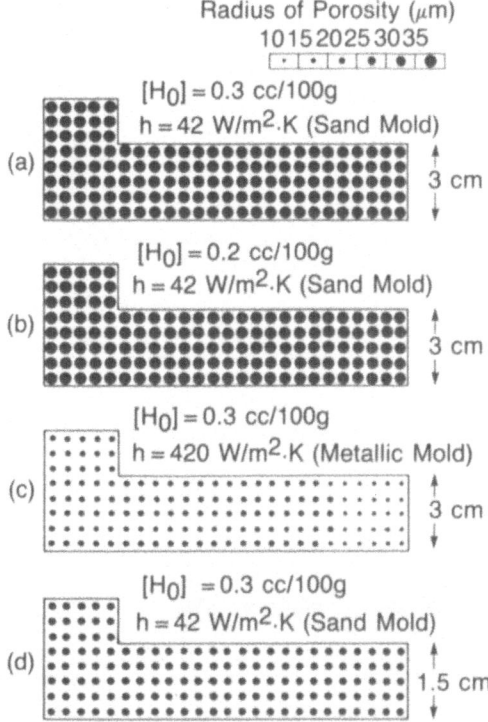

Figure 6. Calculated Radius of Porosity in Al-4.5% Cu Plate Castings.[9]

form, they spread near the riser as shown in Figure 7c. Pellini[18] reported that the maximum feeding distance for plates is $4.5D$ which is composed of an edge contribution of $2.5D$ and a riser contribution of $2D$. Pellini's empirical rules have been used widely for casting designs.[19] The present modeling results are coincident with the experimental results obtained by Pellini.

The temperature and metal pressure distributions calculated at a later stage in solidification of these steel plate castings are shown in Figure 8. These isotherms are 2 K apart, ranging from 1761 to 1755 K, and correspond to 5.7 to 2.0% liquid, respectively, where porosity defects are likely to occur. When the porosity defects are absent $(L/D = 3)$, the temperature gradient toward the riser is more than about 2 K/cm. On the other hand, when the porosity defects occur, e.g., at $(L/D = 6)$, the temperature gradient toward the riser is less than 1 K/cm. This result supports the opinion of Niyama, et al.[3] that the critical temperature gradient for occurrence of a shrinkage cavity in steel castings is 2 K/cm.

Local pressures showed a low value (50 kPa) at the final freezing front in the large $(L/D = 6)$ aspect ratio relative to 80 kPa for $L/D = 3$. This lower local pressure promotes CO evolution which enhances formation of porosity.

*References p. 440*

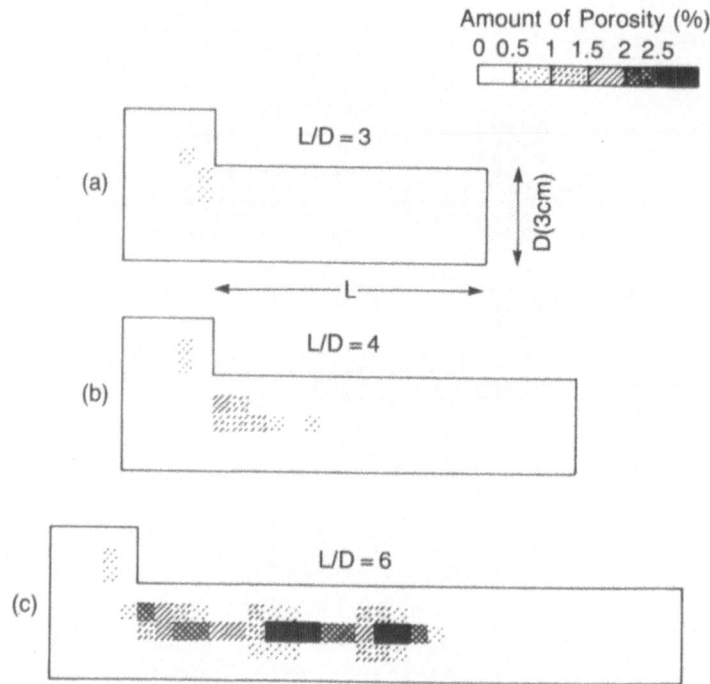

Figure 7. Calculated Amount of Porosity in Steel Plate Castings (0.14% C, 0.30% Si) poured into Sand Molds.[9]

Interdendritic flow velocities calculated at a later stage of solidification of the steel plate casting poured into a sand mold are shown in Figure 9.[9] This is for a case of occurrence of severe porosity defects. The direction of the vector of velocity is reversed near the point farthest from the riser, i.e., shrinkage during solidification is compensated by porosity growth at the point farthest from the liquid metal source.

## DISCUSSION

Formation of porosity in solidifying alloys is influenced by gas evolution and by interdendritic flow in the later stages of solidification. Empirical rules for assuring sound castings closely match the results presented for modeling studies.

The modeling results presented have considered the simultaneous influence of gas evolution, interdendritic fluid flow and shrinkage. The practical implementation of these results requires that gas content in the liquid metal be minimized before casting, and that degassing be utilized as appropriate. Design of the molding system to increase cooling characteristics and to assure directional solidification is critical in minimizing microporosity.

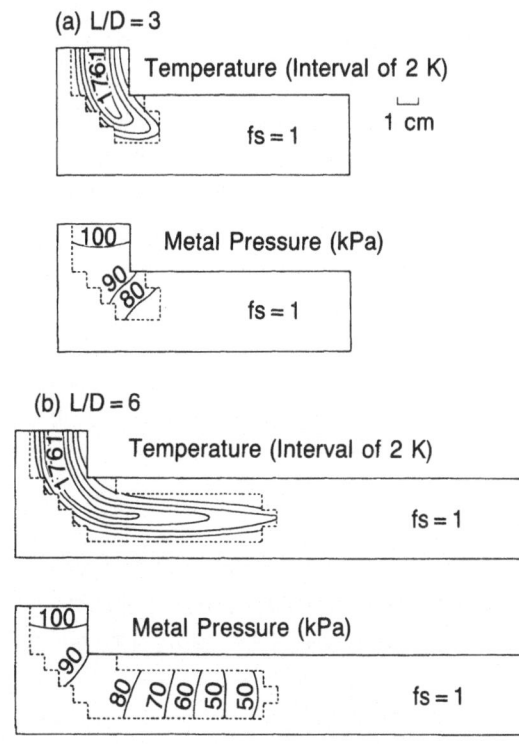

Figure 8. Calculated Temperature and Metal Pressure Distribution at a Later Stage of Solidification of Steel Plate Castings Poured into Sand Molds.[9]

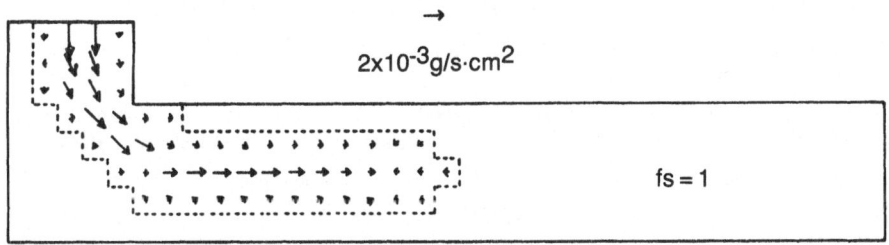

Figure 9. Interdendritic Flow Velocities Calculated at a Later Stage in Solidification of a Steel Plate Casting Poured into a Sand Mold.[9]

## SUMMARY

1. Key factors influencing porosity formation in solidifying alloys have been identified, and their synergism defined.

2. Modeling results which take into account gas evolution and interdendritic fluid flow during solidification have been reviewed. These results are shown to be in good qualitative agreement with experiments on Al-4.5%Cu castings.

3. The modeling results have provided a basis for more specifically defining empirical rules for predicting soundness of steel castings.

4. Minimization of gas content by degassing and increasing mold chilling power are recommended to produce sound castings. However, limitations of gas content and increasing mold chilling power depend on casting shape, alloy composition, required mechanical properties, and other foundry variables which require more detailed investigation for further quantification.

## REFERENCES

1. A. Jeyarajan and R. D. Pehlke, *Trans. Amer. Foundrymen's Soc.*, 1978, *86*, 457–464.
2. W. C. Erickson, *AFS Int. Cast Metals J.*, 1980, *5*, 1, 30–41.
3. E. Niyama, T. Uchida, M. Morikawa, and S. Saito, *AFS Int. Cast. Metals J.*, 1981, *6*, 2, 16–22.
4. I. Imafuku, *Trans. Japan Soc. of Mech. Engrs. Ser. C.*, 1981, *47*, 918–926.
5. T. S. Piwonka and M. C. Flemings, *Trans. TMS-AIME*, 1966, *236*, 1157–1165.
6. J. Campbell, *AFS Cast Metals Res. J.*, 1969, *5*, 1, 1–8.
7. V. de L. Davies, *AFS Cast Metals Res. J.*, 1975, *11*, 2, 33–44.
8. I. Ohnaka, Y. Mori, Y. Nagasaka, and T. Fukusako, *J. Japan Foundrymen's Soc.*, 1981, *53*, 673–679.
9. K. Kubo and R. D. Pehlke, *Met. Trans. B*, 1985, *16B*, 359–366.
10. M. C. Flemings, *Solidification Processing*, McGraw-Hill Book. Co., New York, NY, 1974, 34, 148, 207, and 234.
11. K. Kubo and R. D. Pehlke, and T. Fukusako*, The University of Michigan, Ann Arbor, MI and *Osaka University, Suita, Osaka, Japan, unpublished research, 1984.
12. P. C. Carman, *Trans. Inst. Chem. Eng.*, 1937, *15*, 150–166.
13. E. T. Turkdogan, *Trans. TMS-AIME*, 1965, *233*, 2100–2112.
14. B. R. Deoras and V. Kondic, *Foundry Trade J.*, 1956, *100*, 361–364.
15. S. Nishi and T. Kurobuchi, *Keikinzoku*, 1974, *24*, 245–254.
16. S. Nishi, Y. Shinada, and T. Kurobuchi, *Keikinzoku*, 1974, *24*, 130–134.
17. K. Kubo, T. Fukusako, and I. Ohnaka, *J. Japan Foundrymen's Soc.*, 1979, *51*, 586–591.
18. W. S. Pellini, *Trans. Amer. Foundrymen's Soc.*, 1953, *61*, 61–80.
19. P. R. Beeley, *Foundry Technology*, Butterworths, London, 1972, 112.

## DEFINITION OF SYMBOLS

| | |
|---|---|
| $a$ | constant in Eq. [9] |
| $C$ | specific heat (kJ/kg·K) |
| $[C_0]$ | initial carbon content in steel (wt pct) |
| $[C_L]$ | carbon content in liquid (wt pct) |
| $[C_S]$ | carbon content in solid (wt pct) |
| $D$ | thickness of plate casting (m) |
| $d$ | dendrite cell size (m) |
| $f_L$ | liquid fraction |
| $f_S$ | solid fraction |
| $f_V$ | porosity |
| $\Delta f_S$ | increase of solid fraction |
| $\Delta f_V$ | increase of porosity |
| $g$ | acceleration of gravity (9.80 m/s$^2$) |
| $[H_0]$ | initial hydrogen content in aluminum alloy (cc/100 g) |
| $[H_L]$ | hydrogen content in liquid (cc/100 g) |
| $[H_S]$ | hydrogen content in solid (cc/100 g) |
| $h$ | heat transfer coefficient (W/m$^2$·K) |
| $K_{CO}$ | equilibrium constant of Eq. [15] ((wt pct)$^2$/atm) |
| $K_{LH}$ | equilibrium constant of Eq. [13] (cc/100 g·atm$^{1/2}$) |
| $K_{SH}$ | equilibrium constant of Eq. [12] (cc/100g·atm$^{1/2}$) |
| $K_{SiO_2}$ | equilibrium constant of Eq. [21] ((wt pct)$^3$) |
| $k$ | permeability (m$^2$) |
| $k_0$ | equilibrium partition ratio of copper in aluminum alloy |
| $K_{Fe-C}$ | equilibrium partition ratio of carbon in steel |
| $k_{Fe-O}$ | equilibrium partition ratio of oxygen in steel |
| $L$ | latent heat of fusion (kJ/kg) |
| $m_L$ | liquidus slope (K/wt pct) |
| $n$ | constant in Eq. [10] |
| $[O_0]$ | initial oxygen content in steel (wt pct) |
| $[O_L]$ | oxygen content in liquid (wt pct) |

| | |
|---|---|
| $[O_S]$ | oxygen content in solid (wt pct) |
| $P$ | metal pressure (Pa) |
| $P_g$ | gas pressure (Pa) |
| $r$ | radius of porosity (m) |
| $[Si]$ | silicon content of steel (wt pct) |
| $[Si_0]$ | initial silicon content of steel (wt pct) |
| $\Delta SiO_2$ | content of deoxidation product by silicon (wt pct) |
| $T$ | temperature (K) |
| $T_0$ | room temperature (K) |
| $T_m$ | melting point of pure metal (K) |
| $T_p$ | pouring temperature (K) |
| $t$ | time (s) |
| $\Delta t$ | time increment (s) |
| $u$ | velocity in $x$ direction (m/s) |
| $V$ | volume of porosity (m$^3$) |
| $v$ | velocity in $y$ direction (m/s) |
| $x$ | coordinate (m) |
| $y$ | coordinate (m) |
| $\alpha_C$ | constant in Eq. [16] $\left(\dfrac{0.146}{f_L\rho_L + f_s\rho_s}\right)$ |
| $\alpha_H$ | constant in Eq. [11] $\left(\dfrac{27300}{f_L\rho_L + f_s\rho_s}\right)$ |
| $\alpha_O$ | constant in Eq. [17] $\left(\dfrac{0.390}{f_L\rho_L + f_s\rho_s}\right)$ |
| $\beta$ | constant in Eq. [17] (0.467) |
| $\Delta\theta_f$ | local solidification time (s) |
| $\lambda$ | thermal conductivity (W/m·K) |
| $\mu$ | viscosity of liquid metal (Pa·s) |
| $\rho$ | average density. $f_L\rho_L + f_s\rho_s$ (kg/m$^3$) |
| $\rho_L$ | density of liquid (kg/m$^3$) |
| $\rho_S$ | density of solid (kg/m$^3$) |
| $\sigma_{LG}$ | gas-liquid interfacial energy (N/m) |

## DISCUSSION

**B. Thomas** *(University of Illinois)*

Please tell me more about the heat flow model you used? Did you say it includes fluid flow?

**A.** Actually, no, the model as formulated does not include fluid flow in the heat transfer analysis because the thermal gradients in the range where solidification is being completed are extremely small, so this term was not included. However, fluid displacements in the final stages of solidification are considered.

**B. Thomas** *(University of Illinois)*

Were the heat flow computations performed with a finite element package or did you write the program yourself?

**A.** Dr. Kubo wrote the program. It was a finite difference program.

**S. Asai** *(Nagoya University)*

You have done a good job of modeling. You mentioned the model assumes a dendritic structure. How would you modify the model to account for an equiaxial structure?

**A.** You need some guideline or equation which would describe fluid flow through such a mass of crystallites.

**S. Asai** *(Nagoya University)*

Does that mean that for the dendritic case you use Darcy's equation with permeability dependent on direction, while for the equiaxial structure the permeability would be assumed to be isotropic?

**A.** Yes. The value of the permeability was based on a correlation taken from castings of similar shape.

**S. Asai** *(Nagoya University)*

What are the criteria for determining the porosity?

**A.** One calculates the pressure gradient in the liquid. Knowing the permeability, one determines if it will feed, and what local shrinkage has taken place? Next, the distribution of gas between liquid and solid is calculated and the pressure is determined as well. By iterating, one can determine the extent of the local porosity.

**Jay Janowak** *(Amax Metals Groups)*

You did not talk about cast iron, but how would you measure the microporosity in gray or ductile iron, recognizing that the density variation caused by graphite volume in gray cast iron may be larger than the percentage of porosity?

A. The microporosity in these structures could be determined by metallographic examination of sections. A possible check on these measurements could be made by density measurements, although as you suggest, graphite volume variations may rule out these measurements for this purpose.

**P. Wieser** *(Case Western University)*

To what do you attribute reduced porosity in a chilled mold?

A. With the more rapid solidification obtained in a chilled mold, there is a steeper pressure gradient in the liquid. So, even in those areas where dendritic solidification may start to form, you have a strong thermal gradient and should get local feeding. Your solidification zone is not as extended.

**Y. Sahai** *(Ohio State University)*

How were the grid sizes determined?

A. This is a standard procedure, to reduce both the grid sizes in space and time until one gets a consistent result. I think the local pore sizes could be identified as unit elements and so you're looking at something like two or three millimeters as the space step. The time step would be half a minute, or several seconds, at least.

**Y. Sahai** *(Ohio State University)*

What was the criterion for choosing the time step?

A. Convergence. You need to satisfy three aspects in your model. You need consistency, which just means that you use the right equations and then you need stability, particularly in the explicit finite difference method. That speaks to the ratio between the time step and the grid size. With respect to the two dimensions, if you decrease the grid size in space by a half, then the time step has to come down by a quarter to maintain the same ratio of $\Delta t / \Delta x^2$. And then finally, the question of convergence which is to get a convergent solution for the set of equations which can be established by decreasing time and space steps until the solution is unchanged.

**N. El-Kaddah** *(University of Alabama)*

What is the CO pressure in the pores?

A. On the order of 1.5 to 2.5 atmospheres.

**K. G. Davis** *(CANMET)*

I'm wondering if the Hatachi people had success in correlating microporosity in steel with the critical value of temperature gradients.

**A.** I'm aware of that work but we haven't made any direct comparison on that basis. However, I have cited the gradient. In fact, Niyama is at Hatachi and his guideline is two degrees per centimeter or more. Of course, they're looking at relatively large castings for steam power plants.

**S. Jacobs** *(CTIF)*

I have a comment and a question. The comment is: we have made porosity measurements on aluminum castings with the same dimensions as yours and our measurements are in total agreement with your predictions. The question is: could you comment on the role of mass feeding on microporosity formation?

**A.** I think that mass feeding certainly can be important in casting solidification, but it doesn't play a critical role in microporosity formation.

**S. Jacobs** *(CTIF)*

Is it possible for your model to take into account the effects of the increasing concentrations of copper in the liquid phase as solidification progresses?

**A.** Certainly it could. However, we did not do that, and I suspect that the effect would be a relatively small adjustment in the gas solubility distribution between liquid and solid. This would not affect the overall results that much.

# ON THE DETECTION, BEHAVIOUR AND CONTROL OF INCLUSIONS IN LIQUID METALS

### RODERICK I. L. GUTHRIE

*Macdonald Professor of Mining and Metallurgy*
*McGill University*
*Montreal, H3A 2A7*

## ABSTRACT

A technique resting on the electric sensing zone principle has been developed for monitoring liquid metal quality. The instrumentation developed makes use of very heavy electrical currents to compensate for the low resistivity of liquid metals. It is capable of providing information on the number density, and size distribution, of inclusions in liquids such as molten aluminum, magnesium, zinc, copper and steel. Industrial tests in aluminum show that metal quality is a dynamic phenomenon, and that melt hydrodynamics play a very important role in the entrainment and removal of inclusions. Particles in the size range 20-300 microns can be routinely measured, but smaller, or larger, particles can be detected through suitable adjustments of the electric sensing zone aperture size and the magnitude of the applied current. It is shown that inclusions can be controlled to some extent through the phenomena of sedimentation, flotation and/or filtration.

## INTRODUCTION

The presence of unwanted second phase particles in solidified metals and alloys such as non-metallic inclusions or undesirable intermetallic precipitates can seriously flaw the performance and value of the final product. For critical applications such as aluminum or steel beverage cans, steel corded tires, aerospace landing gears, turbine blades or thin walled auto-engine blocks, etc., it is important that the size and number of inclusions be minimised, and/or kept below known, and acceptable limits. The problem until recently, has been that none of the inclusion detection techniques were able to directly record liquid metal quality.

For instance, in the steel industry automatic image analysis currently represents 'state of the art' technology. This entails cutting up a continuously cast ingot taking samples from strategic locations within the ingot and preparing them for microstructural examination. Automatic image analysis (i.e. counting and sizing) must then be carried out on as large a surface as reasonably possible. The main problem with such an approach is obviously the time, labour and effort needed to discover, *post facto*, possible liquid metal processing problems.

In the aluminum industry, the assessment of melt quality has developed along similar lines, the Porous Disc Filtration Analysis (PoDFA) method developed by Alcan being typical state of the art technology. In this approach, about 5 kg of liquid aluminum are spooned out of the melt into a heated crucible fitted with a porous disc set in its base. Molten aluminum, being typically much cleaner than commercial steels, is filtered through the disc so as to collect, by cake filtration phenomena, second phase particles such as titanium diboride clusters, alumina films, and/or silicates in a small heel of residual aluminum. Again, the sample is allowed to solidify and a cross section of the porous disc and overlaying sample is cut for micrographic analysis. The operation can typically provide answers within two days. It is labour intensive and tends to be subjective, i.e., it is also advisable to have the same micrographer assess inter-plant samples.

For the iron, zinc, copper and magnesium foundry industries, similar needs for melt quality assessment are thought to exist. For instance, copper-beryllium alloys used for electrical contacts, springs, etc., are prone to oxidation, any beryllia inclusions being detrimental to surface quality, as well as to electrical and physical properties. On-line detection of second phase particles could be a boon.

Similarly, in magnesium foundry operations, where solute iron is to be eliminated for 'corrosion free' magnesium auto-wheels to meet Canadian Winter Performance Criteria, vanadium additions are used to precipitate iron as an iron-vanadium intermetallic. These inclusions must be allowed to settle to the bottom of a crucible before the purified magnesium metal is cast. Again, an on-line quality assessment device is needed for effective process control, particularly as these represent small batching operations.

## DETECTION OF INCLUSIONS IN MELTS

To date, only two techniques have been developed which provide metallurgists with the opportunity to assess liquid metal quality on-line. Historically, the ultrasonic probe was the first to be studied, the Alcan UPILM (Ultrasonic Probe in Liquid Metals), started in 1966 and the 4M probe of Reynolds (Mansfield Molten Metal Monitor) being typical.[1] Although these techniques bear the promise of providing on-line results concerning metal cleanliness, their sensitivity appears to be extremely limited by existing transducer technology.[2]

Following twenty years of development therefore, the UPILM approach was abandoned by Alcan, in favour of a new technique, known as the LiMCA (Liquid Metal Cleanliness Analysis) method.

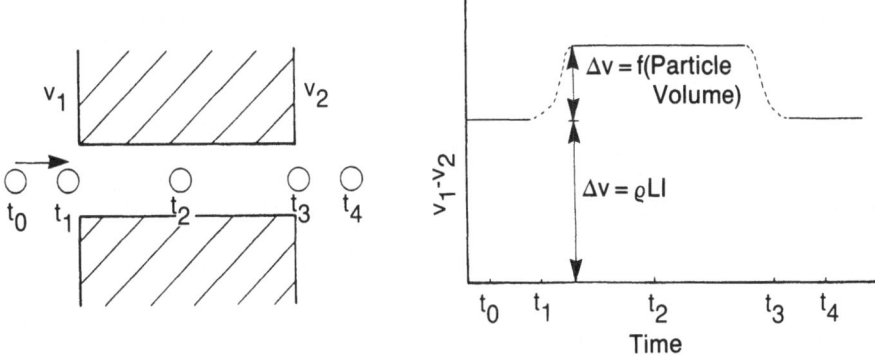

Figure 1. Operating principle of the electric sensing zone (E.S.Z.), or resistive pulse technique (R.P.T.).

**Principle of Particle Detection by the Electric Sensing Zone Technique**—The principle of the resistive pulse/electric sensing zone technique is shown in Figure 1. The particles to be analysed are suspended in an electrically conducting fluid as the latter is drawn through an orifice into an electrically insulating vessel. At the same time a constant current is maintained through the orifice which completes the circuit between two terminals of a power supply. When a particle enters the orifice it displaces its volume of the conducting fluid, causing a temporary rise in the electrical resistance of the orifice. This resistance change in the presence of the applied current produces a voltage pulse of a duration that is approximately equal to the transit time of the particle across the orifice. Alternatively, a constant voltage source can be used, in which case current pulses are observed. The amplitude of the voltage (or current) change is, as a first approximation, proportional to the volume of the particle. Thus, by measuring and counting each pulse the particle size distribution and concentration of the suspended particles can be deduced.

The novelty of the technique in its application to liquid metals, lies in the extremely heavy currents used for generating small signals for measuring second phase particles. The central theoretical problem of the resistive pulse technique is to predict the magnitude of the change in electrical resistance, $\Delta R$, caused when a non-conducting particle of equivalent diameter, $d$, enters a sensing volume (often a right circular cylinder of diameter, $D$, and length, $L$). Owing to distorted electrical fields, there is no simple analytical solution that is valid for all values of $(d/D)$ ranging from 0 to 1.

Deblois and Bean[3] have shown that for $(d/D) < 0.4$, a relation first proposed by J. C. Maxwell can be used to derive an explicit expression for $\Delta R$. Maxwell showed for a dilute suspension of insulating spheres that the effective resistivity could be expressed as:

$$\rho_{eff} = \rho \left( 1 + \frac{3}{2} f + ... \right) \tag{1}$$

*References p. 465*

where $f$ is the volume fraction occupied by the spheres. For a cylindrical sensing volume of length, $L$, and diameter, $D$, filled with a fluid of resistivity, $\rho$, the resistance is:

$$R_1 = \frac{\rho L}{A} = \frac{4\rho L}{\pi D^2} \qquad (2)$$

When a sphere of diameter, $d$, is contained within the sensing volume, the value of $f$ becomes

$$f = \frac{V_{sphere}}{V_{cylinder}} = \frac{2d^3}{3D^2 L} \qquad (3)$$

substituting Equation 3 into Equation 1 and substituting Equation 1 for $\rho$ in Equation 2 gives the resistance of a cylinder with a small non-conducting sphere contained within it:

$$R_2 = \frac{4\rho L}{\pi D^2}\left(1 + \frac{d^3}{D^2 L} + ...\right) \qquad (4)$$

Subtracting Equation 2 from Equation 4 then gives the desired expression for $\Delta R$:

$$\Delta R = \frac{4\rho d^3}{\pi D^4} \qquad (5)$$

This expression correctly predicts that the resistive pulse is independent of the length of the sensing volume. Recasting it in terms of the particle volume, $V_p$:

$$\Delta R = K V_p \qquad (6)$$

where the constant K is determined by the resistivity of the fluid and the diameter of the sensing volume:

$$K = \frac{24\rho}{\pi^2 D^4} \qquad (7)$$

Smythe[4,5] analyzed the problem of predicting $\Delta R$ using a numerical technique and has published correction factors (with a stated accuracy of 1 part in $10^7$) for Equation 5 of the form:

$$\Delta R = \frac{4\rho d^3}{\pi D^4} f(d/D) \qquad (8)$$

Smythe's values of $f(d/D)$ are shown in Table 1.

It can be seen that for $(d/D) = 0.4$, the error in the use of Equation 5 is in the order of 5% and decreases to insignificance when $(d/D)$ is less than 0.2. Deblois et al[6] further showed that the correction factor $f(d/D)$ could be expressed as:

$$f(d/D) = [1 - 0.8(d/D)^3]^{-1} \qquad (9)$$

with a relative error of $3 \times 10^{-4}$% at $(d/D) = 0.1$, and less than 1% at $(d/D) = 0.8$. Thus Equation 8 can be written explicitly as:

$$\Delta R = \frac{4\rho d^3}{\pi D^4 [1 - 0.8(d/D)^3]} \qquad (10)$$

## TABLE 1

Correction Factors, f(d/D) for Particle to Orifice Diameter Ratios (d/D) for Use with Equation (8).[4,5]

| d/D | f(d/D) | d/D | f(d/D) |
|---|---|---|---|
| 0.1 | 1.000798 | 0.6 | 1.208438 |
| 0.2 | 1.006416 | 0.7 | 1.379503 |
| 0.3 | 1.021988 | 0.8 | 1.710860 |
| 0.4 | 1.053744 | 0.9 | 2.556723 |
| 0.5 | 1.110694 | 0.95 | 3.861028 |

valid for $(d/D)$ ranging from 0 to 0.8 with an accuracy of 99%.

In practice, voltage rather than resistance changes are measured. Figure 2 shows a typical diagram of an electric circuit for resistive pulse counting in liquid metals, where a battery with an emf of $V_T$ drives a current $I_O$, while a ballast resistance $R_B$, regulates this current. The voltage pulses, $\Delta V_{AB}$, caused by the passage of particles, riding on a steady potential difference $V_{AB}$, are detected at the ends of feeder electrodes, $A$ and $B$. The value of this voltage pulse is a product of the change in resistance, $\Delta R$ and the electric current, $I_O$. The pulse can also be a function of the ballast resistance as the following derivation shows. Referring to Figure 2:

$$V_{AB} = \frac{R_E}{R_E + R_B} V_T \qquad (11)$$

$$\Delta V_{AB} = \left(\frac{\partial V_{AB}}{\partial R_E}\right)_{R_B} \Delta R_E \qquad (12)$$

$$\Delta V_{AB} = \frac{(R_E + R_B) - R_E}{(R_E + R_B)^2} V_T \Delta R_E \qquad (13)$$

and since

$$I_O = \frac{V_T}{R_E + R_B} \qquad (14)$$

then:

$$\Delta V_{AB} = \frac{R_B}{R_E + R_B} \Delta R_E I_O \qquad (15)$$

where $I_O$ is the current passing through the orifice in the absence of a particle in the electric sensing zone. Finally, and in the limit, for cases where the ballast

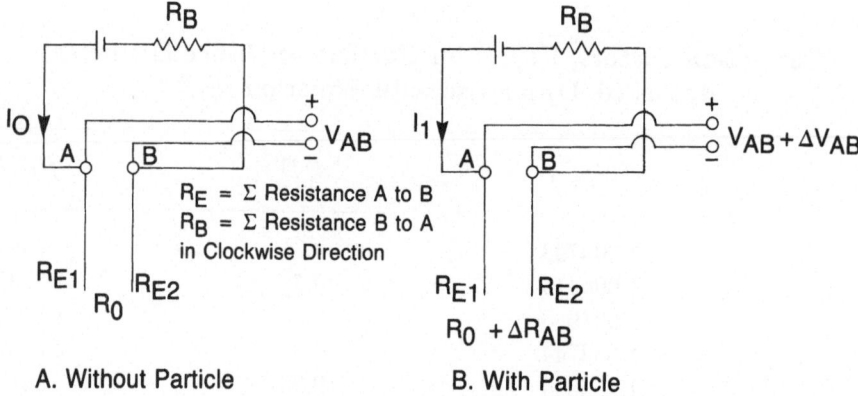

Figure 2. Typical circuit diagram illustrating the flow of current through two electrodes, $R_{E1}$ and $R_{E2}$, the electric sensing zone, $R_O$, and ballast resistor, $R_B$, as driven by a heavy duty lead-acid battery.

resistance, $R_B$ is much greater than the electrodes plus the orifice resistance, $R_E$, this expression simplifies to:

$$\Delta V_{AB} = I_O \Delta R = \frac{4 I_O \rho}{\pi} \frac{d^3}{D^4} f(\frac{d}{D}) \tag{16}$$

provided the electrode resistances are small and can be neglected.

To conclude therefore, a knowledge of pulse height voltages allows one to deduce individual particle diameters, according to Equations 15 or 16, as appropriate.

Figure 3 presents a plot of $\Delta V_{AB}$ as a function of particle diameter under the following conditions: $\rho = 25 \times 10^{-8} \Omega m$ (molten aluminum), $I = 60$ amperes.

It can be seen that the amplitude of the signals caused by particles of a size likely to be encountered in molten aluminum (20 to 100 μm) are quite small (tens of microvolts to millivolts), making highly sensitive measuring equipment mandatory. Equations 15 and 16 predict that the amplitude of the voltage pulses produced by the passage of non-conductive particles is: (i) proportional to the applied current, (ii) inversely proportional to the fourth power of the orifice diameter, (iii) proportional to the particle volume.

For aluminum, the decision was made to aim for a particle detection limit of 20 μm (equivalent spherical diameter) while maximizing in so far as possible the orifice diameter (to maximise the sampling rate and to reduce the chance of plugging). The conditions ultimately chosen thus represented a compromise between the detection limit, the sampling rate, and the size of the current supply system. For the aluminum work presented in this paper, therefore, the following conditions applied: $D = 300 \mu m$, $I = 60$ amperes, pre-amplifier gain $= 1000$. Under these

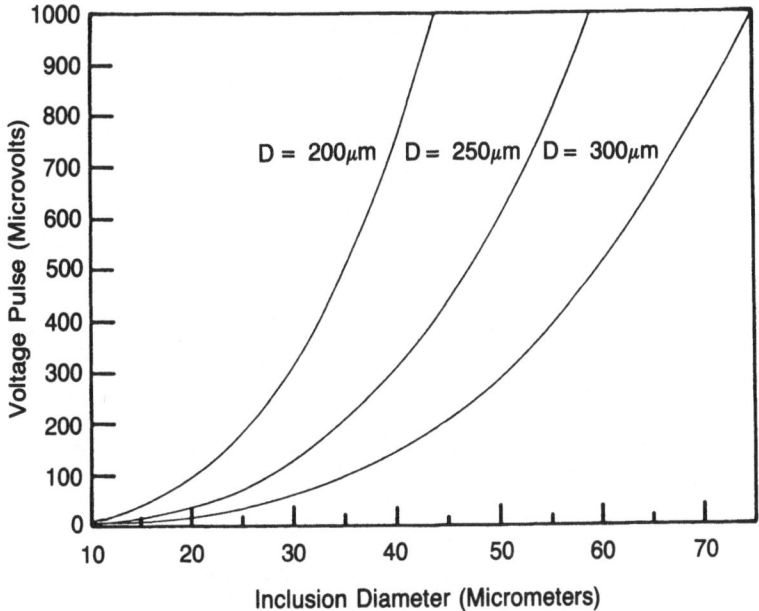

Figure 3. Predicted voltage pulses (microvolts) generated by non-conducting inclusions passing through 200, 250 and 300 μm diameter sensing zones versus equivalent spherical diameter. (I = 60 amperes, metal = aluminum, 700°C).

conditions, a particle 20 μm in diameter produced a voltage pulse of 19 μV compared to a typical background noise level of 5 μV (peak to peak).

**LiMCA Equipment**—Typical LiMCA equipment is shown schematically in Figure 4.[7] It consists of an insulating sampling cell containing a small orifice through which the metal to be sampled is drawn under reduced pressure. The flow of current through the orifice is established with a constant power supply and two electrodes, one inside the sampling tube and the other immersed in the metal outside. The voltage across the two electrodes can be observed using an oscilloscope. The peaks caused by the passage of particles through the orifice are amplified and recorded using a multichannel analyser device which measures and records the amplitude of each peak. After each test, a histogram is provided, displaying the number of peaks observed as a function of their magnitude.

For applications in aluminum, heat resistant glass sampling tubes are used and the electrodes are made from mild steel. In addition to aluminum, the method has been successfully used in molten zinc, lead, copper, magnesium, mercury, gallium, iron and steel. For copper, iron and transformer steel melts, quartz, as well as boron-nitride sampling tubes, have proved successful. For liquid magnesium, boron nitride performs satisfactorily, silica (quartz) being reduced. Quartz tubes in molten aluminum are also chemically reduced, leading to erosion and expansion of the

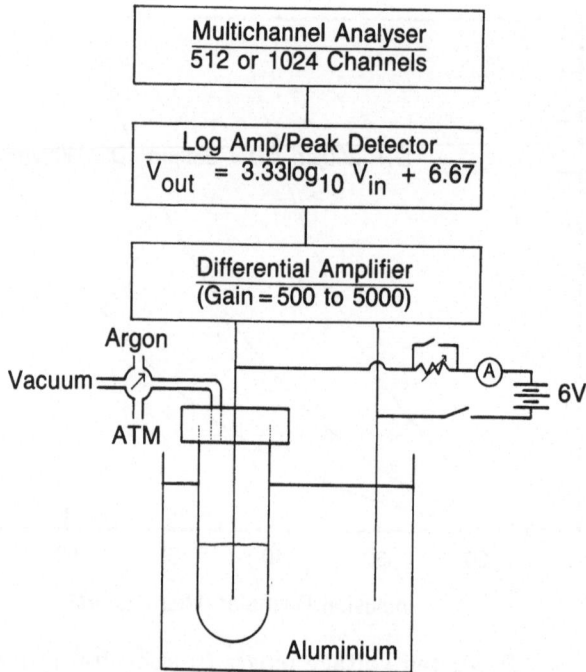

Figure 4. Schematic view of apparatus and equipment for detecting inclusions in liquid metals.

electric sensing zone region.

Table 2 lists the resistivities of a number of molten metals. As the signal amplitude is directly proportional to electrical resistivity, detection limits for most light metals should be at least as low as those for molten aluminum (i.e. $\simeq 20~\mu m$) for the system described. Nonetheless, smaller orifices ($\simeq 100~\mu m$) can be used for monitoring particles.

It is also worth noting the resistivity of a saline solution in Table 2 in comparison to the much lower resistivities of liquid metals. This aggravated the problem of detecting particles in liquid metals, making sensitive electronic equipment mandatory for signal detection.

## BEHAVIOR OF INCLUSIONS IN MELTS

Much data have been gathered in relation to the quality of molten aluminum during various processing operations carried out in Alcan's casting centres.[8] Two examples are now presented to illustrate these.

### EXAMPLE I. Precipitation and Settlement of (Ti/V)$B_2$ Particles for Electrical Conductivity Grade Aluminum.

The presence of titanium and vanadium in solid solution in aluminum reduces the latter's electrical conductivity. To circumvent this problem, boron additives are

## TABLE 2

### Resistivities of Molten Metals and Aqueous Electrolytes[7]

| Fluid | Resistivity $\Omega$ m | Temperature, (°C) |
|---|---|---|
| Cu | $2.2 \times 10^{-7}$ | 1084 |
| Ga | $2.48 \times 10^{-7}$ | 30 |
| In | $2.9 \times 10^{-7}$ | 157 |
| Fe | $13.9 \times 10^{-7}$ | MP~1540 |
| Pb | $9.79 \times 10^{-7}$ | 340 |
|  | $10.1 \times 10^{-7}$ | 400 |
| Mg | $2.74 \times 10^{-7}$ | 650 |
| Hg | $9.84 \times 10^{-7}$ | 50 |
|  | $10.3 \times 10^{-7}$ | 100 |
| Rb | $2.3 \times 10^{-7}$ | 122 |
| Zn | $3.53 \times 10^{-7}$ | 413 |
| K | $1.36 \times 10^{-7}$ | 64 |
| Sn | $4.5 \times 10^{-7}$ | 232 |
| Cd | $3.4 \times 10^{-7}$ | 400 |
| Bi | $12.9 \times 10^{-7}$ | 300 |
| Ca | $3.3 \times 10^{-7}$ | 839 |
| Al | $2.5 \times 10^{-7}$ | 700 |
| .9% NaCl (Physiological Saline) | $7.1 \times 10^{-1}$ | 25 |
| 0.1 N KCl | $8.3 \times 10^{-1}$ | 25 |

frequently made to molten aluminum destined for electrical conductivity (E.C.) grade. The boron reacts with dissolved titanium and vanadium to form insoluble complex borides (Ti/V)$B_2$ in which Ti and V can occur in all proportions. The density of the product formed lies between that of the two pure compounds i.e. in the range $\rho = 4,500$kg/m$^3$ (Ti$B_2$) and $\rho = 5,100$kg/m$^3$ (V$B_2$) and is considerably higher than that of the molten aluminum: $\rho = 2,300$kg/m$^3$. Precipitated particles therefore tend to settle under the influence of gravity. Having first demonstrated the ability of the LiMCA technique to detect precipitated (Ti/V)$B_2$ particles in the laboratory, tests were then conducted in a production plant which produces E.C. grade wire by the Properzi process.

*References p. 465*

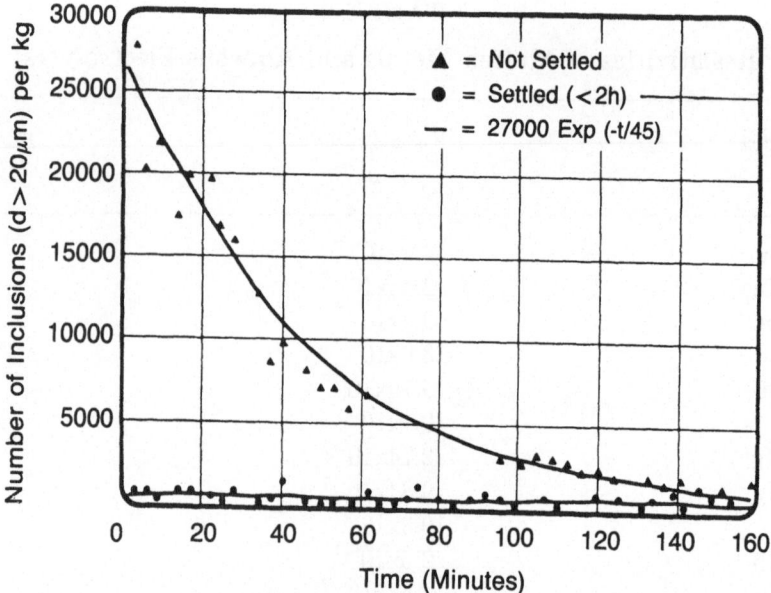

Figure 5. The sedimentation of $(Ti/V)B_2$ particles from conductivity grade aluminum melts following boron additions to a holding furnace. Plot of total number of inclusions greater than 20 microns, and less than 300 microns per kilogram of metal, versus time (minutes).

The standard practice at this installation is to allow for a period of quiescent settling as the last step in the batching procedure prior to casting. The lower curve in Figure 5 shows the results obtained using the LiMCA technique to monitor the metal cleanliness at the furnace outlet following standard batching practice.

To investigate the kinetics of the settling process one batch was followed in which the metal was cast immediately after the molten aluminum in the furnace had been mechanically stirred. This resulted in a high number of inclusions initially, the metal cleanliness following the upper curve given in Figure 5. It was found that the settling curve in this latter case could, as a reasonable approximation, be represented by a simple exponential decay function.

## EXAMPLE II. Melt Quality during Continuous Casting Operations in Aluminum

The practice of continuous casting, as opposed to semicontinuous direct chill casting, is becoming increasingly widespread throughout the aluminum industry. Alcan used this process for the continuous production of sheet at one of its Arvida complex plants. Due to the nature of the metal preparation process, several tandem furnaces are employed in which the batching operations including charging, melting,

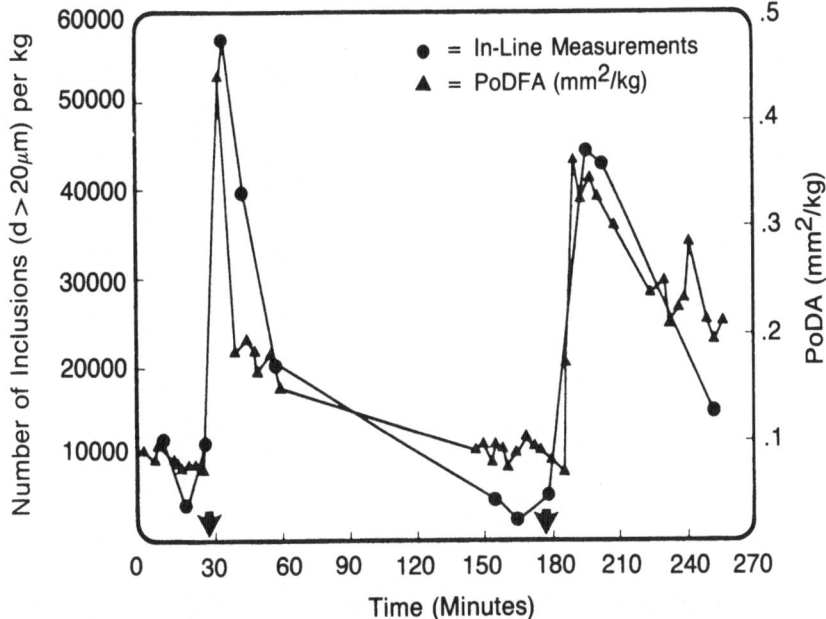

Figure 6. LiMCA and PoDFA analyses for second phase particulates in a transfer trough as a function of time into cast. Arrows indicate times at which furnace changes took place.

fluxing and holding, are carried out in essentially 'batch' mode, each of which is then used, in turn, to provide an uninterrupted metal feed to the casting machine.

In order to examine on-line metal cleanliness during several feeding cycles, the LiMCA analyser was installed up-stream of the caster at a location within the metal launder common to all furnaces.

Figure 6 shows the results obtained as a function of time starting near the end of one furnace batch, continuing for the entire duration of a second batch and showing the first 60 minutes of a third batch. Parallel metal filtration samples (PoDFA) were also taken throughout the test period. As seen, an excellent correlation between the two techniques is apparent.

Figure 7 shows the particle size distribution as measured by LiMCA at three successive times following the first furnace change-over. Here time zero corresponds to the first arrow marker at the 28th minute in Figure 6.

In discussing these data, perhaps the most striking observation metallurgically, is the fact that molten aluminum contains a considerable number of suspended inclusions even after melt cleaning treatments such as settling, fluxing and filtration. For instance, in the laboratory, and even using pure metal and prolonged (days) settling periods, "zero" count rates were seldom observed. This is not particularly surprising considering that the detection limit of the device, based on detecting one 16 $\mu$m particle in 16 ml of aluminum, corresponds to 100 parts per trillion ($10^{12}$) on a volume basis!

References p. 465

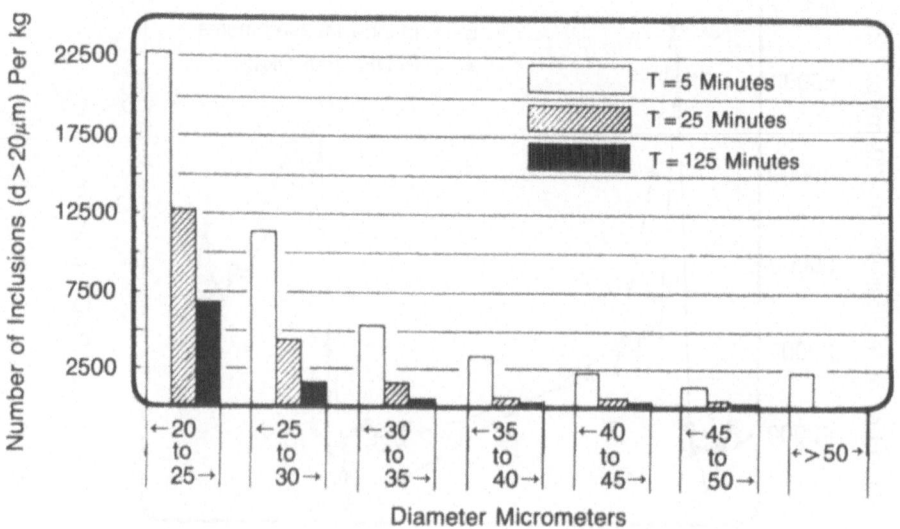

Figure 7. Sedimentation of titanium diboride particles expressed in terms of size distribution, 5, 25 and 125 minutes after a furnace change over. Note that large particles (e.g. 50μm) settle out more rapidly than smaller particles.

The ability of the technique to quantitatively detect and measure titanium and vanadium diborides is, at first glance, surprising since the reported electrical resistivities of these compounds at 700-720°C are of the same order of magnitude as that of molten aluminum. Theoretically the effect of passing particles that are more conductive than the supporting fluid through the sensing zone should be to produce inverted signals (i.e. a net decrease in the orifice resistance). Indeed, this phenomenon has been observed by, for instance, allowing the metal to partially solidify while sampling. The resistive pulses produced as a result of boron addition were however all positive indicating that the precipitated particles behaved as though they were less conductive than molten aluminum. If the conductivity of these particles is comparable to, but somewhat lower than that of the molten metal, then Equation 15 would have to be modified to allow for particle conductivity. However, the metallographic results that were obtained by the inspection of polished sections also revealed the existence of particle clusters in the size range predicted by the LiMCA results.

The exponential settling curve exhibited by the boron treated metal in Figure 5 was measured in a tilting-type furnace, in which metal is withdrawn from the top-most zone. It is postulated that natural convection currents resulting from heat losses through the upper part of the furnace are of sufficient magnitude to maintain particles in the range $d = 20$ to 50 μm in suspension. Particles that reach the bottom surface of the furnace are however removed, or remain trapped in a quiescent boundary layer. If the overall metal circulation remains relatively constant throughout the cast, then the rate of particle drop out to the furnace bottom should be proportional to the concentration of suspended particles. This leads directly on to an exponential decay curve of particle concentration versus time of the type observed in Figure 5.

However, at the present time there are too many unknowns, including the absolute size and effective densities of the $(Ti/V)B_2$ particles, as well as the patterns of metal circulation within a tilting reverberatory furnace, to speculate further.

The foregoing comments also apply to the results obtained for the example involving continuous casting (Figure 6). Again the metallographic results indicated that the predominant inclusion type is a metal boride ($TiB_2$). Inspection of Figure 6 clearly shows that the LiMCA and PoDFA results are in excellent agreement. After each furnace change there occured a large increase in the level of suspended inclusions which decayed throughout the cast and again rose sharply upon subsequent changeover.

There were two sources of $TiB_2$ inclusions: the grain refining addition that was made continuously upstream of the sampling point, and recycled grain refined scrap (edge trim, damaged coils). Since the grain refiner was added continuously it should, in principle, have contributed a constant level of suspended particulates into the metal. The scrap on the other hand, was charged to the furnaces where the boride particles have the opportunity to agglomerate and settle. Such sedimentation is believed to be the cause of the cyclical variation observed in the inclusion content of these melts. The possibility that the "bursts" are the result of entrained refractory particles washed off the fresh sections of launder exposed after furnace changeover can be ruled out on the basis of the PoDFA results and the fact that stirring the metal in the transfer trough upstream of the sampling point had no measurable effect on the LiMCA results.

Figure 7 was included to demonstrate the ability of the present technique to provide particle size distribution data. It is seen that there is a large preponderance of smaller particles and that after a furnace change, the large particles sediment faster than the smaller ones.

### EXAMPLE III. Melt Quality in an Inductively Stirred Melt of Steel

While no work has so far been conducted at commercial steel making plants, there has been considerable activity in McGill University, attempting to perfect a probe for low carbon steel applications.[9] So far, tests have been carried out successfully over temperatures ranging between 1300°C and 1400°C, using a 'metallic glass' iron alloy.

Thus, Figure 8 illustrates a test on a 5%B, 3%Si steel, at 1400°C, in which the number density of borosilicate inclusions within the melt following melt down in an inductively stirred furnace was raised by passing air through the melt for one minute, at a flowrate of 4 Nl/min. As anticipated, further oxidation of boron

*References p. 465*

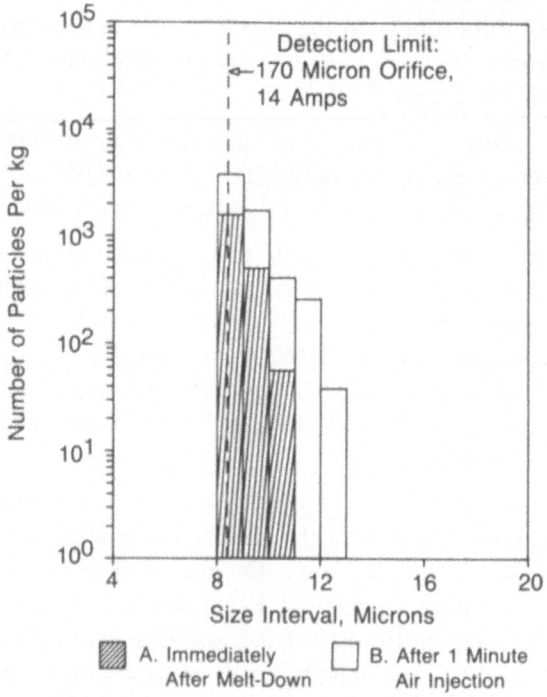

**Figure 8.** Size distribution of inclusions immediately after melt down, and following one minute of air injection, in an inductively stirred transformer steel melt.

and silicon led to an increased number of particles in all size ranges, with a shift towards larger inclusions.

There is obvious interest in detecting the presence of small inclusions for any thin sheet casting operations. Thus, the smallest orifice size used to date with the present E.S.Z. instrument on this alloy steel has been 120 μm. With the melt being sampled using a current of 5 amperes, the calculated detection limit was for particles of 7.8 μm diameter. As orifice sizes are reduced, the detection limit diameter is lowered and the inclusion population density generally rises sharply. For such conditions, particles enter the E.S.Z. in rapid succession. This is illustrated in Figure 9. If the particle population is too high, coincidence can be a problem. This is measured as the likelihood of two or more particles entering the sensing zone at the same time. Inclusion coincidence may cause an increase or a decrease in the true counts depending on whether it is predominantly of a primary or of a secondary nature.[10] For instance, if two inclusions, both below the detection limit pass through the sensing zone simultaneously, they can augment their signals and be counted as a single larger particle. This is generally not a problem much beyond

# DETECTION, BEHAVIOUR AND CONTROL

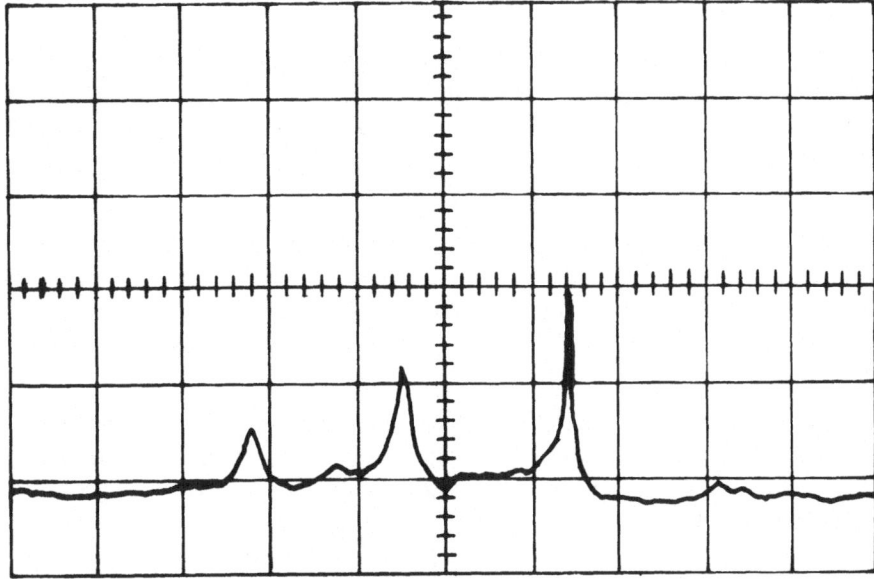

Figure 9. A train of inclusions passing through a 120μm orifice set within a quartz sensing probe, as detected by an oscilloscope. The horizontal intervals represent a 1 millisecond, the vertical intervals 50 microvolts.

the lower limit of detection. As seen, unambiguous pulses of the correct time period can be clearly distinguished in Figure 9.

## THE CONTROL OF INCLUSIONS IN MELTS

In order that inclusions within melts be controlled, the various reasons for their presence must be known. As illustrated in the previous text, inclusions in melts tend to be ubiquitous, and an inevitable consequence of deoxidation, desulfurisation, alloying and other refining and metal transfer operations.

For instance in the iron foundry industry, the use of magnesium-ferrosilicon alloys (e.g. the Inmold process) for modifying cast iron structures will naturally lead to the formation of magnesium silicate and sulfide type inclusions. Similarly, in the refining of aluminum melts, molten salt droplets (inclusions) are generated when fluxing with argon-chlorine gas. The gas mixture bubbling through the melt reacts with dissolved sodium, potassium and magnesium to form mixed salts. These are presumably sheared off the bubble surfaces to form minute salt droplets, or inclusions, within the melt.

Other inclusions derive from the chemical, or physical, degradation of refractory materials containing molten metals/slags. Thus a fireclay bricked ladle containing an aluminum killed steel can be chemically attacked and reduced by steel, owing to a low oxygen activity within the melt. Large aluminum silicate type

*References p. 465*

inclusions will be found.

Hydrodynamic erosion of the container material can be another factor leading to the presence of second phase particulates within the melt. Yet other sources of inclusions are droplets entrained from overlaying slags or salt layers. These can result during mixing operations such as submerged gas bubbling, or be caused by slag vortexing, and similar hydrodynamic phenomena.

Finally all exposed metal streams coming into contact with air are prone to reoxidation, and inclusion formation. It is for such reasons that streams passing from a ladle to a tundish, and then to an oscillating mold during continuous casting are conventionally protected by tightly sealed, magnesia nozzles/shrouds, with submerged entry ports.

The above remarks show that adventitious inclusions can be controlled by good metal transfer operations, by using stable refractories, and by carefully controlling metal flows. However, for those inclusions which are an inevitable consequence of refining and/or alloying operations, partial removal is generally possible through making use of a particle's natural tendency to sediment or to float. For sedimentation, or flotation, plug flow and an absence of turbulence would generally be expected to achieve the best results, and this is illustrated in Figure 10, where the separation of inclusions within a full scale model tundish[11] are shown to be a function of tundish flow conditions (dams, weirs) and Stokes rising velocity or inclusion diameter. As expected, larger inclusions rise faster to the tundish surface and are therefore removed in greater quantities. However, in practice, anomalies exist.

Solid oxide inclusions in steel melts tend to be extremely ferro-phobic, and turbulent agitation may lead to inclusion coalescence, sintering, and certainly attachment to the refractory side walls of a container vessel. Tundish nozzle blockage by solid inclusions such as alumina, or zirconia, are well known examples of this phenomenon.

Filtration of metals is an obvious answer to the control and elimination of such inclusions. Thus in aluminum processing operations, critical (inclusion sensitive) alloys are commonly cast through an in-line filtration unit as a last line of defense prior to solidification.

Figures 11 and 12 compare the respective efficiencies of a commercially available single use filter with that of a deep bed type of filter. Extensive in-plant testing of both filter systems using LiMCA equipment has shown that the removal efficiency for particles larger than $20\mu$m, is normally 30 to 60% for single use filters, during a typical cast of 75 minutes, whereas a deep bed filter will usually remove more than 90% of all such inclusions over the course of its useful life ($\sim$1-2$\times 10^6$ kg of Al). These results were obtained at a single casting center while producing the same alloy. One should note however, that liquid inclusions will not generally be trapped by such devices, and in some circumstances, there is evidence

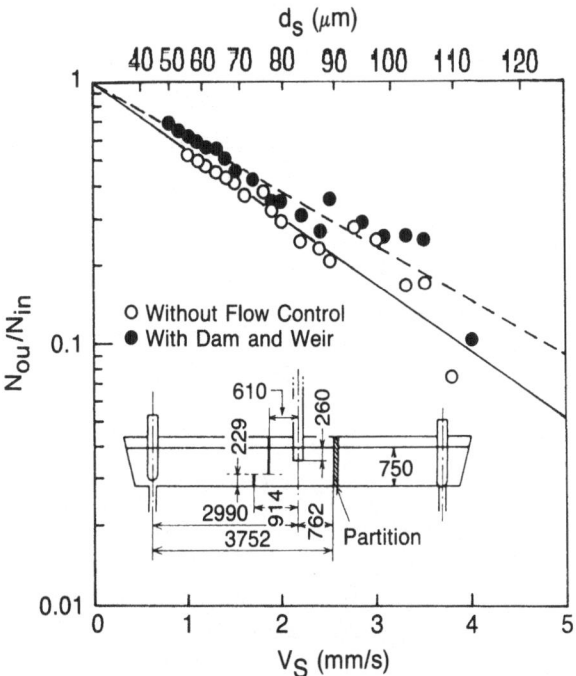

Figure 10. A semi-logarithmic plot showing the number of glass sphere inclusions exiting a full scale water model of a 30 ton twin slab caster arrangement, as a function of inclusion diameter ($d_S$), or Stokes rising velocity($V_S$), with and without the use of dams and weirs. Hydrophyllic glass microspheres (S.G. =0.22), water flowrate $7.6 \times 10^{-3} m^3 s^{-1}$.

that they may actually harm a filter's ability to collect solid particles. Similarly, any flow surges can cause major quality problems.

## CONCLUSIONS

An electric sensing zone method (LiMCA) for particle size analysis has been successfully developed for use in liquid metals. The technique is rapid, quantitative, and extremely sensitive. It provides both the concentration and particle size distribution of suspended inclusions in melts with results available at approximately two minute intervals.

The technique has been successfully demonstrated in a number of molten metal systems at temperatures ranging from ambient to 1600°C. Experiments carried out in the aluminum industry have proven that the technique can generate significant results in a harsh industrial environment.

On-line metal quality measurements show metal cleanliness to be a dy-

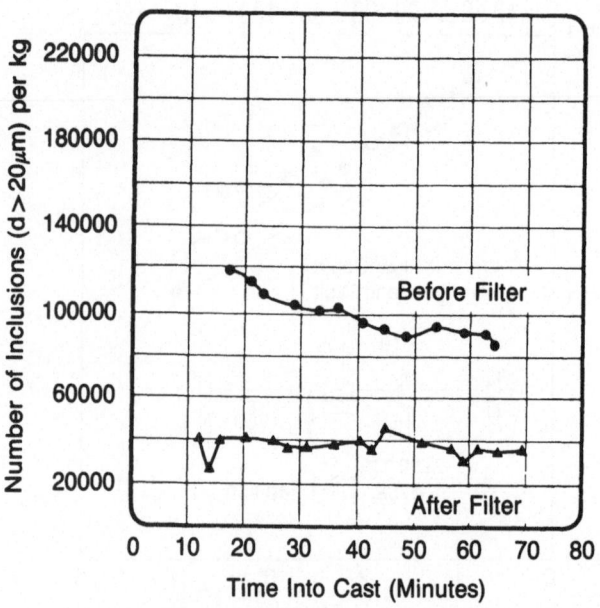

Figure 11. Inclusion number density on entry and exit to a single use filter versus time into cast.

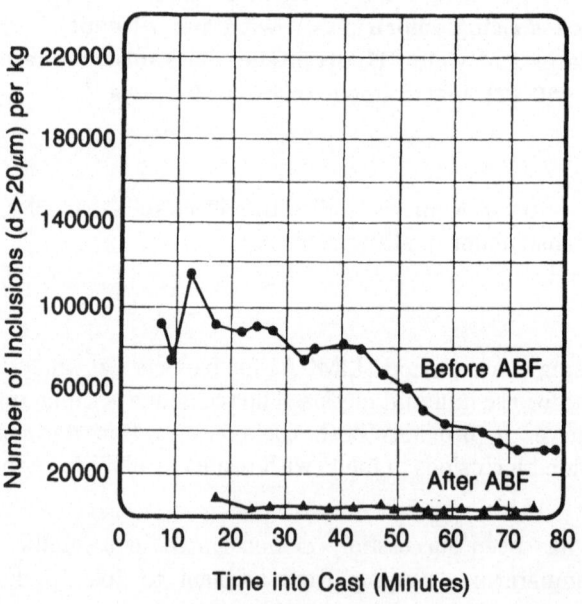

Figure 12. Inclusion concentrations on entry and exit to a deep bed filter (Alcan Bed Filter) versus time.

namic phenomenon that will require continuous monitoring in order to guarantee consistency.

Inclusion levels within melts can be controlled through eliminating all adventitious sources, by good hydrodynamic design of slag/metal flow systems, and by filtration techniques.

## REFERENCES

1. D. A. Doutre,; "The Development and Application of a Rapid Method of Evaluating Molten Metal Cleanliness," Ph.D. Thesis, McGill University, 1984.
2. R. I. L. Guthrie, "The Detection of Non-Metallic Inclusions in Liquid Metals by a Direct Method," *Steel Industry/Researcher Sensor Research Workshop*, May 1984, CSIRA/CANMET, Burlington, Ontario, 33–38.
3. R. W. DeBlois and C. P. Bean, "Counting and sizing in of sub-micron particles by the resistive pulse technique," *Rev. Sci. Instr.*, *41*, 7, (1970), 909–915.
4. W. R. Smythe, "Flow Around a Sphere in a Circular Tube" *The Physics of Fluids*, *4*, 6, (1961), 756–759.
5. W. R. Smythe, "Flow Around a Sphere in a Circular Tube" *The Physics of Fluids*, *17*, 5, (1964), 633–638.
6. R. W. DeBlois, C. P. Bean, and R. K. A. Wesley, "Electrokinetic Measurements with Sub-micron Particles and Pores by the Resistive Pulse Technique", *J. of Colloid and Interface Science*, *61*, 2, (1977), 323–335.
7. D. Doutre and R. I. L. Guthrie, "Method and Apparatus for the Detection and Measurement of Particulates in Molten Metal," *U.S. Patent* 4,555,662, Nov. 1985.
8. D. Doutre and R. I. L. Guthrie, "On-Line Measurements of Inclusions in Liquid Metals," *Int'l Symposium on the Refining and Alloying of Liquid Aluminum and Ferro-Alloys*, The Norwegian Instit. of Tech., Trondheim, Aug. 1985, 147–163.
9. S. Kuyucak, H. Nakajima and R. I. L. Guthrie, "On-Line Measurements of Inclusions in Liquid Steel," *Fifth International Iron and Steel Congress, Process Technology Proceedings*, Washington DC, (1986) 193–198.
10. H. Bader, J. R. Gordon, and O. B. Brown, "Theory of Coincidence Count Correction for Optical and Resistive Pulse Particle Counters," *Review of Scientific Instruments*, *43*, 10, (1972), 1407–1412.
11. H. Nakajima, S. Tanaka, F. Sebo, L. Dumitru, D. Harris and R. I. L. Guthrie, "On the Separation of Non-Metallic Inclusions from Tundishes in Continuous Casting Operations—a Water Model Study." *Fifth International Iron and Steel Congress, Steelmaking Proceedings*, Washington DC, (1986).

## DISCUSSION

**D. M. Stefanescu** *(University of Alabama)*

Did you examine particulate material in liquid cast iron by your method?

**A.** We have looked at molten cast irons, but have only carried out two or three tests, to date. We've melted 14 kg charges inductively, and detected a number of small inclusions in the size range 10-15$\mu$m in an iron containing 4.6%C, 1%Si,

0.15%Mn, 0.012%S and 0.035%P and temperatures ranging between 1250 and 1400°C.

### D. M Stefanescu *(University of Alabama)*

Since you have not examined cast iron in any detail, I would like to offer the following comment. In cast iron the main inclusion is graphite and it is present in much larger amounts than any other insoluble entity. About ten years ago I studied the electrical resistivity of solid and liquid iron. The interesting thing was that there was a remarkable difference between the electrical resistivity of spheroidal graphite in liquid cast iron and flake graphite in liquid cast iron due to the crystallographic orientation of the graphite.

This suggests that you could use this method to determine the degree of nodularity of a certain iron. I suggest you may want to look into this further because it could be of great practical value for assessing the effectiveness of the nodularizing treatment for cast iron before castings are poured.

### Y. Sahai *(Ohio State University)*

In your steel work did you find a nozzle clogging problem? The danger of clogging seems great because the opening in your probe is only about 200 microns in diameter.

A. That's right. One might anticipate nozzle clogging in view of the fact that much larger diameter tundish nozzles can become blocked by solid inclusions. Sometimes it can also happen with the LiMCA probe, although it's not a major problem. The last part of the work has been done by Mr. Selcuk Kuyucak and he has encountered the phenomenon. For instance, a 0.1% Al addition to a boron steel can sometimes precipitate sufficient alumina to block a 200$\mu$m diameter sensing zone. One of our solutions is to increase the sensing zone diameter to 400$\mu$m or to allow most of them to float out prior to melt sampling.

### G. Kipouros *(General Motors Research Laboratories)*

Do you have any problems with detecting $TiB_2$ in aluminum?

A. No. Theoretically one might expect not to detect them because the conductivity of $TiB_2$, is higher than the conductivity of aluminum. For some reason the current will be deflected by the interface, so that even if we have a conducting particle or an intermetallic one, our experiences to date show that we can still detect it.

# METAL REFINING BY FILTRATION

## DIRAN APELIAN and KYUNG K. CHOI
*Materials Engineering Department*
*Drexel University*
*Philadelphia, Pennsylvania 19104*

## ABSTRACT

Melt purification and removal of non-metallic inclusions from the melt prior to casting has much merit. The size, shape, type and distribution of non-metallic inclusions present in the finished metal product are the performance fingerprints of the cast shop. It is advantageous to remove these non-metallic particles as they are known to reduce fluidity of the melt, increase internal porosity in the casting, reduce strength, ductility and fatigue resistance of the product and also result in poor machinability and surface finish. In general, the larger the inclusions are, the greater are their deleterious effects. Non-metallic inclusions act as stress-raisers, and can cause premature failure of a specific component. They provide not only initiation sites for fracture but also play a significant role in the propagation of the crack. In this paper the emphasis will be on inclusion removal by filtration of the melt prior to casting.

The filtration process is not one of physical separation as in screening and separation, but rather melt filtration is a two step serial transport process. First, as a result of bulk fluid flow the inclusions are transported to the filter surface. In the second step inclusion capture occurs due to interfacial or surface forces. In general, particle transport can occur by impingement, interception, sedimentation, diffusion, and other hydrodynamic effects. Particle attachment can be a result of forces developed through pressure, chemical, or Van der Waal effects. The relative dominance of each mechanism is a function of particulate type and size, fluid approach velocity, as well as temperature and media characteristics. As an example, at high temperatures (steel melts) the inclusions sinter to the filter surface, whereas lower temperatures (aluminum melts) the inclusions are attached at the medium by secondary forces.

Accurate mathematical representations of filtration processes can be powerful tools for system design and process optimization. The theoretical basis for inclusion removal by filtration is developed and applied.

In addition, inclusion detection techniques will be briefly reviewed since the effect of the filtration process can only be verified by such experimental procedures: by chemical analysis, quantitative metallography, volumetric analysis, and non-destructive ultrasonic techniques.

The experimental work which is described and discussed in this paper includes filtration trials with aluminum, steel, and superalloy melts. In all of the experimental work, the inclusions being removed are well characterized. A distinction is made between solid and liquid inclusions in terms of inclusion capture mechanisms. The solid inclusions (such as $Al_2O_3$ in steel) are non-deformable, whereas the liquid inclusions ($CaO \cdot SiO_2$ in steel) are deformable. The effect of filter wetting characteristics with respect to filtration performance are discussed. Table 1 lists the melt systems investigated; filtration results for aluminum, superalloy and steel melts are presented and discussed.

## INTRODUCTION

Melt purification and removal of non-metallic inclusions from the melt prior to casting has much merit. The size, shape, type and distribution of non-metallic inclusions present in the finished metal product are the performance fingerprints of the cast shop. It is advantageous to remove these non-metallic particles as they are known to reduce fluidity of the melt, increase internal porosity in the casting, reduce strength, ductility and fatigue resistance of the product and also result in poor machinability and surface finish.[1,2,3] In general, the larger the inclusions are, the greater are their deleterious effects. Non-metallic inclusions act as stress-raisers, and can cause premature failure of a specific component. They provide not only initiation sites for fracture but also play a significant role in the propagation of the crack. Various methods of inclusion separation and removal have been reported.[4] In this paper the emphasis will be on inclusion removal by filtration of the melt prior to casting.

In general, the non-metallic inclusions present in most melts are an inevitable feature of the production route and can be classified as follows:

- *Exogeneous inclusions*: foreign particulates which have been added to the melt, inadvertently, such as coarse clusters of grain refining agent, or refractory particles from the vessel or reactor.

- *Deoxidation products and salts*: these particulates are suspended in the melt as a result of the prior metal treatment process, such as $Al_2O_3$ or $SiO_2$ deoxidation products in the steel melt or light metal halide salts in aluminum melts. The latter are liquid while the former are solid at casting temperatures.

- *Oxides of the melt*: these oxides are either floating on the melt surface and/or entrapped within the melt due to turbulence, e.g., $Al_2O_3$ in aluminum melts and entrapped slag particulates in bottom poured steelmaking ladles.

Molten metals contain solid and liquid inclusions. The former are rigid while the latter are deformable and may coalesce into larger globules or even form conduits in the filter. The operative capture mechanisms for the removal of liquid inclusions are distinctly different than those for solid inclusions.[5]

Filtration is not a new processing method for inclusion removal. Specifically, aluminum alloys have been commercially filtered for over 16 years.[6,7] The initial impetus for the genesis of this process was the need for inclusion free metal for manufacturing aluminum foil. Steels are not commercially filtered today, however the feasibility of inclusion removal from steel melts by filtration has been demonstrated in the laboratory and in the pilot plant.[6,7]

Today, with the drive for automation, flexible manufacturing and process control, the need is more than ever to establish the fundamentals of inclusion removal from molten metal. Process control, scale-up and optimization is only possible with an understanding of the operating mechanisms.

## BACKGROUND

The filtration process is not one of physical separation as in screening and separation, but rather melt filtration is a two step serial transport process. First, as a result of bulk fluid flow the inclusions are transported to the filter surface. In the second step inclusion capture occurs due to interfacial or surface forces. At high temperature, these inclusions sinter to the filter surface. Figure 1 shows a schematic representation of the inclusion size with respect to the average filter pore diameter. For example, compare the size of the 2.5 $\mu$m inclusion *vis a vis* a diameter of 1270 $\mu$m for the 10 ppi (pores per inch) ceramic foam filter. It is quite evident that physical separation is not the mode of capture.

In general, particle transport can occur by impingement, interception, sedimentation, diffusion, and other hydrodynamic effects. Particle attachment can be a result of forces developed through pressure, chemical, or Van der Waal effects. The relative dominance of each mechanism is a function of particulate type and size, fluid approach velocity, as well as temperature and media characteristics.

**Theory**—Accurate mathematical representations of deep-bed filtration processes can be powerful tools for systems design and process optimization. Two types of mathematical models are commonly used to describe deep-bed filtration:

1) parametric or functional models based on fluid flow mechanics through porous media. The objective of these models is to develop relationships between filter media and operating variables, such as filtration pressure and flow velocity of the melt.

2) phenomenological or kinetic models predicted on kinetic rate expressions that are developed for a particular filtration system. Kinetic models are useful in predicting filter performance.

The literature is abundant with phenomenological models of various investigators in low temperature systems. Of particular interest are the discussions by Ives,[8] Herzig, *et al.*,[9] Tien, *et al*,[10] and Kraj.[11] Herzig and Tien provide simplifying transformations that result in easily solved ordinary differential equations.

*References pp. 490–492*

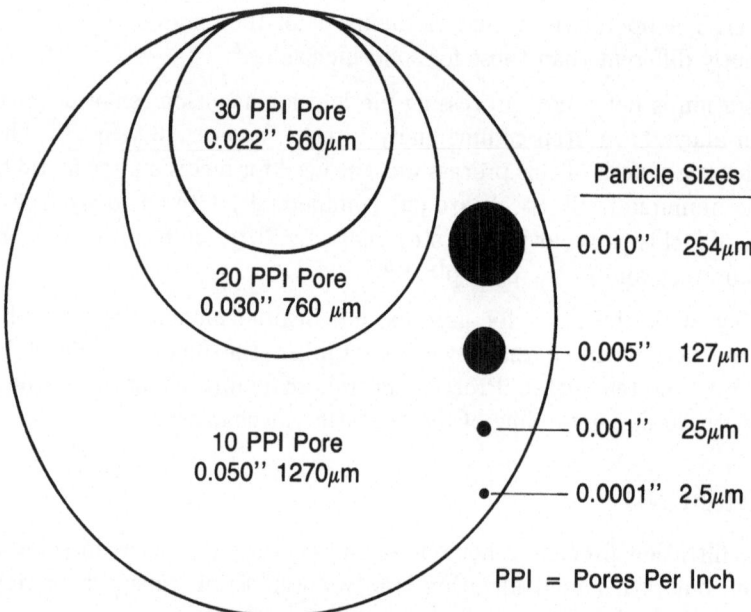

Figure 1. Schematic representation of the inclusion size with respect to the filter pore diameter.

The first application of kinetic models to melt deep-bed filtration was made by Apelian and Mutharasan.[12,13] The fundamental equation

$$\frac{\partial \sigma}{\partial \tau} + U_m \frac{\partial C}{\partial Z} = 0 \tag{1}$$

was obtained by constructing a mass balance over a differential volume element of the filter bed. Where:

$\sigma$ = volume fraction of retained inclusions in the filter

$\tau$ = filtration time, compensated for initial time delay for particles to reach filter depth, $Z$.[10]

$U_m$ = melt superficial velocity, volume melt · time$^{-1}$ · filter frontal area$^{-1}$

$C$ = volume fraction of inclusions in the melt

$Z$ = flow direction dimension

With use, the inclusions are entrapped on the surface of the filter grains. The rate of change of entrapped inclusions per unit of filter volume, $\partial \sigma / \partial \tau$, is a function of inclusion concentration in the melt.[14] That is:

$$\frac{\partial \sigma}{\partial \tau} = KC \tag{2}$$

where K, a kinetic parameter, is a function of the melt physical properties, melt flow rate, filter properties and the shape and size of the inclusions. K is also a function of $\sigma$,[13] obeying the relationship

$$K = K_o \left[1 - \frac{\sigma}{\sigma_m}\right] \tag{3}$$

where $K_o$ is the kinetic parameter coefficient and $\sigma_m$ is the inclusion retention capacity of the filter bed. Combining Equations 1-3, results in a first order partial differential equation which gives the concentration of inclusions in the melt as a function of time and distance along the filter.[13] If it is assumed that the inclusion retention capacity of the filter is large compared to the entrapped inclusions; that is $\sigma/\sigma_m \approx 0$, the solution to the partial differential equation simplifies to [13]

$$\frac{C_z}{C_i} = \exp\left[-\frac{K_o Z}{U_m}\right] \tag{4}$$

where: $C_i$ and $C_z$, respectively, are the volume fractions of inclusions in the melt at the filter entrance and at a depth, $Z$, in the filter.

Applying Equation 4 to conditions at the exit of the filter gives

$$\frac{C_o}{C_i} = \exp\left[-\frac{K_o L}{U_m}\right] \tag{5}$$

where $C_o$ = the outlet volume fraction of inclusions in the melt

$L$ = the height of the filter bed.

Experiments can be designed to measure $C_i$, $C_o$, $U_m$ and $L$, and thus $K_o$ can be calculated from Equation 5. Once $K_o$ is experimentally evaluated, filter efficiency can be predicted. Filter efficiency $\eta$, is defined

$$\eta = \frac{C_i - C_o}{C_i} \times 100\% \tag{6}$$

Use of Equation 5 in the above yields

$$\eta = \left[1 - \exp\left[-\frac{K_o L}{U_m}\right]\right] \times 100\% \tag{7}$$

The preceding suggests that inclusion removal efficiency can be improved by either increasing filter length or by decreasing melt velocity. The model assumes that the inclusions are solid and non-deformable; the term $K_o$ is unique to a particular filter-melt-inclusion system.

**Filter Performance Evaluation**—An important experimental detail is the manner in which filter performance is evaluated. There are essentially three generic methods of measuring metal cleanliness.

*References pp. 490-492*

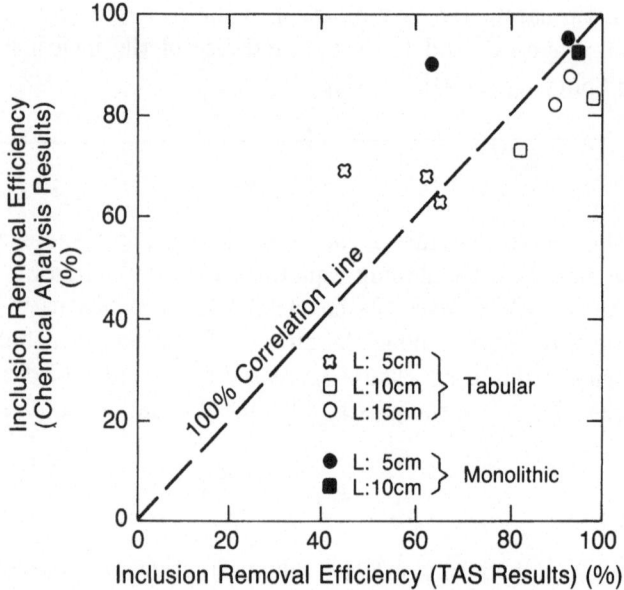

Figure 2. Comparison of inclusion removal efficiency data based on TAS and chemical analysis results indicating a strong correlation between the two methods of analysis.

- Chemical Analysis

Elemental composition is determined either *via* wet or spectographic analysis in unfiltered and filtered samples. Filter performance is assessed by the efficiency parameter, $\eta$ given by Equation 6.

- Quantitative Metallography

With the advent of computers and image analysis, equipment now exists which differentiates between 100 shades of gray. Quantitative metallography, once a tedious task, is now carried out fairly rapidly yielding data on the number of inclusions observed per square millimeter. Figure 2 shows a comparative analysis between metallographic data (TAS) and chemical analysis.

- Volumetric Analysis

The previous two techniques have the disadvantage that the results may depend on sampling method and sectioning technique (metallographic preparation). These vulnerabilities are circumvented by remelting a given volume of metal and forcing the inclusions to settle or to float depending on their density. Subsequently, the inclusions which are now "concentrated" are analyzed by conventional methods. There are two methods of analysis based on this concept: (i) electron beam melting (EB), and (ii) centrifugal separation.

The EB method involves drip melting of a sample under vacuum using an electron beam, and collecting the molten metal in a hemispherical, water-cooled mold. During the melting and solidification of the EB-specimen (button), the nonmetallic particles rise to the surface and are subsequently concentrated in a central floating raft. The particles in the raft can be removed chemically for quantitative analysis, or can be observed directly using optical, SEM, image-analysis, or other techniques. Estimates of the nonmetallic particle content are made from area measurements of magnified photos of the rafts. Cleanliness is determined by the size of the specific raft area (i.e., cm² per kg of sample); the smaller the specific area, the cleaner the sample. Figure 3 shows the methodology of the EB technique. Figure 4 shows the button and the central raft of concentrated inclusions.

The centrifuge method involves melting of the whole sample and centrifuging the heated crucible. The inclusions will separate under centrifugal force. This technique has been used successfully by Simensen[15] and his colleagues.

**Solid Versus Liquid Inclusions**—As pointed out earlier, solid inclusions in general are non-deformable whereas liquid inclusions are deformable. In addition, the wettability of the filter by the melt and the inclusion phase must be taken into account. In this paper, we give an overview of inclusion removal by filtration in aluminum, superalloy and ferrous melts. However, in our laboratory at Drexel University, we have investigated a variety of model systems (hot metal systems as well as low-temperature model systems) as shown in Table 1.

In the following sections, an overview of these results are presented.

**Filtration of Aluminum Melts**—Traditionally filtration of aluminum has been associated with products such as aluminum foil. More recently the increasing use of aluminum in drawn thin walled beverage cans, computer memory discs, demanding aerospace situations and decorative applications such as automotive trim, *etc.* has placed a renewed emphasis on melt cleanliness. Impurities present in the melt either as solid or liquid particles are known to reduce the fluidity of the melt, cause internal porosity in castings, reduce mechanical strength, ductility and fatigue resistance of the product and also result in poor machinability, surface finish, *etc.*[16-21]

The particles present in molten metal prior to casting are an inevitable feature of the production route. It has been found from industrial observation that the majority of the above particles which are deleterious to product quality lie in the range of 1-30 μm and are dilutely suspended. Removal of these non-metallic inclusions may be accomplished by filtration.[6,7]

Inclusions deposit onto the grains of the filter medium due to diffusion, direct interception, gravity, and/or surface forces. Mechanical entrapment has been observed to be responsible for filtration of inclusions larger than 30 μm, whereas, it is believed that surface forces are responsible for the retention of inclusions smaller than 30 μm. In a depth filter the inclusions are dispersed through part or all of its volume (depth). It thus has the advantage of having a large surface area for entrapment and can trap particles much smaller than the pores in the filter bed.

We have studied two types of filters which have wide scale industrial applica-

## TABLE 1
### Hot Metal Systems and Low-Temperature Model Systems Investigated at Drexel University

| Melt | Inclusion Phase | Inclusion Size Range, $\mu$m | Filter Material and Geometry | Melt Velocity, cm/s |
|---|---|---|---|---|
| Aluminum at 1020 K | $TiB_2$ | 1-10 | Tabular Alumina (0.1-0.5 cm size) | 0.05-0.78 |
| | | | Ceramic Foams (30 ppi) | 0.06-0.78 |
| | | | Monolithic Alumina 300,400 cells/in$^2$ | 0.16-1.83 |
| | | | Monolithic Cordierite 400 cells/in$^2$ | |
| Superalloys (Ni-based) | $Al_2O_3$ $HfO_2$ | 0.25* 5.7* | Ceramic Foams (20, 30 ppi) Monolithic Mullite 240 cells/in$^2$ | 0.08-1.30** |
| Steel at 1873 K | $Al_2O_3$ | 0-5 | Tabular Alumina 6.2-0.6 cm size | 0.03-10 |
| | Manganese Silicates (liquid) | 1-10 | Ceramic Foams Alumina (10,30,50 ppi) | 0.2-0.10 |
| | | | Monolithic Alumina 400 cells/in$^2$ | 0.04-0.15 |
| Hydrocarbon (Diesel fuel) at 298 K | $CaCO_3$ | 2-3 | Tabular Alumina (0.1-0.8 cm size) | 0.02-0.2 |
| Distilled Water | 3-3' dimethyl diphenyl (0.01 %) | 3-7 | Glass Beads (3,5 mm) | 0.015-0.055 |
| | | | Silicone Oil Coated Glass Beads (3,5 mm) | 0.015-0.055 |
| Chloroform +Ethyl Phthalate | Distilled Water + Dye | 100-500 | Glass Bead (3,6 mm) | 0.025-0.309 |
| | | | Silanized Glass Beads (3,6 mm) | 0.025-0.309 |
| | | | (3,6 mm) | 0.025-0.309 |
| | | | Teflon Balls (6mm) | |
| Distilled Water | Iodine + Kerosene | 100-500 | Glass Beads (3,6 mm) | 0.025-0.309 |
| | | | Silanized Glass Beads (3,6 mm) | 0.025-0.309 |
| | | | Teflon Balls (6 mm) | 0.025-0.309 |

\* units are cm$^2$/kg    \*\* units are kg/s

# METAL REFINING BY FILTRATION

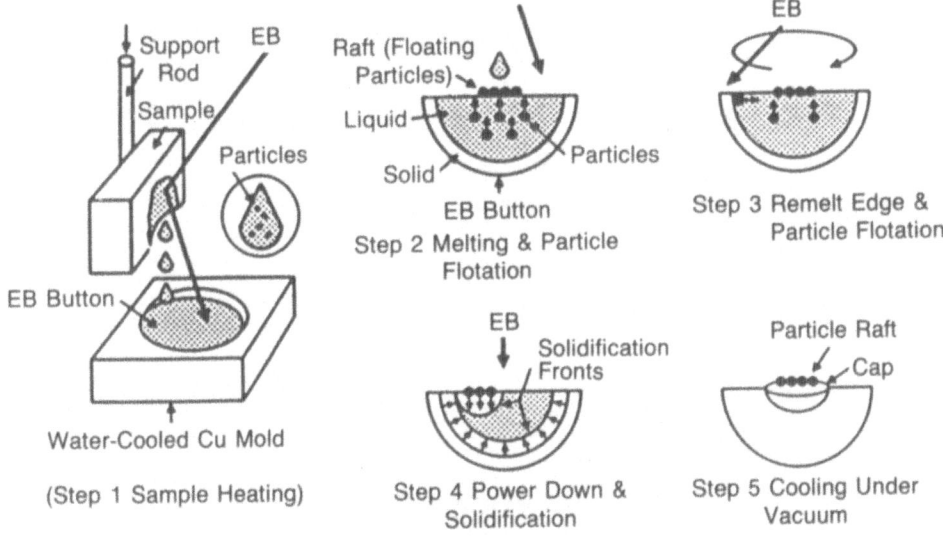

Figure 3. Procedures for producing EB-button specimen for cleanliness evaluation.

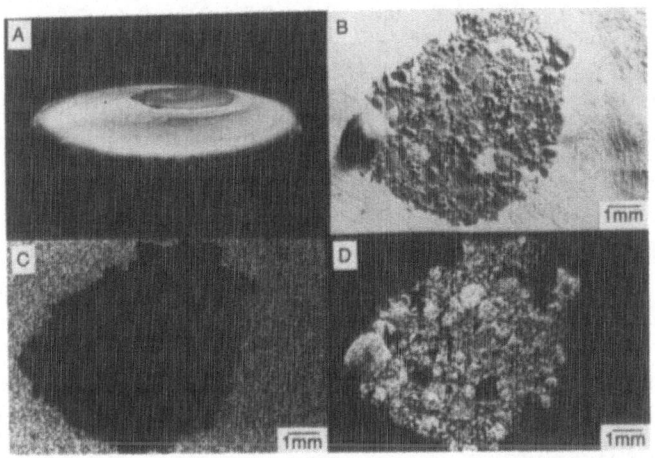

Figure 4. Photos of particle rafts on top of IN-718 EB-buttons (A). Photos (B), (C), (D) show SEM Image, Ti and Al X-ray maps, respectively.

tion for aluminum: the deep bed filter and the ceramic foam filter. The deep bed filter consisted of a packed, 50 mm bed of 1-3 mm tabular alumina particles with a bed void fraction of 0.48. The ceramic foam filter consisted of a sheet of open cell, 30 pores per linear inch, ceramic foam, with 0.85 volume void fraction.

The melt used in these studies was 70 kg of P1020 aluminum at 750°C.

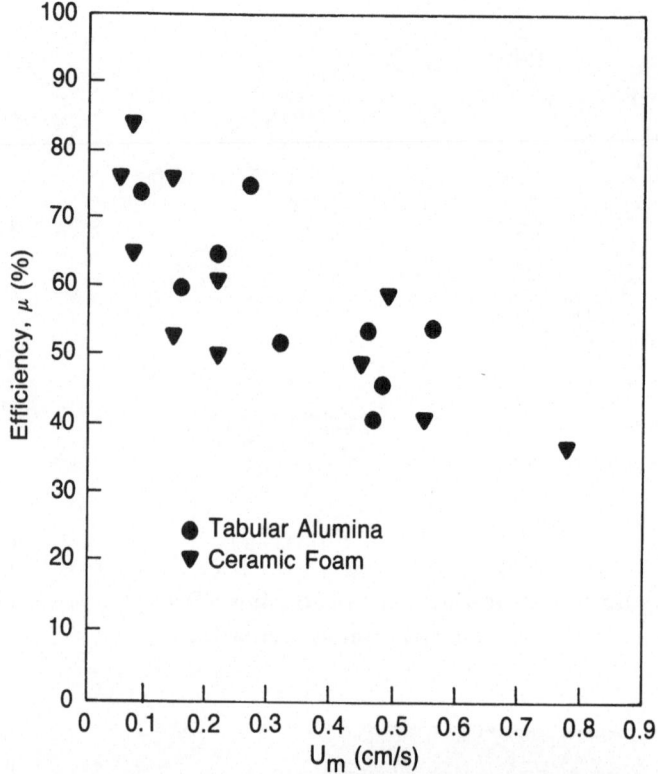

Figure 5. Filtration efficiency as a function of melt velocity for tabular alumina and the ceramic foam filters, each 50 mm thick.

TiB$_2$ was chosen as a tracer inclusion to monitor the filtration efficiency. To form these "synthetic inclusions" a predetermined amount of titanium diboride (Al-5%Ti-1%B) was added to the melt. Titanium diboride "synthetic inclusions" were chosen because: (i) the size of TiB$_2$ particulates is in the critical size range required for this study, i.e. 1 to 30 $\mu$m; (ii) ease of quantitative analysis of Ti and B in aluminum via spectrographic and metallographic techniques; (iii) availability of Al-Ti-B master alloys.

The filtered metal was collected under the exit orifice of the filter using a series of molds. By collecting the metal over a specific period of time and determining its weight the melt flow rate through the filter was calculated.

Figure 5 shows the filtration efficiency of the ceramic foam and deep bed filters evaluated on the basis of superficial velocity. The efficiency of each filter decreased with increasing melt velocity. Filtration efficiencies higher than 35% were observed for both filters at all melt flow rates studied. Figure 5 shows that the ceramic foam and the tabular alumina filters practically had identical efficiency profiles.

To compare the fundamental bed characteristics, the parameter $K_o$, which

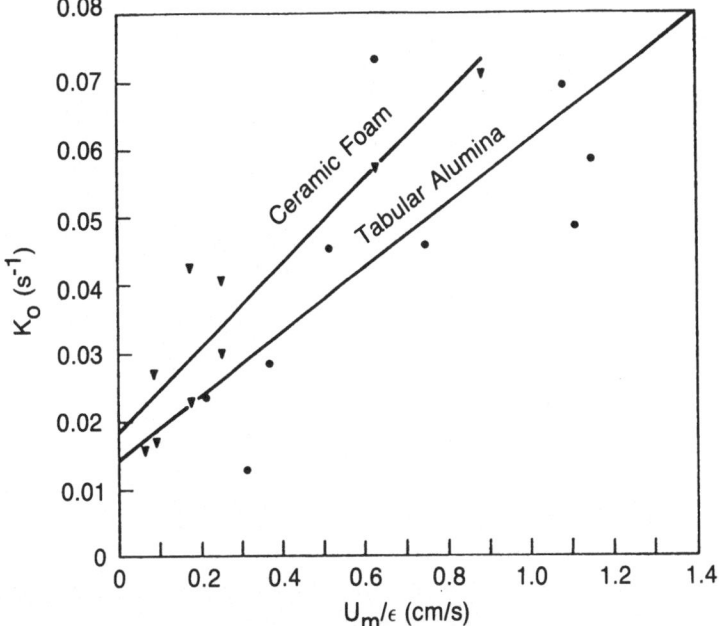

Figure 6. The kinetic parameter, $K_o$, as a function of interstitial melt velocity, $U_m/\epsilon$ for the ceramic foam and tabular alumina filters.

denotes the particle capture kinetics, was determined for each filter as a function of flow rate using Equation 5. The results of this analysis are shown in Figure 6. The linear relationships in Figure 6 were determined by a least squares analysis. The results show that in all cases the inclusion capture kinetics increases with increasing melt flow rate, however the slopes and intercepts vary for the filters evaluated.

The overall trends and variations between the results agree with those predicted by the proposed model. The increase in the parameter $K_o$ with melt velocity, $U_m$, can be explained qualitatively from an inclusion trajectory concept. Figure 7 shows fluid streamlines around a spherical filter grain. The limiting trajectory is the one which passes the spherical filter grain at a distance of $d_i/2$ where $d_i$ is the inclusion diameter. Here, it is hypothesized that the primary mechanism of particle entrapment is interception by the filter grain. As one increases the fluid velocity across the filter grain, the inclusion having a higher mass inertia deviates from the fluid streamlines and tends to impact the filter grain surface. Also, as the fluid velocity increases, the number of inclusions which approach the filter grain also increases in proportion to the velocity. Since $K$ is a measure of inclusion entrapment per unit time (see Equation 2), we find experimentally that $K$ increases with melt velocity.

Although there is considerable scatter in the data in Figure 6, the relative

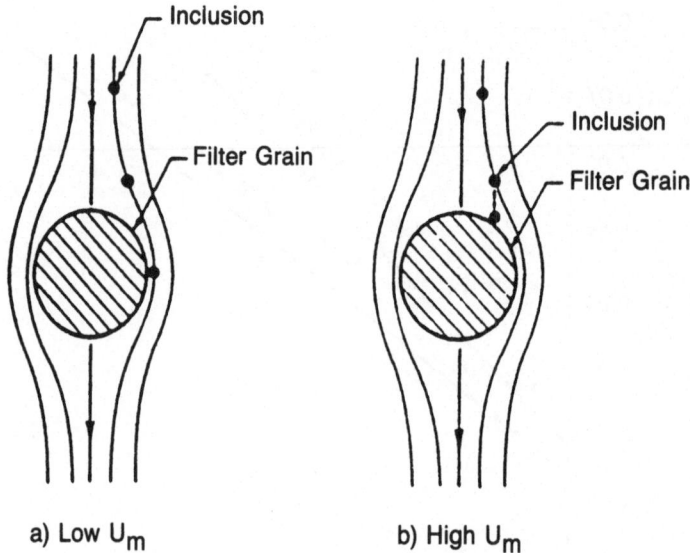

**Figure 7.** The effect of velocity on inclusion trajectory near a spherical filter grain.

positions and slopes of the lines in this figure may also be explained by reference to the proposed model: it was postulated that the intercept of the line with the $K_o$ axis would be a measure of a fundamental bed parameter, intrinsically related to factors such as bed tortuosity, etc.[14] Thus we would expect the ceramic foam filter, with its cellular structure and greater resistance to channeling, to have a higher intercept than the deep bed filter. This is in accord with the experimental results.

Comparison of the slope of the lines for the ceramic foam and the tabular alumina filters in Figure 6 would imply that the ceramic foam, having a somewhat higher slope, has a greater capture kinetics/flow rate dependence than the deep bed filter. This implies that foam filters may perform better than tabular alumina filters at very high melt velocities.

Further evidence suggesting the importance of particle inertia in particle entrapment was obtained from physical modelling studies. Low temperature model system experiments consisting of hydrocarbon (simulating aluminum melt) and calcium carbonate (simulating inclusions in aluminum with sizes of 2-3$\mu$m) was conducted in the velocity range of 0.02-0.08 cm/s. The effect of the Reynolds number for flow in a porous bed on the kinetic parameter is shown in Figure 8 for both Al-TiB$_2$ and the hydrocarbon carbonate system. At low Reynolds numbers, $K_o$ is a weak function of Reynolds number because the inclusion inertia is low due to low fluid velocity.

**Filtration of Superalloy Melts**—The need for cleanliness and tighter control over the content and size of nonmetallic particulates (inclusions) in superalloys is a real and a severe one. The newer, advanced superalloys for jet engines are

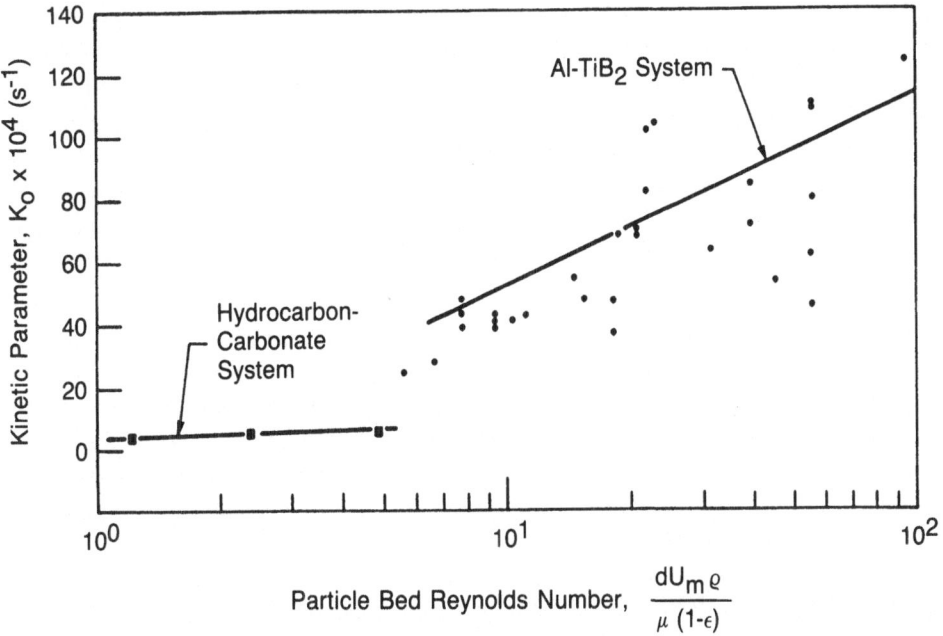

Figure 8. Effect of particle bed Reynolds number on the kinetic parameter.

continually being subjected to significantly higher stress levels and service temperatures than their earlier predecessors. As a result, the tolerance levels for defects in these alloys has markedly decreased in recent years. Defects ranging from 50-750 $\mu$m can be of critical size in highly alloyed superalloys operating under highly stressed low cycle fatigue (LCF) conditions. Such defects are one to two orders of magnitude smaller than those tolerated by the conventionally processed wrought alloys.[22]

Defects in superalloys may manifest themselves as microporosity, intermetallic segregates, nitrides and carbide stringers, and other nonmetallic inclusions.[23-25] The first two types of defects can be controlled by improved processing methods (i.e., better solidification control). The latter types can be better controlled by compositional modifications (i.e., lower carbon contents) and by improved melt and refractory practices. In characterizing the defects in directionally solidified superalloy castings, Narder and Kortovich[25] indicated that the dross (oxide) inclusions were found to be the single most frequent cause for casting rejections. Studies on powder metal, nickel-base superalloy materials have shown that the minimum LCF life is set by the presence of small, nonmetallic oxide inclusions in the material.[26,27] Oxide inclusions having cross sectional areas as small as 0.005 mm,$^2$ or less, have caused LCF failures in test specimens of powder metal alloys.[28]

The need for cleaner superalloys is approaching the limits of cleanliness

*References pp. 490-492*

achievable by the conventional vacuum induction melting processing route. Metal filtration using various types of ceramic filter media has been found to be an effective means of controlling the level and particle size of inclusions. A concern has been the difficulty to quantitatively assess the level of refinement. The electron-beam (EB) cleanliness evaluation test has been found to be a reproducible and a reliable method of assessing the level of melt cleanliness. Ceramic foam filters with a relatively high open-pore volume, have been utilized to produce clean melts of 713LC, 718, 738, and MM200 containing Hf, all of which are nickel-based superalloys.

The specific oxide raft area of filtered and non-filtered samples were measured. The rafts were produced using the EB melting technique. Figure 9 shows the performance results using a variety of 20 pores per linear inch ceramic filters for the four superalloys evaluated. Figure 9 indicates that the application of ceramic filters to superalloy heats can significantly reduce the nonmetallic particle content of the melt. Further, the efficiency of the filter to remove particles from the melt is dependent on alloy composition, on initial particle concentration in the melt, on the pore size of the filter, and on the rate of flow through the filter (viz., at rates of 0.08-0.11 kg/s vs. 0.3-1.3 kg/s). The composition of the foam filter in most cases did not appear to affect filter efficiency in these studies.

**Filtration of Steel Melts**—Nonmetallic inclusions are formed during steelmaking operations as a result of different physical and chemical phenomena, including erosion of refractories, entrainment of slag, inclusions from added materials such as ferroalloys, deoxidation and reoxidation, and precipitation during cooling and solidification of the steel. These inclusions become an integral part of the steel microstructure, participating in the further processing of the steel, and affecting the properties of the final product. The physical properties of the inclusions differ from those of the steel matrix, and modify the stress distribution in the steel under conditions of load-bearing, mechanical deformation, and temperature changes. As a result, they influence nearly every physical property of the steel.

The number of nonmetallic inclusions in a ton of cast carbon steel is extremely large, numbering in the trillions.[29] The vast majority of these inclusions are too small to be seen by optical microscopy. However, the inclusions which can be seen contain nearly all of the oxygen in the steel. Although it may be impossible to remove every inclusion from steel prior to casting, it is desirable to prevent the large, deleterious inclusions from remaining throughout processing to the production of final product.

Much effort is being invested on techniques for removing inclusions from steel melts prior to casting. The observation that tundish nozzles often blocked due to the deposition of inclusions from aluminum-killed steels[30] indicates that filtration is a feasible refining process for steels. Experimental studies[31-33] have shown that nearly all alumina inclusions above $2\mu$m in size, and a large percentage of the inclusions even below $1\mu$m in size, can be removed by passage through high performance filter media composed of aluminum oxide.

The apparatus used to carry out the steel melt filtration experiments is

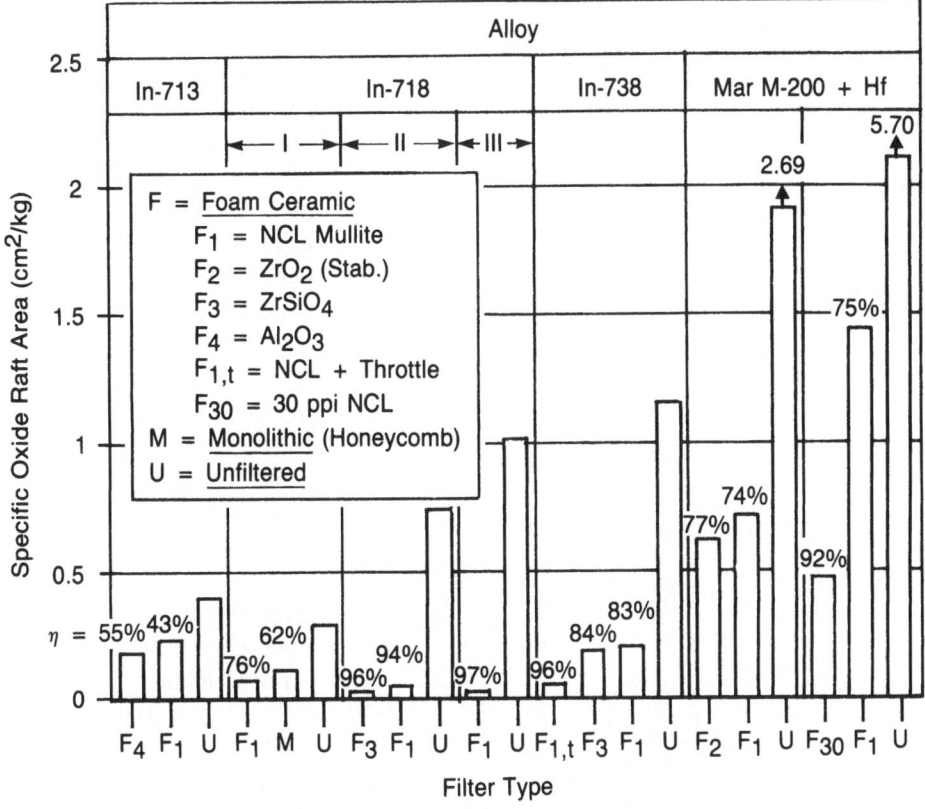

Figure 9. Effect of different 20 pores per linear inch filter materials and initial heat cleanliness on the filtration efficiency of several superalloy melts. Filtration efficiency, $\eta$, is given on top of the bars for filtered samples. Raft areas for unfiltered (U) MAR M-200 + Hf were off scale. Their raft areas are given above their respective bars.

schematically shown in Figure 10. The upper section consists of a stainless steel dome (55 cm ID and 75 cm height) which houses an induction coil, a cylindrical graphite susceptor, and the alumina crucible containing the melt and the filter. The electrical power for heating the crucible assembly is supplied by a 10 kW RF induction unit. The lower section houses the stainless steel melt receiving crucible (5.8 cm ID and 29.2 cm height with a 2° taper), which is water-cooled from the bottom end. This receiving crucible rests on a load cell, which is attached to a strip chart recorder. The upper and lower chambers can be evacuated or backfilled with argon gas simultaneously. Both chambers are equipped with pressure indicators.

Ceramic foam filters were placed within alumina tubes and cemented in place with alumina cement. A stopper rod arrangement was used to prevent melt infiltration into the filter prior to the desired time.

*References pp. 490-492*

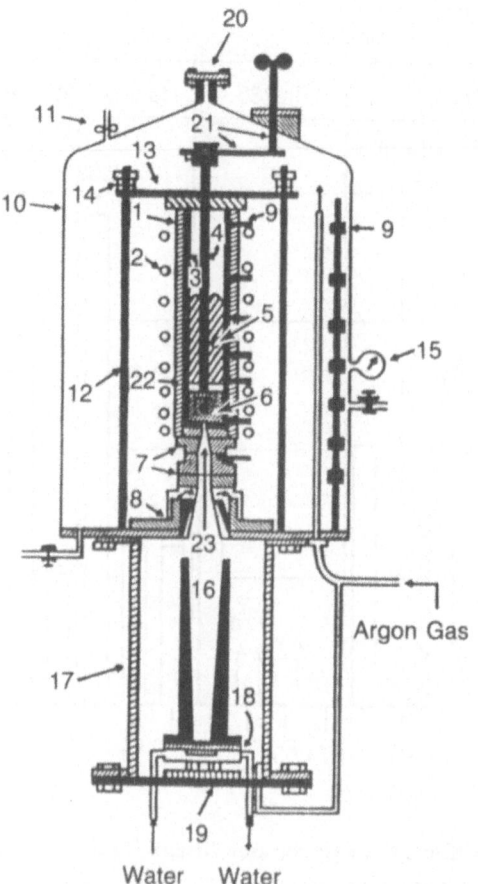

Figure 10. Schematic diagram of steel melt filtration apparatus.

Legend
1. Graphite Susceptor
2. Induction Coil
3. Alumina Crucible
4. Alumina Stopper Rod
5. Steel Melt
6. Alumina Filter Bed
7. Pyrolitic Graphite Rings
8. Water Cooled Stainless Pedestal
9. Thermocouples (W—5Re: W—26Re)
10. Furnace Dome
11. Safety Valve
12. Steel Rod
13. Steel Plate
14. Springs
15. Pressure Gauge
16. Metallic Mold
17. Lower Chamber
18. Water Cooled Brass Plate
19. Load Cell
20. Opening Window
21. Steel Rod
22. Alumina Disc
23. Orifice

A combination of electrolytic iron chips (1.5 kg) and iron powder (0.5 kg) was used as the starting charge material. Deoxidation of the melt was carried out *in situ* by the addition of weighed amounts of electrolytic manganese and ferrosilicon. The unfiltered steel melts were sampled using a fused silica tube and a suction bulb. Samples of filtered steel were cut from the final ingots. All samples were subjected to chemical analyses for manganese, silicon and oxygen. Other pieces of the same unfiltered and filtered steel were mounted, polished, and examined using optical and scanning electron microscopy.

During the experimental run, the temperature at the surface of the alumina crucible assembly was measured at several points with W-5%Re/W-26%Re thermocouples. Once the charge was melted, a sample from the melt was sucked into a fused silica tube (9 mm ID). Measured amounts of electrolytic manganese and ferrosilicon were added to deoxidize the melt, and another sample was taken. Next,

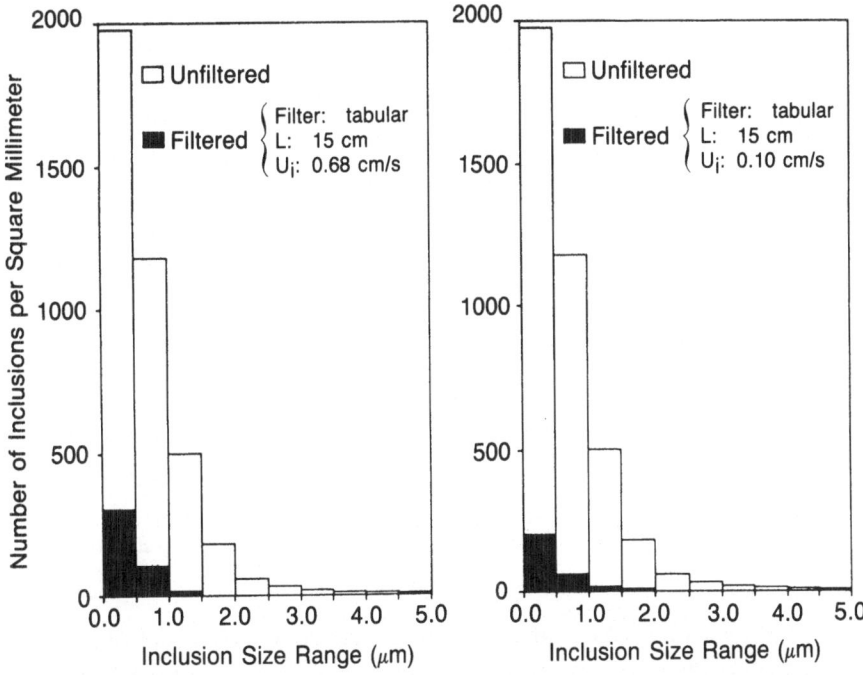

Figure 11. Effect of melt interstitial velocity on filtration performance for removal of alumina inclusions by 15 cm long tabular alumina filters. (a) 0.68 cm/s, (b) 0.10 cm/s.

the top chamber was pressurized with argon and the stopper rod was raised, permitting the melt to flow through the filter. The argon gas pressure in the top dome controlled the melt flowrate during the filtration experiments. The flow rate of the steel was measured using a load cell and a strip chart recorder.

**Removal of Solid Inclusions**—Figure 11 shows the comparison of inclusion size (0-5 $\mu$m) distribution data for unfiltered and filtered steel using 15 cm long tabular alumina filter at two different melt velocities, 0.10 and 0.68 cm/sec. A large reduction in inclusion concentrations can be noted between the filtered and unfiltered samples. In addition, the melt filtered at lower melt velocity contained fewer inclusions as compared to the melt filtered at a higher velocity.

Melt interstitial velocity, $U_i$, is usually varied to control throughput. It is observed from Figure 12 that the inclusion removal efficiency, $\eta$, decreased with an increase in melt interstitial velocity for both tabular and monolithic alumina filters. It is clear from the data that for tabular and monolithic alumina filters, the filtration performance was insensitive to the filter length in the range 5 to 10 cm. Comparing the performance of the filters for a given melt interstitial velocity, the monolithic alumina filter yielded higher inclusion removal than the tabular alumina filter.

Concerning the observed decrease in the inclusion removal efficiency with

*References pp. 490–492*

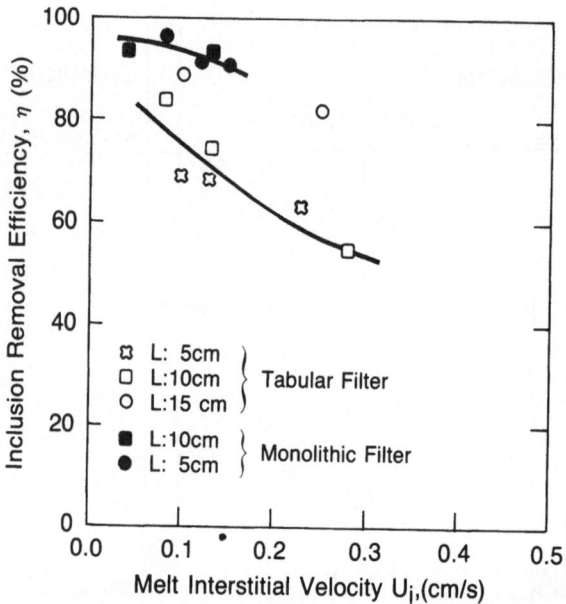

**Figure 12.** Effect of melt interstitial velocity on inclusion removal efficiencies using tabular and monolithic alumina filters.

increasing melt interstitial velocity. It is believed that at higher melt velocities, the interfacial shear is higher and the intercepted inclusion is subjected to a higher dislodgement stress, resulting in a lower efficiency of inclusion removal. Although sintering plays an important role, it is believed that the rate of inclusion sintering is slower than the rate of inclusion dislodgement.

The total surface area of the filter plays an important role in inclusion removal during filtration. In the case of extruded monolithic alumina filters, the total void space for a given filter volume is much greater than that of the tabular alumina filters for the same filter volume. The filter surface for monolithic alumina is rougher than that of the tabular alumina. For a constant filter volume, the total surface area of a monolithic filter is higher than that of the tabular filter as shown in Figure 13. It is observed that the removal efficiency increases with an increase in total surface area.

It is important to consider the effect of turbulence on removal efficiency. The critical Reynolds number for laminar to turbulent flow transition in a packed bed is 1.0, while in the case of the monolithic filter, the critical Reynolds number for laminar to turbulent flow transition is 2300. Figure 14 shows the variation of inclusion removal efficiency as a function of ratio of channel Reynolds number to the respective critical Reynolds number. An important conclusion that may be drawn from the results shown in Figure 14 is that inclusion entrapment is favored when the filter is operated in the laminar region.

# METAL REFINING BY FILTRATION

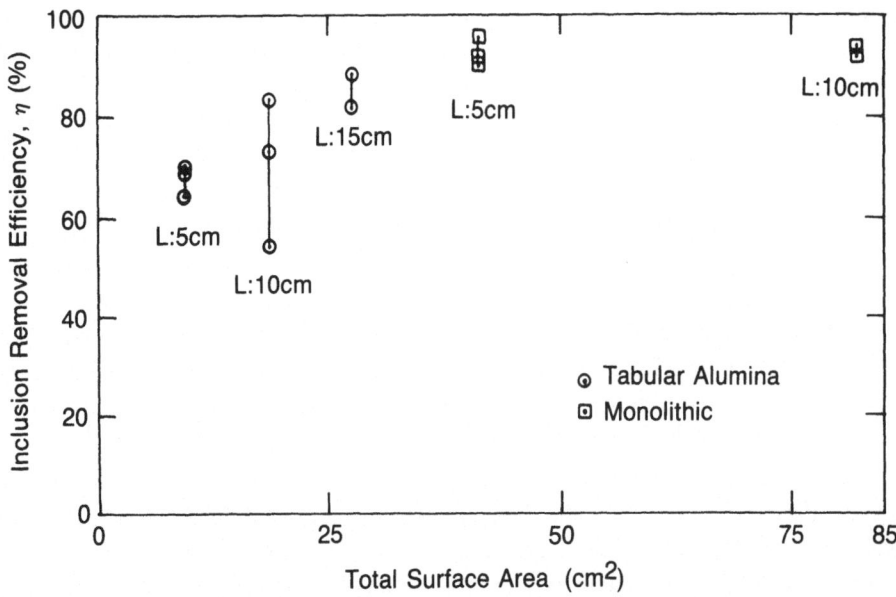

Figure 13. Effect of total surface area of filter on inclusion removal efficiency.

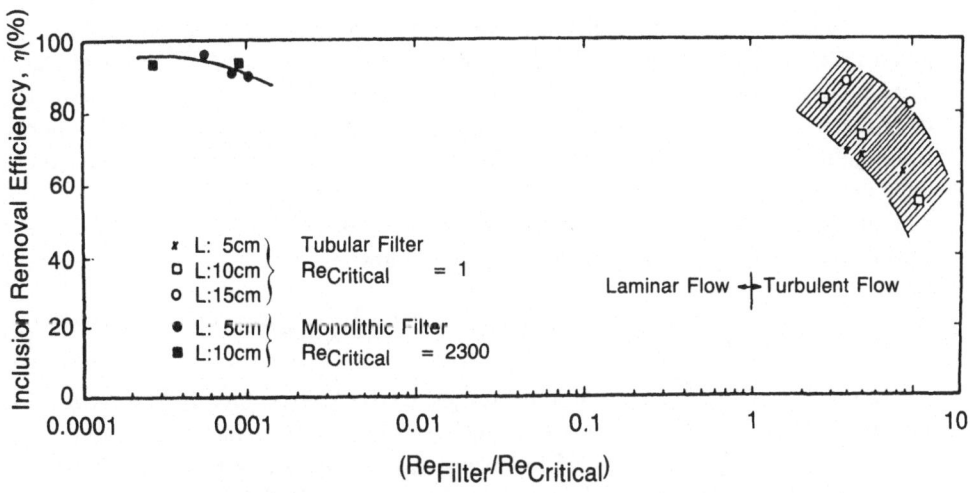

Figure 14. Effect of melt flow behavior on inclusion removal efficiency, laminar flow vs. turbulent flow.

*References pp. 490-492*

**Removal of Liquid Inclusions in Steel**—Examples of common inclusions which are in the liquid state in liquid steel are Ca-Al silicates, Mn silicate, and droplets of entrained slag. These inclusions are usually spherical and deformable at high temperatures in the solid steel, but glassy and brittle at room temperature. During hot working operations, these deformable inclusions are elongated into stringers, which leads to anisotropy of mechanical properties, and is responsible for poor short transverse properties. During cold working operations, these inclusions are non-deformable, and act as local stress intensifiers. During deformation, they may nucleate cracks in the steel, leading to rejection of the product or failure in service. It is highly desirable to reduce the number of these liquid inclusions in steel.

To investigate the process of removal of liquid inclusions, steel melts containing liquid manganese silicate inclusions were prepared and filtered through open cell ceramic foam filters. The steel melts were deoxidized with manganese and ferrosilicon (containing 47.6% Si) to produce the liquid manganese silicate inclusions. The total oxygen contents of the samples were obtained from chemical analysis, and were used for calculating efficiencies of oxygen and inclusion removal.

The manganese and ferrosilicon additions to deoxidize the melt were selected to yield 1% Mn and 0.25% Si at 100% efficiency of addition. The actual Mn and Si levels in the deoxidized liquid steel ranged from 0.41 to 0.87% and 0.05 to 0.20%, respectively. The compositions of the steel melts after deoxidation fell in the range needed for the production of the desired liquid inclusions. Micrographs of inclusions found in samples taken from melts deoxidized with manganese and ferrosilicon are shown in Figure 15. These inclusions are spherical and transparent, indicating that they were liquid in the steel melt, and became glassy during cooling.

The oxygen removal efficiencies, which are related to filtration efficiencies, are given in Table 2. The results clearly show that the oxygen level of the deoxidized steel melt can be significantly reduced by filtration. The oxygen content of the deoxidized steel melt is consistently in the range of 100 to 200 ppm, which is above the equilibrium value. However, the absolute removal efficiency, $\eta$, in Equation 6, does not take into account the equilibrium value of dissolved oxygen in the metal. The appropriate form of the filtration efficiency equation to take into account the equilibrium concentration of oxygen is given below:

$$\eta' = \frac{(C_i - C_E) - (C_o - C_E)}{C_i - C_E} \times 100\% \qquad (8)$$

where $C_E$ is the equilibrium value of oxygen in the filtered metal. The results are shown in Figure 16. As can be seen, when the bulk melt velocity increased from 0.4 to 1.3 cm/s, the efficiency of inclusion capture decreases from about 80% to 45% using alumina filters. The removal efficiency was higher with zirconia filters at 0.9 cm/s, while it was lower at a melt velocity of 0.3 cm/s. More work with zirconia filters is in progress to examine this anomalous behavior. The strong dependence of liquid inclusion capture with velocity is similar to that observed with solid $Al_2O_3$ inclusions.[33] The processes associated with entrapment of liquid inclusions are significantly different from those for solid inclusions. The capture

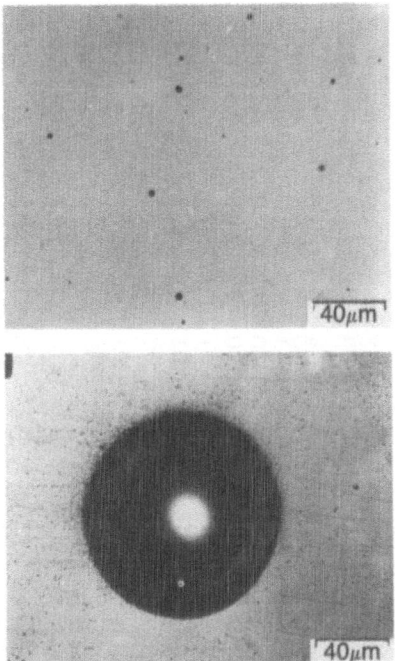

Figure 15. Micrographs of inclusions found in samples taken from melts deoxidized with manganese and ferrosilicon. The inclusions are composed of manganese silicate.

and subsequent retention of inclusions is a function of the properties of the filter surface. In the case of liquid inclusions, the wetting/non-wetting character of the inclusion will determine the propensity of the inclusion to remain entrapped in spite of the shear forces imparted by the flowing melt. Examining the results shown in Figure 16, the decrease in inclusion capture efficiency due to an increase in melt velocity is suggestive of dislodgement caused by the shear forces of the flowing melt.

**Model Studies - Effect of Wettability**—During filtration of steel, the melt does not wet the ceramic filter medium. The inclusions, however, may or may not wet the filter. The role of interfaces in inclusion capture is of paramount importance. The wettability characteristics of the inclusions with respect to the filter has been addressed and studied by appropriately coating the surface of the filter.

Figure 17 shows the dimensionless retained volume as a function of dimensionless time for borosilicate glass balls, coated glass balls and teflon balls at a Reynolds number of 2.3. In the transient mode ($t^* > 0.10$), it can be seen that more inclusions are captured with the teflon balls than with the borosilicate glass balls.

Several observations can be made based on the experimental work. At low Reynolds number, when the bed is nonwetting with respect to the liquid inclusions,

*References pp. 490–492*

## TABLE 2
### Inclusion Removal Efficiencies

| Filter Type | Mn, % | Si, % | O, ppm (equil.) | O, ppm Unfiltered | O, ppm Filtered | $\eta$, % | $\eta'$, % | Bulk melt Velocity, cm/s |
|---|---|---|---|---|---|---|---|---|
| Al$_2$O$_3$ | 0.87 | 0.20 | 70 | 169 | 96 | 43.3 | 73.8 | 0.42 |
|  | 0.74 | 0.13 | 80 | 169 | 96 | 43.3 | 82.1 | 0.42 |
| Al$_2$O$_3$ | 0.81 | 0.18 | 90 | 147 | 116 | 21.4 | 54.8 | 0.89 |
|  | 0.80 | 0.15 | 95 | 147 | 116 | 21.4 | 60.0 | 0.89 |
| Al$_2$O$_3$ | 0.73 | 0.15 | 75 | 142 | 117 | 18.0 | 38.0 | 1.05 |
|  | 0.74 | 0.12 | 80 | 142 | 117 | 18.0 | 41.0 | 1.05 |
| Al$_2$O$_3$ | 0.72 | 0.14 | 77 | 130 | 109 | 16.1 | 39.3 | 1.28 |
|  | 0.73 | 0.12 | 82 | 130 | 109 | 16.1 | 43.3 | 1.28 |
| Al$_2$O$_3$ | 0.70 | 0.13 | 80 | 99 | 103 | −4.0 | — | 1.77 |
|  | 0.70 | 0.10 | 88 | 99 | 103 | −4.0 | — | 1.77 |
| Al$_2$O$_3$ | 0.50 | <0.01 | — | 203 | 197 | 2.7 | — | 69.4 |
| ZrO$_2$ | 0.70 | 0.16 | 90 | 180 | 154 | 14.3 | 28.7 | 0.29 |
|  | 0.68 | 0.11 | 105 | 180 | 154 | 14.3 | 34.4 | 0.29 |
| ZrO$_2$ | 0.71 | 0.13 | 100 | 161 | 116 | 27.8 | 73.1 | 0.89 |
|  | 0.71 | 0.11 | 110 | 161 | 116 | 27.8 | 87.3 | 0.89 |
| ZrO$_2$ | 0.50 | 0.07 | 109 | 155 | 138 | 11.0 | 37.4 | 4 |
|  | 0.50 | 0.05 | 117 | 155 | 138 | 11.0 | 45.3 | 4 |

Notes:
(1) $\eta$=absolute removal efficiency = $100(C_i - C_o)/C_i$ where $C_i$ and $C_o$ are the oxygen content of the unfiltered and filtered melt respectively.
(2) $\eta'$ is defined as a modified oxygen removal efficiency taking into account equilibrium between O, Mn, and Si. Here $\eta' = 100[(C_i - C_E) - (C_o - C_E)]/(C_i - C_E)]$, where $C_E$ is the equilibrium value.

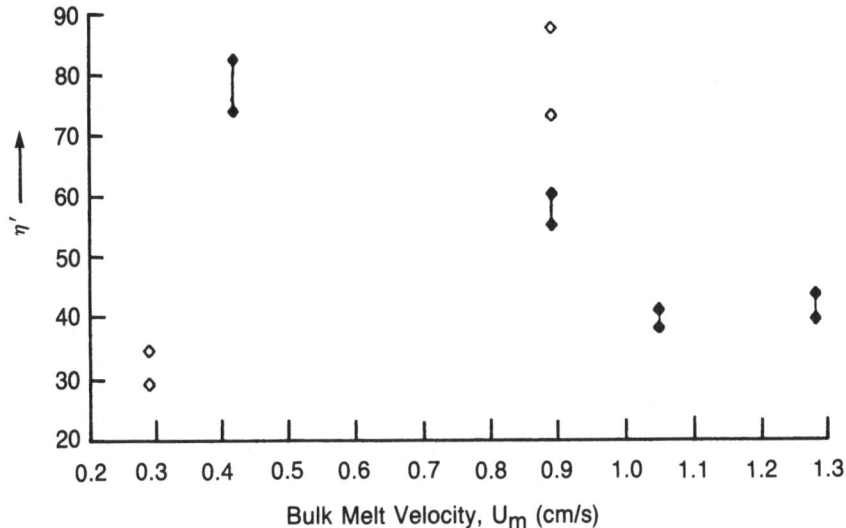

Figure 16. Effect of bulk melt velocity on modified removal efficiency ($\eta'$) of liquid manganese silicate inclusions. The filled data points are for alumina filters; the unfilled ones are for zirconia filters.

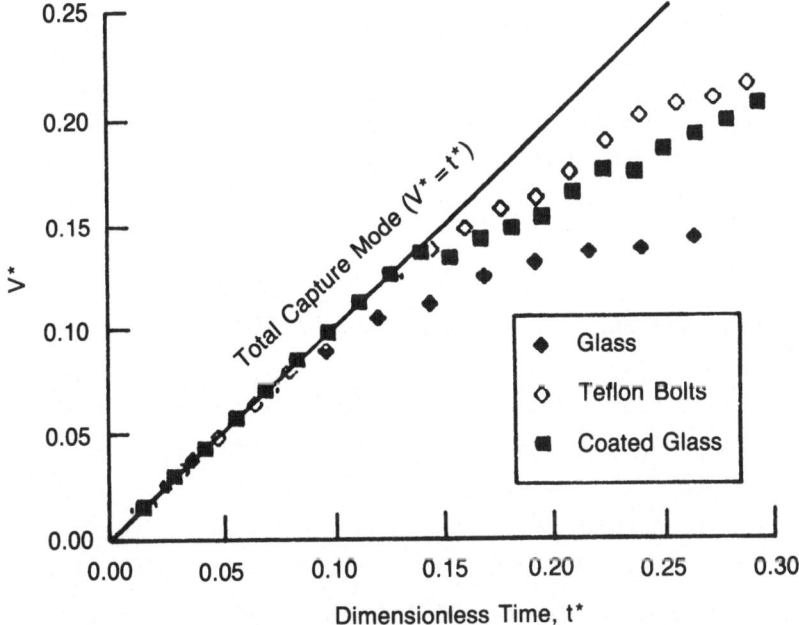

Figure 17. The effect of wettability on dimensionless volume ($V^*$), retained volume of inclusions/initial filter void volume, at $Re = 2.3$. The case of 100% capture is represented by $V^* = t^*$.

*References pp. 490–492*

we observe that:

- During the initial stages of the filter's life the inclusions are randomly captured throughout the filter bed. Due to the large contact angle between the liquid droplets and the filter surface, the inclusion droplets "roll" through the first layers and are captured by direct interception. In contrast, when the filter surface is wetting with respect to the inclusions, then the majority of the droplets are captured in the first two layers of the filter bed.

- At a later stage, large discontinuous globules are captured throughout the bed. When the bed is wetting, conduits are formed.

This behavior is reversed at high Reynolds number. For the case where the inclusions are nonwetting with respect to the filter grain surface, we observe that:

- During the initial stages of the filter's life, due to the high inertial forces, the inclusions are not captured ... what comes in goes out. When the inclusions are wetting to the filter, random capture takes place throughout.

- At a later stage, the inclusions form a liquid layer at the entrance of the bed. For the case where the inclusions are wetting with respect to the filter grain surface, conduits are formed.

## ACKNOWLEDGEMENTS

The contribution of past students who over the years generated the data base for the aluminum, superalloy and steel filtration studies is acknowledged. The author would also like to acknowledge the contribution of colleagues Raj Mutharasan at Drexel University and Will Sutton at United Technologies Corporation. The work discussed in this paper based on the research carried out at Drexel University over the past five years or so, sponsored by NSF, Alcoa, AMMRC, AISI, Special Metals Corp., and Bethelehem Steel Corp.

## REFERENCES

1. S. Ali, D. Apelian and R. Mutharasan: *Canadian Met. Quarterly*, *24*, (1985), 311–318.

2. H. C. Cumming, F. B. Stulen and W. C. Schulte: *Trans, ASM*, *49*, 482–512.

3. J. L. Mihelich, J. R. Bell and M. J. Korchynsky: *J. Iron and Steel Institute*, *209*, (1971), 469.

4. C. E. Eckert, R. E. Miller, D. Apelian, and R. Mutharasan, *Light Metals 1984*, Editor: J. P. McGeer, TMS-AIME, Warrendale, PA, (1984), 1281–1304.

5. D. Apelian, S. Luk, T. Piccone and R. Mutharasan: *5th International Iron and Steel Congress Steelmaking Proceedings*, April 7-9, 1986, Washington D.C., 957–967.

6. K. J. Brondyke and P. D. Hess: *Trans. TMS-AIME*, *230*, (1964), 1553–1556.

7. M. V. Brant, D. C. Bone and E. F. Emley, *Journal of Metals*, *23*, (March 1971), 48–53.

8. K. J. Ives, *Proceedings of the NATO Advanced Study Institute*, (July 1978), 203–224.

9. J. P. Herzig, D. M. LeClerc, and P. LeGoff, *Industrial and Engineering Chemistry*, *62*, 5, (1970), 8–35.

10. C. Tien, R. M. Turian, and H. Penose, *Journal of AIChE*, *25*, (1979), 385–395.

11. W. Kraj, *Bul. Acad. Pol. Sci. Ser. Tech*, *14*, (1966), 8.

12. D. Apelian and R. Mutharasan, "A Theoretical Basis for Depth Filtration of Liquid Metals" submitted to *Trans AIME*, 1985.

13. D. Apelian and R. Mutharasan, *Journal of Metals*, *32*, 9, (1980), 14–20.

14. D. Apelian, and R. Mutharasan, "Modeling of Inclusion Removal of Melt Systems," paper presented at 72nd Annual AIChE Meeting, San Francisco, California.

15. C. J. Simensen, *Met. Trans.*, *13B*, (1982), 31–34.

16. F. R. Mollard, and N. Davidson, "Ceramic Foam - A Unique Method of Filtering Molten Aluminum in the Foundry," presented at the 82nd AFS Casting Congress, Detroit, Michigan, April 1978.

17. R. B. Miclot, *Technical Report* No. 67–1507, Rock Island Arsenal, June 1967.

18. L. A. Alekseev, and I. B. Kumanin, *Izv. Vyssh. Uchebr. Zaved. Tseutn. Metall.* *11*, 1, (1968), 155–159.

19. D. Apelian, Sc.D. Thesis, MIT, 1972.

20. H. E. Miller, *Aluminum*, (1972), 368–371.

21. Dj. Hedjazi, G. H. J. Bennett, and V. Kondic, *Metals Technology*, (Dec. 1976), 537–541.

22. G. W. Meetham, *High Temperature Alloys for Gas Turbines*, Applied Science Pub., London, (1978), 837–859.

23. S. R. Houldsworth, *Superalloys '80, Proc. 4th Int. Symp. of Superalloys*, ASM, Metals Park, OH, (1980), 375–383.

24. E. Bachelet, *Quality Castings of Superalloys*, Applied Science Pub., London, (1978), 665–699.

25. J. M. Narder and C. S. Kortovich, "Characterization of Casting Defects in Typical Castings of a Directionally Solidified Superalloy," Final Rept. AFML-TR-79-4060, TRW Materials Tech. Lab., Cleveland, OH, Contract F33615-76-5373, June, 1979.

26. C. E. Shamblin, R. F. Halten, and W. R. Pfouts, "Manufacturing Methods for Improved Superalloy Powder Production," AFML F33615-78-3-5225, Second Interim Report.

27. E. E. Brown, J. E. Stulga, L. Jennings and R. W. Salkeld, *Superalloys '80, Proc. 4th Int. Symp. on Superalloys*, ASM, Warrendale, PA, (1980), 159–162.
28. J. D. Buzzanell and L. W. Lherbier, *Superalloys '80, Proc. 4th Int. Symp. on Superalloys*, ASM, Warrendale, PA, (1980), 149–158.
29. R. Kiessling, *J. Metals, 21*, 10, (1969), 48–54.
30. G. C. Duderstadt, R. K. Iyengar and J. M. Matesa, *J. Metals, 20*, 4, (1968), 89–94.
31. S. Ali, Ph.D. Thesis, Drexel University, Philadelphia, PA, (June 1984), 136.
32. D. Apelian, R. Mutharasan and S. Ali, *J. Materials Sci. 20*, (1985), 3501–3514.
33. S. Ali, R. Mutharasan and D. Apelian, *Met. Trans. 16B*, (1985), 725–742.

## DISCUSSION

**S. Bahk** *(General Motors Research Laboratories)*

What is the nature of $K_o$? Is it related to, for example, permeability and pore size?

**A.** $K_o$ is a function of the melt system, the size of the inclusions and many other variables; it is a lump parameter. It does not stand for permeability, rather it's a kinetic parameter coefficient that is a function of the porous medium itself. Moreover, $K_o$ is a function of filter geometry. The $K_o$ values for monolithic, foam and tabular filters are all different. Furthermore, for a given geometry the $K_o$ value differs with the type of metal being filtered. Thus, $K_o$ varies from system to system.

**S. Bahk** *(General Motors Research Laboratories)*

So, it essentially depends on the nature of filter and the nature of your molten metal system?

**A.** That's right. For a given melt system, for example: aluminum, I know what inclusions I want to remove. Knowing the $K_o$ value for the different types of filters, I can calculate, on the basis of the model, what the optimum filter size should be. In other words, when will it clog and when it should be replaced. So, we're extrapolating the long term behavior based on short term behavior tests.

**S. Bahk** *(General Motors Research Laboratories)*

Unless you include the effect of temperature on $K_o$, how can you account for changes in $K_o$ due to the change of viscosity with temperature?

**A.** If you examine viscosity data for molten metals over a range of temperatures above the liquidus, there isn't much change to merit further complicating the model.

**S. Bahk** *(General Motors Research Laboratories)*

In molten irons it changes quite a lot.

**A.** Perhaps you are correct. We have looked at the change in viscosity of aluminum with temperature and it doesn't change that much. The more critical thing is the interstitial velocity, $U_m/\epsilon$. Your point is well taken that for steels we should look at temperature.

**D. M. Hanna** *(General Motors Research Laboratories)*

You listed three methods for evaluating filter efficiency. What do you think about a fourth method, the direct analysis of the spent filter?

**A.** Different tools are used for different reasons. To study particle capture mechanisms, for example, the *post-mortem* analysis of the filter is a must, although sample preparation is often difficult. For plant work this is too laborious and some type of on-line technique is necessary.

**Y. Sahai** *(Ohio State University)*

In filtration do wetting and non-wetting inclusions behave the same way?

**A.** No. That is a very good question. In fact, that's the subject of a paper we presented last week. Unfortunately, I didn't present any of those results at this conference.

**Y. Sahai** *(Ohio State University)*

Can you describe the mechanism by which materials that form inclusions stick to the filter material?

**A.** Not yet, but I think we've got some pieces of the puzzle. For example, for the zirconia filters which contain MgO, we find that there is a whole shell of $Al_2O_3 \cdot MgO$ forming on the filters and it acts like these tapes that hang out to catch insects. So the wettability is a very important factor but we don't have the whole story on that yet.

# FILTRATION OF IRONS AND STEELS

### PETER F. WIESER
*Wieser & Associates*
*Cleveland, Ohio 44106*

## ABSTRACT

Filtering liquid metal to remove undesirable suspended phases has been utilized most extensively in direct chill casting of aluminum. In the last few years demand for increased quality of cast products, as well as the availability of filters for higher temperature applications, have led to the use of filters in casting of superalloys, irons, copper-base alloys, stainless steels, and, most recently, carbon and low alloy steels. Because of these trends several graduate students and the writer have worked at Case Western Reserve University on several aspects of iron and steel filtration for several years. These include filtration effectiveness, the sizing of filters, and, to some extent, the effect of filtration on the structure of the filtered metal.

## INTRODUCTION

This paper is based upon research conducted at Case Western Reserve University over the last six years. The research started with the question as to whether newly available filters could be used for irons and steels. Since those initial encouraging trials, filtration work has been carried out to develop the principles involved in filtering irons and steels, and to evaluate ideas for improving the performance of filters.

## FILTER TYPES

The ceramic filters dealt with in this paper are either of the foam type with their unique pore structure, or filters with pores that consist of straight, uniform cross section channels. The latter type is produced either by extrusion or by pressing. The pore size of these filters is typically expressed in terms of cells

Figure 1. Cope surfaces, 25 × 18cm (10 × 7 in.), of unfiltered (top) and filtered (bottom) Inmold ductile iron step block castings.

per square inch, $cpi^2$. Foam filters are produced by coating a polymer foam. The polymer is subsequently removed in the firing process. The pores of these foams consist of slightly elongated bubbles. "Windows" connect each bubble to several others and thereby produce the desired open foam structure. The window diameter is the narrowest diameter in the pore structure and therefore critically affects flow of the liquid metal through the filter. Our research has shown that these windows also have a controlling influence on the capture effectiveness of the foam. The pore size of foams is typically reported in terms of the number of pores per linear inch, ppi.

## SURFACE AND INTERNAL CLEANLINESS

**Exogenous Inclusions**—The surface quality improvement obtainable by placing filters into the gating system of a casting is readily apparent in Figures 1 and 2. These represent respectively the cope surfaces of unfiltered and filtered Inmold ductile iron and medium carbon steel casting plates. The mold configurations used for casting the two materials are shown, respectively, in Figures 3 and 4.

The surface quality improvement was due to the capture of exogeneous particles and of reaction products from the Inmold treatment of iron and from the deoxidation of steel respectively. Exogenous inclusions are frequently larger particles or agglomerates whose diameter may exceed that of the filter pores. These particles are removed by straining. Examples which illustrate the capture of sand,

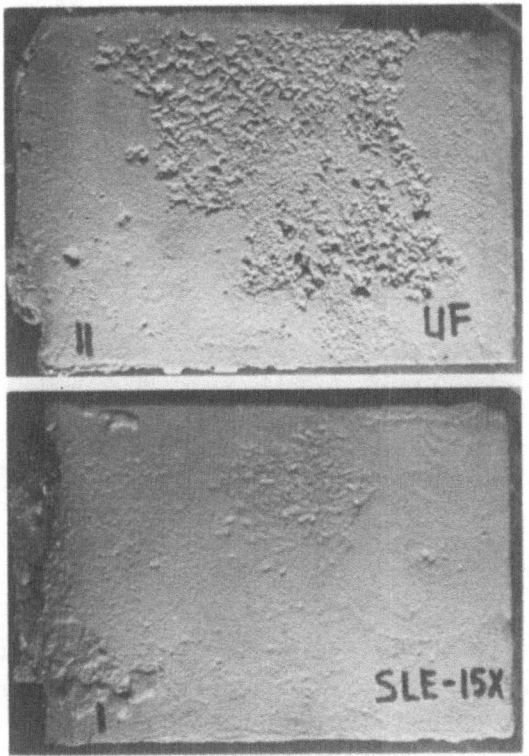

**Figure 2.** Cope surfaces, $28 \times 15$ cm ($11 \times 6$ in.), of unfiltered (top) and filtered (bottom) steel plate castings.

and sand slag mixtures are illustrated in Figures 5 and 6.

**Reaction Products**—The treatment of iron with magnesium or magnesium alloys to produce spheroidal graphite, or ductile iron, is associated with the formation of reaction products such as magnesium sulfide, magnesium oxide, and silicate dross. In the case of the Inmold treatment process insufficient opportunity may exist for these reaction products to be trapped by the gating system. Magnesium sulfide and Type II dross stringers (Figure 7) have been shown to be retained with filters.[1,2] With long runners, the larger Type II dross stringers tend to float to the surface and become trapped by the gating system.[2] In that case, capture of smaller, Type III, dross particles becomes apparent upon inspection of the filters (Figure 8).

Alumina inclusions form upon deoxidation of steel. Their capture inside a foam filter is illustrated in Figure 9. Recent and earlier work[3,4,5] has shown these inclusions to be trapped by interception. This takes place preferentially at section changes as they occur at the filter entrance, and several times in foam filters where windows connect individual "bubbles."

*References p. 511*

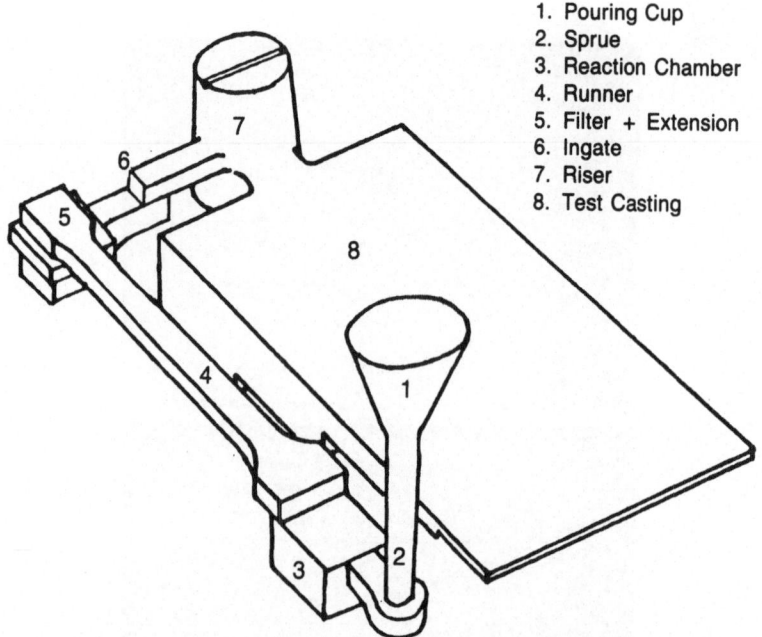

**Figure 3.** Gating arrangement for Inmold ductile iron step block casting. Minimum filter edge to Inmold chamber separation = 15 cm (6 in.).

Since many of the reaction products are 10 to 1000 times smaller than the pore diameter of commercially used filters, capture of these inclusions involves transport as well as attachment to the filter pore wall. Calculations[6] based on the wetting of non-metallics by steel and on surface energies indicate that both the transport of inclusions in the immediate vicinity of the filter pore wall and the initial attachment to the wall is favored for nonwetted inclusions (Figure 10). A stronger bond may subsequently form due to sintering.[7,8,9]

**Casting Cleanliness**—Castings with 30 to 60% lower oxide inclusion contents have been obtained by filtering aluminum-deoxidized carbon and low alloy steels that contained 50 to 100 ppm total oxygen prior to filtration.[4,5] Total oxygen contents 30 to 50% lower have been obtained in tundish filtration tests with aluminum-deoxidized carbon steel.[9,10] Examples of reduced oxygen content are shown in Figure 11 for static castings. For tundish filtration tests, results are shown in Figures 12 and 13.[3,7]

The data for foam filters in Figure 11, and those for filters with straight, constant cross section pores in Figures 12 and 13 demonstrate the important improvement in filtration effectiveness with smaller pore size. Since the pore diameter of the 121 cpin$^2$ filters is of the same order (0.1 - 0.2 cm) as that of the 15 and 25 ppi foam filters the data also indicate the greater filtration effectiveness

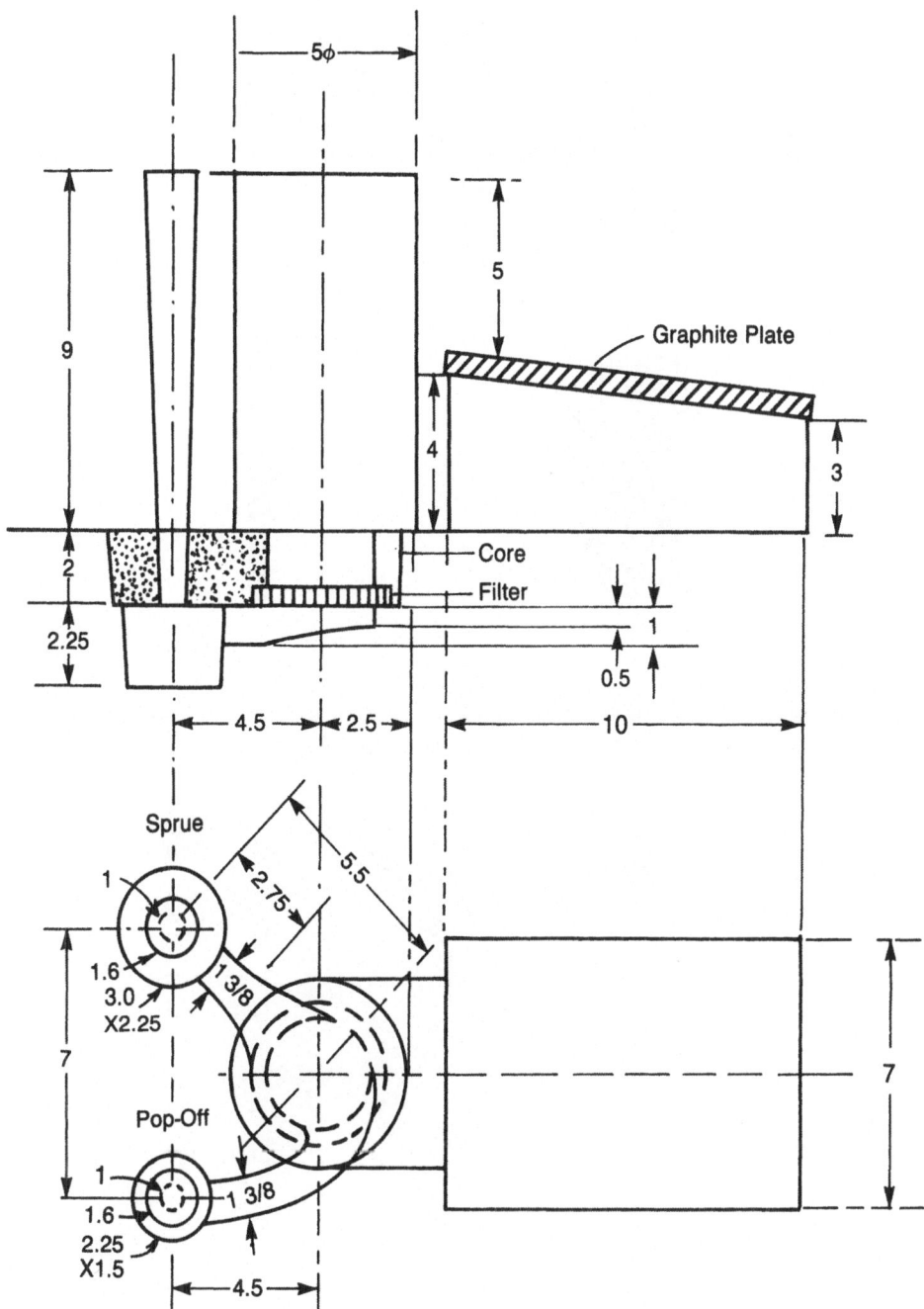

Figure 4. Gating arrangement for tapered steel plate casting. Dimensions are given in inches.

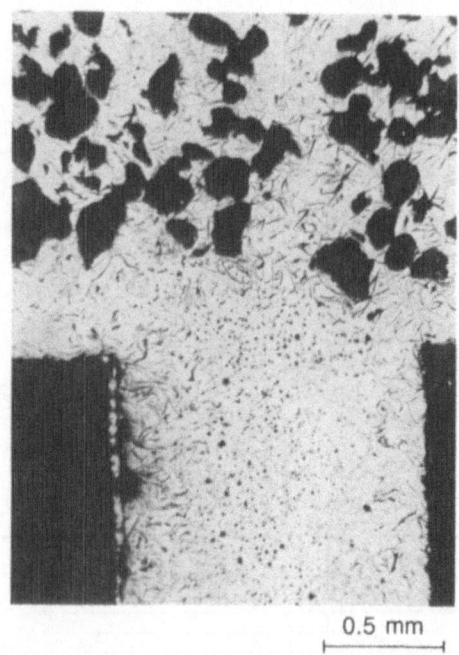

0.5 mm

Figure 5. Eroded sand trapped in front of a filter with straight, constant cross section pores. Inmold ductile iron.

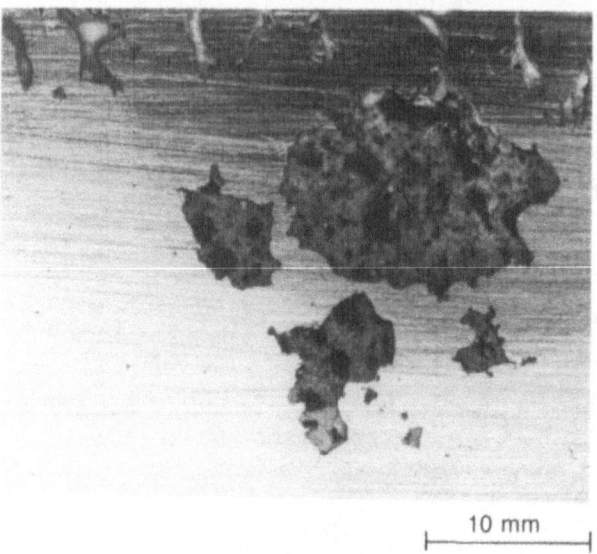

10 mm

Figure 6. Sand-slag particles trapped in front of a foam filter. Carbon steel.

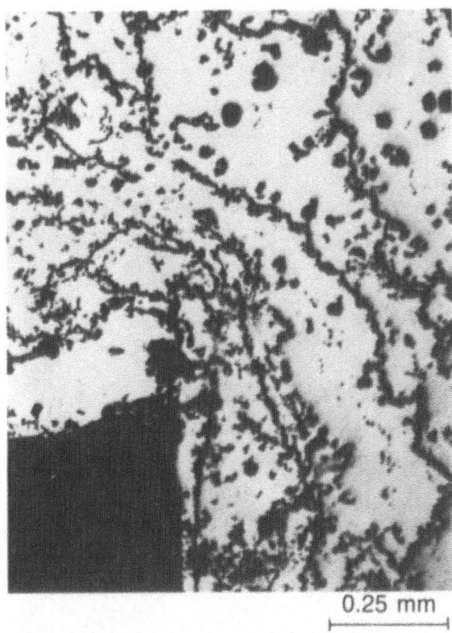

Figure 7. Type II dross stringers trapped at the entrance to a straight, constant cross section pore filter. Inmold ductile iron.

obtained with foams. Sectioning filters after various stages of filtration[3] revealed that this superior capture effectiveness is related to the greater number of preferred deposition sites inside the filter; i.e. the number of "windows" that liquid steel must pass through in traversing the filter thickness.

## NODULE COUNT

Nodule count data from numerous filtered and unfiltered castings shown in Figure 14 indicate an increased nodule count associated with filtration. These results confirm earlier preliminary results[1] and agree with higher cell counts reported for gray iron filtration studies.[11] These results have not yet been explained satisfactorily, they may be based upon abrasion of filter ceramic materials that subsequently act as nuclei, or upon the retention of substances that inhibit the effectiveness of existing nuclei.

## FILTER PRIMING AND FLOW

Filters currently used for cast irons are generally primed without difficulty. For steel castings, carefully designed gating systems will ensure priming.[4,6] One reason for the resistance to priming is the wetting behavior of steel relative to filter ceramics (Figure 15). Another reason is the heat loss and possible freezing when the liquid metal first contacts the filter and enters the filter pores.[12] Figure

*References p. 511*

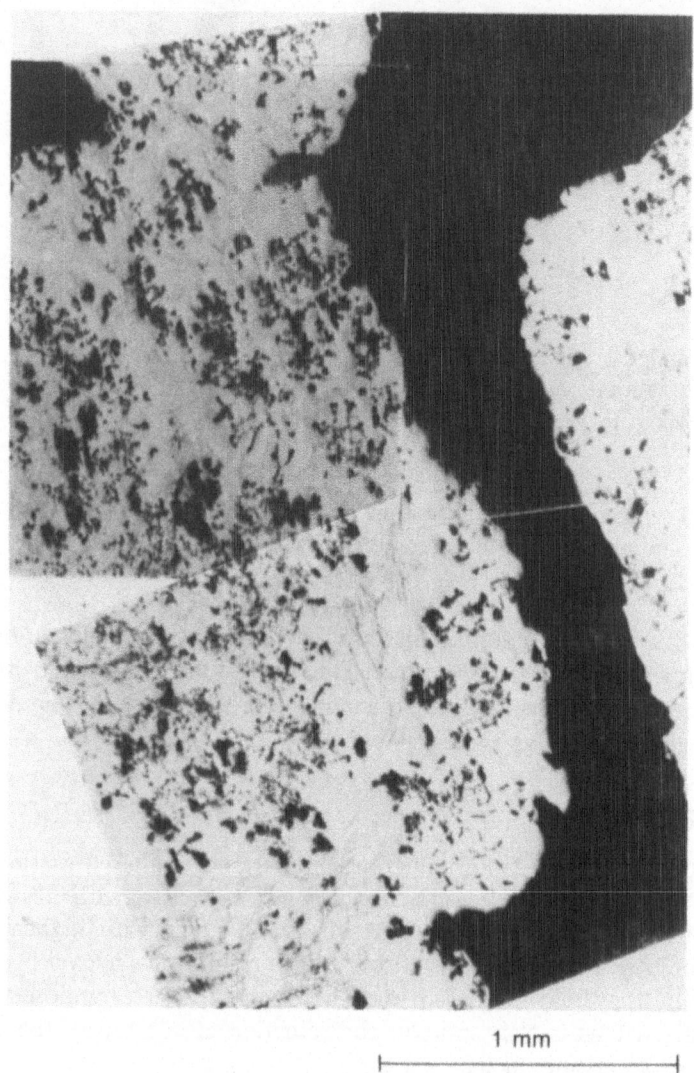

Figure 8. Fine Type III dross particles trapped inside a foam filter. Inmold ductile iron.

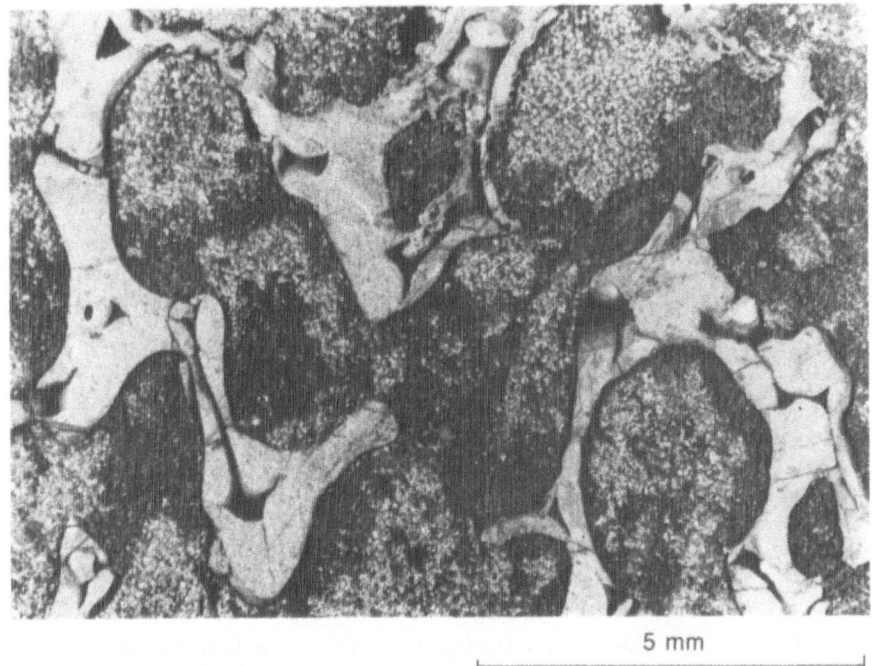

5 mm

Figure 9. Aluminum deoxidation products in carbon steel trapped inside the pores of a foam filter. The filter and the deoxidation products (dark) stand in relief due to partial removal of the metal by deep etching.

16 illustrates sample calculations of metal superheat effects on the initial freezing followed by subsequent remelting of the metal inside a filter pore.[12]

The filter geometry and pore size affects the resistance to the initial flow of metal. Experimental data in Figure 17 [13] also suggests a significant effect of the filter ceramic. This has been attributed to the wetting characteristics of the liquid metal.[13]

## FILTER SIZING

As non-metallic particles are trapped by the filter the resistance to flow increases and eventually leads to complete stoppage of liquid metal flow. The total amount of metal that can flow through a filter before this stoppage occurs has been termed filter life.

The process of sizing filters involves selecting the minimum size filter (cost) that achieves mold filling within the desired time limit without premature filter

*References p. 511*

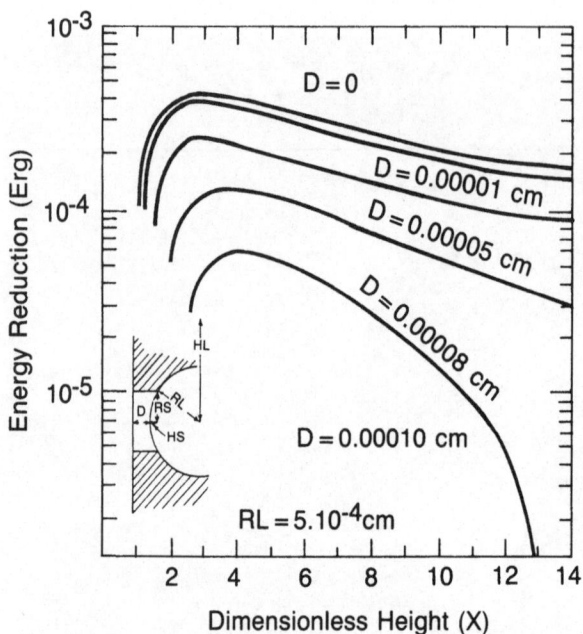

Figure 10. Calculated energy reduction as a function of the dimensionless metal withdrawal height, X = RS/HS, and the separation distance D.
Assumption: 10 micron diameter spherical alumina particles in steel.

clogging. This also involves a compromise between the desire to achieve fast metal flow rates and high filtration effectiveness. Larger diameter pores permit higher flow rates at the expense of lower inclusion capture effectiveness.

Filter sizing could be accomplished with flow calculations on the basis of the resistance of a clean filter and suitable modelling of the resistance that develops with inclusion capture. The latter varies significantly with process variables. In practice therefore a filter is commonly selected empirically with an area that is a multiple of the runner cross section to achieve rapid mold filling. Adjustments to this filter size are subsequently considered on the basis of the casting weight and available filter life data.

Experimental work and calculations have shown that filters for bottom gated steel castings can be sized on the basis of flow calculations alone without considering the effects of the inclusion build-up on the filter (Figure 18).[13] Ignoring the inclusion build-up provides realistic results in this instance because the filter size required to match the flow rate of an unfiltered casting becomes less as the flow rate diminishes while the metal level rises in the mold cavity. This provides the extra filter area needed to compensate for the effects of inclusion deposition from steels with normally encountered inclusion concentrations.

FILTRATION OF IRONS AND STEELS

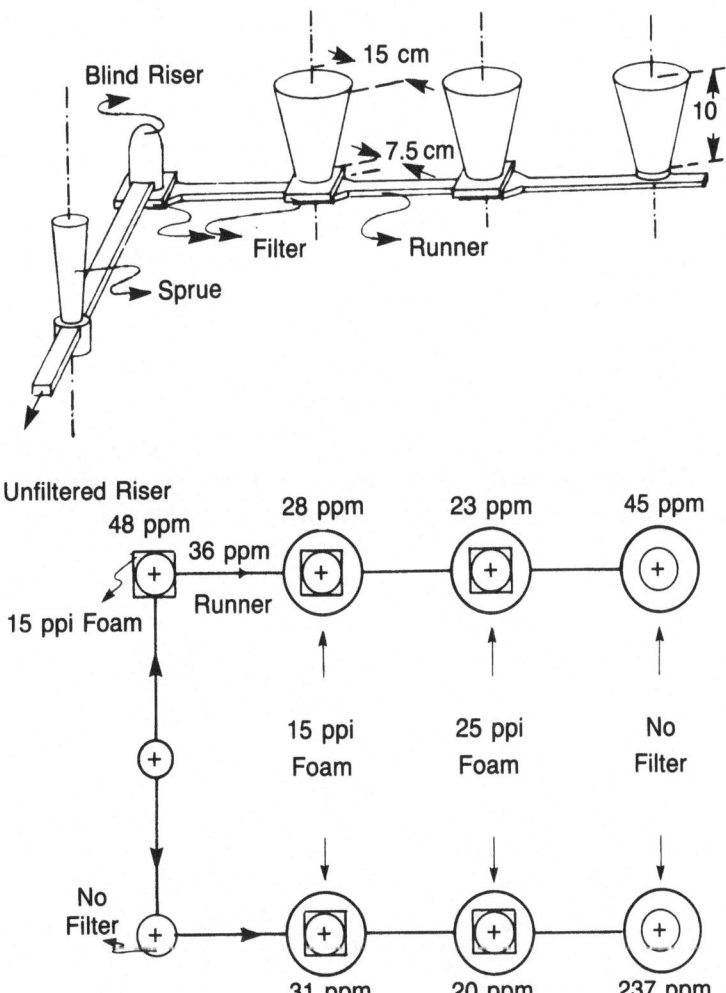

Figure 11. Total oxygen content in filtered and unfiltered static castings. Aluminum-deoxidized low alloy steel.

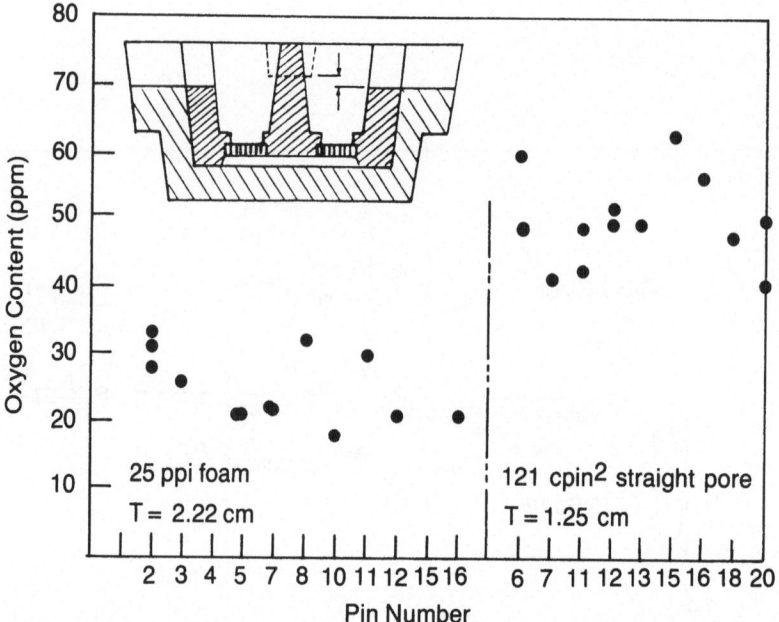

Figure 12. Effect of the filter type on the total oxygen content of filtered carbon steel. Two separate filters performing independently in a counter gravity flow tundish. Aluminum deoxidation. Pin samples taken at various times.

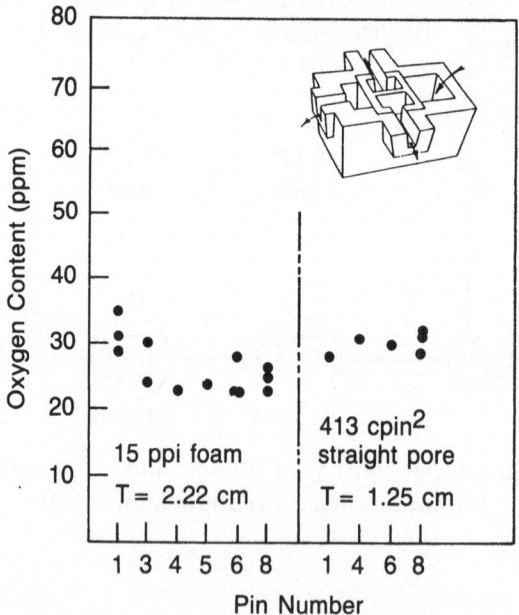

Figure 13. Effect of the filter type on the total oxygen content of filtered carbon steel. Two separate filters performing independently in a counter gravity flow tundish. Aluminum deoxidation. Pin samples taken at various times.

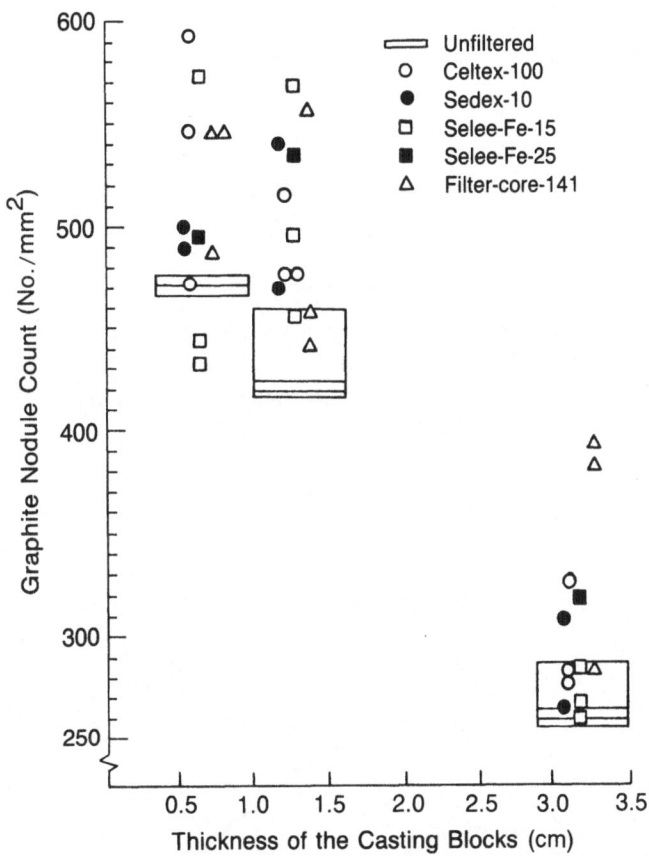

Figure 14. Graphite nodule counts of filtered and unfiltered step block iron castings. Foam and straight pore filters.

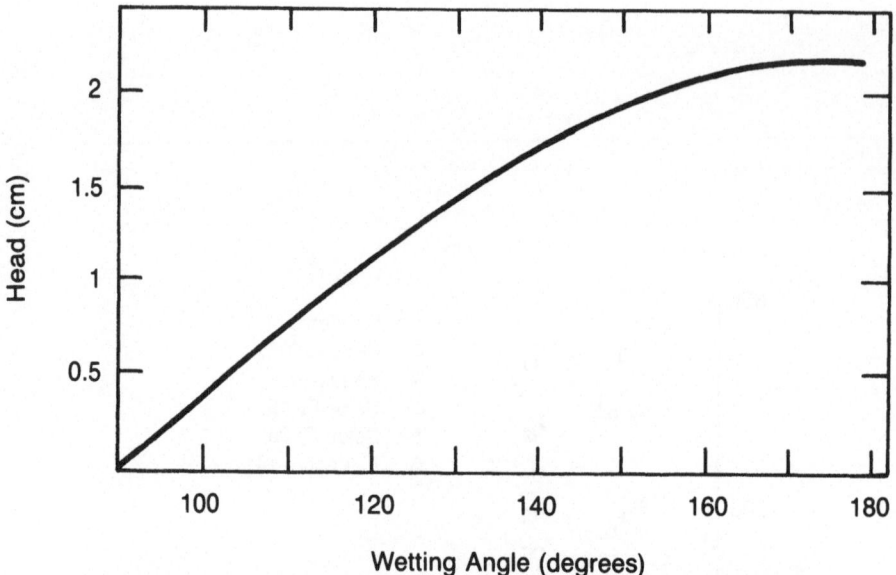

Figure 15. The effect of the wetting angle of steel on the static priming head. Assumption: circular 0.2 cm (0.5 in.) diameter pores, filter at liquid metal temperature.

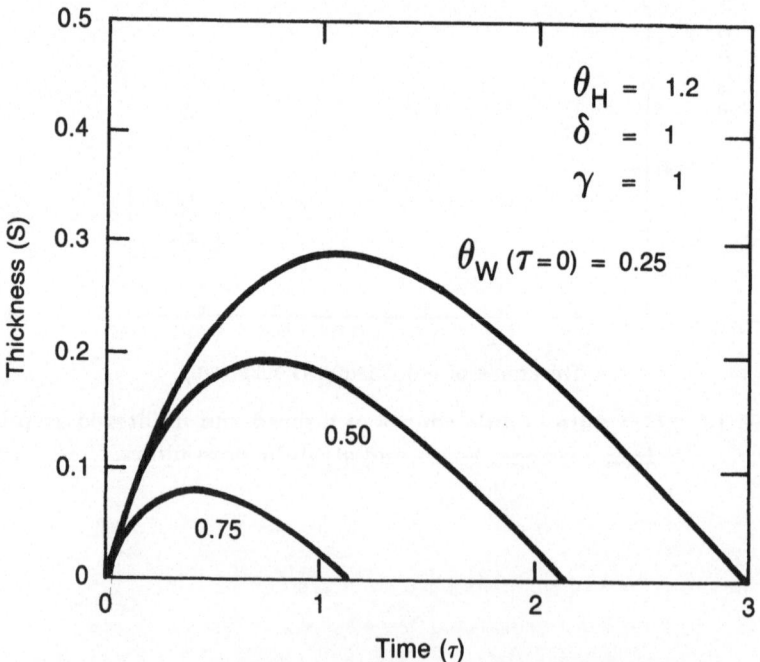

Figure 16. Time, $\tau$ (dimensionless), dependence of frozen metal thickness, S (dimensionless). Liquid metal superheat effect on freezing and remelting of steel within a filter pore.

# FILTRATION OF IRONS AND STEELS

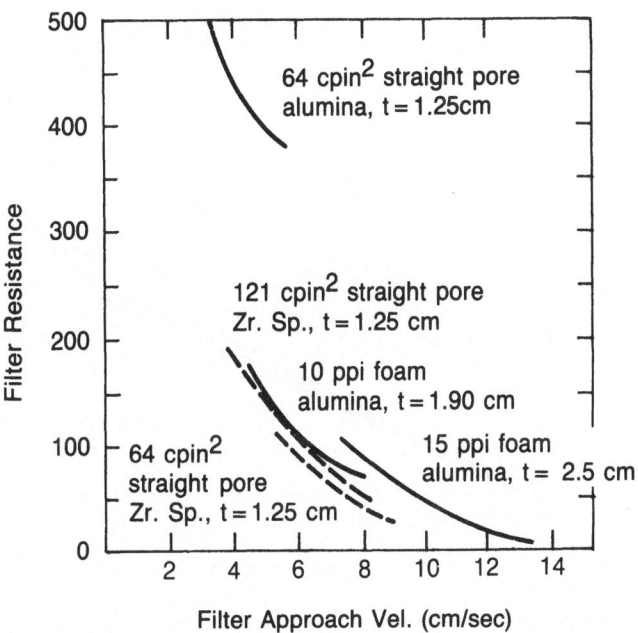

Figure 17. Effect of approach velocity on the resistance (dimensionless) of several filter types to flow of carbon steel. Zr Sp signifies zirconia spinel (60% $ZrO_2$, 30% $Al_2O_3$, 10% MgO).

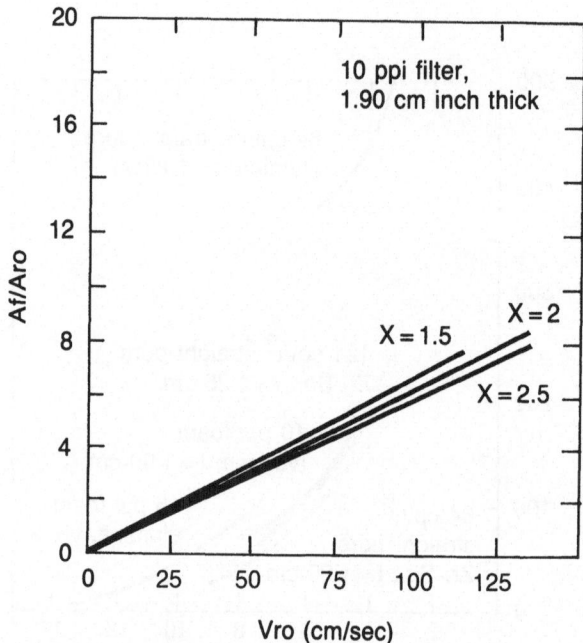

Figure 18. Dependence on the relative filter size, Af/Aro (ratio of filter to runner cross sectional area), on the metal velocity, Vro, in the runner of the filterless system, and the dimensionless runner size, X. Bottom gated system for medium carbon steel. X = ratio of runner cross sections, filtered / unfiltered casting.

## REFERENCES

1. J. W. Brockmeyer, and P. F. Wieser, "Dross Removal from Ductile Iron with Ceramic Foam Filters," *Trans. AFS 93*, 1985, 171.
2. Unpublished Data, S. Edwin, and P. F. Wieser, Filtration of Ductile Iron.
3. Unpublished Data, I. Liu, and P. F. Wieser, Filtration of Steel.
4. A. Ilhan, I. Dutta, J. W. Brockmeyer, P. F. Wieser, "Cast Steel Quality Improvement by Filtration with Ceramic Foam Filters," *Trans. AFS 93*, 1985, 177.
5. J. Dutta, A. Ilhan, P. F. Wieser, "Filtration of Liquid Steel," *Proceedings of the First International Steel Foundry Congress*, Steel Founder's Society of America, Chicago, IL, Nov. 1985, 27.
6. P. F. Wieser, "Separation Processes and Fluid Mechanics of Filtration," *Filtration of Ferrous Metals*, AFS, Des Plaines, IL, 1985, 87–114.
7. R. L. Coble, "Initial Sintering of Alumina and Hematite," *J. Amer. Cer. Soc.*, *41*, 2, 1958, 55–62.
8. S. N. Singh, "Mechanism of Alumina Buildup in Tundish Nozzles During Continuous Casting of Aluminum Killed Steel," *Met. Trans.*, 5, 1974, 2165.
9. P. Galeta, D. Apelian and R. Mutharasan, "Assessment of Tundish Nozzle Blockage Mechanisms - Mathematical Modeling Approach," *Conference Proceedings on Modeling of Casting and Welding Processes*, H. D. Brody, D. Apelian, Editors, AIME, Warrendale, PA, 1981.
10. P. F. Wieser, "Fundamental Considerations in Filtration of Liquid Steel," *Fifth International Iron and Steel Congress Steelmaking Proceedings*, April 1986, to be published.
11. A. Voros, G. Gyorock, "The Filtering of Cast Irons," BISIS Translation 21434, *Banyasz Lapok*, No. 3, 1982, 53.
12. D. L. Feke and P. F. Wieser, "Freezing and Melting Effects in Filter Priming," Third AIME Modelling of Casting-Welding Conference, U.C.S.B. CA., Jan. 1986, Casting Process Modelling Session, Proceeding to be published.
13. J. Duta, P. F. Wieser, "Priming and Flow Through Filters," *Trans. AFS. 94*, 1986, 85–92.

## DISCUSSION

**S. Katz** *(General Motors Research Laboratories)*

You said approximmately 50% of the particles are removed in practical cases of steel filtration. Are the larger particles being trapped?

**A.** It is quite obvious that the larger particles are trapped first. In the case of steel particles that look like dendrites or coral type structures get attached to certain preferred locations on the filter. These particles stick out into the flowing metal and act as a new filter. So, at this preferred location of deposition, one can observe a gradual increase in density of trapped particles.

**S. Katz** *(General Motors Research Laboratories)*

Is there available data on the effect of a 50% reduction in the inclusion population on the properties of cast iron?

**A.** No, we were unable to obtain such data for filtered cast iron. We do however have data on the improvement of the ductile iron casting cleanliness with respect to surface appearance. Also we have data on the fatigue life improvement in high strength steel, the tensile strength and ductility improvement in the case of aluminum. No similar work was done for cast iron, but people have noted improved machining performance, longer tool life, and longer rolling mill roll life due to fewer inclusions.

**R. Loper, Jr.** *(University of Wisconsin)*

What do you think is the reason for the increase in nodule counts after filtration?

**A.** Well, we haven't figured it out yet. We have two simple possibilities to look at. One is that some of the filter refractory is abraded by metal flowing through the filter and that these serve as nucleation sites. The interesting thing is that we saw the increase in the nodule count no matter whether the filter was made of zirconia, partially stabilized zirconia, alumina-zirconia mixtures, or alumina. It didn't make any difference. This would indicate that all these materials can act as nuclei, or that the responsible mechanism is entirely different. One other possibility is that some substances were retained on the filter and that these substances act as poisons to existing nuclei. We have lots of sulfur prints that show the retention of sulfur by the fiber. We think that the sulfur was present as magnesium sulfide. I think more work needs to be done in this area.

# SYMPOSIUM PARTICIPANTS

Agarwal, R. L.
   GM Advanced Engineering Staff
   Manufacturing 'A' Building - 36
   GM Technical Center
   Warren, MI 48090-9040

Apelian, D.
   Drexel University
   Department of Materials Engineering
   Philadelphia, PA 19104

Argyropoulos, S. A.
   University of Toronto
   Department of Metallurgy and
     Materials Science
   Toronto, Ontario M5S 1A4
   CANADA

Asai, S.
   Nagoya University
   Department of Iron & Steel
   Furocho Chikusaku Nagoya 464
   JAPAN

Bachowski, R.
   Ingot Casting Division
   Aluminum Company of America
   Alcoa Center, PA 15069

Backer, G. P.
   GM Advanced Engineering Staff
   Manufacturing 'A' Building - 36
   GM Technical Center
   Warren, Michigan 48090-9040

Bahk, S.
   GM Research Laboratories
   Metallurgy Department
   30500 Mound Road
   GM Technical Center
   Warren, Michigan 48090-9055

Bartram, R.
   Airco Carbide Division
   Airco, Inc.
   P.O. Box 10037
   Louisville, Kentucky 40210

Bates, C. E.
   Southern Research Institute
   2000 Ninth Avenue South
   Birmingham, AL 35255

Berry, J. T.
   University of Alabama
   P.O. Box G
   University, AL 35486

Blaser, D. A.
   GM Research Laboratories
   Engineering Mechanics Dept.
   30500 Mound Road
   Warren, Michigan 48090-9055

Bradley, F. J.
   University of Wisconsin
   1509 University Ave.
   Madison, WI 53706

Bralower, P.
   American Foundrymen's Society
   Golf and Wolf Roads
   Des Plaines, Illinois 60016-2277

Buhr, R. K.
   Energy, Mines and Resources-CANMET
   568 Booth Street
   Ottawa, Ontario K1A 0G1
   CANADA

Caulk, D. A.
   GM Research Laboratories
   Engineering Mechanics Dept.
   30500 Mound Road
   Warren, Michigan 48090-9055

Cho, B. K.
   GM Research Laboratories
   Physical Chemistry Dept.
   30500 Mound Road
   Warren, Michigan 48090-9055

Cole, R.
  General Motors of Canada
  St. Catharines Foundry
  Box 3002 / 570 Glendale
  St. Catharines, Ontario  L2R 7B3
  CANADA

Cornie, J. A.
  Massachusetts Institute of Technology
  Room 8-405
  Cambridge, MA 02139

Creese, R. C.
  West Virginia University
  P.O. Box 6101
  Morgantown, WV  26506-6101

Davis, K. G.
  Energy, Mines and Resources - CANMET
  568 Booth Street
  Ottawa, Ontario K1A 0G1
  CANADA

Day, W.
  Columbus Foundries Inc.
  P.O. Box 4201
  Columbus, GA  31995

DebRoy, T.
  The Pennsylvania State University
  Department of Materials Science &
    Engineering
  University Park, PA  16802

Devers, M. W.
  Central Foundry Division
  1805 Veterans Memorial Hwy.
  Saginaw, Michigan  48605

Doat, R.
  Bois-Franc
  Jarnioux F 69640-Denice
  FRANCE

Doshi, B. K.
  CPC Headquarters
  Room 228-25
  30001 Van Dyke Avenue
  Warren, Michigan  48090-9020

Draper, A. B.
  Penn State University
  207 Hammond Bldg., 103 Eng. E.
  University Park, PA  16802

Edwards, I. L.
  Central Foundry Division
  1805 Veterans Memorial Hwy.
  Saginaw, Michigan  48605

El-Kaddah, N.
  The University of Alabama
  P.O. Box G
  University, AL  35486

Eppich, R. E.
  Dalton Foundries
  Box 1388
  Warsaw, IN  46580

Evans, W.
  Ford Motor Co.
  Manufacturing Development Center
  24500 Glendale Ave.
  Dearborn, MI  48239

Feuring, K.
  Buderus AG
  Wetzlar
  FRG

Fine, H. A.
  University of Kentucky
  Anderson Hall
  Lexington, KY  40506-0046

Fosbinder, L. L.
  Saturn Corp.
  Metal Casting Team
  434 W. 12 Mile Road
  Madison Heights, MI  48071

Fray, D. J.
  Dept. of Metallurgy
    & Materials Science
  University of Cambridge
  Pembroke Street
  Cambridge CB2 3QZ
  ENGLAND

# PARTICIPANTS

Frosch, R. A.
Vice President, GM Research Laboratories
Executive Office/Research
  Administration Bldg.
30500 Mound Road
Warren, Michigan  48090-9055

Fruehan, R. J.
Carnegie-Mellon University
Schenley Park
Pittsburgh, PA  15213

Gaskell, D. R.
Purdue University
School of Materials Engineering
West Lafayette, IN  47907

Gaye, H.
IRSID
B. P. 13
F-57210 Maizieres-les-Metz
FRANCE

Goodrich, G. M.
Taussig Assoc. Inc.
7530 Frontage Road
Skokie, IL  60077

Goto, K. S.
Tokyo Institute of Technology
2-12 Ookayama
Meguro-ku, Tokyo 152
JAPAN

Goundan, K.
AME, Manufacturing Services
General Motors Bldg.
Detroit, MI  48202

Green, R. J.
General Motors of Canada
St. Catharines Foundry
Box 3002 / 570 Glendale
St. Catharines, Ontario  L2R 7B3
CANADA

Gundlach, R. B.
AMAX Metals Group
1600 Huron Parkway
Ann Arbor, MI  48105

Guthrie, R. I. L.
McGill University
Department of Mining & Metallugrical
  Engineering
3450 University Street
Montreal, Quebec H3A 2A7
CANADA

Hacetoglu, A. P.
Cyanamid Canada, Inc.
P. O. Box 240
Niagra Falls, Ontario  L2E 6T4
CANADA

Hanna, M. D.
GM Research Laboratories
Metallurgy Department
30500 Mound Road
Warren, Michigan  48090-9055

Harvey, D. J.
General Motors Research Labs
Metallurgy Department
30500 Mound Road
Warren, MI  48090-9055

Heine, R. W.
University of Wisconsin
1509 University Avenue
Madison, WI  53706

Henein, H.
Carnegie Mellon University
5000 Forbes Ave.
Pittsburg, PA  15213

Hlinka, J. W.
Bethlehem Steel Corp.
Research Department
Bethlehem, PA  18016

Houchens, A. F.
GM Advanced Engineering Staff
Manufacturing 'A' Building - 36
GM Technical Center
Warren, MI  48090-9040

Howell, J.
Deere & Co. Technical Center
3300 River Drive
Moline, IL 61265

Howell, L. J.
GM Research Laboratories
Engineering Mechanics Dept.
30500 Mound Road
Warren, MI 48090-9055

Irons, G. A.
McMaster University
1280 Main Street West
Hamilton, Ontario
CANADA

Jacob, S.
Centre Techniques des Industries
 de la Fonderie
Departement de la Metallurgie
12 Avenue Raphael - 75016 Paris
FRANCE

Janke, D.
Max-Planck-Institut für Eisenforschung
 GmbH
Max-Planck-Str. 1
4 Düsseldorf
FRG

Janowak, J.
AMAX Metals Group
P.O. Box 397
Arlington Hghts., IL 60006

Jarrett, N.
Alcoa Laboratories
Aluminum Company of America
Alcoa Center, PA 15069

Johnson, J. N.
GM Research Laboratories
Metallurgy Department
30500 Mound Road
Warren, MI 48090-9055

Kagawa, A.
Osaka University
Mihogaoka 8-1, Obaraki
Osaka 567
JAPAN

Kaiser, R. H.
Elkem Metals Co.
P.O. Box 1344
Niagra Falls, NY 14302

Kamal, M. M.
Technical Director
GM Research Laboratories
30500 Mound Road
Warren, MI 48090-9055

Kato, E.
Waseda University
4-1 Okubo 3-Chome
Shinjuku-ku, Tokyo 160
JAPAN

Katz, S.
GM Research Laboratories
Metallurgy Department
30500 Mound Road
Warren, MI 48090-9055

Kilpatrick, D.
General Motors of Canada
St. Catharines Foundry
Box 3002 / 570 Glendale
St. Catharines, Ontario L2R 7B3
CANADA

Kim, C. W.
GM Advanced Engineering Staff
Manufacturing 'A' Building - 36
GM Technical Center
Warren, MI 48090-9040

Kipouros, G. J.
GM Research Laboratories
Electrochemistry Dept.
30500 Mound Road
Warren, MI 48090-9055

# PARTICIPANTS

Kundrat, D. M.
Armco, Inc.
703 Curtis St.
Middletown, OH 45043

Kusakawa, T.
Waseda University
4-1 Okubo 3-Chome
Shinjuku-ku, Tokyo 160
JAPAN

Landefeld, Craig F.
GM Research Laboratories
Metallurgy Department
30500 Mound Road
Warren, MI 48090-9055

Lange, K. W.
Institut für Eisenhüttenkunde
Technische Hochschule Rhein-Westf.
Aachen
WEST GERMANY

Langner, E. E. Jr.
American Foundrymen's Society, Inc.
American Cast Iron Pipe Co.
P. O. Box 2727
Birmingham, AL 35202

Lillybeck, N. P.
Deere & Company
3300 River Drive
Moline, IL 61265

Lobenhofer, R. W.
American Foundrymen's Society, Inc.
Golf & Wolf Roads
Des Plaines, IL 60016-2277

Loper, C. R. Jr.
Metallurgical & Mining Eng.
University of Wisconsin
1509 University Avenue
Madison, WI 53706

Martin, K. L.
GM Advanced Engineering Staff
Manufacturing 'A' Building - 36
GM Technical Center
Warren, MI 48090-9040

Mattavi, J. N.
Assistant Head, Fluid Mechanics Dept.
GM Research Laboratories
30500 Mound Road
Warren, MI 48090

McCluhan, T. K.
Elkem Metals Co.
P. O. Box 1344
Niagara Falls, NY 14302

McLean, A.
University of Toronto
Department of Metallurgy & Materials
    Science
Toronto, Ontario M5S 1A4
CANADA

Meyst, P. A.
Iron Casting Research Institute
870 W. 37th Ave.
Columbus, OH 43212

Mikelonis, P. J.
General Castings
P.O. Box 828
706 E. Main Street
Waukesha, WI 53187

Miller, W. K.
General Motors Research Labs.
Metallurgy Department
30500 Mound Road
Warren, MI 48090-9055

Miller, J. C.
Reynolds Metals Company
P.O. Box 27003
Richmond, VA 23261

Mobley, C. E.
Dept. of Metallurgical Eng.
Ohio State University
Columbus, OH 43210

Nafziger, R.
Albany Research Center
US Bureau of Mines
P.O. Box 70
Albany, OR 97321

Naro, R. L.
Ashland Chemical Co.
P.O. Box 221
Columbus, OH 43216

Niedringhaus, J. C.
Armco Inc.
703 Curtis Street
Middletown, OH 45043

Osborne, R.
GM Advanced Engineering Staff
Manufacturing 'A' Building - 36
GM Technical Center
Warren, MI 48090-9040

Ozturk, B.
Department of Metallurgy and
  Materials Science
Shenley Park
Pittsburgh, PA 15213

Peck, W. J.
Central Foundry Division
State Road Route 281
Defiance, OH 43512

Pehlke, R. D.
The University of Michigan
3062 Dow, North Campus
Ann Arbor, MI 48109

Plutshack, L. A.
Foseco, Inc.
20200 Sheldon Rd.
Brookpark, OH 44142

Powers, P. J.
P.J.P. Engineering Consulting
6 Maya Lane
Hot Springs Village, AR 71909

Pritchett, T. R.
Kaiser Aluminum Co.
P.O. Box 877
Pleasanton, CA 94566

Prucha, T. E.
CMI International Inc.
14638 Apple Drive
Fruitport, MI 49415

Puckett, J.
Nelson Metal Products Corp.
2950 Prarie Street
Grandville, MI 49418

Rashid, M. S.
GM Research Laboratories
Metallurgy Department
30500 Mound Road
Warren, MI 48090-9055

Rehder, J. E.
University of Toronto
36 Castle Frank Road, Apt. 309
Toronto, Ontario M4W 2Z7
CANADA

Reinemann, G.
American Foundrymen's Society
Morris Bean & Co.
P.O. Box 108
Yellow Springs, OH 45387

Robertson, D.
Generic Mineral Center for
  Pyrometallurgy
University of Missouri
Rolla, MO 65401-0249

Robinson, G. H.
Head, Metallurgy Department
GM Research Laboratories
30500 Mound Road
Warren, MI 48090-9055

Robinson, S. W.
Airco Carbide Division
Airco Inc.
P.O. Box 10037
Louisville, KY 40210

Rooy, E. L.
Aluminum Company of America
1501 Alcoa Building
Pittsburgh, PA 15219

Ruff, G. F.
Central Foundry Division
1805 Veterans Memorial Hwy.
Saginaw, MI 48605

# PARTICIPANTS

Ryntz, E. F.
 GM Research Laboratories
 Metallurgy Department
 30500 Mound Road
 Warren, MI 48090-9055

Sahai, Y.
 Department of Metallurgical Engineering
 The Ohio State University
 Columbus, OH 43210

Sano, N.
 University of Tokyo
 Department of Metallurgy & Materials
  Science
 Bunkyo-ku, Tokyo
 JAPAN

Selines, R. J.
 Union Carbide Corporation
 Tarrytown Technical Center
 Tarrytown, NY 10591

Shea, M. M.
 GM Research Laboratories
 Metallurgy Department
 30500 Mound Road
 Warren, MI 48090

Shulhof, W. P.
 Central Foundry Division
 1805 Veterans Memorial Hwy.
 Saginaw, MI 48605

Sigworth, G. K.
 Cabot Corporation
 P.O. Box 1296
 Reading, PA 19603

Singh, R.
 GM Advanced Engineering Staff
 Manufacturing 'A' Building - 36
 GM Technical Center
 Warren, MI 48090-9040

Smith, E.
 Research and Technology Sector
 Energy, Mines and Resources
 580 Booth Street
 Room 2055
 Ottawa, Ontario K1A 0E4
 CANADA

Sommerville, I. D.
 University of Toronto
 184 College Street
 Toronto, Ontario M5S 1A4
 CANADA

Sovran, G.
 GM Research Laboratories
 Fluid Mechanics Department
 30500 Mound Road
 Warren, MI 48090-9055

Sponseller, D. L.
 AMAX Materials Research Center
 1600 Huron Parkway
 Ann Arbor, MI 48106

St. Pierre, G. R.
 Ohio State University
 116 West 19th Avenue
 Columbus, OH 43210

Stefanescu, D. M.
 The University of Alabama
 P.O. Box G
 University, AL 35486

Stroom, P. A.
 Saturn Corp.
 Metal Casting Team
 434 12 Mile Road
 Madison Heights, MI 48071

Svoboda, J.
 Steel Founders' Society of America
 455 State St.
 Des Plaines, IL 60016

Szekely, J.
 Massachusetts Institute of Technology
 Department of Materials & Science
  Engineering
 Cambridge, MA 02139

Thomas, B.
 Department of Metallurgy
 University of Illinois
 Mechanical Engineering Bldg.
 Room 140
 1206 W. Green St.
 Urbana, IL 61801

Tippin, R. B.
　The University of Alabama
　P.O. Box G
　University, AL　35486

Tiwari, B. L.
　GM Research Laboratories
　Electrochemistry Dept.
　30500 Mound Road
　Warren, MI　48090-9055

Trojan, P. K.
　University of Michigan
　3939 Vorheis Road
　Ann Arbor, MI　48105

Tuesday, C. S.
　Technical Director
　GM Research Laboratories
　30500 Mound Road
　Waren, MI　48090-9055

Turkdogan, E. T.
　Mail Station 68
　United States Steel Corporation
　One Tech Center Drive
　Monroeville, PA　15146

Vernia, P.
　Metallurgy Department
　GM Research Laboratories
　30500 Mound Road
　Warren, MI　48090-9055

Wada, H.
　Metallurgy Department
　University of Michigan
　3062 Dow, North Campus
　Ann Arbor, MI　48109

Wall, E.
　Central Foundry Division
　Veterans Memorial Highway
　Saginaw, MI　48605

Warrick, R. J.
　Lynchburg Foundry
　Box 5200
　Lynchburg, VA　24505

Weiss, J. C.
　GM Advanced Engineering Staff
　Manufacturing 'A' Building - 36
　GM Technical Center
　Warren, MI　48090-9040

White, J. R.
　Department of Sociology and
　　Anthropology
　Youngstown State University
　Youngstown, OH　44555

Wieser, P.
　Case Western University
　Wieser & Associates
　14738 East River Road
　Columbia Station, OH　44028

Winter, B.
　Ford Motor Company
　Manufacturing Development Center
　24500 Glendale Ave.
　Dearborn, MI　48239

Zurecki, Z.
　Air Products & Chemicals, Inc.
　P.O. Box 538
　Allentown, PA　18105

# SUBJECT INDEX

Activity
  of elements in aluminum, 32, 33
  of elements in iron, 32, 57-60, 85-89,
    136-160, 227, 228, 237
  slags, 124, 127-131, 365
Activity coefficients
  C in Fe-O-X, 227-228
  Li in AL, 399-401
  O in Fe-C-X, 227-228, 237
Alloy
  additions to ferrous melts, 15-17
  elements, 11
  oxidation, 11
  production, 10
Aluminum, 26
  additions to cupola, 8, 9, 19
  analysis in iron, 32
  degassing of, 29
  filtration of, 41, 473
  gas porosity in, 30, 432, 435-437
  hydrogen in, 29, 31, 432, 433, 435-437
  impurities in, 27, 28, 31, 396-398
  lithium in, 33, 399-401
  magnesium in, 29, 30, 33
  oxidation, 30
  purification, 28, 396-398
  sodium in, 28
  strontium in, 28
  tramp elements in, 27-31
  water, reaction with, 30
Aluminum, impurity removal by
  $AlF_3$, 398
  chlorine, 29, 31, 397
  precipitation, 28, 398
  premelt separation, 28, 396-397
  separation during melting, 397
Aluminum oxide
  filtration, 462
  inclusions, 39, 41, 448, 455

Basicity of slags, 102, 113, 117, 120-121
Beta-alumina, 249-250
Bismuth
  effects in iron, 375-376, 379
  neutralization by REM, 379-390
  recovery % in iron, 376
Blow holes, 33
Boudouard reaction (see Coke gasification)

$CaC_2$, 17, 24, 72, 310-324
$Ca_2Sn$ formation free energy, 370
$CaH_2$ electrolyte, 247
Calcium oxide
  activity, 128, 129, 365

  formation free energy, 356
Capacity coefficient, 262
Carbide capacity of slags, 107, 111, 113
Carbon
  additions to cast iron, 15
  analysis in iron, 32
  dissolution in iron, 77
  nucleation and growth in cast iron, 42, 60-64
  pickup in cupola, 77, 171-175
Carbon monoxide
  defects in cast iron, 55, 56, 433, 436
  solubility in steel, 433, 434
Carbonate capacity of slags, 102
Cast iron, 32, 36
  activity of alloy elements, 85
  carbon additions to, 15, 77
  chemical analysis, 32, 227, 228, 237
  desulfurization of, 17, 18, 24, 64-74,
    105-110, 125-127, 364-365
  eutectic temperatures, 22, 142-145
  filtration of, 496-510
  gas porosity in, 33, 35, 36, 136, 153-159
  graphite growth kinetics, 22, 60, 62
  graphite nucleation, 42, 63, 64
  graphitization, 145-149
  hydrogen in, 35
  inclusions in, 37, 38, 39
  interfacial energy, 54
  magnesium addition to, 16
  magnesium solubility, 86
  MgS solubility, 87
  neutralization of tramp elements, 23, 379-390
  nitrogen in, 35, 55-59, 111-115, 495-512
  nitrogen partition, 149-153
  oxidation of, 37
  partition coefficients, 136-142
  purification of, 17
  TiC solubility, 89
  tramp element removal, 17-24, 57-64,
    358-370, 375-390, 415-419
Casting defects, 33, 55-59, 495-512
Casting, cleanliness, 41, 447-512
Ceramic foam filter, 475, 476, 478
Coke
  coal chars, 9
  formcoke, 8-10
  gasification of, 5, 10, 74-76, 180-186
  size, 4, 9
  pore diffusion, 75, 76
  properties, 9
Cokeless cupola, 6
Cupola
  alloy formation in, 10, 11
  aluminum additions to, 8, 9, 11

burden height, 3
carbon pickup, 77, 171-175
CO/CO$_2$ ratio, 164
coke, 6, 8, 9, 74-76, 179-186
desulfurization, 11
divided blast, 6, 7
energy efficiency, 3, 176
formcoke, 8
heat balance, 176
heat losses, 175-178
heat transfer, 3, 6, 175-178
hydrogen removal, 29
iron oxidation in, 164
off-gas latent heat, 3, 5
off-gas recuperation, 6
off-gas sensible heat, 6
off-gas temperature, 3-4
oxygen injection, 7
preheat zone, 6
refractory linings, 5
sampling, 164-167
secondary air, 6, 7
sensible heat losses, 3, 175-178
shell heat flux, 3, 5, 6, 176-178
slag-metal reactions, 77, 78, 167-171
Cupola slag
   formation, 164, 166-167
   iron emulsion, 168-171
   reactions in 167-171
Cyanide capacity of slags, 107, 111, 113, 120

Degassing
   aluminum, 20, 24, 29, 31
   cast iron, 20, 420-425
   kinetics, 24
Dephosphorization, 121
Desulfurization of iron and steel
   atmosphere effects, 352
   by CaO and CaC$_2$, 341-344
   by CaO-Al$_2$O$_3$, 65, 67, 68
   by CaO-CaF$_2$-SiO$_2$, 18, 69-74, 364-365
   by Mg, 340-341
   by Na$_2$CO$_3$, 339-340
   by injected CaC$_2$, 310-324
   kinetics of, 70, 337-339, 347-351
   partition coefficient, 345
   plant tests, 72, 73
   sulfide capacity, 105-110, 125-127, 346-347
   stirring energy, 74
   thermochemistry of, 334-337
Ductile iron
   graphite growth, 60-63
   graphite nucleation, 42, 61-64

Electric furnaces, 2
Electric Sensing Zone Technique, 449-454
Electrochemical measurement
   C in Fe-O-X, 220-230, 236
   H in Al, 33, 247-249, 254
   Li in Al, 249-250
   Mg in Al, 33
   Na in Al, 33, 250
   O in Fe-C-X, 220-230, 236
   reference electrodes for Al, 245-246
   Si in iron, 230-235
   Zn in Pb, 251
Electrolytes
   beta-alumina, 33, 249-250
   CaH$_2$, 33, 247
   Li$_{3.6}$Si$_{0.6}$P$_{0.4}$O$_4$, 33, 249
   liquid silicate, 230-231
   MgCl$_2$-CaCl$_2$, 33
   polarization of, 236
   selection criteria, 245, 254-255
   SrCe$_{0.9}$5Yb$_{0.05}$O$_3$, 257
   table of, 33, 258-259
   thoria, 222, 225
   uranyl phosphate, 247
   zirconia, 222, 229
Electromagnetic stirring, 202-208
Energy efficiency in cupola, 3-7, 74-76, 175-178
Equilibrium, attainment in Fe-C-Si-Mn-S, 239-240

Filter
   ceramic foam, 469, 475, 476, 478, 480
   cross sectional area, 503
   deep bed, 475
   efficiency, 40, 470-472, 476-484, 488, 489
   evaluation of, 472, 475
   opening size, 40, 470, 478, 479, 503, 504
   performance, 471-512
   surface area, 484
   tabular alumina, 476, 478
   types of, 495, 496
Filtration
   aluminum melts, 462, 463, 473
   cast iron melts, 496-500
   effect of melt velocity, 509-510
   effect on nodule count, 501, 507
   filter priming, 501
   filter type, 495, 496, 506
   inclusions, 462, 463
   kinetic parameter coefficient, 471
   mechanisms, 473
   models, 469, 470, 471, 487, 504, 508
   Reynolds number, 484
   solid inclusions, 483

# SUBJECT INDEX

steel melts, 481, 485, 497, 498
superalloy melts, 478-480
Fluid flow modelling
   by electromagnetic stirring, 202-208
   by gas stirrring, 200
   by natural convection, 199
   general considerations, 193-197
   lumped parameters, 196-199
   mold filling, 208-213
Formcoke, 8, 10
Free energy of formation
   $Ca_2Sn$, 370
   $CaO$, 356

Gas dissolution kinetics, 413-418
Gas evolution kinetics, 420, 421
Gas porosity
   aluminum, 29-33, 57, 58, 430, 436
   cast iron, 20, 21, 55-60, 136, 153-159, 419-422
   carbon monoxide defects, 34, 55, 56, 337, 433, 436
   chromium, effect of, 58
   cooling rate, effect on, 435, 436
   formation, 428, 435-439
   hydrogen defects, 33, 35, 36, 55, 56, 419-422
   model, 427-445
   mold geometry, effect on, 436
   mold roughness, 55, 56
   nitrogen defects, 20, 21, 35, 36, 55, 56,107, 111-116, 149-161, 419-422
   resin sand-binders, effect of, 35
   steel, 34, 433, 436-437
   surface tension, effect on, 55, 56, 414, 415
   titanium, effect of, 59, 60
Gas purging
   aluminum, 29-31
   cast iron, 20, 21, 420-422
Gas stirring, 24, 200-202, 262-270
Gasification of coke
   coke density, 9
   coke size, 5, 9
   effect on cupola, 180
   interphase mass transfer, 75, 76
   intrinsic kinetics, 75, 76, 183-186
   modelling, 180-186
   pore diffusion, 75, 67
Graphite
   growth kinetics, 22, 60-62
   nucleation, 42, 63, 64
Graphitization, 22, 145-149

Hydrogen, measurement in Al, 246-249

Heat flux to cupola shell, 176-178
Hydrogen
   gas defects, 35, 36, 55, 56, 432, 435-437
   gas entrainment, 36
   measurement in Al, 246-249
   purging, 29-31, 420
   solubility in aluminum, 31, 432, 433
   solubility in iron, 57-60, 117, 153-161, 412-413, 419-421

Impurities, 21
   in aluminum, 26-31
   in cast iron, 21-25, 58-60, 64-74, 105-132, 136-160, 412-423
   in steel, 24
   slag extraction, 24, 105-132
Inclusions
   aluminum melts, 39, 40, 473-478
   analyzer for liquid metal, 449-454
   cast iron melts, 37-39, 461, 496-500
   control, 462, 463
   detection, 41, 447, 448-454
   filtration, 37-41, 462, 463, 467-512
   liquid, 473, 485, 486
   measurement, 41
   removal efficiency, 38, 40, 488, 489
   sedimentation, 454-456, 458, 460
   sources, 37, 40, 468, 469, 496-498
   steel melts, 461, 497, 498
   superalloy melts, 478-480
Injection of solids with gas
   angled injection, 310, 329, 331
   $CaC_2$, 310-324
   critical loading, 307
   jet penetration, 307-309
   jetting and bubbling, 304-310
   nozzle clogging, 330
   particle wetting, 328
   solids/gas ratio, 320-322
   with top slags, 313-316
Interdendritic feeding, 428
Interfacial energy, cast iron, 54

Liquid velocity, 263-264, 470, 471, 474, 476-484, 502-504
Lithium in Aluminum
   effects on Al properties, 399
   measurement, 249-250, 402-404
   removal by $AlF_3$, 401
   removal by chlorine, 400
   removal by $CO_2$, 401, 404-405, 408

Magnesium
   addition to cast iron, 16, 497, 502

in aluminum, 29, 30
MgS solubility in cast iron, 87
removal, 28
solubility in cast iron, 85, 86
Manganese
in blast furnace, 77-80
in cupola, 11, 13
Mass transfer in gas-stirred vessels
capacity coefficient, 262
controlling step, 262
gas-liquid, 270-273
liquid-liquid, 273-280
entrapment criterion, 279-280
gas rate effect, 275-278, 295
injection depth effect, 298
physical modelling, 294-300
tuyere diameter effect, 298
tuyere placement effect, 278, 298
solid-liquid, 281-284
Mixing of metal
by natural convection, 199-200
by gas stirring, 24, 200-202, 262-270
Mixing time, 197-199, 264-270, 292
and mixing power, 265-267
scale-up, 269
Mold filling, modelling, 208-213

Nitride capacity of slags, 107, 111, 113, 120
Nitrogen
capacity of slags, 21, 107, 111, 113, 120
defects, 20, 21, 34-36, 55, 56
dissolution kinetics, 411-419
gas entrainment, 36, 55-60
partition in freezing iron alloy, 149-159
purging, 420
slag extraction, 21, 107, 111, 113, 120
solubility in iron, 20, 35, 57-60, 412-413

Optical basicity
definition, 103, 104
steel desulfurization slags, 346
of slags, 103, 104, 107-110, 112, 116, 121-126, 128-132
Oxygen
in cast iron, 32
in steel, 34, 433
Oxygen activity
in Fe-C-Si-Al-CaC$_2$, 317-319
in Fe-C-Si-CaC$_2$, 316-317

Partition coefficients
in freezing cast iron
eutectic-austenite, theory, 138
liquid-austenite, theory, 136, 137

multi-component systems
effect on eutectic temperature, 142-145
effect on graphitization, 145-149
nitrogen, 149-153
in iron-slag
As, Pb, Sn with CaO-CaF$_2$, 367
Cu with CaO-CaF$_2$, 367-370
Cu with Na$_2$S-FeS, 361
Sb with CaO-CaF$_2$, 365-367
Sb, Mn, V, Nb with Na$_2$O-SiO$_2$, 358-361
Phosphate
activity, 124, 127
capacity of slags, 106, 108, 117, 121, 123
Pinhole defects, 33, 55, 56, 136, 153-159, 411, 419, 428, 432, 433, 436
Porous plug ladles, 24, 25

Rare earth neutralization, 23, 379-390
Recycling, Al cans, 394
Reynolds number, 484

Scrap separation, 28
Secondary air, 6, 7
Sensors
composition, 11, 32, 33, 34, 220-236, 247-250
surface-to-volume ratio, 12
liquid metal cleanliness, 37-42
Silicon
analysis in iron, 32, 230-235
behavior in blast furnace, 78-80
in cast iron, 32
modification in Al, 253
oxidation in cupola, 167-168
recovery in cupola, 11, 12, 15
surface tension in iron, 92
Slag-metal
equilibrium, 8, 11, 14, 37-39 78, 81, 239-240
reactions, 8, 11-20, 28, 29, 37-39, 64-74, 77-82, 239-240
Slags
basicity, 102, 113, 117, 120-121
CaO activity, 128, 129
carbide capacity, 107, 111, 113
carbonate capacity, 103
cyanide capacity, 107, 111, 113, 120
Na$_2$O activity, 130-131
nitride capacity, 107, 111, 113, 120
optical basicity, 103, 104, 107-110, 112, 116, 121-126, 128-132
P$_2$O$_5$ activity, 124, 127
phosphate capacity, 106, 108, 117, 121, 123
sulfide capacity, 105, 107, 108, 116, 120, 126

# SUBJECT INDEX

water capacity, 106, 110, 117
Sodium oxide activity, 130, 131
Solution loss reaction (see Coke gasification)
Stanton number, 262
Stokes law, 38
Stratification in ladle, 199-200
Sulfide capacity
    $CaO-CaF_2-SiO_2$, 364
    slags, 105, 107, 108, 116, 120, 126
    steel ladle slags, 346-347
Sulfur
    behavior in blast furnace, 78-80
    behavior in cupola, 12, 15, 17
    graphite growth, 61, 63, 64
    nitrogen dissolution, 415, 416
    solubility of MgS, 87, 88
    surface tension, 91-96
    superalloys, 479
Surface
    tension, 55, 56, 58-64, 90-96
    kinetics, 413-415
    thermodynamics, 90
Surface-to-volume ratio sensor, 12

Thoria electrolyte, 222, 225
Titanium, TiC solubility in iron, 87
Tramp elements
    in aluminum, 28-31, 396-401
    in cast iron, 17-24, 57-64, 358-370, 375-390,
        415-419
    in steel, 24
    neutralization, 23, 379-390
    slag extraction, 24, 105-132

Water capacity of slags, 106, 110, 117
Wire injection, 329

Zirconia electrolyte, 222, 229